Essential Mathcad for Engineering, Science, and Math ISE

Essential Mathcad for Engineering, Science, and Math ISE

Brent Maxfield

ELSEVIER

AMSTERDAM • BOSTON • HEIDELBERG • LONDON
NEW YORK • OXFORD • PARIS • SAN DIEGO
SAN FRANCISCO • SINGAPORE • SYDNEY • TOKYO

Morgan Kaufmann Publishers is an Imprint of Elsevier

Academic Press is an imprint of Elsevier
30 Corporate Drive, Suite 400, Burlington, MA 01803, USA
525 B Street, Suite 1900, San Diego, California 92101-4495, USA
84 Theobald's Road, London WC1X 8RR, UK

Library of Congress Cataloging-in-Publication Data
MATLAB and Simulink are registered trademarks of The MathWorks, Inc. See www.mathworks.com/trademarks for a list of additional trademarks. The MathWorks Publisher Logo identifies books that contain "MATLAB®" and/or "Simulink®" content. Used with permission. The MathWorks does not warrant the accuracy of the text or exercises in this book. This book's use or discussion of "MATLAB®" and/or "Simulink®" software or related products does not constitute endorsement or sponsorship by The MathWorks of a particular use of the "MATLAB®" and/or "Simulink®" software or related products.

British Library Cataloguing-in-Publication Data
A catalogue record for this book is available from the British Library.

ISBN: 978-0-12-374846-1

For information on all Academic Press publications
visit our Web site at www.elsevierdirect.com

Transferred to Digital Printing 2012

Contents

Preface xix
Acknowledgement xxiii

PART I BUILDING YOUR MATHCAD TOOLBOX

CHAPTER 1 AN INTRODUCTION TO MATHCAD 3

Before You Begin 3
- Mathcad Basics 4

Creating Simple Math Expressions 5
Editing Lines 6
Editing Expressions 8
- Selecting Characters 8
- Deleting Characters 8
- Deleting and Replacing Operators 9

Wrapping Equations 9
Toolbars 10
- Calculator Toolbar 11
- Greek Toolbar 12

Regions 13
- Math Regions 14
- Text Regions 14

Functions 14
- Built-In Functions 14
- User-Defined Functions 14

Units 15
- Assigning Units to Numbers 16
- Evaluating and Displaying Units 17

Arrays and Subscripts 18
- Creating Arrays 18
- Origin 20
- Subscripts 20
- Range Variables 22

Plotting: X-Y Plots 25
- Graphing Toolbar 25
- Setting Plotting Ranges 28

Programming, Symbolic Calculations, Solving, and Calculus 29
Resources Toolbar and My Site 30

Summary ... 30
Practice ... 32

CHAPTER 2 VARIABLES AND REGIONS ... 33
Variables ... 33
Types of variables ... 33
Rules for Naming Variables ... 34
Case and Font ... 34
Characters that Can Be Used in Variable Names ... 34
Literal Subscripts ... 35
Special Text Mode ... 35
Chemistry Notation ... 36
String Variables ... 37
Why Use Variables? ... 37
Regions ... 39
Using the Worksheet Ruler ... 39
Tabs ... 39
Selecting and Moving Regions ... 39
Aligning Regions and Alignment Guidelines ... 40
Text Regions ... 41
Changing Font Characteristics ... 41
Inserting Greek Symbols ... 42
Controlling the Width of a Text Region ... 42
Moving Regions Below the Text Region ... 42
Paragraph Properties ... 43
Text Ruler ... 45
Spell Check ... 45
Additional Information About Math Regions ... 46
Math Regions in Text Regions ... 46
Math Regions that Do Not Calculate ... 46
Find and Replace ... 47
Find ... 47
Replace ... 49
Inserting and Deleting Lines ... 50
Summary ... 51
Practice ... 51

CHAPTER 3 SIMPLE FUNCTIONS ... 53
Built-in Functions ... 53
User-Defined Functions ... 55

Why Use User-Defined Functions? 58
Using Multiple Arguments 58
Variables in User-Defined Functions 59
Examples of User-Defined Functions 60
Passing a Function to a Function 61
Custom Operator Notation 63
Prefix Operator 63
Postfix Operator 63
Infix Operator 63
Treefix Operator 64
Warnings 65
Engineering Example 3.1: Column Buckling 67
Engineering Example 3.2: Torsional Shear Stress 67
Summary 68
Practice 69

CHAPTER 4 UNITS! 71
Introduction 71
Definitions 72
Changing the Default Unit System 73
Using and Displaying Units 73
Derived Units 76
Custom Default Unit System 76
Units of force and units of mass 78
Creating Custom Units 79
Units in Equations 80
Do Not Redefine Built-in Units 81
Units in User-Defined Functions 83
Units in Empirical Formulas 83
SIUnitsOf() 85
Custom Scaling Units 87
Fahrenheit and Celsius 87
Change in Temperature 89
Degrees Minutes Seconds (DMS) 89
Hours Minutes Seconds (hhmmss) 90
Feet Inch Fraction (FIF) 92
Money 94
Creating Your Own Custom Scaling Function 94
Dimensionless units 95
Limitation of Units 96

Summary ... 96
Practice ... 97

PART II HAND TOOLS FOR YOUR MATHCAD TOOLBOX

CHAPTER 5 ARRAYS, VECTORS, AND MATRICES ... 101
Review of Chapter 1 ... 101
Range Variables ... 102
Using Range Values to Create Arrays ... 104
Using Units in Range Variables ... 105
Calculating Increments from the Beginning and Ending Values ... 108
Displaying Arrays ... 109
Table Display Form ... 110
Using Units with Arrays ... 112
Calculating with Arrays ... 113
Addition and Subtraction ... 113
Multiplication ... 114
Division ... 117
Array functions ... 120
Creating Array Functions ... 120
Size Functions ... 120
Lookup Functions ... 120
Extracting Functions and Operators ... 120
Sorting Functions ... 121
Calculation Summary ... 121
Converting a range variable to a vector ... 121
Engineering Examples ... 122
Engineering Example 5.1: Using Vectors in a User-Defined Function ... 123
Engineering Example 5.2: Using Vectors in Expressions ... 124
Engineering Example 5.3: Using Element by Element Multiplication (Vectorize Operator) ... 125
Engineering Example 5.4: Charge on a capacitor ... 126
Summary ... 127
Practice ... 127

CHAPTER 6 SELECTED MATHCAD FUNCTIONS ... 129
Review of Built-In Functions ... 130
Toolbars ... 130

Selected Functions 131
max and *min* Functions 131
mean and *median* Functions 132
Truncation and Rounding Functions 137
Summation Operators 141
if Function 145
linterp Function 146
Miscellaneous Categories of Functions 151
Curve Fitting, Regression, and Data Analysis 151
Namespace Operator 151
Error Function 151
String Functions 151
Picture Functions and Image Processing 153
Complex Numbers, Polar Coordinates, and Mapping Functions 154
Mapping Functions 154
Polar Notation 154
Angle Functions 155
Reading From and Writing To Files 156
Summary 157
Practice 158

CHAPTER 7 PLOTTING 161
Creating a Simple X-Y QuickPlot 161
Creating a Simple Polar Plot 162
Using Range Variables 162
Setting Plotting Ranges 164
Graphing with Units 168
Graphing Multiple Functions 169
Formatting Plots 172
Axes Tab 172
Traces Tab 173
Number Format Tab 174
Labels Tab 174
Defaults Tab 176
Zooming 177
Plotting Data Points 179
Range Variables 179
Data Vectors 180
Numeric Display of Plotted Points (Trace) 180

Using Plots for Finding Solutions to Problems 183
Parametric Plotting 183
Plotting Over a Log Scale 184
Plotting Conics 185
Plotting a Family of Curves 185
3D Plotting 188
Engineering Example 7.1: Shear, Moment, and Deflection Diagrams 192
Engineering Example 7.2: Determining the Flow Properties of a Circular Pipe Flowing Partially Full 192
Input Parameters 195
Summary 196
Practice 197

CHAPTER 8 SIMPLE LOGIC PROGRAMMING 199
Introduction to the Programming Toolbar 199
Creating a Simple Program 200
Return Operator 202
Boolean Operators 204
Adding Lines to a Program 207
Using Conditional Programs to Make and Display Conclusions 211
Engineering Examples 212
Engineering Example 8.1 212
Engineering Example 8.2 215
Summary 216
Practice 216

PART III POWER TOOLS FOR YOUR MATHCAD TOOLBOX

CHAPTER 9 INTRODUCTION TO SYMBOLIC CALCULATIONS 219
Getting Started with Symbolic Calculations 219
Evaluate 222
Float 222
Expand, Simplify, and Factor 227
Explicit 232
Using More than one Keyword 233
Units with Symbolic Calculations 233

Additional Topics to Study 237
Summary 237
Practice 238

CHAPTER 10 **SOLVING ENGINEERING EQUATIONS** **241**
Root Function 241
Polyroots Function 242
Solve Blocks Using Given and Find 247
Isolve Function 250
Solve Blocks Using *Maximize* and *Minimize* 250
TOL, CTOL, and Minerr 252
Using Units 253
Engineering Examples 253
Engineering Example 10.1: Object in Motion 253
Engineering Example 10.2: Electrical Network 256
Engineering Example 10.3: Pipe Network 257
Engineering Example 10.4: Chemistry 259
Engineering Example 10.5: Determining the Flow Properties of a Circular Pipe Flowing Partially Full 261
Engineering Example 10.6: Box Volume 262
Engineering Example 10.7: Maximize Profit 262
Summary 263
Practice 264

CHAPTER 11 **ADVANCED PROGRAMMING** **265**
Local Definition 265
Looping 266
For Loops 267
While Loops 271
Break and *Continue* Operators 273
Return Operator 276
On Error Operator 277
Engineering Example 11.1 278
Development of Intensity Duration Frequency (IDF) Curves 278
Step 1: Compute the Rainfall Intensities (cm/hr) 280
Step 2: Sort Data and Assign a Rank 280

Step 3: Calculate Return Periods Using the Wiebull Formula 282
Step 4: Plot the Rainfall Intensity vs Return Period 282
Step 5: Create a data matrix 283
Step 6: Plot a 5, 10, and 25 Year IDF Curve (Using Data Points) 283
Step 7: Model IDF Curves to Fit a Function of the Form: $i = \frac{a}{t+b}$ 284
Step 8: Plot the IDF Curves Using the Mathematical Formula 286
Summary 287
Practice 287

CHAPTER 12 CALCULUS AND DIFFERENTIAL EQUATIONS 289
Differentiation 289
Integration 291
Differential Equations 292
Ordinary Differential Equations (ODEs) 294
Partial Differential Equations (PDEs) 294
Engineering Example 12.1 298
Integration 298
Shear 298
Moment 298
Slope 298
Deflection 299
Engineering Example 12.2 300
2 Cell, Well Mixed, Lagoon System 300
Mass Balance Equation 301
Input Variables 301
Method 1: *Odesolve* 302
Method 2: *Radau* 304
Practice 308

PART IV CREATING AND ORGANIZING YOUR ENGINEERING CALCULATIONS WITH MATHCAD

CHAPTER 13 PUTTING IT ALL TOGETHER 311
Introduction 311
Guidelines for Naming Variables 312

Naming Guideline 1: Use Descriptive Variable Names 312
Naming Guideline 2: Use a Combination of Uppercase and Lowercase Letters to Help Make Your Variable Names Easier to Read 313
Naming Guideline 3: Use Underscores to Separate Different Names in Your Variable Names 313
Naming Guideline 4: Make Good Use of Subscripts in Your Variable Names 313
Naming Guideline 5: Use the (') Key If You Need to Use a "Prime" (Single Apostrophe) in Your Variable Name 314
Mathcad Toolbox 314
Variables 314
Editing 315
User-Defined Functions 315
Units! 315
Mathcad Settings 316
Customizing Mathcad with Templates 316
Hand Tools 316
Power Tools 316
Let's Start Building 317
What is Ahead 317
Summary 317
Practice 318

CHAPTER 14 MATHCAD SETTINGS 319
Preferences Dialog Box 319
General Tab 320
File Locations Tab 321
HTML Options Tab 322
Warnings Tab 323
Script Security Tab 325
Language 326
Save Tab 326
Summary of the Preference Tab 327
Worksheet Options Dialog Box 327
Built-in Variables Tab 327
Calculation Tab 328

Display Tab 330
Unit System Tab 331
Dimensions Tab 331
Compatibility Tab 331
Result Format Dialog Box 331
Number Format Tab 332
Display Options Tab 333
Unit Display Tab 334
Tolerance Tab 335
Individual Result Formatting 335
Automatic Calculation 336
Summary 337
Practice 338

CHAPTER 15 CUSTOMIZING MATHCAD 341
Default Mathcad Styles 341
Default Math Styles 342
Default Text Styles 342
Additional Mathcad Styles 343
Additional Math Styles 344
Additional Text Styles 347
Changing and Creating New Math Styles 348
Changing the Variables Style 349
Changing the Constants Style 349
Creating New Math Styles 351
Changing and Creating New Text Styles 351
Changing Text Styles 352
Creating New Text Styles 355
Headers and Footers 357
Creating Headers and Footers 357
Information to Include in Headers and Footers 359
Examples 359
Margins and Page Setup 361
Toolbar Customization 362
Summary 363
Practice 363

CHAPTER 16 TEMPLATES 365
Information Saved in a Template 365
Mathcad Templates 366
Review of Chapters 4, 14, and 15 367
Creating Your Own Customized Template 367
EM Metric 368
EM US 376
Normal.xmct File 376
Summary 377
Practice 378

CHAPTER 17 ASSEMBLING CALCULATIONS FROM STANDARD CALCULATION WORKSHEETS 379
Copying Regions from Other Mathcad Worksheets 380
XML and Metadata 380
Provenance 384
Creating Standard Calculation Worksheets 385
Protecting Information 387
Locking Areas 388
Protecting Regions 390
Advantages and Disadvantages of Locking Areas versus Protecting Worksheets 392
Potential Problems with Inserting Standard Calculation Worksheets and Recommended solutions 393
Guidelines 394
How to Use Redefined Variables in Project Calculations 395
Resetting Variables 396
Using User-Defined Functions in Standard Calculation Worksheets 397
Using the Reference Function 400
When to Separate Project Calculation Files 403
Using *Find* and *Replace* 403
Summary 404

CHAPTER 18 IMPORTING FILES FROM OTHER PROGRAMS INTO MATHCAD 405
Introduction 405
Object linking and embedding (OLE) 406
Bringing Objects into Mathcad 406
Drawing Tools 408
Use and Limitations of OLE 408
Common Software Applications that Support OLE 408
Microsoft Excel 409
Microsoft Word or Corel WordPerfect 409
Adobe Acrobat 409
AutoCAD 409
Multimedia 410
Data Files 410
Summary 411

CHAPTER 19 COMMUNICATING WITH OTHER PROGRAMS USING COMPONENTS 413
What is a component? 413
Application Components 414
Microsoft Excel 414
Microsoft Access and Other Open DataBase Connectivity (ODBC) Components 415
MathWorks MATLAB® 418
Data Components 420
Data Import Wizard Component 421
File Input Component 424
Data Table 425
File Output Component 426
Read and Write Functions 427
Data Acquisition 428
Scriptable Object Component 428
Controls 429
Inserting Mathcad into Other Applications 429
Pro/ENGINEER 430
Summary 430

CHAPTER 20 MICROSOFT EXCEL COMPONENT 433
Introduction 433
Inserting New Excel Spreadsheets 434
Multiple Input and Output 435
Hiding Arguments 436
Using Excel Within Mathcad 437
Using Units with Excel 437
Input 437
Output 439
Inserting Existing Excel Files 439
Mechanics 439
Embedding versus Linking 444
Printing the Excel Component 444
Getting Mathcad Data From and Into Existing Excel Spreadsheets 445
Example 447
Summary 447
Practice 448

CHAPTER 21 INPUTS AND OUTPUTS 449
Emphasizing Input and Output Values 449
Input 449
Output 450
Project Calculation Input 451
Variable Names 452
Creating Input for Standard Calculation Worksheets 452
Inputting Information from Mathcad Variables 454
Data Tables 454
Microsoft Excel Component 456
Which Method Is Best to Use? 457
Summarizing Output 457
Controls 457
Text Box 458
Check Box 460
List Box 461
Radio Box 461
Submit Box 462
Summary 462

CHAPTER 22 HYPERLINKS AND TABLES OF CONTENTS 465
Hyperlinks 465
Linking to Regions in Your Current Worksheet 466
Linking to Region in Another Worksheet 467
Notes about Hyperlinks 467
Creating a Pop-Up Document 469
Tables of Contents 470
Mathcad calculation E-book 472
Summary 473

CHAPTER 23 CONCLUSION 475
Advantages of Mathcad 475
Creating Project Calculations 476
Additional Resources 476
Conclusion 477

Appendix 479
Index 495

Preface

This book is a result of feedback from many readers of the book *Engineering with Mathcad: Using Mathcad to Create and Organize your Engineering Calculations*.

The goal of *Engineering with Mathcad* was to get readers using Mathcad's tools as quickly as possible. This was accomplished by providing a step-by-step approach that enabled easy learning. As a result of reader feedback, *Essential Mathcad* makes it even easier to learn Mathcad. We added a new Chapter 1 that quickly introduces many useful Mathcad concepts. By the end of Chapter 1 you should be able to create and edit Mathcad expressions, use the Mathcad toolbars to access important features, understand the difference between the various equal signs, understand math and text regions, know how to create a user-defined function, attach and display units, create arrays, understand the difference between literal subscripts and array subscripts, use range variables, and plot an X-Y graph. Readers felt that the discussion of Mathcad settings and templates in Part 1 slowed down their learning of Mathcad. As a result of this feedback, the chapters Mathcad Settings, Customizing Mathcad, and Templates have been moved to Part IV. These chapters will have more meaning after readers have a greater understanding of Mathcad. Most of the material from *Engineering with Mathcad* is included in this book, but it has been rearranged in order to allow quicker access to Mathcad's tools.

Readers asked for more applied examples of using Mathcad from various disciplines. *Essential Mathcad* provides many additional examples from fields such as: Chemistry, water resources, hydrology, engineering mechanics, sanitary engineering, and taxes. These examples help illustrate the concepts covered in each chapter.

A challenge with any book is to hit a balance between too little material and too much material. Based on feedback from *Engineering with Mathcad*, I feel that we have achieved a good balance in *Essential Mathcad*. Some have said that the first edition did not cover enough advanced topics for their math, physics or advanced engineering courses. Others asked for coverage of some essential engineering topics. On the other hand, some said that the book was too long and covered too much material. Essential Mathcad is an attempt to achieve an even better balance. By adding the new Chapter 1, An Introduction to Mathcad, and rearranging other chapters, I think we have helped make learning Mathcad even easier. By adding discussion of some requested topics, I think we have satisfied the desires of many readers who wanted discussion of more topics. This book cannot and does not include a discussion of all the many Mathcad functions and features. It does attempt to focus on the functions and features that will be most useful to a majority of the readers.

BOOK OVERVIEW

This book uses an analogy of teaching you how to build a house. If you were to learn how to build a house, the final goal would be the completed house. Learning how to use the tools would be a necessary step, but the tools are just a means to help you complete the house. It is the same with this book. The ultimate goal is to teach you how to apply Mathcad to build comprehensive project calculations.

In order to begin building, you need to learn a little about the tools. You also need to have a toolbox where you can put the tools. When building a house, there are simple hand tools and more powerful power tools. It is the same with Mathcad. We will learn to use the simple tools before learning about the power tools. After learning about the tools, we learn to build.

This book is divided into four parts:

Part I—Building Your Mathcad Toolbox. This is where you build your Mathcad toolbox—your basic understanding of Mathcad. It teaches the basics of the Mathcad program. The chapters in this part create a solid foundation upon which to build.

Part II—Hand Tools for Your Mathcad Toolbox. The chapters in this part will focus on simple features to get you comfortable with Mathcad.

Part III—Power Tools for you Mathcad Toolbox. This part addresses more complex and powerful Mathcad features.

Part IV—Creating and Organizing Your Project Calculations with Mathcad. This is where you start using the tools in your toolbox to build something—project calculations. This part discusses embedding other programs into Mathcad. It also discusses how to assemble calculations from multiple Mathcad files, and files from other programs.

ADDITIONAL RESOURCES

This book is written as a supplement to the Mathcad Help and the Mathcad User's Guide. It adds insights not contained in these resources. You should become familiar with the use of both of these resources prior to beginning an earnest study of this book. To access Mathcad Help, click **Mathcad Help** from the **Help** menu, or press the F1 key. The Mathcad User's Guide is a PDF file located in the Mathcad program directory in the "doc" folder.

In addition to the Mathcad Help and the Mathcad User's Guide, the Mathcad Tutorials provide an excellent resource to help learn Mathcad. The Mathcad Tutorials are accessed by clicking **Tutorials** from the **Help** menu. Take the opportunity to review some of the topics covered by the tutorials.

This book (if sold in North America) includes a CD containing the full, non-expiring version of Mathcad v.14. The software is intended for educational use only. The book along with CD provides a complete introduction to learning and using Mathcad. A companion website is provided along with the text and includes links to additional exercises and applications, errata, and other updates related to the book. Please visit www.elsevierdirect.com/9780123747839.

TEMINOLOGY

There are a few terms we need to discuss in order to communicate effectively.

The terms, "click," "clicking" or "select" will mean to click with the left mouse button.

The terms "expression" and "equation" are sometimes used interchangeably. "The term "equation" is a subset of the term "expression." When we use the term "equation," it generally means some type of algebraic math equation that is being defined on the right side of the definition symbol ":=". The term "expression" is broader. It usually means anything located to the right of the definition symbol. It can mean "equation" or it can mean a Mathcad program, a user-defined function, a matrix or vector or any number of other Mathcad elements.

Acknowledgement

Grateful acknowledgement is given to Dixie M. Griffin, Jr., PhD, Professor of Civil Engineering at Louisiana Tech University. A significant number of the engineering examples in this book were adapted from worksheets provided by Dr. Griffin. He has accumulated thousands of Mathcad worksheets over the years. Many of these worksheets are posted on his Mathcad webpage—http://www2.latech.edu/~dmg.

Acknowledgement is also given to the users and reviewers of the first edition who provided feedback that has been incorporated into this new edition, including Colin Campbell, University of Waterloo, Denis Donnelly, Siena College, Robert Newcomb, University of Maryland, John O'Haver, University of Mississippi, and A.J. Wilkerson, University of Hull.

Many engineers have improved my career. Indirectly, they have helped create this book. There are too many to mention by name, but I wish to express thanks to the engineers in the Structural Engineers Association of Utah. I also wish to thank my colleagues at work who have provided continual improvement to our use of Mathcad at work. It is a pleasure to work with them. Special thanks goes to Leon Williams—a good friend and mentor for nearly two decades.

This list of thanks would not be complete without special thanks to my wife Cherie, who put up with the late nights, early mornings, and Saturdays spent working on this book. She has been a stalwart supporter of this effort. I could not have done it without her support. I also want to thank my five sons and two daughter-in-laws who are the joy of my live. They have been very understanding and patient with the time spent away from them.

Brent Maxfield

PART

Building Your Mathcad Toolbox

1

Just as you store tools in a toolbox, you store Mathcad tools in your Mathcad toolbox. Your Mathcad toolbox is the place where you will store your Mathcad skills—the tools that will be discussed in Parts II and III. You build your Mathcad toolbox by learning about the basics of the Mathcad program and the Mathcad worksheet. The chapters in Part I teach about variables, expression editing, user-defined functions, and units. These chapters create a foundation upon which to build. They create your Mathcad toolbox.

CHAPTER

An Introduction to Mathcad

1

This chapter is intended to quickly teach you some fundamental Mathcad concepts. We will only touch the surface of many Mathcad concepts. In later chapters, we will get into more depth, and build on the concepts covered in this chapter. This chapter also teaches techniques to create and edit Mathcad expressions.

Chapter 1 will:

- Show how to do simple math in Mathcad.
- Teach how to assign and display variables.
- Explain how to create and edit math expressions.
- Demonstrate the editing cursor and the different forms it takes.
- Discuss the use of operators.
- Demonstrate how to wrap a math region.
- Briefly discuss the Mathcad toolbars.
- Introduce and define math and text regions.
- Introduce built-in and user-defined functions.
- Introduce units.
- Introduce arrays and subscripts.
- Discuss the variable ORIGIN.
- Describe the difference between literal and array subscripts.
- Introduce range variables.
- Introduce X-Y plots.
- Encourage completing several Mathcad tutorials.

BEFORE YOU BEGIN

If you don't already have Mathcad installed on your computer, take a few minutes and install the included version of Mathcad 14. This is the full unexpiring version of Mathcad. This will allow you to follow along and practice the concepts discussed in this book. It will also give you access to Mathcad Help and Mathcad Tutorials.

Essential Mathcad is based on the US version of Mathcad. It is also based on the US keyboard. There may be slight differences in Mathcad versions sold outside of the United States.

We suggest that you read and do the exercises in the Mathcad tutorial before or just after reading this chapter. You can open the Mathcad tutorial by clicking **Tutorials** from the **Help** menu. This opens a new window called the Mathcad Resources window. In this window you will see a list of Mathcad tutorials. Click the **Getting Started Primers**. Each of these primers is excellent. You may choose to do them all, but for the purpose of this chapter, focus on the following topics: Entering Math Expressions, Building Math Expressions, Editing Math Expressions, First Things First, and Adding Text and Images. This chapter cannot replace the experience gained by completing the Mathcad tutorials.

Mathcad Basics

Whenever you open Mathcad, a blank worksheet appears. You can liken this worksheet to a clean sheet of calculation paper waiting for you to put information on it.

Let's begin with some simple math. Type `5+3=`. You should get the following:

$$5 + 3 = 8 \ \blacksquare$$

Now type `(2+3)*2=`. You should get the following:

$$(2 + 3) \cdot 2 = 10 \ \blacksquare$$

You can also assign variable names to these equations. To assign a value to a variable, type the variable name and then type the colon `:` key. For example, type `a1:5+3`.

$$a1 := 5 + 3$$

Now type `a1=`. This evaluates and displays the value of variable a1.

$$a1 = 8 \ \blacksquare$$

Let's assign another variable. Type `b1:(2+3)*2`.

$$b1 := (2 + 3) \cdot 2$$

Now type `b1=`. This displays the value of variable b1.

$$b1 = 10 \ \blacksquare$$

Now that values are assigned to variable a1 and variable b1, you can use these variables in equations. Type `c1:a1+b1`.

c1 := a1 + b1

Now type `c1=`. You should get the following result:

c1 = 18▪

> **Tip!** As you begin using variables, it is important to understand the following Mathcad protocol. In order to use a previously defined variable, the variable must be defined above or to the left of where it is being used. In other words, Mathcad calculates from left to right, top to bottom.

As you can see, Mathcad does not require any programming language to perform simple operations. Simply type the equations as you would write them on paper.

CREATING SIMPLE MATH EXPRESSIONS

There are two ways to create a simple expression. The first way is to just type as you would say the expression. For example, you say 2 plus 5, so you would type the following `2+5` . You say 2 to the 4th power, so you would type `2^4`. You say the square root of 100, so you type `\100`.

The second way to create a simple expression is to type an operator such as +, —, *, or /. This will create empty placeholders (black boxes) that you can then click to fill in the numbers or operands. For example, if you press the + key anywhere in your worksheet, you will get the following:

▪ + ▪

Click in the first placeholder and type `2`, then press `TAB` or click in the second placeholder and type `5`. Your expression should now look like this:

2 + 5

In this example, 2 and 5 are operands of the + operator.

You can use this procedure with any operator. Let's try the exponent operator. Press `^` to create the exponent operator. You can also click x^Y on the calculator toolbar. You should have the following:

▪^▪

Click in the lower placeholder and type `2`, then press `TAB` or click in the upper placeholder and type `4`. Your expression should now look like this:

$$2^4$$

These methods of creating expressions work very well for creating simple expressions. As your expressions become more complex, there are a few things we must learn.

EDITING LINES

Creating more complex math expressions is very easy once you learn the concept of the editing lines. These are similar to a two-dimensional cursor with a vertical and a horizontal component. There is a vertical editing line and a horizontal editing line. As an expression gets larger, the editing lines can grow larger to contain the expanding expression. Notice how in the previous examples the editing lines just contained a single operand. Pressing the spacebar will cause the editing lines to grow to hold more of the expression. For example, if you type `2+5 spacebar`, you get the following:

$$2 + 5$$

Whatever is held between the editing lines becomes the operand for the next operator. So, if you type `2+5 spacebar^3`, you get the following:

$$(2 + 5)^3$$

In this case (2+5) is the x operand for the operator x to the power of y. Notice how the editing lines now contain only the number 3. This means that if you type any operator, the number 3 is the operand for the operator. Thus, if you type `+ 4`, you get the following:

$$(2 + 5)^{3+4}$$

But, if you press the `spacebar` first, the editing lines expand to enclose the whole expression. This expression becomes the operand for the next operator. Thus, if you now type `+ 4`, you get the following:

$$(2 + 5)^3 + 4$$

The whole expression became the operand for the addition operator.

It is very important to understand this concept of using the editing lines to determine what the operand is of your next operator. You can also use parentheses to set the operand for operators. Pressing the single quote (`'`) adds a pair of opposing parentheses.

The following example will help reinforce these concepts. Let's create the following expression:

$$\frac{\left(\frac{1}{2}-\frac{1}{3}\right)^2}{\sqrt{\frac{4}{5}+\frac{2}{7}}}$$

To create this expression, use the following steps:

1. Type `1/2 spacebar`. The editing lines now hold the fraction 1/2. This becomes the operand for the subtraction operator.

$$\frac{1}{2}$$

2. Type `- 1 / 3 spacebar spacebar`. The editing lines should now hold both fractions. This becomes the operand for the power operator.

$$\frac{1}{2}-\frac{1}{3}$$

3. Type `^2 spacebar`. The editing lines should now hold the entire numerator. This becomes the operand for the division operator.

$$\left(\frac{1}{2}-\frac{1}{3}\right)^2$$

4. Type `/\(or use the square root icon on the math toolbar) 4/5 spacebar spacebar`. This makes everything under the radical the operand for the addition operator.

$$\frac{\left(\frac{1}{2}-\frac{1}{3}\right)^2}{\sqrt{\frac{4}{5}}}$$

5. Type **+ 2 / 7**. This completes the example.

$$\frac{\left(\frac{1}{2}-\frac{1}{3}\right)^2}{\sqrt{\frac{4}{5}+\frac{2}{7}}}$$

Notice how during each step, the spacebar was used to enlarge the editing lines to include the operand for the following operator.

The Mathcad tutorial has additional examples that provide worthwhile practice.

EDITING EXPRESSIONS

Another important concept to know is how to edit existing expressions. In order to understand this concept, it is important to understand how to move the vertical editing line. This vertical editing line can be moved left and right using the left and right arrow keys. You can also toggle the vertical editing line from the right side to the left side and back by pressing the **INSERT** key. For expressions that are more complex you can also use the up and down arrows to move both editing lines.

Selecting Characters

If you click anywhere in an expression and then press the spacebar, the editing lines expand to include more and more of the expression. How the editing lines expand depend on where you begin and on what side the vertical editing line is on. The editing lines work differently in different versions of Mathcad. The best way to understand how they work is to experiment and to follow the examples in the Mathcad tutorial.

> **Tip!** I have found that if you begin with the vertical editing line on the right side of the horizontal editing line, the expansion of the editing lines makes more sense. The general rule is that as the editing lines expand and cross an operator, the operand for that operator is then included within the lines.

Deleting Characters

You can delete characters in your expressions by moving the vertical editing line adjacent to the character. If the vertical editing line is to the left of the character, press the **DELETE** key. If the vertical editing line is to the right of the character, press the **BACKSPACE** key.

$\frac{1}{2}+\frac{1}{4}$ $\frac{1}{2}\square\frac{1}{4}$ $\frac{1}{2}-\frac{1}{4}$

FIGURE 1.1 Replacing an operator

To delete multiple characters, drag-select the portion of the expression you want to delete. If the vertical editing line is to the left of the highlighted area, press the **DELETE** key. If the vertical editing line is to the right of the highlighted area, press the **BACKSPACE** key.

Deleting and Replacing Operators

To replace an operator, place the editing lines so that the vertical editing line is just to the left of the operator. Next, press the **DELETE** key. This will delete the operator, usually leaving a hollow box symbol where the operator used to be. Now, type a new operator, and it will replace the box symbol. See Figure 1.1.

You may also have the vertical editing line to the right of the operator and use the **BACKSPACE** key to delete and replace the operator.

The best way to understand this concept is to experiment with it.

WRAPPING EQUATIONS

There are times when a very long expression might extend beyond the right margin. If this is the case, the entire expression will not print on the same sheet of paper.

There is a way to wrap your equations so that they are contained on two or more lines; however, you are only able to wrap equations at an addition operator.

To wrap an equation, press **CTRL+ENTER** just prior to an addition operator. Mathcad inserts three dots indicating that the expression is to be continued on a following line. On the following line, Mathcad inserts the addition operator with a placeholder box. Because Mathcad automatically inserts the addition operator, you are not able to wrap an equation at other operators.

> **Tip!** You may wrap an equation at a subtraction operator by making the following operand a negative number (in essence adding a negative number).

See Figure 1.2 for examples of wrapping equations.

$$\text{Example}_1 := 1 + 2 + 3 + 4 + 5 + 6 \ldots + 7 + 8 + 9$$ Wrapped Equation

$$\text{Example}_1 = 45$$

$$\text{Example}_2 := 1 + 5 + 3 + 4 \ldots + -3 + 2$$ Wrapped Equation using negative number.

$$\text{Example}_2 = 12$$

$$\text{Example}_3 := \left(1 + 2 + 3 + 4 + 5 + 6 \ldots + 7 + 8 + 9\right)\cdot 2$$ Multiplication using wrapped equation.

$$\text{Example}_3 = 90$$

FIGURE 1.2 Wrapping equations

TOOLBARS

Now that you understand how to create and edit Mathcad expressions, let's start exploring some of Mathcad's features.

One of the easiest ways to access many of Mathcad's features is by the use of toolbars. You access Mathcad toolbars by clicking **Toolbars** from the **View** menu. For our discussion it is important to have the following toolbars turned on: Standard, Formatting, and Math. See Figure 1.3 to see these toolbars.

The Math toolbar allows you to quickly access many of the other toolbars. From this toolbar you will be able to open the following toolbars: Calculator, Graph, Vector and Matrix, Evaluation, Calculus, Boolean, Programming, Greek Symbol, and Symbolic Keyword. Hover your mouse above each icon on the Math toolbar to see a tooltip reminding you which toolbar each icon opens.

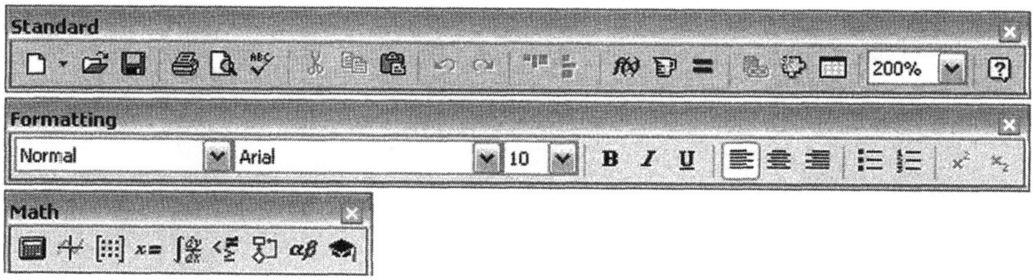

FIGURE 1.3 Standard, Formatting, and Math toolbars

Calculator Toolbar

The Calculator toolbar allows you to quickly access some basic math operators and trigonometric functions. See Figure 1.4. The Calculator toolbar behaves just like a calculator. It inserts the numbers and operators into Mathcad as you click the buttons on the toolbar. If you click an operator prior to entering numbers, Mathcad inserts blank placeholders into the worksheet. Press the TAB key to move between placeholders.

In-Line Division

In-line division is a way to save space when you have several divisions in your expression. It displays division similar to a textbook. To add an in-line division operator to your expression, type CTRL+/ rather than just the /. You can also use the division (÷) icon on the Calculator toolbar. See Figure 1.5.

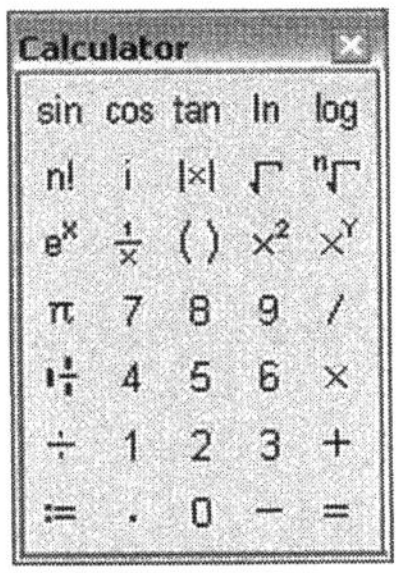

FIGURE 1.4 Calculator toolbar

$$\textbf{Result}_1 := \frac{\dfrac{\dfrac{540}{45-52}}{6}}{\dfrac{7+\dfrac{3}{5}}{\dfrac{45}{6}}} \qquad \textbf{Result}_2 := \frac{540}{\dfrac{45-52}{6}} \div \left(7+\frac{3}{5}\right) \div \frac{45}{6}$$

$$\textbf{Result}_1 = -8.12 \qquad \textbf{Result}_2 = -8.12$$

$$\textbf{Result}_3 := 540 \div [(45-52) \div 6] \div [7+(3 \div 5)] \div (45 \div 6)$$

$$\textbf{Result}_3 = -8.12$$

Use CTRL+ FORWARD SLASH
or the Calculator toolbar to get In-line Division.

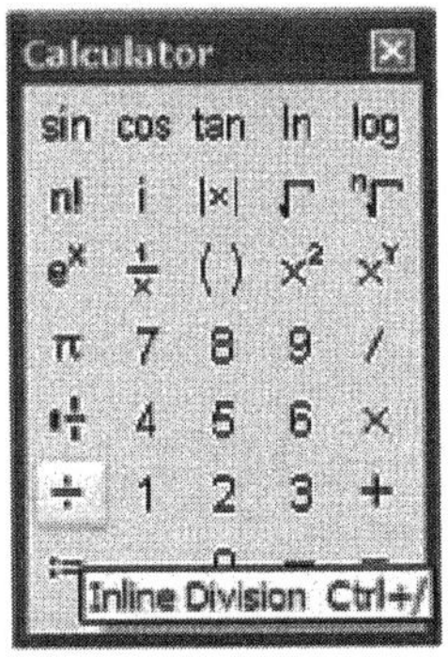

FIGURE 1.5 In-line division

Mixed Numbers

Mixed numbers allow you to input and show values as integers and fractions. To enter a mixed number press CTRL+SHIFT+PLUS or use the icon on the Calculator toolbar. See Figure 1.6.

To display results as mixed fractions, double-click the displayed result. This opens the Result Format dialog box. Select **Fraction** from the Format list, and check the "Use mixed numbers" check box.

Greek Toolbar

The Greek toolbar allows you to quickly enter Greek letters. See Figure 1.7. Chapter 2 will discuss Greek letters in more detail.

Summary of Equal Signs

There are four equal signs used in Mathcad. It is important to understand the difference between them.

- The assignment operator (:=) COLON is used to define variables, functions, or expressions.

Conventional input. Note that first number was not input as 5.667.

$$\mathbf{MixedNumber_1} := 5.666 + 3.333 + 5.625$$

$\mathbf{MixedNumber_1} = 14.62$ Normal Result format

$\mathbf{MixedNumber_1} = \frac{1828}{125}$ Fraction Result format

$\mathbf{MixedNumber_1} = 14\frac{78}{125}$ Fraction Result format with "Use mixed Numbers" checked.

Same equation using mixed number format - CTRL+SHIFT+PLUS.

$$\mathbf{MixedNumber_2} := 5\frac{2}{3} + 3\frac{1}{3} + 5\frac{5}{8}$$

$\mathbf{MixedNumber_2} = 14.63$ Normal Result format

$\mathbf{MixedNumber_2} = \frac{117}{8}$ Fraction Result format

$\mathbf{MixedNumber_2} = 14\frac{5}{8}$ Fraction Result format with "Use mixed Numbers" checked.

FIGURE 1.6 Mixed numbers

FIGURE 1.7 Greek toolbar

- The evaluation operator (=) `EQUAL SIGN` is used to evaluate a variable, function, or expression numerically.
- The Boolean equality operator (=) `CTRL+EQUAL SIGN` is used to evaluate the equality condition in a Boolean statement. It is also used for programming, solving, and in symbolic equations. It will be discussed in more detail in future chapters.
- The global assignment operator (≡) `TILDA ~ or SHIFT+ACCENT` is used to assign a global variable. All global assignment definitions in the worksheet are scanned by Mathcad prior to scanning for normal assignment definitions. This means that global assignments can be defined anywhere in the worksheet and still be recognized. Global assignments should be used with caution.

> **Tip!** The use of global definitions is discouraged because they do not participate in redefinition warnings, and they can create confusing redefinition chains if used in the middle of a document.

The assignment operator, evaluation operator, and global assignment operator are found on the Evaluation toolbar. The Boolean equality operator is found on the Boolean toolbar.

REGIONS

A region is a location where information is stored on the worksheet. Your entire Mathcad worksheet will be comprised of individual regions. You can view the regions in your worksheet by clicking **Regions** from the **View** menu. There are two types of regions—math regions and text regions.

Math Regions

Math regions contain variables, constants, expressions, functions, plots, among others. These regions are basically anything except text regions. These regions are created automatically whenever you create any expression or definition.

Text Regions

Text regions allow you to add notes, comments, titles, headings, and other items of interest to your calculation worksheet. There are several ways to create a text region. The simplest way to create a text region is to start typing text. As soon as you use the spacebar, Mathcad converts the math region into a text region. This is a handy feature, unless you press the spacebar by accident when you are entering a variable name. Once a math region is converted to a text region, it cannot be changed back to a math region. (You can use the **undo** command, if you immediately catch the mistake.) Other ways to create text regions are to use the double quote (") key, or choose **Text Region** from the **Insert** menu.

When you are finished typing the text, if you press the ENTER key, Mathcad inserts a new paragraph in the same text region. In order to exit a text region, click outside the region. You can also press CTRL+SHIFT+ENTER, or you can use the arrow keys to move the cursor outside the text region.

In Chapter 2, we will discuss text regions in much more depth.

FUNCTIONS

Functions will be discussed briefly in Chapter 3 and built upon throughout the book. The following paragraphs will get you started.

Built-In Functions

Mathcad has hundreds of built-in functions. You access these function from the Insert Function dialog box, which is opened by selecting **Functions** from the **Insert** menu. See Figure 1.8. You can also type CTRL+E or click the $f(x)$ icon on the Standard toolbar. The Insert Function dialog box lists categories of functions on the left, and lists function names on the right. The boxes below the function name list the arguments expected and a brief description of each function.

User-Defined Functions

User-defined functions are very similar to built-in functions. They consist of a name, a list of arguments (in parentheses following the name), and a definition giving the relationship between the arguments. The name of the user-defined

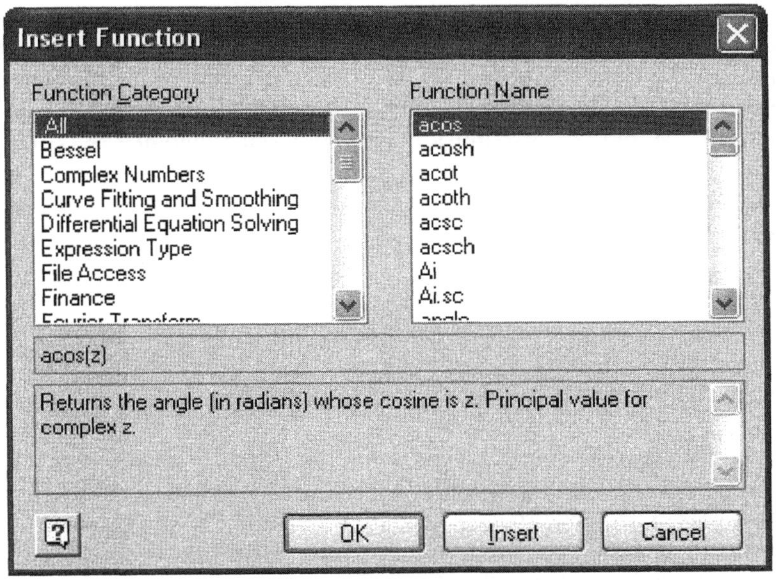

FIGURE 1.8 Insert Function dialog box showing the function categories and function names

$\mathbf{CircleArea(r)} := \pi\,\mathbf{r}^2$	$\mathbf{CircleArea}(5) = 78.54$
$\mathbf{SquareArea(L)} := \mathbf{L}^2$	$\mathbf{SquareArea}(4) = 16.00$
$\mathbf{RectangleArea(L,H)} := \mathbf{L}\cdot\mathbf{H}$	$\mathbf{RectangleArea}(4,5) = 20.00$
$\mathbf{BoxVolume(L,H,W)} := \mathbf{L}\cdot\mathbf{H}\cdot\mathbf{W}$	$\mathbf{BoxVolume}(4,5,10) = 200.00$

FIGURE 1.9 User-defined functions

function is simply a variable name. See Figure 1.9 for an illustration of using some user-defined functions.

UNITS

This section is intended only to get you started with units. Many experienced Mathcad users still do not understand the significant benefits of using Mathcad units, so it is important to read and study Chapter 4, "Units!"

Once a unit is assigned to a variable, Mathcad keeps track of it internally and displays the unit automatically. You will never need to remember the conversion factors for various units. You will never need to convert it from one unit system to another. Mathcad does it all for you. All you need to do is tell Mathcad how you want the unit displayed. For example, you can attach the unit of meters (m) to a variable. You will then be able to tell Mathcad to display this variable in any unit of length such as millimeters (mm), centimeters (cm), kilometers (km), inches (in), yards (yd), or miles (mi). Mathcad does the conversion for you. If Mathcad does not have the unit of measurement built in, you can define it, and use it over and over.

Assigning Units to Numbers

To assign units to a number, simply multiply the number by the name of the unit. If you cannot remember the name of the unit, you can select from a list of over 100 built-in Mathcad units. These are found in the Insert Unit dialog box. See Figure 1.10.

To open the Insert Unit dialog box, select **Unit** from the **Insert** menu. You can also click the measuring cup icon in the Standard Toolbar, or you can use the shortcut CTRL+U. See Figure 1.11. The System shown in the Insert Unit dialog box is the default unit system selected from the Worksheet Options dialog box. Chapter 4 will discuss the various unit systems. If you select **All** from the Dimension box, then all the built-in units available for that system will be shown in the Unit box. Note that some units will be available only for some specific unit systems.

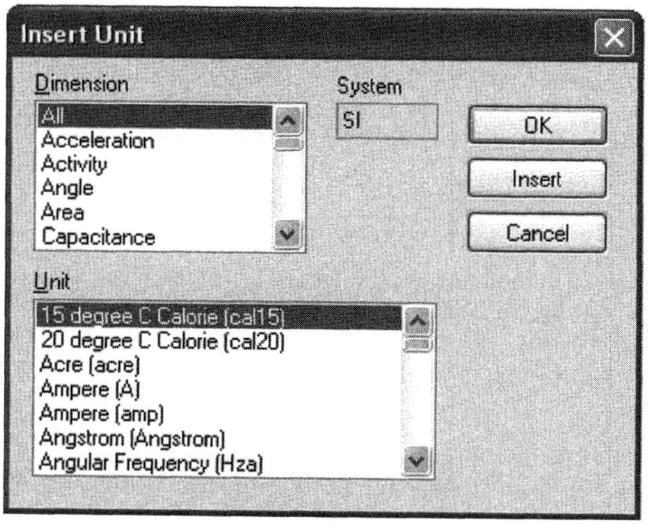

FIGURE 1.10 Insert Unit dialog box

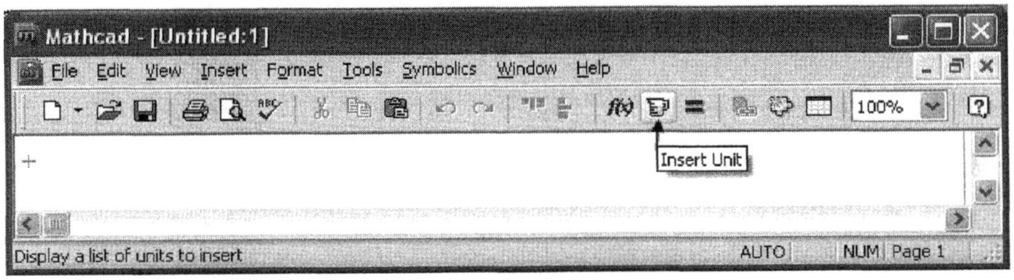

FIGURE 1.11 Icon to insert units

Examples of units attached to numbers and their default results. The Mathcad default unit system is set to SI, so units display in the SI system.

If you don't know the unit names, then use the Insert Units dialog box. Type the number, type "*", and then click "Insert."

$1\text{ft} = 0.305\,\text{m}$	Units of length
$180\text{deg} = 3.142$	Units of angle. The result is in radians.
$1\text{gal} = 3.785\,\text{L}$	Units of volume
$1\text{min} = 60\,\text{s}$	Units of time
$1\text{gm} = 1 \times 10^{-3}\,\text{kg}$	Units of mass

FIGURE 1.12 Examples of units attached to numbers

To assign units from the Insert Unit dialog box, type a number, type the asterisk [*] and select the desired unit from the Unit box, then click **OK**. Figure 1.12 shows some examples of units attached to numbers.

Evaluating and Displaying Units

When you evaluate an expression with a unit attached, the unit Mathcad displays by default is based on the chosen default unit system (see Chapter 4). After evaluating an expression by pressing the [=] key, Mathcad displays the default unit followed by a solid black box. This box is the unit placeholder. If you want Mathcad to display a unit different from the default unit, click the unit placeholder and type the name of the unit you want displayed. You can also double-click the unit placeholder and select a unit from the Insert Unit dialog box. See Figure 1.13 for some examples of displaying units.

$\text{Units}_F := 3\text{ft} + 1\text{m} + 33\text{mm} + 4\text{yd}$

$\text{Units}_F = 5.605\,\text{m}$

Mathcad displays in the default unit system (in this case SI). To display the results in inches, do the following:

Type: Units.F=[tab]in[enter]

$\text{Units}_F = 220.669\,\text{in}$

To display the results in mm, do the following:

Type: Units.F=[tab]mm[enter]

$\text{Units}_F = 5.605 \times 10^{3}\,\text{mm}$

You can display the results in many different units:

$\text{Units}_F = 5.605 \times 10^{-3}\,\text{km}$

$\text{Units}_F = 3.483 \times 10^{-3}\,\text{mi}$

$\text{Units}_F = 0.028\,\text{furlong}$

$\text{Units}_F = 6.13\,\text{yd}$

$\text{Units}_F = 18.389\,\text{ft}$

FIGURE 1.13 Displaying results in different units

> **Tip!** After evaluating an expression, press the **TAB** key to automatically move you to the unit placeholder.

Chapters 14, 15, and 16 will show how to set and keep default unit systems.

ARRAYS AND SUBSCRIPTS

An array is simply a vector or a matrix. A vector is a matrix with only a single column. This section briefly introduces the topic; Chapter 5 will have a much more in-depth discussion.

Creating Arrays

Use the Insert Matrix dialog box to create a matrix. This dialog box can be accessed in three ways: selecting **Matrix** from the **Insert** menu, typing the shortcut **CTRL+M**., or selecting the matrix icon (showing a three-by-three matrix) on the Matrix toolbar. See Figure 1.14.

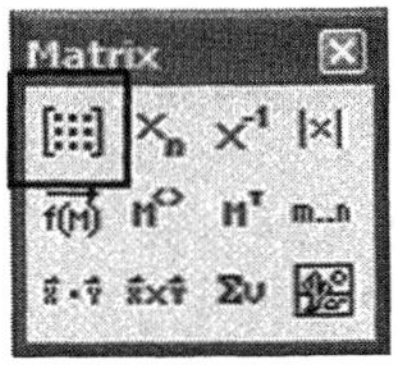

FIGURE 1.14 Insert matrix icon on Matrix toolbar

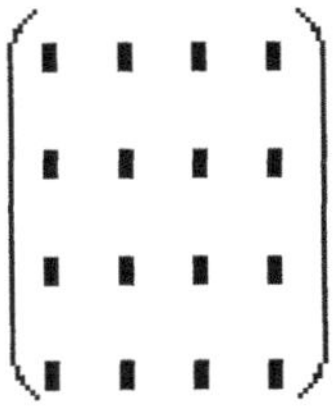

FIGURE 1.15 Blank 4×4 matrix

Once the Insert Matrix dialog box is open, change the number of rows and columns to the desired numbers and click **OK**. For example, if you type 4 and 4 in the Rows and Columns boxes you will get a matrix as shown in Figure 1.15.

Now, simply fill in the placeholders with numbers or expressions. Use the TAB key or arrow keys to move from placeholder to placeholder. See Figure 1.16 for two sample matrix definitions using numbers and expressions.

Once you create a vector or matrix, you can add additional rows or columns by using the Insert Matrix dialog box. To do this, select an element in the vector or matrix, and then open the Insert Matrix dialog box. Mathcad will insert

$$\text{Matrix\_1} := \begin{pmatrix} 1 & 2 & 3 & 4 \\ 5 & 6 & 7 & 8 \\ 9 & 10 & 11 & 12 \\ 13 & 14 & 15 & 16 \end{pmatrix} \qquad \text{Matrix\_1} = \begin{pmatrix} 1.00 & 2.00 & 3.00 & 4.00 \\ 5.00 & 6.00 & 7.00 & 8.00 \\ 9.00 & 10.00 & 11.00 & 12.00 \\ 13.00 & 14.00 & 15.00 & 16.00 \end{pmatrix}$$

$$\text{Matrix\_2} := \begin{pmatrix} 1+2 & 2+3 & 3+4 & 4+5 \\ 5+6 & 6+7 & 7+8 & 8+9 \\ 9+10 & 10+11 & 11+12 & 12+13 \\ 13+14 & 14+15 & 15+16 & 16+17 \end{pmatrix} \qquad \text{Matrix\_2} = \begin{pmatrix} 3.00 & 5.00 & 7.00 & 9.00 \\ 11.00 & 13.00 & 15.00 & 17.00 \\ 19.00 & 21.00 & 23.00 & 25.00 \\ 27.00 & 29.00 & 31.00 & 33.00 \end{pmatrix}$$

FIGURE 1.16 Sample matrix definitions

additional rows below the selected element, and insert additional columns to the right of the selected element. Tell Mathcad how many additional rows and/or columns you want to add. If you want to add one additional row, but not an additional column, then type 1 for row and 0 for column. If you want to add rows above or columns to the left, then select the entire vector or matrix prior to using the Insert Matrix dialog box. After entering the number of rows and/or columns, click **OK** or **Insert**. If you select **Insert** first, be sure to click **Close** to close the box. If you click **OK** to close the box, additional rows and/or columns will be added.

You can also use the Insert Matrix dialog box to remove rows and columns. To do this, select an element in the row or column you want to delete. Tell Mathcad how many rows and/or columns you want to delete, and then click **Delete**. Mathcad will delete the row and/or column of the selected element and additional rows below the element and additional columns to the right of the element. Be sure to click **Close** to close the box. If you click **OK**, additional rows and/or columns will be added.

Origin

The value of the variable name ORIGIN tells Mathcad the starting index of your array. The Mathcad default for this variable is 0. This means that a vector or matrix begins indexing with zero. In other words, the first element is the 0th element. Thus, in Matrix_1 of Figure 1.16, the value of the 0th element of the matrix (Matrix_1(0,0)) would be 1.

> **Tip!** I find it awkward to begin array numbering with 0. I like the first variable in an array to be labeled 1 rather than 0. I set the built-in variable ORIGIN to the value of 1.

For most scientific and engineering calculations, it is suggested that you change the value of ORIGIN from 0 to 1. With the value of ORIGIN set at 1, the first element of a matrix is the 1st element. Thus, in Matrix_1 of Figure 1.16, the value of the first element of the matrix (Matrix_1(1,1)) would be 1. For the remainder of this book, the value of ORIGIN will be set at 1.

To change the value of ORIGIN, use the Built-In Variables tab in the Worksheet Options dialog box. You open this dialog box by clicking **Worksheet Options** from the **Tools** menu. On the Built-in Variables tab, change the value of Array Origin (ORIGIN) from 0 to 1. See Figure 1.17.

Subscripts

A discussion of arrays would not be complete without a discussion of subscripts. It is critical to understand the difference between two types of subscripts because they behave very differently. These two types of subscripts are called literal subscripts and array subscripts.

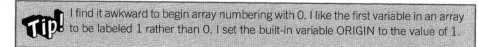

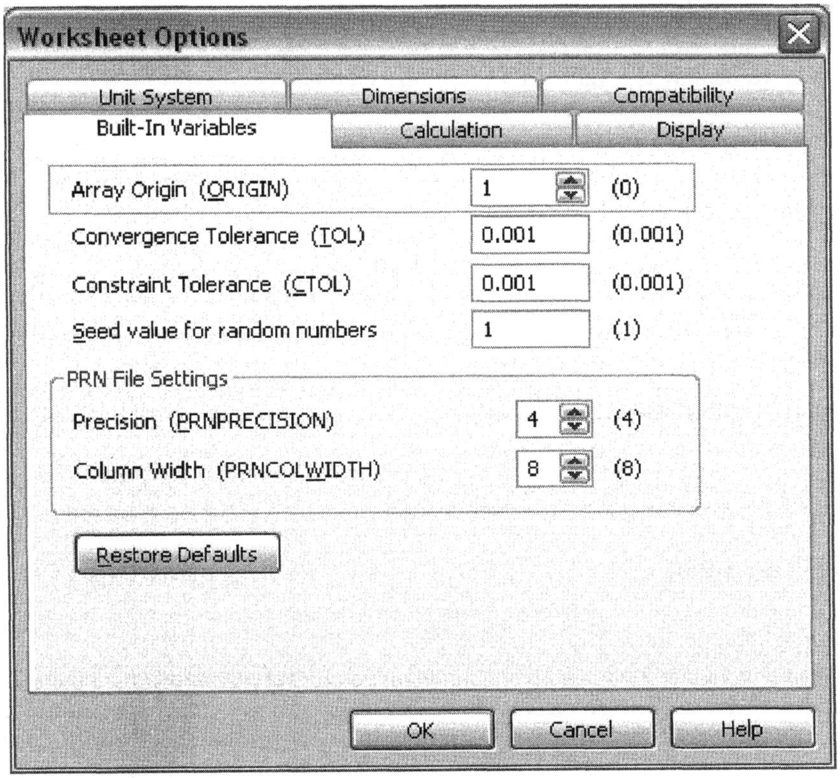

FIGURE 1.17 Built-In Variables tab of Worksheet Options dialog box

Literal Subscripts

Literal subscripts are part of a variable name. They allow you to have variable names such as F_s or f_y. To type a literal subscript, type the first part of the variable name, and then type a period. The insertion point will drop down half a line. All characters typed after this point will be part of the subscript. (See Chemistry Notation in Chapter 2 for an exception.) See Figure 1.18 for an example of variable names using literal subscripts.

Example_1

Sample_2

f_y

f'_c

FIGURE 1.18 Example of variable names using literal subscripts

Array Subscripts

An array subscript is not part of the variable name. An array subscript allows Mathcad to display the value of a particular element in an array. It is used to refer to a single element in the array. The array subscript is created by using the `[` key. This is referred to as the subscript operator. Thus, if you want Mathcad to display the value of the first element in Matrix_1 in Figure 1.16 you would type: `Matrix_1[1,1=`. $\textbf{Matrix\_A}_{1,1} = 1.00$ (remember we changed ORIGIN from 0 to 1). If you want Mathcad to display the value of the element in the 3rd row, 4th column, you would type `Matrix_1[3,4=` $\textbf{Matrix\_A}_{3,4} = 12.00$.

In this example, the variable name was Matrix_1. The variable contains a 4 row–4 column matrix. The array subscript is not part of the variable name. It is used only to display an element of the array.

You can also use an array subscript to assign elements of an array. If you type `Matrix_1[1,1:20` then the value of the 1st element in Matrix_1 will be changed from 1 to 20. See Figure 1.19.

Figure 1.20 shows how to use array subscripts for a vector. Figure 1.21 shows how to use array subscripts to assign new values to vectors and arrays.

Range Variables

Range variables will be used extensively in later chapters, but this section will only introduce the concept.

A range variable is similar to a vector in that it takes on multiple values. It has a range of values. The range of values has a beginning value, an ending value, and uniform incremental values between the beginning and ending values. Range variables can be used to iterate a calculation over a specific range of values, or to plot a function over a specific range of values. They often are used as integer subscripts for defining arrays. A range variable looks like this: RangeVariableA:=1, 1.5 .. 5. This range variable begins with 1.0. The second number in the range variable sets the increment value. Mathcad takes the difference between the first and second numbers and uses this as the incremental value. In this case, the increment is 0.5. The last number in this range is 5.0. Thus, this range variable has the values 1.0, 1.5, 2.0, 2.5, 3.0, 3.5, 4.0, 4.5, and 5.0.

To define a range variable, type the variable name followed by a colon `:`. This creates the variable definition. In the placeholder, type the beginning value, and

$$\textbf{Matrix\_1}_{1,1} := 20$$

$$\textbf{Matrix\_1} = \begin{pmatrix} 20.00 & 2.00 & 3.00 & 4.00 \\ 5.00 & 6.00 & 7.00 & 8.00 \\ 9.00 & 10.00 & 11.00 & 12.00 \\ 13.00 & 14.00 & 15.00 & 16.00 \end{pmatrix}$$

FIGURE 1.19 Changing the value of a single array element

$$\text{Vector\_1} := \begin{pmatrix} 2 \\ 22 \\ 222 \\ 2222 \end{pmatrix} \qquad \text{Vector\_1} = \begin{pmatrix} 2.00 \\ 22.00 \\ 222.00 \\ 2222.00 \end{pmatrix}$$

$\text{Vector\_1}_1 = 2.00$

$\text{Vector\_1}_2 = 22.00$

$\text{Vector\_1}_3 = 222.00$

$\text{Vector\_1}_4 = 2222.00$

$\text{Vector\_1}_0 = \blacksquare$ With ORIGIN set to one, there is no zero element

There are only 4 elements. The 5th is not recognized.

$\text{Vector\_1}_5 = \blacksquare$ $\boxed{\text{Vector\_1}_5 = \blacksquare\,\blacksquare}$

This array index is invalid for this array.

FIGURE 1.20 Using array subscripts

You can add additional elements to an array, by defining them with array subscripts.

$\text{Vector\_1}_6 := 22222$ $\qquad$ $\text{Matrix\_1}_{5,5} := 3333$

$$\text{Vector\_1} = \begin{pmatrix} 2.00 \\ 22.00 \\ 222.00 \\ 2222.00 \\ 0.00 \\ 22222.00 \end{pmatrix} \qquad \text{Matrix\_1} = \begin{pmatrix} 20.00 & 2.00 & 3.00 & 4.00 & 0.00 \\ 5.00 & 6.00 & 7.00 & 8.00 & 0.00 \\ 9.00 & 10.00 & 11.00 & 12.00 & 0.00 \\ 13.00 & 14.00 & 15.00 & 16.00 & 0.00 \\ 0.00 & 0.00 & 0.00 & 0.00 & 3333.00 \end{pmatrix}$$

FIGURE 1.21 Using array subscripts

then type a comma [,]. This adds a second placeholder in the expression. Now enter the second value in the placeholder. The second value sets the incremental value. Now type a semicolon [;]. This places two dots in the worksheet, and adds a third placeholder. Enter the ending value in the placeholder. If the second value is less than the beginning value, the range variable will be decreasing, and the last value sets the lower limit to the range variable. See Figure 1.22 for sample range variables and their displayed results.

Comparing Range Variables to Vectors

Because range variables and vectors are similar, it is important to understand the difference between them. Table 1.1 is a comparison of range variables and vectors.

The second value sets the increment.

RangeVariable_A := 5,5.1..6

RangeVariable_A

5
5.1
5.2
5.3
5.4
5.5
5.6
5.7
5.8
5.9
6

Range variables may also decrease

RangeVariable_B := 10,8..−10

RangeVariable_B

10.00
8.00
6.00
4.00
2.00
0.00
-2.00
-4.00
-6.00
-8.00
-10.00

If a second value is not given, then an increment of 1 is used.

RangeVariable_C := 1..10

RangeVariable_C

1.00
2.00
3.00
4.00
5.00
6.00
7.00
8.00
9.00
10.00

FIGURE 1.22 Sample range variables

Table 1.1 Comparing Range Variables and Vectors

Range Variables	Vectors
Range variables must increment (up or down) in uniform steps.	Vectors may have numbers in any order.
Range variables must be real.	Vectors may use real or complex numbers.
You cannot access individual elements of range variables.	Each element of a vector can be accessed by using array subscripts.
When using range variables in calculations, the results are displayed, but the individual results are not accessible.	When using vectors in calculations, the results are also displayed, but each individual result is accessible. See Chapter 5 for details.
Range variables can be used to iterate calculations over a range of values. The calculation is performed once for each value in the range.	Vectors can also be used as arguments for calculations. The calculation is performed once for each value in the vector.
Range variables often are used as subscripts to write or access data in vectors and matrices.	Range variables (starting at ORIGIN and incrementing by 1) can be used to create a vector of values.
Range variables begin at the defined beginning value.	Vectors use ORIGIN as the first element.

PLOTTING: X-Y PLOTS

Graphing Toolbar

The Graphing toolbar is shown in Figure 1.23. The Graphing toolbar allows you to quickly insert two-dimensional X-Y plots, Polar plots, and three-dimensional plots. Plotting will be discussed at length in Chapter 7, but let's take a quick look at how to create some simple plots.

To create a simple X-Y QuickPlot, click the icon on the Graphing toolbar. You may also type @, or hover the mouse over **Graph** on the **Insert** menu and click **X-Y Plot**. This places a blank X-Y plot operator on the worksheet.

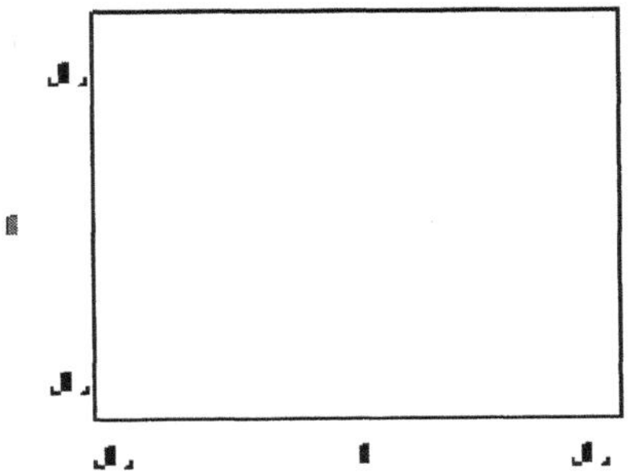

Click the bottom middle placeholder. This is where you type the x-axis variable. Type the name of a previously undefined variable. The variable is allowed to be x, but can be any Mathcad variable name. Next, click the middle left placeholder, and type an expression using the variable named on the x-axis. Click outside the operator to view the X-Y plot. Mathcad automatically selects the range for both the x-axis and the y-axis. Another shortcut is to type only the expression in the left placeholder. Mathcad automatically adds the independent variable in the bottom placeholder. See Figure 1.24.

Another way to create a QuickPlot is to define a user-defined function prior to creating the plot. Open the X-Y plot operator by typing @. Click the bottom placeholder and type a variable name for the x-axis. This variable name does not need to be the same one used as the argument to define the function. On the left placeholder, type the name of the function. Use the variable name from the x-axis as the argument of the function. Here again, Mathcad selects the range for both the x-axis and the y-axis.

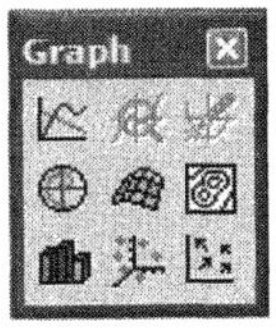

FIGURE 1.23 Graphing toolbar

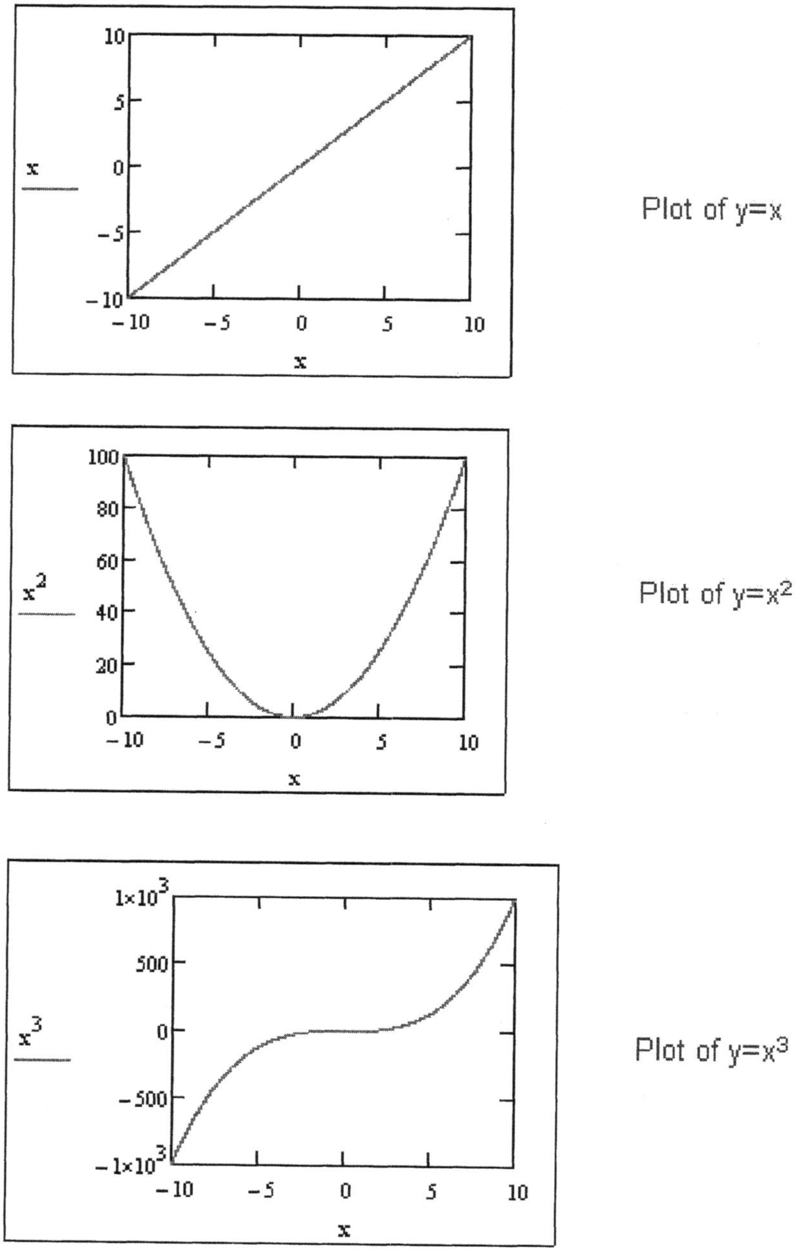

FIGURE 1.24 X-Y QuickPlot of equations

You can skip the step of typing a variable for the x-axis. Mathcad will automatically add the argument used in the y-axis function. See Figure 1.25.

If you use a previously defined variable, Mathcad will not plot a graph over a range of values. It will plot only the value of the variable used. In some cases, this might be only a single point. For a QuickPlot, it is important to use only undefined variables. We will discuss the use of range variables in plots in Chapter 7. This is a case where a previously defined variable can be used.

$ff(h) := h$ $\quad gg(i) := i^2$ $\quad hh(j) := j^3$ Define functions

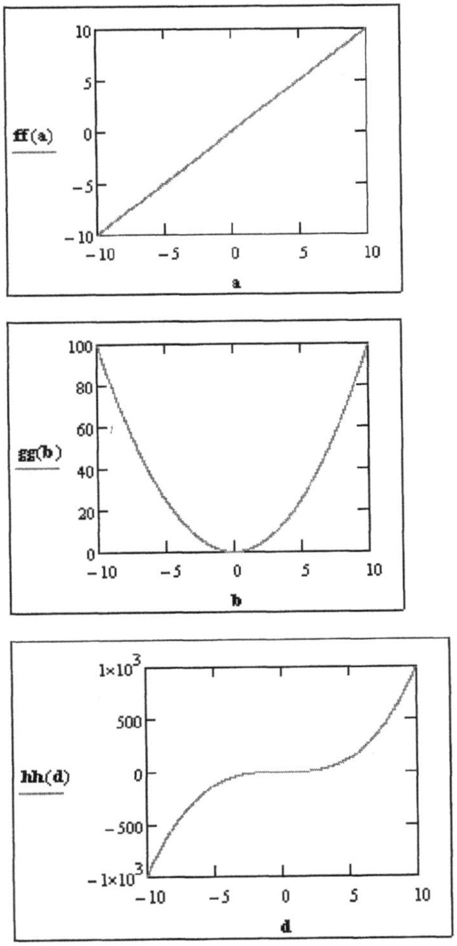

Notice how the variables on the x-axis do not need to match the arguments used to define the function.

Note: The variable "c" could not be used because "c" is a built in variable for the speed of light. The variable needs to be a previously undefined variable.

FIGURE 1.25 X-Y QuickPlot of functions

Setting Plotting Ranges

Mathcad automatically sets the plotting range, but there is a way to change it. You might have noticed additional placeholders when you opened an X-Y plot. These placeholders set the lower and upper limits of the plot.

The placeholders on the bottom set the lower and upper limits on the x-axis. The placeholders on the left set the lower and upper limits on the y-axis. Once you create a QuickPlot, these placeholders will have default values added. To change the default values, click the limit placeholder and delete the value. Next, add a new plot limit. You can tell which plot limits still have the default values because there will be small brackets on the bottom sides of the default values. Once you change the default values, the brackets are no longer displayed. See Figures 1.26 and 1.27.

$$D(t) := \frac{1}{2} \cdot 9.81 \cdot t^2$$

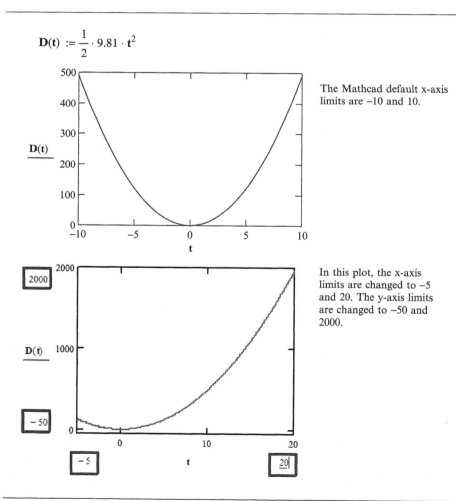

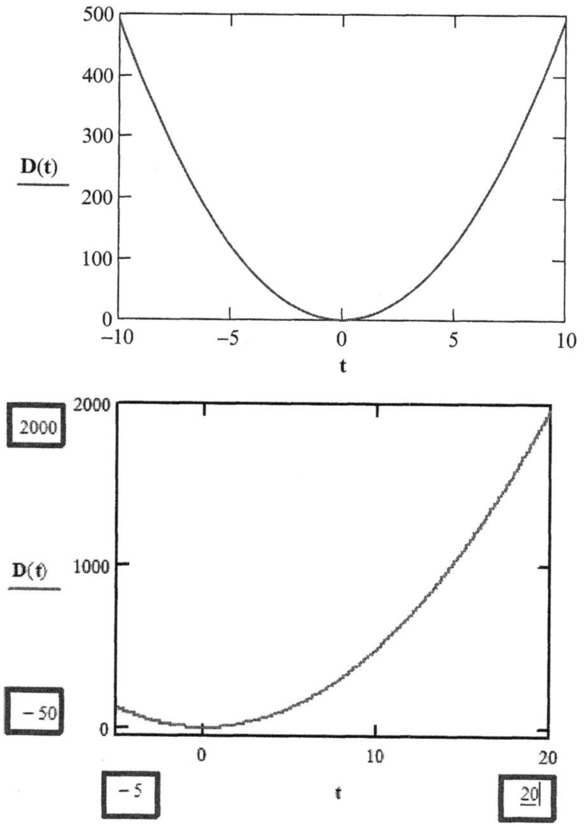

The Mathcad default x-axis limits are −10 and 10.

In this plot, the x-axis limits are changed to −5 and 20. The y-axis limits are changed to −50 and 2000.

FIGURE 1.26 Setting plot range

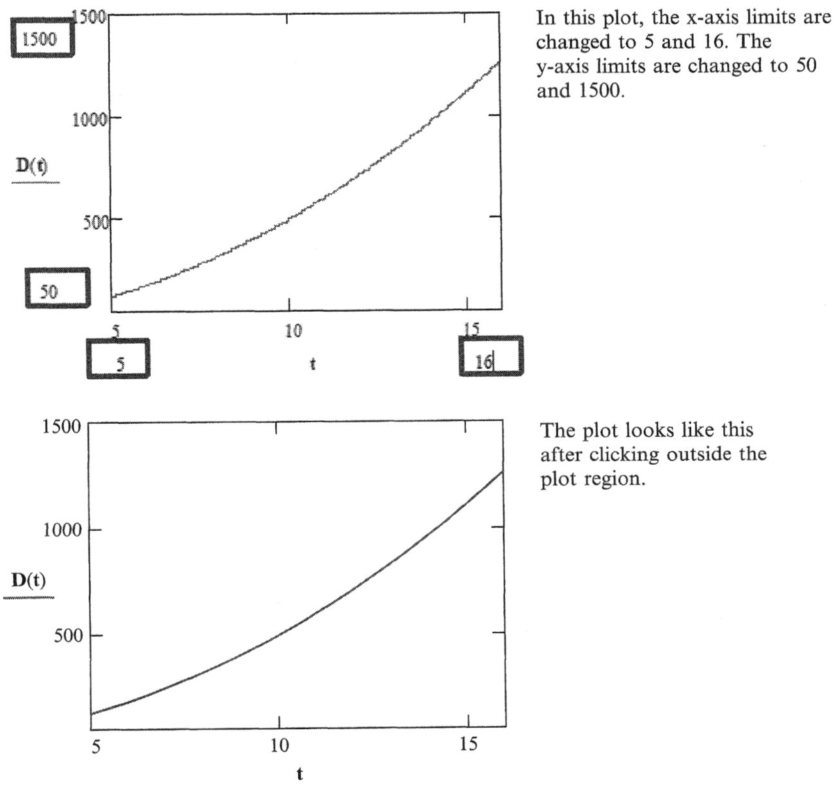

FIGURE 1.27 Setting plot range

PROGRAMMING, SYMBOLIC CALCULATIONS, SOLVING, AND CALCULUS

There are many wonderful Mathcad features that we have not covered in this chapter, but this chapter is an introduction. If we covered all the features, then we would need a book to discuss them. That is what the rest of this book is about, teaching you some of the essential features of Mathcad.

In future chapters we will build on the concepts learned in this chapter. We will also discuss how to use Mathcad programming to create useful and powerful functions. We will discuss the use of symbolic calculations to return algebraic results rather than numeric results. Chapter 10 will discuss some of Mathcad's powerful solving features. In Chapter 12, we will demonstrate how Mathcad can solve calculus and differential equation problems. Part IV will discuss how to use Mathcad to create and organize scientific and engineering calculations.

RESOURCES TOOLBAR AND MY SITE

The Resources toolbar is your one-stop place to access Mathcad information. See Figure 1.28. From the drop-down box, you can access Mathcad tutorials, QuickSheets, reference tables, and any Mathcad e-books or extension packs you have installed. The My Site is a treasure chest of information. You can use the default site, or you can select a different site from the Preferences dialog box on the **Tools** menu. From My Site you can access the PTC web site, the Mathcad User Forum, the Mathcad Web Resource Center, Mathcad Download Site, and the Mathcad Knowledge Base. If you are looking for additional information on a Mathcad topic, My Site is the place to begin your search. See Figure 1.29.

SUMMARY

The intent of this chapter was to get you up and running with Mathcad by introducing key Mathcad features. It is also intended to whet your appetite for the information covered in future chapters. The best way to gain an understanding of the concepts introduced in this chapter is to practice. If you have not done so already, open the Mathcad tutorials and go through the **Getting Stared Primers** mentioned at the beginning of this chapter.

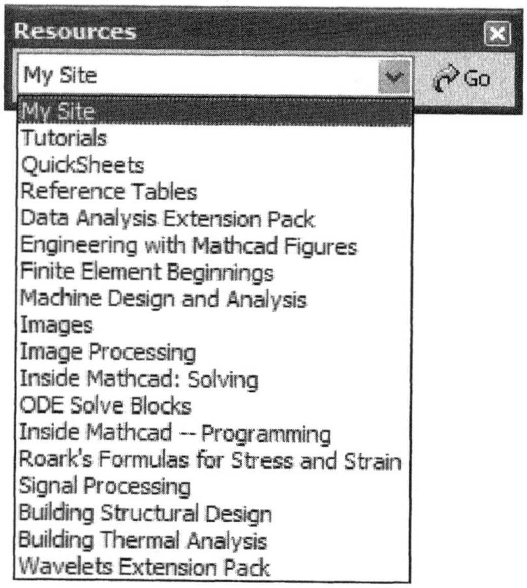

FIGURE 1.28 Resourcest toolbar

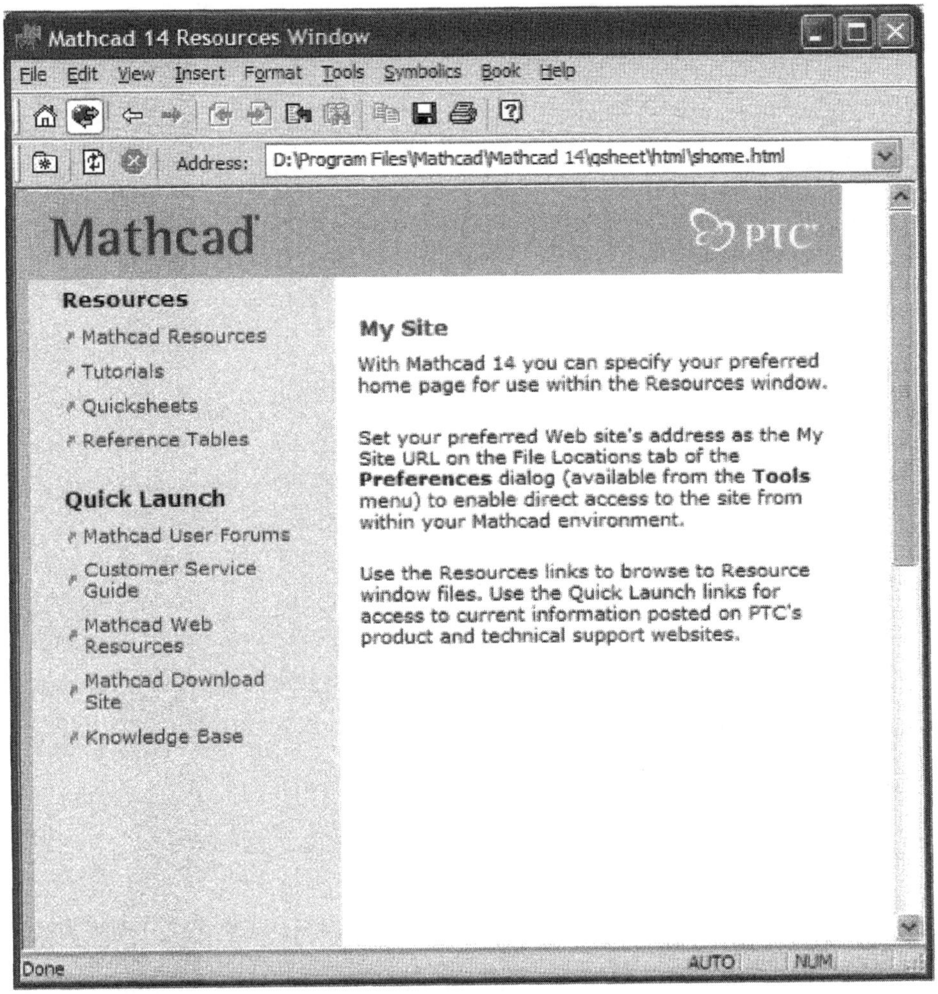

FIGURE 1.29 My Site

In Chapter 1 we:

- Showed how to create and edit Mathcad expressions using the editing lines.
- Described the Mathcad toolbars.
- Differentiated between the different Mathcad equal signs.
- Discussed regions.
- Introduced functions, units, arrays, and plotting.
- Introduced range variables.

- Emphasized the difference between literal subscripts and array subscripts.
- Described the variable ORIGIN.

PRACTICE

Additional problems and applications can be found on the companion site: www.elsevierdirect.com/9780123747839.

1. Enter the following equations into a Mathcad worksheet:

$$\frac{-b + \sqrt{b^2 - 4 \cdot a \cdot c}}{2 \cdot a}$$

$$\frac{F_0}{\sqrt{m^2 \cdot \left(\omega_0^{\ 2} - \omega^2\right)^2 + b^2 \cdot \omega^2}}$$

$$\left(\frac{2}{3} - \frac{F_y \left(\frac{1}{r_T}\right)^2}{1530 \mathrm{x} 10^3 \cdot C_b}\right) \cdot F_y$$

$$\frac{-1}{2} d_2 + \sqrt{\frac{2 \cdot v_2^{\ 2} \cdot d_2}{g} + \frac{d_2^{\ 2}}{4}}$$

2. Give each of the preceding equations a variable name. Assign variable names and a value to the variables used in the equations. These variable assignments will need to be made above the equation definition. Show the result. Change some of the input variable values and see the impact they have on the results.

3. Choose 10 equations from your field of study (or from a physics book) and enter them into a Mathcad worksheet. Assign the variables the equation needs prior to entering the equation. Select appropriate variable names. Don't select easy equations—pick long complicated formulas that will give you some practice entering equations.

4. Choose some of these equations and change some of the operators in the equation.

5. Choose some of these equations and make the equation wrap at an addition or subtraction operator.

CHAPTER

Variables and Regions

2

VARIABLES

Variables are one of the most important features of Mathcad. As in algebra, variables define constants and create relationships. As we saw in Chapter 1, your Mathcad worksheet will be full of variables. It is therefore important to quickly gain a solid foundation in their use.

Chapter 2 will:

- Discuss types of variables.
- Give rules for naming variables.
- List characters that can be used in variable names.
- Introduce string variables.
- Discuss the worksheet ruler and tabs.
- Tell how to move, align, and resize regions.
- Discuss the use of Find and Replace.

TYPES OF VARIABLES

Variables can consist of numbers or constants such as A := 1 or B := 67. They can consist of equations such as C := A + B or D := A + 3. You can set one variable equal to another such as E := A. Variables can also consist of strings of characters such as F := "This is an example of a string variable." Variables can even have logic programs associated with them so that the value of the variable depends on the outcome of Boolean logic. As you go through this book you will see that variables can be very simple or very complex. For the purpose of this chapter we will stay with simple examples. More detailed examples will follow in later chapters.

RULES FOR NAMING VARIABLES

Case and Font

The first important thing to remember about variable names is that they are case, font, size, and style sensitive. Thus the variable "ANT" is different from the variable "ant" (uppercase versus lowercase), and the variable "Bat" is different from the variable "**Bat**" (normal font versus bold font). The variable "Cat" is different from the variable "Cat" (different font size), and the variable "Dog" is different from the variable "Dog" (different font style). If at some point in your worksheet Mathcad isn't recognizing your variable, check to make sure your variables are exactly the same in case, font, size, and style.

Characters that Can Be Used in Variable Names

There are some rules for naming variables:

- Variable names can consist of upper- and lowercase letters.
- The digits 0 through 9 can be used in a variable name, except that the leading character in a variable name cannot be a digit. Mathcad interprets anything beginning with a digit to be a number and not a variable.
- Variable names may consist of Greek letters. The easiest way to insert Greek letters is to use the Greek letter toolbar. Find this toolbar by highlighting **Toolbars** from the **View** menu and then clicking **Greek**. Select the desired Greek letter from the toolbar, and it will be inserted into your worksheet. Another way to insert Greek letters is to type the equivalent roman letter and then type CTRL+G. See Figure 2.1 for a table of equivalent Greek

α	a	η	h	ο	o	ϖ	v
β	b	ι	i	π	p	ω	w
χ	c	φ	j	θ	q	ξ	x
δ	d	κ	k	ρ	r	ψ	y
ε	e	λ	l	σ	s	ζ	z
ϕ	f	μ	m	τ	t		
γ	g	ν	n	υ	u		
Α	A	Η	H	Ο	O	ς	V
Β	B	Ι	I	Π	P	Ω	W
Χ	C	ϑ	J	Θ	Q	Ξ	X
Δ	D	Κ	K	Ρ	R	Ψ	Y
Ε	E	Λ	L	Σ	S	Ζ	Z
Φ	F	Μ	M	Τ	T		
Γ	G	Ν	N	Υ	U		

FIGURE 2.1 Table of equivalent Greek letters

letters. Search "Greek toolbar" in the Index of Mathcad Help for this table of Greek equivalent letters.

- The infinity symbol ∞ can be used only as the beginning character in a variable name. To insert the infinity symbol, type CTRL+SHIFT+Z.

Literal Subscripts

Literal subscripts were discussed in Chapter 1. Remember, to type a subscript, type the first part of the variable name, and then type a period. The insertion point will drop down half of a line. All characters typed after this point will be part of the subscript. To remove a subscript, delete the period that occurs just before the subscript. Also remember that a literal subscript looks similar to an array subscript, but it behaves much differently. Array subscripts will be discussed in Chapter 5.

Special Text Mode

Keyboard symbols can be used in variable names, but Mathcad operators can not. However, most keyboard symbols are also Mathcad shortcuts that insert a Mathcad operator or perform another Mathcad function. (See the Appendix for a list of keyboard shortcuts.) This prevents you from using most keyboard symbols in your variable names. When you try to use a symbol that is also a Mathcad shortcut, Mathcad inserts the operator or executes the command referenced by the shortcut. For example, if you type A $ B, Mathcad inserts the range sum symbol because the $ symbol is the keyboard shortcut for the range sum. A variable name cannot use the addition operator. If you type A + B : 6, Mathcad will not recognize the variable name, and will give an error. See Figure 2.2.

Mathcad provides a way to use both symbols and operators in variable names by providing a special text mode. To activate the special text mode, begin the variable name by typing a letter, then type CTRL+SHIFT+K. Once the special text mode is entered, the editing lines turn from blue to red. You are now free to enter

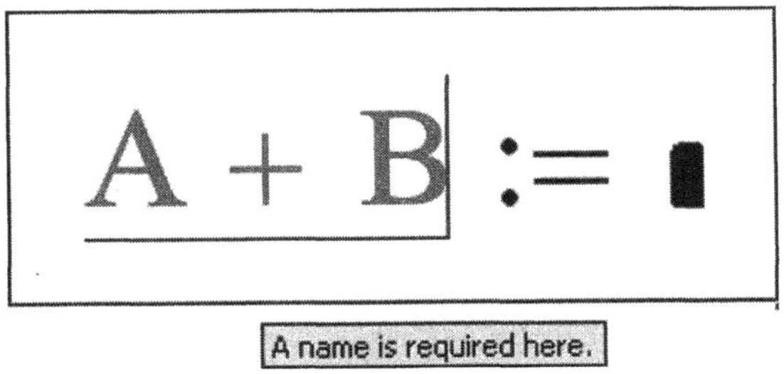

FIGURE 2.2 Operators cannot be used in a variable name

A+B := 1	A+B = 1
B$C: := 2	B$C: = 2
x^2 := 4	x^2 = 4
{Test} := 100	{Test} = 100

FIGURE 2.3 Examples of variable names using the special text mode

any keyboard symbols. If you want your variable name to begin with a symbol move the cursor back to the beginning of the variable and type the symbol. When you are done entering symbols, type **CTRL+SHIFT+K** again to return to normal math editing mode. See Figure 2.3 for examples of variable names using keyboard symbols.

Chemistry Notation

Mathcad provides a means to have your variable names look like an expression or an equation. It is a special mode called Chemistry Notation. To activate the Chemistry Notation type **CTRL+SHIFT+J**. This inserts a pair of brackets with a placeholder between them. You are now free to insert whatever letters, numbers, and operators you want between the brackets. Using this mode you can make your variable name look like an equation. You are not limited to staying within subscripts like you are when you name a normal variable. Chemistry Notation is useful when you have a long equation with many parts. You might want to separate the equation into smaller parts. In order to do this you need to give each part of the equation a variable name. Sometimes it is difficult to determine what to name each part. With Chemistry Notation, each part can be given the variable name to match the part of the equation. See Figure 2.4.

$a := 1 \quad b := 5 \quad c := 6$

$$\left[\sqrt{b^2 - 4\cdot a\cdot c}\right] := \sqrt{b^2 - 4\cdot a\cdot c}$$

This variable name was entered using Chemistry Notation. The variable name looks just like the defined equation.

$$\left[\sqrt{b^2 - 4\cdot a\cdot c}\right] = 1$$

$$\left[\frac{NaOH}{Na_2\cdot CO_3}\right] := \frac{80}{106}$$

An example of a chemistry ratio entered using Chemistry Notation.

$$\left[\frac{NaOH}{Na_2\cdot CO_3}\right] = 0.755$$

FIGURE 2.4 Variable names using Chemistry Notation

STRING VARIABLES

A string is a sequence of characters between double quotes. It has no numeric value, but it can be defined as a variable. To create a string variable, type the variable name followed by pressing the colon **:** key. Type the double quotes key **"** in the placeholder. You will see an insertion line between a pair of double quotes. You can then type any combination of letters, numbers, or other characters. When you are finished with the string, press **ENTER**.

String variables are useful to use as error messages. If you need a certain input to be a positive number, you can assign a string variable to have the value, "Input must be positive." If a number less than zero is entered, Mathcad can display this string variable as an error message. String variables are also useful as a means of displaying whether a certain condition is met. You can assign one variable to have the value "Yes" and another variable to have the value "No." If a specific condition is met, Mathcad can display the string variable associated with "Yes." If the specific condition is not met, Mathcad can display the string variable associated with "No." These logic programs will be discussed in Chapter 8, "Simple Logic Programming." See Figure 2.5 for some examples of text strings.

WHY USE VARIABLES?

Figure 2.6 shows three different ways to get similar results. The first method shown is direct. If you were not going to be saving the worksheet and you needed a quick answer, the first method works fine. Just type the numbers to get an answer. Use the result of the first equation, and type it into the second equation. Use the result of the second equation, and type it into the third equation.

The second method shown is to assign a variable name to the intermediate answers. The benefit of this method is that you will always have the result of each expression available to use in other expressions in your worksheet. The equations shown are very simple and basic, but in your scientific or engineering calculations the equations or expressions can be very complex. Once the result of an

$TextString_1$:= "This is a text string"

$TextString_1$ = "This is a text string"

$TextString_2$:= "Yes" $TextString_2$ = "Yes"

$TextString_3$:= "No" $TextString_3$ = "No"

FIGURE 2.5 Examples of text strings

The following example shows three methods of getting the same results.

Use direct numbers

$5 + 7 = 12.00$

$12 \cdot 3 = 36.00$

$36 - 8 = 28.00$

Use intermediate results

$\mathbf{Answer_1} := 5 + 7 \qquad \mathbf{Answer_1} = 12.00$

$\mathbf{Answer_2} := \mathbf{Answer_1} \cdot 3 \qquad \mathbf{Answer_2} = 36.00$

$\mathbf{Answer_3} := \mathbf{Answer_2} - 8 \qquad \mathbf{Answer_3} = 28.00$

Assign variable names to input and output values

$\mathbf{Input_1} := 5 \quad \mathbf{Input_2} := 7 \quad \mathbf{Input_3} := 3 \quad \mathbf{Input_4} := 8$

$\mathbf{Answer_4} := \mathbf{Input_1} + \mathbf{Input_2} \qquad \mathbf{Answer_4} = 12.00$

$\mathbf{Answer_5} := \mathbf{Answer_4} \cdot \mathbf{Input_3} \qquad \mathbf{Answer_5} = 36.00$

$\mathbf{Answer_6} := \mathbf{Answer_5} - \mathbf{Input_4} \qquad \mathbf{Answer_6} = 28.00$

FIGURE 2.6 Using variables in calculations

expression is calculated by Mathcad, you want to capture it for future use. You do this by assigning the result to a variable.

The third method shown is to assign all values to variable names. There are four input values and three output results. You may be saying to yourself, "Why would I type all those extra key strokes? It is much more time consuming to type Input1+Input2 than just typing 5+7." Well, let's assume that the numbers 5, 7, 3, and 8 represent some type of engineering input. You now use these numbers over and over in your calculations. If you keep using just the numbers 5, 7, 3, and 8 in your many different equations, what happens if at some point the input value 7 is changed to 9? If you have used the variable Input2=7, then you just change the value of Input2 from 7 to 9 and you are done. Mathcad does the rest. If you didn't use the input variable, you will need to go through your worksheet and change (or attempt to change) every instance where the number 7 represented the input variable, and change it from 7 to 9. This could be an impossible task if you have a complex worksheet.

Remember that the goal of this book is to teach you how to use Mathcad as a tool for creating scientific and engineering calculations. Because of this, it is recommended that you get into the habit of using the third method illustrated in Figure 2.6. Most of the examples used in this book will use this method. We will assign the input values to variable names; assign a variable name to the expression; and then display the results of the expression.

Chapter 13 provides some useful naming guidelines for variables to be used in your scientific and engineering calculations.

REGIONS

In Chapter 1 we discussed how a Mathcad worksheet is comprised of many different regions. This section will now discuss how to manipulate and organize regions.

Using the Worksheet Ruler

The worksheet ruler at the top of your worksheet can help you align regions and set tabs. To make the ruler appear, click **Ruler** from the **View** menu. Repeat the procedure for hiding the ruler.

You can change the measurement system used on the ruler by right-clicking the ruler and selecting from the list of measurements: inches, centimeters, points, or picas. Remember that there are 72 points per inch and 6 picas per inch. When you change the ruler measurements, some of the dialog box measurement systems change to the new ruler measurement system.

Tabs

You can use tabs to help align regions in your worksheet. If you press the TAB key prior to creating a math or text region, the region will be left-aligned with a tab stop.

Mathcad defaults to tab stops of one-half inch. You can set tab stops on the worksheet ruler by clicking the worksheet ruler at the location where you want to set the tab stop. Once a tab stop is shown on the ruler, you can adjust the tab by clicking the tab and dragging it along the ruler. You can clear the tab stop by clicking the tab and dragging it off the worksheet toolbar. Another way to set tab stops is by choosing **Tabs** from the **Format** menu. This opens the Tabs dialog box. From this dialog box you can clear all tab stops and set new tab stops at exact tab stop locations.

Selecting and Moving Regions

You can select and move a single region or multiple regions. To move a single region, click within the region, and then place your cursor near the perimeter of the region until the cursor changes from an arrow to a hand. Now left-click and hold the mouse button. Drag the region to where you want it.

To move multiple regions, drag-select the regions. To do this, click outside of a region, and then hold down the left mouse button and drag it across several regions and release the mouse button. All regions within this area will now be selected, and each will have a dashed line surrounding the region. To select non-adjacent regions, hold the CTRL key and click within each desired region. To move the selected regions, place the cursor in one of the regions, left-click and hold the mouse button. Drag the regions to a new location. You can also use the arrow keys to move the selected regions.

Aligning Regions and Alignment Guidelines

There will be times when you want to align different regions either vertically or horizontally. Aligned regions appear much more professional. To align regions, select the desired regions, highlight **Align Regions** from the **Format** menu, and then select either **Across** or **Down**. You can also use the alignment icons on the Standard toolbar.

If your selected regions are roughly aligned in a horizontal row, the Across alignment will place the top of each region in a horizontal line. If your selected regions are roughly aligned in a vertical column, the Down alignment will place the left side of each region in a vertical line.

Using the **Align Regions** feature may cause regions to overlap. If a vertical line will pass through more than one of your selected regions, the Across alignment will cause these regions to overlap. If a horizontal line will pass through more than one of your selected regions, the Down alignment will cause these regions to overlap. In order to prevent regions from overlapping, it is important to select only regions in a roughly horizontal or roughly vertical layout. If this is not possible, move some of the regions prior to using the alignment feature.

Mathcad warns you to check your regions prior to executing the requested alignment. If you accidentally align regions that cause an overlap, you can undo the alignment, or you can select the overlapping regions and click **Separate Regions** from the **Format** menu. This will separate the regions vertically.

It was mentioned earlier that if you press the TAB key prior to creating a region, the new region will be left-aligned to the tab stop. If you did not use the tab stop when creating regions, you can still align your regions to the tab stop. Mathcad has a feature called alignment guidelines. These are green lines that extend down from the tab stops. See Figure 2.7.

You can move your regions to align with these alignment guidelines. To turn on guidelines for all the tab stops, open the Tabs dialog box by clicking **Tabs** from the **Format** menu. Then place a check in the Show Guide Lines For All Tabs box. This will place the green guidelines at all existing tab stops. If you add additional tab stops after checking this box, you will need to repeat the procedure. To set a guideline for an individual tab stop, right-click the tab stop and select **Show Guideline**. To remove an existing guideline, right-click the tab stop and select

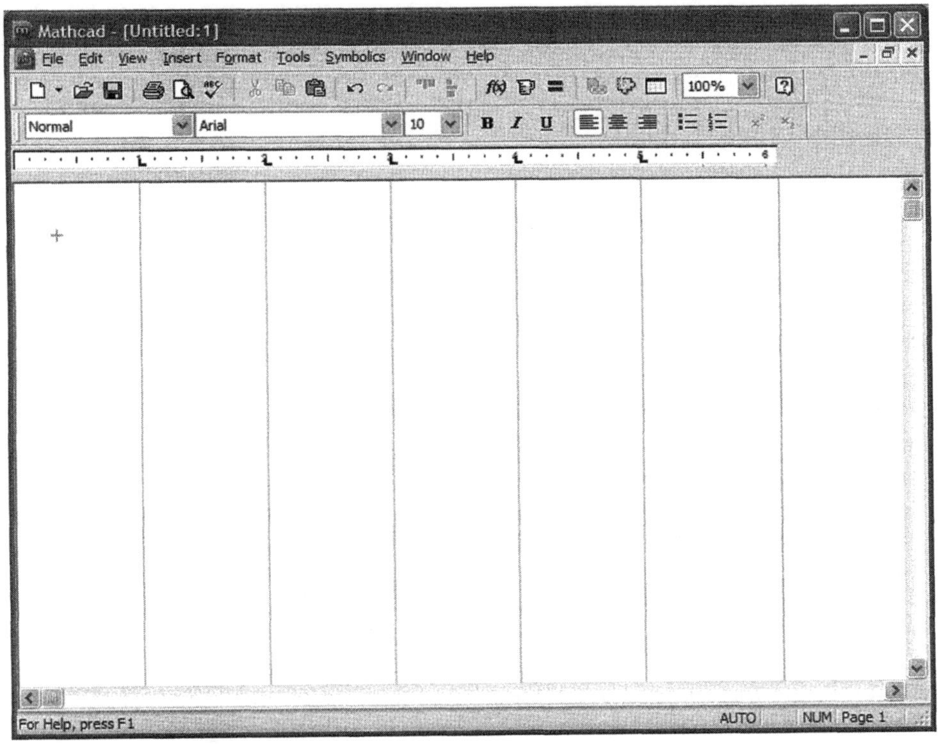

FIGURE 2.7 Alignment guide lines

Show Guideline (there should be a check next to it). You can also remove all guidelines at once by unchecking the "Show Guide Lines For All Tabs" box in the Tabs dialog box.

After you have done a Down alignment of a selected group of regions, you can move this group of regions to align them with one of the guidelines.

TEXT REGIONS

In Chapter 1 we learned how to create a text region by using the double quote (") key, or by choosing **Text Region** from the **Insert** menu. This chapter will focus on how to modify and edit text regions.

Changing Font Characteristics

Once you create a text region you can type text just as you would in a word processor. You can also use tabs or change font characteristics such as font type, font size, or font color. You can also use such things as bold, italic, underline, strike-out, subscript, and superscript. To change the font characteristics while in a text region, highlight the text and choose **Text** from the **Format** menu.

Inserting Greek Symbols

To insert Greek letters, use the Greek Symbol toolbar. You can open this toolbar by choosing **Toolbars** from the **View** menu and then selecting **Greek**. If the Math toolbar is open, you can click the icon representing the Greek letters. You can also type a Roman letter and immediately type CTRL+G. This converts the alphabetic character to its Greek symbol equivalent. See Chapter 1 and the Appendix for tables of equivalent Greek letters.

Controlling the Width of a Text Region

When you start typing in a text region, the region grows to the right until it reaches the right margin. At that point the text wraps to a new line. There are times when you do not want the text region to grow all the way to the right margin. To force the region to wrap before it reaches the right margin, press CTRL+ENTER at the point where you want the text region to wrap. The text may not immediately wrap, but when you begin typing the next word, the cursor will move to the next line. Do not use the ENTER key to change the width of the text region. The ENTER key is used to add a new paragraph to a text region.

To change the width of an existing text region, place the cursor at the point where you want the text region to wrap and press CTRL+ENTER. You can also click within the text region and then move the handle on the right side of the text region. The text will wrap according to the new text region width. If the original text region used the ENTER key to set the width of the text region, the new text will not align with the new width.

Moving Regions Below the Text Region

As you add text to a new text region, the region grows. You might find that the growing text region begins to overlap on top of other regions. There is a way to prevent this from occurring. To do this, right-click inside the text region, click **Properties**, select the Text tab, and place a check in the box adjacent to Push Regions Down As You Type. Now as you type new text, or modify the width of existing text, any regions below the text region will move down or up depending on how you size the text region. See Figure 2.8.

This feature must be set for every text region. There is not a way to set it globally. Be cautious about using this feature. If it is set, some of your math regions might be moved downward. This could cause some of your variable definitions to change or not be recognized. This would occur if a variable definition is moved downward and an adjacent expression (to the right) uses the values from the variable definition.

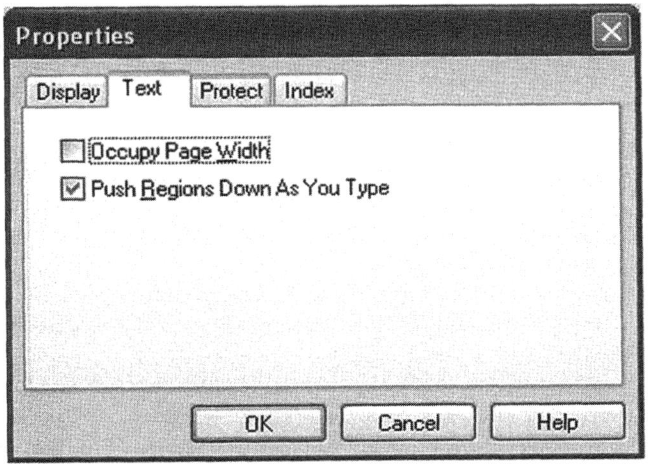

FIGURE 2.8 Push Regions Down As You Type check box

Paragraph Properties

A text region is similar to a simple word processor. You are able to format the text region in much the same way as you would in a word processor. We discussed earlier how to change the font characteristics of text in the text region. You can also set many paragraph characteristics such as margins, alignment, first line indent, hanging indent, bullets, automatic numbering, tabs, and more.

To set the paragraph characteristics, click the text region and select **Paragraph** from the **Format** menu. You can also right-click in a text region and select **Paragraph** from the drop-down menu. See Figure 2.9.

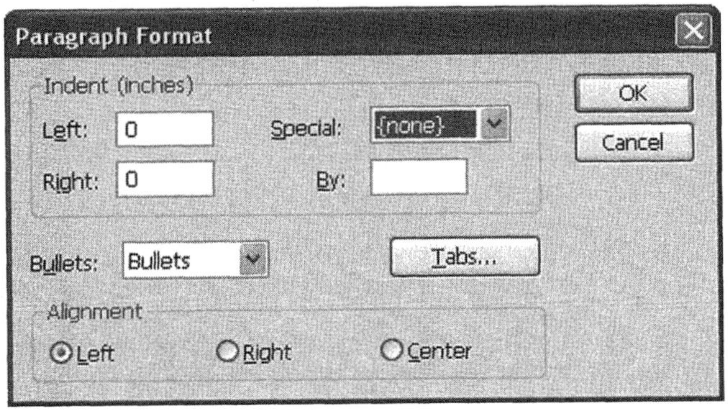

FIGURE 2.9 Paragraph Format dialog box

The indent boxes for left and right are based on the edges of the text region, not the page margins of your worksheet. If you set the margins to be one inch from both left and right, then as you change the size of your text region, the text will always remain one inch from the edges of your text region.

Clicking Special will allow you to indent the first line or allow you to have a hanging indent on your first line. After selecting **First Line** or **Hanging Indent**, tell Mathcad how much you want to indent by changing the number in the **By** box.

The Bullets box allows you to use bullets or automatic numbering to each paragraph in your text region.

The **Tabs** button allows you to set tab locations just as in a word processing program. The tabs are measured from the left edge of the text region, not from the left edge of the page or page margin.

Each paragraph in the text region can have different paragraph settings. See Figure 2.10 for an example of how a text region will look with specific settings. The first paragraph has different features than the last three paragraphs.

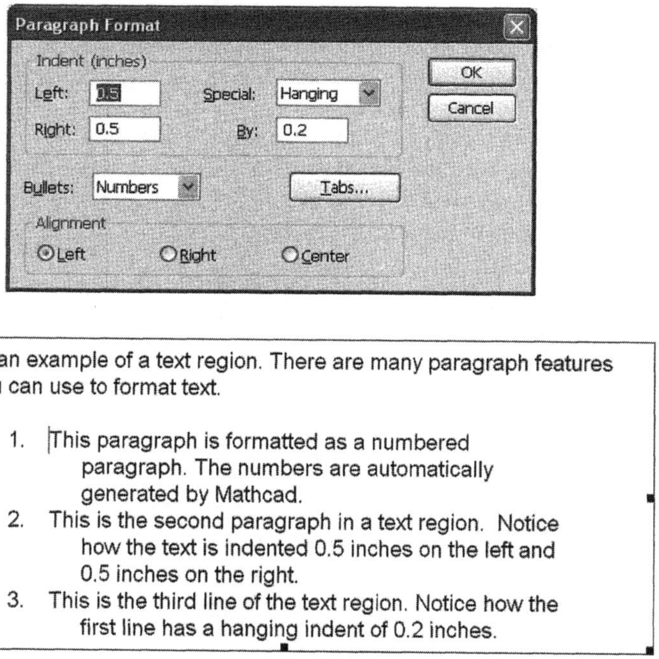

FIGURE 2.10 Paragraph formatting for bottom three paragraphs of text region

Text Ruler

When the worksheet ruler is showing, you will also have a ruler when you are working in a text region. This text ruler changes width to match the width of the text region. The ruler begins at the left edge of the text region and extends to the right edge of the text region. From the ruler, you can set left and right margins, indents, hanging indents, and tabs. To do this, slide the left and right indent markers to the desired positions. You can also add tabs to the text ruler by clicking the ruler. See Figure 2.11.

Spell Check

Mathcad has a built-in spell checker. The spell checker checks the spelling only in text regions, not math regions. To activate the spell checker, click **Spelling** on the **Tools** menu. If a misspelled word is found, you have the option to change to one of the suggested replacement words, ignore the suggestions, add the word to your personal dictionary, or have Mathcad offer additional suggestions.

Mathcad can check several different languages. It can also check several different dialects. For example, you can tell Mathcad to use the British English instead of the American English. To select a different language or dialect, select **Preferences** from the **Tools** menu, and then click the Language tab. From this tab, under the Spell Check Options, you can select a specific language, and some languages will allow you to select a specific dialect.

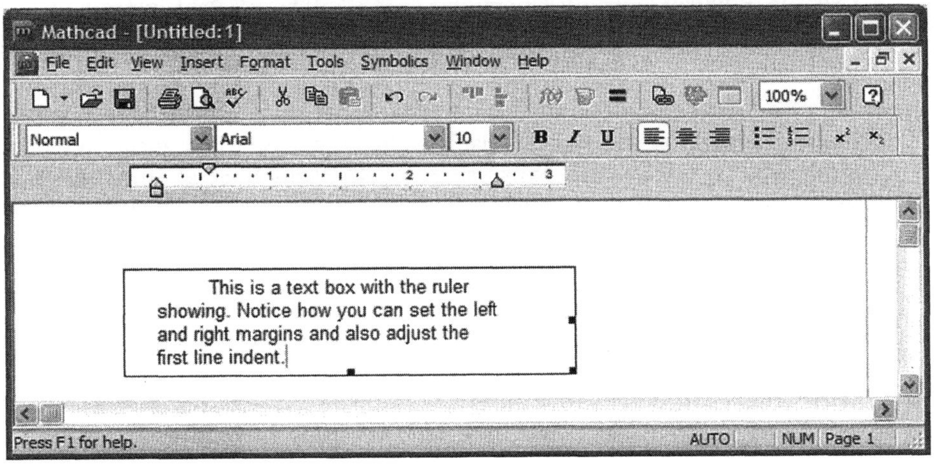

FIGURE 2.11 Text region with ruler turned on

ADDITIONAL INFORMATION ABOUT MATH REGIONS

Math Regions in Text Regions

To insert a math region in a text region, click **Math Region** from the **Insert** menu. This places a blank placeholder in the text region, where you can type a math expression. After you are finished with the expression, use the right arrow to move back into the text box, where you can continue typing text. See Figure 2.12.

> **Tip!** Before inserting a math region, I like to add one or two spaces after the insertion point. This makes it easier to continue typing after I insert the math region. It is not necessary, but it makes it easier to see the cursor after leaving the math region.

Math Regions that Do Not Calculate

There will be many times when you want to display an equation prior to the point where the variables are defined. When you try to do this, Mathcad will give you an error message.

There are several ways to work around this:

- Type the expression using variable names that have not yet been defined. Before clicking out of the Mathcad region, right-click and choose **Properties**. Click the Calculation tab and place a check mark in the Disable Evaluation box. This will place a solid black box in the upper-right corner of your math region and will prevent Mathcad from evaluating the expression.
- Use the Boolean equality operator `CTRL+EQUAL SIGN` instead of the assignment operator `COLON`. This does not make a variable assignment, but it allows you to display what you want without an error.
- After you define your variables and expression, copy the math region that has the expression you want to display. Then move up to the location where you want to display the expression. Click **Paste Special** from the **Edit** menu and select **Picture (Metafile)**. This displays a graphic image of the math region definition. This method is not recommended. You can imagine

You can include a math region $\mathbf{Example} := \frac{600\mathrm{N}}{3\mathrm{m} \cdot 2\mathrm{m}}$ in a text region by using the Math Region command from the Insert Menu.

$\mathbf{Example} = 100.00\,\mathrm{Pa}$

FIGURE 2.12 Including math regions in text regions

$NoVariables_1 := a \cdot x^2 + b \cdot x + c$	Using "Disable Evaluation" from the Properties dialog box.
$NoVariables_1 = \blacksquare$	This value does not exist because the above expression was disabled.
$NoVariables_2 = a \cdot x^2 + b \cdot x + c$	Using a Boolean equal
$NoVariables_2 = \blacksquare$	This value does not exist because the Boolean equal sign does not define a variable.
$NoVariables_3 := a \cdot x^2 + b \cdot x + c$	The image to the left is a pasted graphic image. It is not a math region. This method is not recommended, because it can be very confusing when trying to check the calculations. If you use this method, be sure to make it clear that it is a graphic and not a math region.
$a := 1 \quad b := 5 \quad c := 6 \quad x := 4$	
$NoVariables_3 := a \cdot x^2 + b \cdot x + c$	This math region was copied and pasted as a graphic above.
$NoVariables_3 = \blacksquare$	

FIGURE 2.13 Displaying math regions without having variables defined

the confusion it can cause when trying to check a calculation. If you choose to use this method, make it very clear that the region is a graphic image and not a Mathcad math region.

> **Tip!** My favorite way is to use the Boolean equality operator.

See Figure 2.13 for an example of using these different methods of displaying expressions.

FIND AND REPLACE

Find

The Find and Replace features can easily help you to either find or replace variables or text in your worksheets. To use these features click the **Edit** menu.

Let's first look at the Find dialog box. See Figure 2.14. Type what you want to find in the "Find What" box.

> **Tip!** If you are searching for Greek letters, type \ followed by the Roman equivalent letter. If you are searching for a tab, use ^t. If you are searching for a return, use ^p.

In Figure 2.14, we typed "Ex." Now let's look at how the different check boxes affect how Mathcad finds things.

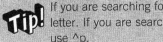

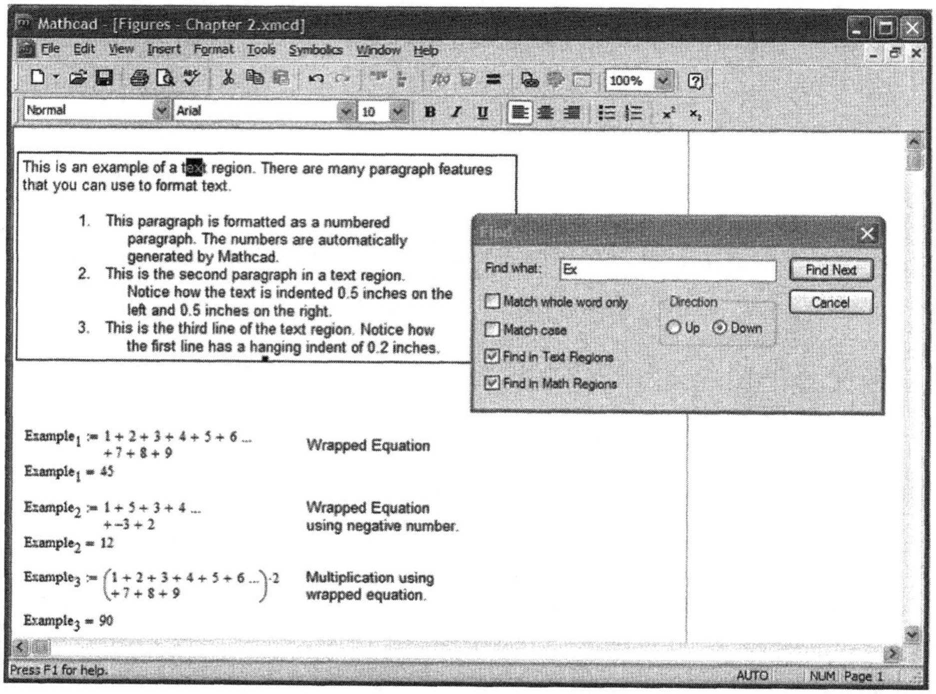

FIGURE 2.14 Using the Find dialog box

You can make Mathcad look in only text regions or only math regions. In order to look in both text regions and math regions it is important to place a check in both the "Find in Text Regions" and "Find in Math Regions" check boxes. The other check boxes tell Mathcad to match the whole word or to match the case.

In Figure 2.14 we checked only the bottom two check boxes. This means that Mathcad will find all instances of "Ex"—uppercase and lowercase—in both the text regions and math regions. In Figure 2.14, Mathcad found the "ex" in "text." It will also find the "ex" in "example" and "Ex" in "Example."

If we checked the top box, "Match whole word only," Mathcad will not find any instances of "Ex" because all instances of "ex" are contained in other words. If we uncheck the top box and check the next box, "Match case", Mathcad will find only the variables using the word "Example" in the math regions.

If we change the Find What from "Ex" to "ex," and leave the "Match Case" box checked, then Mathcad will find the instances of "ex" in the text regions, but will not find the variables using the word "Example" in the math regions.

Tip! The Find in Math Regions is unchecked by default in older versions of Mathcad. Make sure to check this box if you are searching for variables.

Replace

The Replace dialog box is identical to the Find dialog box, except that there is a new line, "Replace with." In Figure 2.15, we will find all instances of "Example" and replace them with "WrappingExample." It is important to make sure that the top box, "Match whole word only", is not checked, otherwise Mathcad will not find the variables such as "Example1." We also checked the Match case box, but in our case, it really would not matter because the "Find in Text Regions" box is unchecked.

Figure 2.16 shows what happened after we replaced all instances of the variable "Example" with "WrappingExample,"—Mathcad quickly replaced these. Because we replaced the variable name with a longer variable name, the math regions are now overlapping the text regions. If this happens, you can quickly select the text regions and move them to the right as we discussed earlier in the chapter.

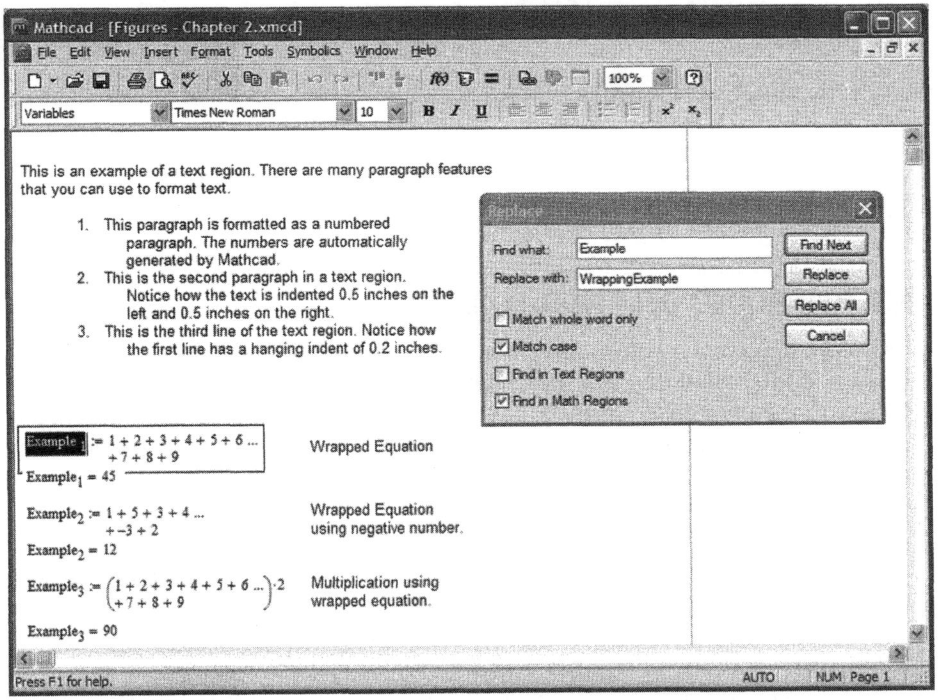

FIGURE 2.15 Using the Replace dialog box

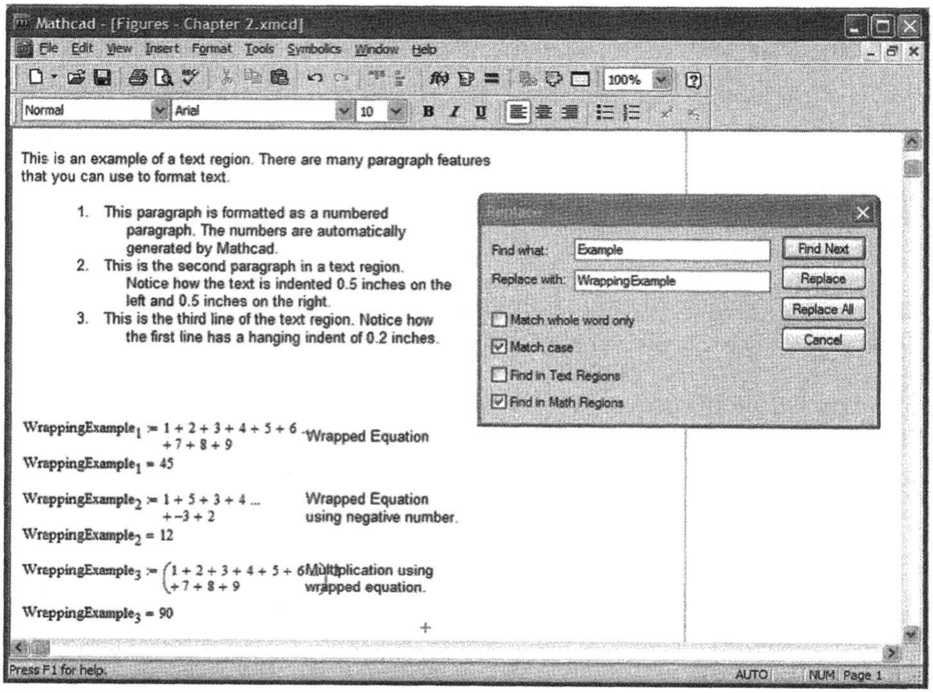

FIGURE 2.16 After using Replace

INSERTING AND DELETING LINES

It is possible to move regions down and create more blank space for new regions. You can do this by holding down the ENTER key, or you can right-click above the region you want to move and select "Insert Lines." This opens the Insert Lines dialog box, where you can input the number of lines to insert. When pasting new information into Mathcad, it is wise to insert blank spaces between the area where you are pasting the information.

If you have extra space between regions, you can delete the blank lines. To do this, place your cursor at the top of the blank space and use the DELETE key to delete the blank lines. You can also right-click and select Delete Lines. Mathcad knows how many blank lines are between the cursor and the next region. Clicking **OK** will delete the blank spaces and move the regions together.

If you want to force a page break in your worksheet, press CTRL+ENTER. This places a horizontal line in your worksheet indicating the location of the page break. You can drag this page break indicator up or down in your worksheet.

SUMMARY

Variables are an important part of engineering calculations in Mathcad. Learning how to use variables effectively will make it much easier to create scientific and engineering calculations. Knowing how to format text regions can also make your calculation worksheets look better and be easier to follow.

In Chapter 2 we:

- Learned that variables are case, font, size, and style sensitive.
- Discussed which characters can be used in variable names.
- Learned about the special text mode and Chemistry Notation.
- Discussed string variables.
- Emphasized the importance of defining variables in Mathcad calculations.
- Discussed moving and aligning regions.
- Described the attributes of text regions.
- Explained paragraph properties in text regions.
- Demonstrated the use of find and replace.

PRACTICE

Additional problems and applications can be found on the companion site: www.elsevierdirect.com/9780123747839.

1. Open a new Mathcad worksheet.
2. Add 10 simple variable definitions from your field of study; include some subscript names.
3. Add 10 variable definitions that are two words or more. Use two different methods for differentiating between words.
4. Add 10 variable definitions that include characters requiring the use of the special text mode.
5. Add 10 variable definitions that use the Chemistry Notation.
6. Add 10 string variable definitions.
7. Create a text region and write two or three paragraphs about the things you learned in this chapter. After creating the paragraphs, change some of the paragraph characteristics and font characteristics. Use the spell checker to check for spelling errors.
8. Use the Find and Replace features to search for and replace certain text and math characters.

CHAPTER

Simple Functions

3

This chapter introduces Mathcad's built-in functions and describes their basic use. It also discuses the use of simple user-defined functions.

The power of user-defined functions is not realized even by many long-time Mathcad users. In some instances, user-defined functions can be confusing and complicated, causing many users to ignore them. After briefly discussing built-in functions, this chapter will focus on simple user-defined functions. The goal of this chapter is to make you comfortable with their concept and use.

Chapter 3 will:

- Introduce built-in functions.
- Discuss what an argument is.
- Introduce user-defined functions.
- Show different types of arguments.
- Give examples of different function names and argument names.
- Tell when to use a user-defined function.
- Describe how to use variables in user-defined functions.
- Give examples of user-defined functions in technical calculations.
- Provide warnings about the use of functions.

BUILT-IN FUNCTIONS

Built-in functions range from very simple to very complex. Examples of simple built-in Mathcad functions are ***sin()***, ***cos()***, ***ln()***, and ***max()***. Every Mathcad function is set up in a similar way. The function name is given, followed by a pair of parentheses. The information that is typed within the parentheses is called the argument. Every function has a name and an argument. The function takes the information from the argument (contained within the parentheses) and processes the information based on rules that are defined for the specific function, returning a result.

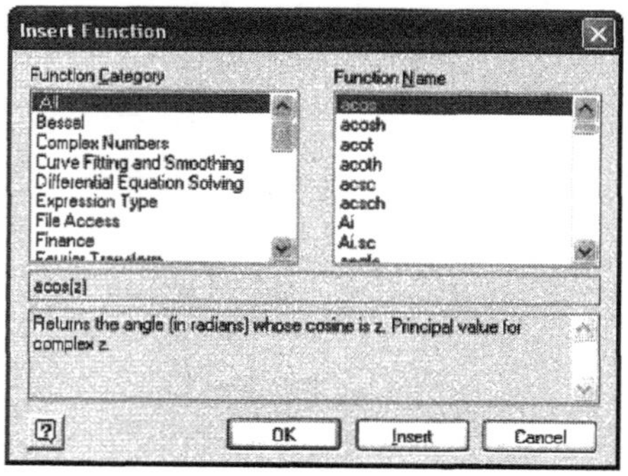

FIGURE 3.1 Insert Function dialog box showing the function categories and function names

To see a list of all the built-in functions Mathcad has, select **Function** from the **Insert** menu. A dialog box will appear that lists all the built-in functions of Mathcad. See Figure 3.1. The functions are grouped by category in the left column. The functions assigned to the highlighted category appear on the right. If you know the function name, you can search for the function by clicking **All** in the left column and then clicking in the right column and typing the function name. The function will then be highlighted.

Once a function is selected in the Insert Function dialog box, you will see some useful information in boxes at the bottom of the dialog box. The upper box shows a list of arguments (within the parentheses) the function is expecting.

Some functions expect only a single argument. Other functions expect two arguments. Some functions require multiple arguments. Some functions can have a variable number of arguments. These will be indicated by three dots following the listed arguments. Figure 3.2 shows four functions with different numbers of arguments.

The lower box contains a description of what the function does. This describes what Mathcad will return when the arguments are included in the function list. It also describes what type of information the function is expecting (such as if the argument must be in radians, or whether it must be an integer).

Take a moment to scan the complete list of Mathcad's built-in functions from the Insert Function dialog box. In later chapters, we will discuss selected functions. If you are interested in knowing about a specific function, refer to Mathcad Help. Figure 3.3 shows examples of using four different Mathcad functions requiring various argument lengths.

A Mathcad expression can have an unlimited number of functions included. To insert a function within an expression, simply highlight the placeholder and use the Insert Function dialog box to insert the function. If you are familiar with the function and its arguments, you do not need to use the dialog box. You simply

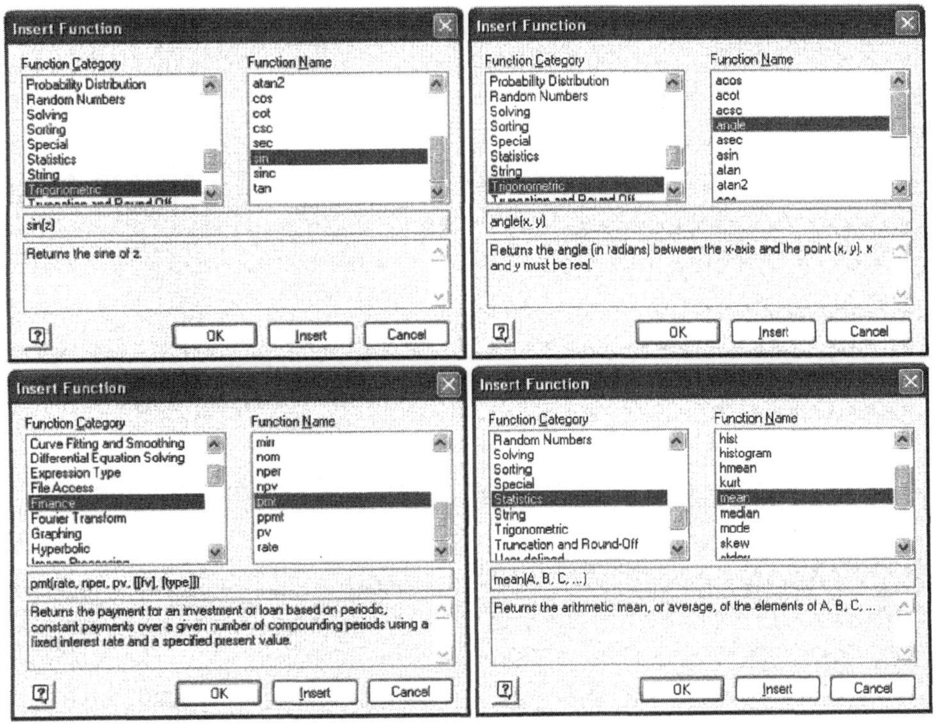

FIGURE 3.2 Note the two boxes below the Function Name. The first box shows the arguments. The second box describes the function. Note that some functions require only one argument. Other functions require multiple arguments. Some arguments allow unlimited arguments.

type the name of the function and include the required arguments between parentheses. Remember that function names are case sensitive. If you type a function name and Mathcad does not recognize it, use the Insert Function dialog box to see if the first letter is upper- or lowercase. Figure 3.4 shows an expression using multiple built-in functions.

USER-DEFINED FUNCTIONS

Remember from Chapter 1 that user-defined functions are very similar to built-in functions. User-defined functions consist of a variable name, a list of arguments, and a definition giving the relationship between the arguments. The same rules that apply to naming variables also apply to naming functions. Refer to Chapter 2 for a listing of the naming rules. The arguments used in the user-defined function do not need to be defined previously in your worksheet. The function is simply

$\mathbf{Example_{1a}} := \mathbf{sin}\left(\frac{\pi}{4}\right)$

$\mathbf{Example_{1a}} = 0.71$

$\mathbf{Example_{1b}} := \mathbf{angle}(3,4)$

$\mathbf{Example_{1b}} = 0.93$ $\quad \mathbf{Example_{1b}} = 53.13 \cdot \mathbf{deg}$

In the second result "deg" was typed in the placeholder

$\mathbf{Example_{1c}} := \mathbf{pmt}\left(\frac{10\%}{12}, 36, 10000, 0, 0\right)$

See Mathcad Help for information about the pmt function

$\mathbf{Example_{1c}} = -322.67$

$\mathbf{Example_{1d}} := \mathbf{mean}(1,3,5,7,10)$

$\mathbf{Example_{1d}} = 5.20$

The list of arguments for this function can be any length.

FIGURE 3.3 Examples of using various types of arguments

$\mathbf{Example_2} := 1 + \mathbf{cos}\left(\frac{\pi}{2} + 7\,\mathbf{tan}\left(\frac{\pi}{4}\right)\right) + \mathbf{sin}(\pi)$

$\mathbf{Example_2} = 0.34$ This is the default result

$\mathbf{Example_2} = 0.34 \cdot \mathbf{rad}$ Result with "rad" typed in placeholder

$\mathbf{Example_2} = 19.65 \cdot \mathbf{deg}$ Result with "deg" typed in placeholder

FIGURE 3.4 Example of using built-in functions within an expression

telling Mathcad what to do with the function arguments, thus the arguments are not defined prior to using the function.

The following is a simple user-defined function:

$$\mathrm{SampleFunction}(x) := x^2$$

In this example, SampleFunction is the name of the function, and x is the argument. When you type `SampleFunction(2)=`, Mathcad takes the value of the argument (2), and applies it everywhere there is an occurrence of the argument in the definition. So in this example, Mathcad replaces x with the number 2 and squares the number, returning the value of 4. See Figure 3.5 for examples of this function with various arguments. Notice that the argument can also be the result of an expression. Any expression is allowed as long as the result of the expression is a value that is expected by the function. The beauty of this is that you don't need to calculate the value of the function argument if it is the result of another equation.

$\mathbf{SampleFunction_1(x)} := x^2$	Function definition
$\mathbf{SampleFunction_1}(2) = 4.00$	Simple positive argument
$\mathbf{SampleFunction_1}(-4) = 16.00$	Simple negative argument
$\mathbf{SampleFunction_1}(\sqrt{16}) = 16.00$	Argument using the result of a built-in function
$\mathbf{SampleFunction_1}\left(\frac{1}{2}\right) = 0.25$	Argument using the result of an expression
$\mathbf{SampleFunction_1}(2 \cdot 4) = 64.00$	Argument using the result of an expression
$\mathbf{SampleFunction_1}\left(4 \cdot \sin\left(\frac{\pi}{4}\right)\right) = 8.00$	Argument using the result of an expression with a built-in function.

FIGURE 3.5 SampleFunction with various arguments

Let us look at some sample user-defined functions. See Figure 3.6. Notice the different types of function names and function arguments. It doesn't matter what letter or combination of letters you use for the argument.

You can even use characters as arguments if you switch to the special text mode discussed in Chapter 1 (CTRL+SHIFT+K). See Figure 3.7.

$\mathbf{f(a)} := a^{\frac{1}{3}}$ $\quad \mathbf{f}(27) = 3.00$	Function names can be a single letter.
$\mathbf{F_2(a)} := 2 \cdot a^2 + a + 3$ $\mathbf{F_2}(3) = 24.00$	Note that the "a" in this function is totally independent of the "a" in the function above. Each argument applies only to the defined function.
$\mathbf{ExampleFunction_2(Dog)} := 2^{\mathbf{Dog}}$ $\mathbf{ExampleFunction_2}(4) = 16.00$	Arguments are not limited to single letters
$\mathbf{AnotherFunction(x2)} := \sqrt{x2 + 8}$ $\mathbf{AnotherFunction}(8) = 4.00$	Arguments may also consist of letters and numbers. This is not recommended as it appears that the argument is "x" multiplied by 2, rather than "x2".

FIGURE 3.6 Examples of various function names and arguments

$\mathbf{SampleFunction_3(@)} := @^2 + 2$ $\mathbf{SampleFunction_3}(10) = 102.00$	To get this function type the following: SampleFunction3 (**[Shift]+[Ctrl]+k** @ **[Shift]+[Ctrl]+k**): **[Shift]+[Ctrl]+k** @ **[Shift]+[Ctrl]+k** ^2 + 2" This takes many key strokes. The advantage of using a symbol as an argument must outweigh the extra effort to add the symbol.

FIGURE 3.7 Example of using a symbol as the argument

Why Use User-Defined Functions?

Once a user-defined function is defined, it can be used over and over again. This makes user-defined functions very useful and powerful. For example, user-defined functions are useful if you are repeatedly doing the same steps in your calculations. This is illustrated in Figure 3.8.

Using Multiple Arguments

Until now, we have been using a single argument in the argument list. Mathcad allows you to have several arguments in the argument list, as illustrated in Figure 3.9.

Suppose that you need to find the result of $x^3 + 2 \cdot x^2 - x + 3$ for many different values of x. You can setup this expression: $y := x^3 + 2 \cdot x^2 - x + 3$. You can then define x:=1 and then type y=. You could then redefine x to be x:=5 ,and then type y= to get another value of y. This method works, but it takes time and is not convenient because you must continuously redefine "x".

$x := 1$

$y := x^3 + 2 \cdot x^2 - x + 3 \qquad y = 5.00$

You would need to redefine "x" to be another number in order to get a result for a different value of x. Your original answer using x:=1 will also be lost.

(In later chapters we will discuss range variables and arrays which will make the above scenario easier.)
Another way solve the above situation is to define a function. After the function is defined, you can type the function using numerous values for the argument.

$y(x) := x^3 + 2 \cdot x^2 - x + 3$

$y(1) = 5.00 \qquad y(10) = 1193.00$

$y(5) = 173.00 \qquad y(100) = 1.02 \times 10^6$

Note the wavy line beneath the "y" indicating that the variable "y" is being redefined. In this case it is changed from an expression to a function.

FIGURE 3.8 Using a function

$$\mathbf{MultipleArgument_1(a, b) := a^2 + b^2}$$

$$\mathbf{MultipleArgument_1(2, 3)} = 13.00$$

$$\mathbf{MultipleArgument_2(dog, cat, goat) := \frac{\sqrt{dog} + 2^{cat}}{goat}}$$

$$\mathbf{MultipleArgument_2(4, 3, 2)} = 5.00$$

FIGURE 3.9 Using multiple arguments

Variables in User-Defined Functions

You are also allowed to use a variable in your function definition that is not a part of your argument list; however, this variable must be defined prior to using it in your user-defined function. The value of the variable—at the time you define your user-defined function—becomes a permanent part of the function.

If you redefine the variable below this point in your worksheet, the function still uses the value at the time the function was defined. This case is illustrated in Figures 3.10 and 3.11.

When you evaluate a previously defined user-defined function, the values you put between the parentheses in the argument list may also be previously defined variables. This is illustrated in Figures 3.12 through 3.14.

$\mathbf{Input_1} := 4$

$\mathbf{VariableExample(V)} := \mathbf{Input_1} \cdot \mathbf{V}$

$\mathbf{VariableExample(3)} = 12.00$

The value of "Input1" in the above function remains the value at the point at which the function was defined (4).

$\mathbf{Input_1} := 8$ Redefine the value of "$Input_1$".

$\mathbf{VariableExample(3)} = 12.00$

In the above example, "$Input_1$" was redefined from 4 to 8. This did not affect the result of the function. The value of "$Input_1$" in the function depends on the value of "$Input_1$" at the point at which the function was defined (4), not the new value of "$Input_1$" (8).

FIGURE 3.10 Using variables in user-defined functions

$\mathbf{Input_1} := 4$ (Note: The wavy lines are because the variable names were used previously in Figure 3.10.)

$\mathbf{Input_1} := 8$ Redefine the value of "$Input_1$".

$\mathbf{VariableExample(V)} := \mathbf{Input_1} \cdot \mathbf{V}$

$\mathbf{VariableExample(3)} = 24.00$

In this example, the value of "$Input_1$" in the function IS changed. This is because the value of "$Input_1$" was changed from 4 to 8 PRIOR to the definition of the user-defined function.

FIGURE 3.11 Changing the value of a variable in a user-defined function

$$\mathbf{Distance}(\mathbf{a},\mathbf{b}) := \sqrt{\mathbf{a}^2 + \mathbf{b}^2}$$

$$\mathbf{Distance}(3,4) = 5.00$$

$x_1 := 3$ $y_1 := 4$

$x_2 := 5$ $y_2 := 12$

$x_3 := 20$ $y_3 := 25$

$$\mathbf{Distance}(x_1, y_1) = 5.00$$

$$\mathbf{Distance}(x_2, y_2) = 13.00$$

$$\mathbf{Distance}(x_3, y_3) = 32.02$$

The argument list can use previous defined variables

FIGURE 3.12 Using variables in the argument list

$\mathbf{B} := 10$	This defines variable "B".
$\mathbf{Function_1}(\mathbf{B}) := \mathbf{B}^2$	The argument "B" in this function is independent of the value of variable "B" above. In this case, "B" defines the argument of the function "$Function_1$". It does not depend on the value of variable "B" above.
$\mathbf{Function_1}(3) = 9.00$	Uses 3 as the argument of "$Function_1$". Result of "$Function_1$" is 3^2.
$\mathbf{Function_1}(\mathbf{B}) = 100.00$	Uses variable "B" as the argument of "$Function_1$" Result of "$Function_1$" is 10^2
$\mathbf{B} := 12$	Redefines variable "B"
$\mathbf{Function_1}(\mathbf{B}) = 144.00$	Result of the function changes because the value of variable "B" has changed. Result of "$Function_1$" is 12^2

FIGURE 3.13 Note the difference between variable B and argument B

EXAMPLES OF USER-DEFINED FUNCTIONS

Let's now look at two examples of how user-defined functions can be used in technical calculations. See Figures 3.15 and 3.16.

$C_1 := 2$	This defines variable "C_1".
$\text{Function}_2(D) := C_1{}^D$	In this function "C_1" is a fixed variable from outside the function, and "D" is the argument of the function (taken from the argument list).
$\text{Function}_2(3) = 8.00$	Result of "Function_2" is 2^3. C_1=2, Argument D=3.
$\text{Function}_2(C_1) = 4.00$	Result of "Function_2" is 2^2. The variable "C_1" becomes the value of the argument in "Function_2". C_1=2, Argument D=2 (Value of variable "C_1").
$C_1 := 4$	Redefine variable "C_1".
$\text{Function}_2(C_1) = 16.00$	Result of "Function_2" is 2^4. C_1=2 (for the function), Argument D=4 (New value of variable "C_1"). Remember that for the function, "C_1" remains the original value at the time that the function was defined, not the redefined value of "C_1".

FIGURE 3.14 Note how C1 is captured in the function and how it also becomes the argument D

Write a function to calculate the area of a trapezoid.

$$\text{Trap}_{\text{Area}}(B1, B2, h) := \frac{1}{2} \cdot h \cdot (B1 + B2) \qquad \text{Trap}_{\text{Area}}(5, 10, 18) = 135.00 \qquad \text{Trap}_{\text{Area}}(8, 0, 9) = 36.00$$

Write a function to calculate the center of gravity of a trapezoid.

$$\text{Trap}_{\text{CG}}(B1, B2, h) := \frac{h \cdot (2 \cdot B2 + B1)}{3 \cdot (B1 + B2)} \qquad \text{Trap}_{\text{CG}}(5, 10, 18) = 10.00 \qquad \text{Trap}_{\text{CG}}(8, 0, 9) = 3.00$$

In order to use these functions and results in technical calculations, you may want to assign the input values and results to variable names such as these:

$T1_{B1} := 5 \qquad T1_{B2} := 10 \qquad T1_{\text{Height}} := 18$

$T2_{B1} := 8 \qquad T2_{B2} := 0 \qquad T2_{\text{Height}} := 9$

Input values to be used in functions.

$T1_{\text{Area}} := \text{Trap}_{\text{Area}}(T1_{B1}, T1_{B2}, T1_{\text{Height}}) \qquad T1_{\text{Area}} = 135.00$

$T1_{\text{CG}} := \text{Trap}_{\text{CG}}(T1_{B1}, T1_{B2}, T1_{\text{Height}}) \qquad T1_{\text{CG}} = 10.00$

$T2_{\text{Area}} := \text{Trap}_{\text{Area}}(T2_{B1}, T2_{B2}, T2_{\text{Height}}) \qquad T2_{\text{Area}} = 36.00$

$T2_{\text{CG}} := \text{Trap}_{\text{CG}}(T2_{B1}, T2_{B2}, T2_{\text{Height}}) \qquad T2_{\text{CG}} = 3.00$

Results are assigned to variable names. These results can now be used later in the calculations.

FIGURE 3.15 Finding the area and center of gravity of a trapezoid

PASSING A FUNCTION TO A FUNCTION

It is possible to use a previous user-defined function in a new user-defined function. Figure 3.17 gives an example of a user-defined function SectionModulus that calculates the section modulus of a rectangular beam. The arguments for this

Write a function to calculate the surface area of a right circular cylinder.

$$\mathbf{Cylinder_{Area}(r,h) := 2\cdot\pi\cdot r\cdot h + 2\cdot\pi\cdot r^2}$$

$\mathbf{Cylinder_{Area}(40\cdot cm, 1.2\cdot m)} = 4.02\ \mathbf{m}^{2.00}$ $\mathbf{Cylinder_{Area}(30cm, 10m)} = 19.42\ \mathbf{m}^{2.00}$

Write a function to find the volume of a right circular cylinder.

$$\mathbf{Cylinder_{Volume}(r,h) := \pi\cdot h\cdot r^2}$$

$\mathbf{Cylinder_{Volume}(40cm, 1.2m)} = 0.60\ \mathbf{m}^{3.00}$ $\mathbf{Cylinder_{Volume}(30cm, 10m)} = 2.83\ \mathbf{m}^{3.00}$

In order to use these functions and results in technical calculations, you may want to assign the input values and results to variable names such as these:

$\mathbf{r_1 := 40cm}$ $\mathbf{h_1 := 1.2m}$

$\mathbf{r_2 := 30.cm}$ $\mathbf{h_2 := 10\cdot m}$

Input values to be used in functions.

$\mathbf{Area_1 := Cylinder_{Area}(r_1, h_1)}$ $\mathbf{Area_1} = 4.02\ \mathbf{m}^{2.00}$

$\mathbf{Volume_1 := Cylinder_{Volume}(r_1, h_1)}$ $\mathbf{Volume_1} = 0.60\ \mathbf{m}^{3.00}$

$\mathbf{Area_2 := Cylinder_{Area}(r_2, h_2)}$ $\mathbf{Area_2} = 19.42\ \mathbf{m}^{2.00}$

$\mathbf{Volume_2 := Cylinder_{Volume}(r_2, h_2)}$ $\mathbf{Volume_2} = 2.83\ \mathbf{m}^{3.00}$

Results are assigned to variable names. These results can now be used later in the calculations.

FIGURE 3.16 Finding the surface area and volume of a right circular cylinder

$$\mathbf{SectionModulus(b,d) := \frac{1}{6}\cdot b\cdot d^2}$$

$$\mathbf{Stress(b,d,M) := \frac{M}{SectionModulus(b,d)}}$$

$\mathbf{Stress(.5m, 1.3m, 800000N\cdot m)} = 823.88\cdot \mathbf{psi}$

The above example used a fixed function. The next example shows the included function to be variable.

$\mathbf{F_1(a) := a^2}$ $\mathbf{G_1(b) := \frac{1}{b^2}}$ Define two user-defined functions.

$\mathbf{Sample(x,H) := \frac{3}{x}\cdot H(x)}$ This user-defined function uses two arguments. They are "x" - a variable and "H" a function name to be defined later. The function actually used will be given as an argument of the user-defined function "Sample".

$\mathbf{Example_1 := Sample(2, F_1)}$ $\mathbf{Example_1} = 6.00$ $\frac{3}{2}\cdot 2^2 = 6.00$ This uses the user-defined function F_1.

$\mathbf{Example_2 := Sample(2, G_1)}$ $\mathbf{Example_2} = 0.38$ $\frac{3}{2}\cdot\frac{1}{2^2} = 0.38$ This uses the user-defined function G_1.

$\mathbf{Example_3 := Sample(2, sin)}$ $\mathbf{Example_3} = 1.36$ $\frac{3}{2}\cdot \mathbf{sin(2)} = 1.36$ This uses the Mathcad function sin().

$\mathbf{Example_4 := Sample(2, ln)}$ $\mathbf{Example_4} = 1.04$ $\frac{3}{2}\cdot \mathbf{ln(2)} = 1.04$ This uses the Mathcad function ln().

FIGURE 3.17 Function in a function

function are b and d. The new function, Stress, calculates the stress in the beam for a given moment M. This function uses the SectionModulus function. Thus, the arguments for Stress must include all the arguments needed for both functions.

The second example in Figure 3.17 is a bit more complicated, but much more powerful. In this example, the user-defined function Sample uses a function H(x) as an argument. The function used will not be defined until the user-defined function Sample is executed. This means that when you execute the user-defined function Sample, you need to include a variable argument for x, and also include a function argument for H. The function can be a user-defined function or a Mathcad function that uses a single argument.

CUSTOM OPERATOR NOTATION

The custom operators allow you to display the names of functions to appear as operators in different forms. There are four different custom operators: prefix, postfix, infix, and treefix. The typical function is displayed as f(x) or f(x,y). The custom operators allow a function to be displayed as fx, xf, xfy, and $x^f y$. The custom operators are located on the Evaluation toolbar.

Prefix Operator

The prefix operator is similar to the typical function notation except there are no parentheses. When you click the prefix operator you get two placeholders. The first placeholder is for the name of the function. The second placeholder is for the name of the function's argument.

Postfix Operator

The postfix operator is similar to the prefix operator, except it places the function last and the argument first.

Infix Operator

The infix operator needs two arguments. It places the function between the x and y arguments. This allows you to define custom functions that can behave like operators. A simple example would be to define a function

Divided By. (Remember that you can include a space in a variable name by using the special text mode, CTRL+SHIFT+K.) You can then have displayed "6 divided by 2=3."

Treefix Operator

The treefix operator needs two arguments as well. This operator places the function name on top with lines extending down to the arguments.

Figure 3.18 gives simple examples of these four different operators. These operators can be much more complex. An excellent discussion of this topic is found in the December 2001 Mathcad Advisor Newsletter. There is also additional information in Mathcad Help.

Prefix operator (fx)

sin(45**deg**) = 0.71 Normal function notation

sin 45**deg** = 0.71 Prefix notation

Postfix operator (xf)

This example uses the FIF (Feet, Inch, Fraction) function/unit.

FIF("5' 4 1/2"") = 64.5000 · **in** Normal function notation

FIF("5' 4 1/2"") = "5' 4-1/2"" · **FIF**

"5' 4 1/2"" **FIF** = 64.5000 · **in** Postfix notation

Infix operator (xfy)

Create a new function called, "to the power of".
Remember, to include spaces in your variable name,
use the special text mode by typing
SHIFT+CTRL+K.

to the power of(**a**, **b**) := $\mathbf{a}^{\mathbf{b}}$

to the power of(2, 5) = 32.00 Normal function notation

2 **to the power of** 5 = 32.00 Infix notation

Treefix operator (xfy)

to the power of = 32.00 Treefix notation

2 5

FIGURE 3.18 Custom operators

WARNINGS

Several warnings will help make the use of functions more effective:

- Be careful not to redefine a user-defined function. User-defined function names are similar to variables. If you redefine a function, it no longer works. See Figure 3.19.

- Be careful not to redefine a built-in function. See Figure 3.20.

- Remember that if you use a variable in your function definition, the value of the variable does not change, even if you later rename the variable.

- Once you define your user-defined function, you will not be able to display the function definition again in your worksheet (unless you use symbolics). If you type the name of the function and an equal sign, Mathcad does not display the function. See Figure 3.21. This can make it difficult to remember exactly what the function definition was. You must also remember what order the arguments go in. One option for displaying the user-defined function definition later in your worksheet is to copy the function definition and paste the definition where you want it displayed. This essentially redefines the function to the same definition. Now right-click within the pasted version of the definition and click **Disable Evaluation**. This will disable the redefinition of the user-defined function. You may wonder, why worry about it? You get the same answer as before because it is the same

$h(x) := x^2 + x + 1$ Defines function "h".

$h(2) = 7.00$

$h := 3$ Redefines "h" as a variable

$h(2) = \blacksquare$ Function "h" is no longer recognized, because it was redefined as variable "h".

$h(x) := x^2 + x + 1$ Defines function "h".

$h(2) = 7$

$h := 3$ Redefines "h" as a variable

$h(2) = \blacksquare\blacksquare$ Function "h" is no longer recognized, because it was redefined as variable "h".

This value must be a function, but has the form: Unitless.

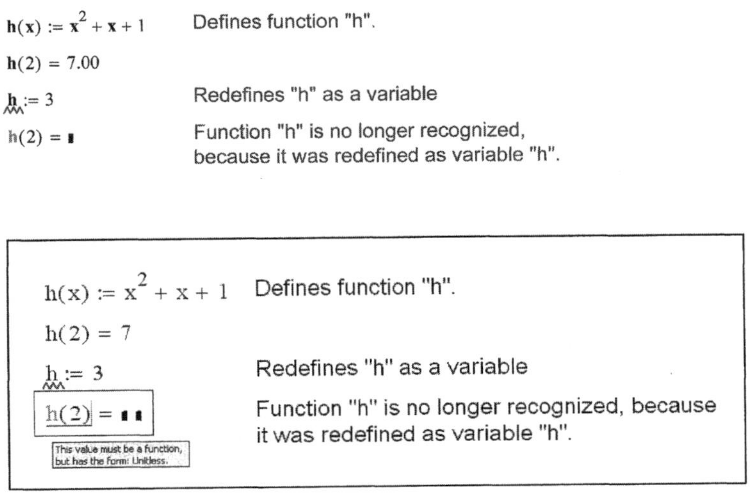

FIGURE 3.19 Be careful not to overwrite user-defined functions

$\mathbf{sin(60deg)} = 0.87$

$\mathbf{sin} := 0.707$ Redefines sin as a variable

$\mathbf{sin(60deg)} = \blacksquare$ Built-in function "sin" is no longer recognized, because it was redefined as variable "sin".

$\mathbf{sin} = 0.71$

$\sin(60deg) = 0.866$

$\sin := 0.707$ Redefines sin as a variable

$\sin(60\cdot deg) = \blacksquare\blacksquare$ Built-in function "sin" is no longer recognized, because it was redefined as variable "sin".

This value must be a function, but has the form: Unitless.

$\sin = 0.707$

FIGURE 3.20 Be careful not to overwrite built-in functions

$g_1(x,y) := x^2 + \frac{y}{2}$ Function definition

$g_1 = f(\mathrm{any1}, \mathrm{any1}^\wedge 2) \rightarrow \mathrm{any1}^\wedge 2$ Mathcad does not display the original definition

In order to display the function definition at a later point in your worksheet, copy the origial definition and paste it where you want it displayed. Right click within the definition and click "Disable Evaluation." WARNING: BE SURE TO DISABLE THE PASTED FUNCTION! You do not want to have two definitions of the same function in your worksheet. If you change the original function, you may forget to change the other copies of the function. This will result in incorrect results in your calculations.

$g_1(x,y) := x^2 + \frac{y}{2}$ ∎ This is a disabled definition. Mathcad ignores the definition. If you want to display the function after it is defined be sure to disable the definition. WARNING: If the original function is changed, this display will be WRONG.

$g_1(3,4) = 11.00$ The result of this fuction is based on the original definition, not the disabled definition. If the original function is changed, this result will also change.

To display the original function using Symbolics, you use the Symbolic toolbar. Follow these steps:

- Type the name of the function and its arguments.
- Click on "explicit" on the Symbolic toolbar.
- Type a comma and type the name of the function (without the argument)
- Click outside of the region.

$g_1(x,y)\ \mathbf{explicit}, g_1 \rightarrow x^2 + \frac{y}{2}$ You may use your original arguments,
or
you may use new arguments.

$g_1(a,b)\ \mathbf{explicit}, g_1 \rightarrow a^2 + \frac{b}{2}$

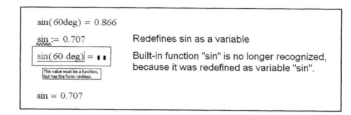

FIGURE 3.21 How to display a function definition later in your calculations

definition. The problem comes if you later redefine the original user-defined function. Mathcad will use the value of the changed function until it comes to the point in your worksheet where you pasted the original function. From this point on, Mathcad will use the old function definition. By disabling the function definition, Mathcad continues to use the new version of the function. However, Mathcad still displays an old version of the function. This may lead to some confusion in your calculation, but not an incorrect answer.

- There is a way to display the original function using Mathcad's symbolic processor. Symbolic calculations will be discussed in Chapter 9. Figure 3.21 shows how to use Symbolics to show a function after it has been defined.

ENGINEERING EXAMPLE 3.1: COLUMN BUCKLING

The Euler column formula predicts the critical buckling load of a long column with pinned ends. The Euler formula is $\mathbf{P_{cr}} = \frac{\pi^2 \cdot \mathbf{E} \cdot \mathbf{I}}{\mathbf{L}^2}$ where E is the modulus of elasticity in (force/length^2), I is the moment of inertia (length^4), L is the length of the column.

Create a user-defined function to calculate the critical buckling load of a column.

$$\mathbf{P_{cr}}\,(\mathbf{E},\mathbf{I},\mathbf{L}) := \frac{\pi^2 \cdot \mathbf{E} \cdot \mathbf{I}}{\mathbf{L}^2}$$

Calculate the critical buckling load for:

1. A steel column with E=29,000ksi, I=37 in^4, and L=20ft
2. A wood column with E=1,800,000 psi, I=5.36in^4, and L=10ft

$\mathbf{P_1} := \mathbf{P_{cr}}(29000\mathbf{ksi}, 37\mathbf{in}^4, 20\mathbf{ft})$ $\quad\quad$ $\mathbf{P_1} = 183.86 \cdot \mathbf{kip}$

$\mathbf{P_2} := \mathbf{P_{cr}}(1800000\mathbf{psi}, 5.36\mathbf{in}^4, 10\mathbf{ft})$ $\quad\quad$ $\mathbf{P_2} = 6.61 \cdot \mathbf{kip}$

ENGINEERING EXAMPLE 3.2: TORSIONAL SHEAR STRESS

The torsional shear stress of a linear elastic homogeneous and isotropic shaft is given by the equation: $\tau = \frac{\mathbf{T} \cdot \mathbf{p}}{\mathbf{J}}$, where T=torque, p=distance from the shaft center, and J=polar moment of inertia with respect to the longitudinal axis of the shaft. For a solid circular shaft, J=π*D^4/32.

Create a user-defined function to calculate the shear stress in a circular shaft.

$$J(D) := \frac{\pi \cdot D^4}{32} \qquad \tau(D, T) := \frac{T \cdot \frac{D}{2}}{J(D)}$$

$$\tau_1 := \tau(1.5in, 8000in \cdot lbf) \qquad \tau_1 = 12.07 \cdot ksi$$

$$\tau_2 := \tau(4in, 300in \cdot kip) \qquad \tau_2 = 23.87 \cdot ksi$$

The twist in a shaft can be calculated by the equation $\theta = \frac{T \cdot L}{J \cdot G}$, where L is the length of the shaft and G is the shear modulus of elasticity or modulus of rigidity (for steel G=11,200 lbf).

Create a user-defined function to calculate the twist of a circular shaft.

$$G := 11200 \cdot ksi \qquad \theta(D, T, L) := \frac{T \cdot L}{J(D) \cdot G}$$

$$\theta_1 := \theta(1.5in, 8000in \cdot lbf, 50in) \quad \theta_1 = 4.12 \cdot deg \quad \theta_1 = 0.07 \cdot rad$$

$$\theta_1 := \theta(4in, 300in \cdot kip, 80in) \quad \theta_1 = 4.89 \cdot deg \quad \theta_1 = 0.09 \cdot rad$$

SUMMARY

We have just scratched the surface of user-defined functions. The intent of this chapter was to introduce the concepts of simple built-in functions, and to get you comfortable with the concept and use of user-defined functions. There is much more to learn. User-defined functions will be discussed in much more detail in later chapters of this book.

In Chapter 3 we:

- Discussed Mathcad's built-in functions.
- Expanded the discussion of user-defined functions.
- Explained the benefits of using user-defined functions rather than using expressions.
- Showed how to include multiple arguments and variables in user-defined functions.
- Issued several warnings about using built-in and user-defined functions.

PRACTICE

Additional problems and applications can be found on the companion site: www.elsevierdirect.com/9780123747839.

Note: Save your worksheet with the following user-defined functions for use with the practice exercises in Chapter 4.

1. Write user-defined functions to calculate the following. Choose your own descriptive function names. These functions will have a single argument. Evaluate each function for two different input arguments.

   a. The volume of a circular sphere with a radius of R1
   b. The surface area of a sphere with a radius of R1
   c. Converting degree Celsius to degree Fahrenheit
   d. Converting degree Fahrenheit to degree Celsius
   e. The surface area of a square box with length L1

2. Write user-defined functions to calculate the following. Choose your own descriptive function names. These functions will have multiple arguments. Evaluate each function for two sets of input arguments.

   a. The inside area of a pipe with an outside diameter of D1 and thickness of T1
   b. The material area of a pipe with an outside diameter of D1 and thickness of T1
   c. The volume of a box with sides L1, L2, and L3
   d. The surface area of a box with sides L1, L2, and L3
   e. The distance traveled by a free-falling object (neglecting air resistance) with an initial velocity, acceleration, and time: V0, a, and T

3. From your area of study (or from a physics book) write 10 user-defined functions. At least five of these should be with multiple arguments. Evaluate each function for two sets of input arguments.

4. The bending moment (at any point x) of a simply supported beam with a uniformly distributed load is defined by the following formula: M_x=w*x*(L-x)/2, where w=force per unit of length, L=length of beam, and x is the distance from the right support. Write a user defined function to calculate the bending moment at any point x. Use the arguments w, L, and x. When testing the formula be sure to input your values in constant units of length.

5. The maximum deflection of a simply supported beam with a uniformly distributed load is defined by the following formula: Deflection=5*w*L^4/(384*E*I). Write a user defined function to calculate the maximum deflection of the beam. Use the arguments w, L, E, and I. When testing the formula, be sure to input your values in constant units of force and length. The units are as follows: w=load/unit of length; L=length; E=force/length^2; and I=length^4.

6. The moment of inertia I of a rectangle about its centroidal axis is defined by the following formula: I=b*d^3/12. Write a user defined function to calculate the moment of inertia. Use the arguments b and d.

CHAPTER

Units!

4

One of the most powerful features of Mathcad is its capability to attach units to numbers. Mathcad units will be one of your best friends as a scientist or engineer. Not only will Mathcad units simplify your work, they are one of the best means you have of catching mistakes. You might remember the 1999 Mars Climate Orbiter that ended in disaster by burning up in the atmosphere of Mars. Why? Because engineers failed to convert English measures of rocket thrust into metric measures. Using Mathcad units will help prevent problems like this from occurring. Once units are attached to values, Mathcad can display the units in any desired unit system. It does this easily and automatically. You are not left wondering what units the numbers represent. If you use units consistently, Mathcad will alert you to problems in your expressions. For example, if you expected the results to be in lbf/ft^2, but they are in lbf/ft, then something is missing from your expression or from your input.

Chapter 4 will:

- Discuss unit dimensions, units, and the default unit systems.
- Show how to assign units to numbers.
- Discuss units of force and units of mass.
- Explain how to create custom units.
- Describe how to use units in equations and functions.
- Present ways of dealing with units in empirical formulas.
- Demonstrate how to use custom scaling units.
- Illustrate the use of custom dimensionless units.

INTRODUCTION

Remember from Chapter 1 that to assign units to a number you should multiply the number by the name of the unit. If you do not know the name of the unit, you can use the Insert Unit dialog box by selecting **Unit** from the **Insert** menu, or by clicking the measuring cup icon on the Standard toolbar.

To change the displayed unit, add the desired unit name to the unit placeholder box, or change the unit name in the unit placeholder. You can double-click on the unit placeholder to select a unit from the Insert Unit dialog box.

DEFINITIONS

For our discussion of units, we will use the following terminology:

Unit dimension: A physical quantity, such as mass, time, or pressure, that can be measured.

Unit: A means of measuring the quantity of a unit dimension.

Base unit dimension: One of seven basic unit dimensions: length, mass, time, temperature, luminous intensity, substance, and current or charge. Mathcad version 14 added currency as a base dimension as well.

Base unit: A default unit measuring one of the seven base unit dimensions. For example, the following are base units: meter (m), kilogram (kg), second (s), Kelvin (K), candela (cd), mole (mol), and Ampere (A).

Derived unit dimension: A unit dimension derived from a combination of any of the seven base unit dimensions. For example, the following are derived unit dimensions: area (length^2), pressure (mass/(length*time^2), energy (mass*length^2/time^2), and power (mass*length^2/time^3).

Derived unit: A unit measuring a derived unit dimension. For example, the following are derived units: pounds per square inch (psi), Joules (J), British Thermal Unit (BTU), and Watts (W).

Unit system: A group of units used to measure base unit dimensions and derived unit dimensions. There are four unit systems available in Mathcad:

- SI, with base units of meter (m), kilogram (kg), second (s), Kelvin (K), candela (cd), mole (mol), and Ampere (A)
- MKS, with base units of meter (m), kilogram (kg), second (s), Kelvin (K), candela (cd), mole (mol), and Coulomb (C)
- CGS, with base units of centimeter (cm), gram (gm), second (s), Kelvin (K), candela (cd), mole (mol), and statampere (statamp)
- U.S., with base units of feet (ft), pound mass (lb), second (s), Kelvin (K), candela (cd), mole (mol), and Ampere (A)

Default unit system: The unit system you tell Mathcad to use. Mathcad defaults to the SI unit system, unless you change it.

Custom unit system: Beginning with Mathcad version 13, you can choose a unit system and then change the default base units (i.e., from feet to inches). You can also select which derived units to use.

CHANGING THE DEFAULT UNIT SYSTEM

When you open Mathcad, the SI unit system is the default unit system. To change to another default unit system in your current worksheet, select **Worksheet Options** from the **Tools** menu. Click the Unit System tab, and select the desired default unit system. See Figure 4.1. You can also choose None as the default unit system. However, if you select None, Mathcad will not recognize any of the built-in units. The custom unit system will be discussed shortly.

Tip! Do not choose None as the default unit system. In engineering calculations, units are critical.

USING AND DISPLAYING UNITS

To add numbers with units attached, the units must all be from the same unit dimension. For example, you can add any units of length, or you can add any units of time, but you cannot add units of length to units of time. Figure 4.2 shows some examples of adding simple units. If you attempt to add different unit dimensions, Mathcad warns that the unit dimensions do not match.

The units Mathcad displays by default are based on the chosen default unit system (or custom unit system), but you can change the displayed units. Let's now look at how to display different units. See Figures 4.3 and 4.4.

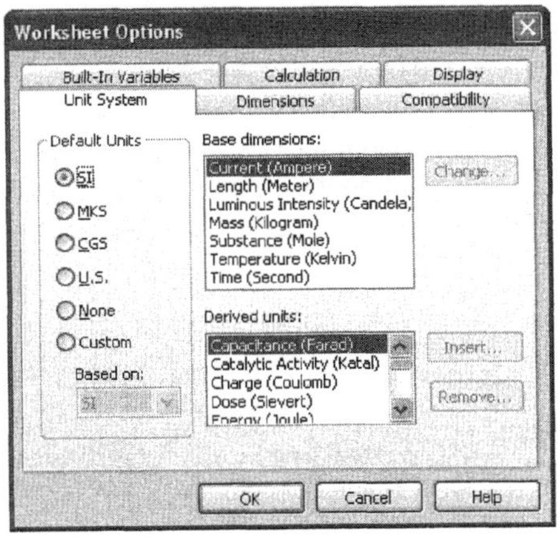

FIGURE 4.1 Setting default unit system

Examples of unit addition.

Attached units can be from any unit system as long as the unit dimensions are the same (such as length, force, etc).
If you don't know the unit names then use the Insert Units dialog box.

$\mathbf{Units_A := 6m + 4mm}$ Type: 6*m+4*mm:

$\mathbf{Units_A} = 6.00\,\mathbf{m}$

$\mathbf{Units_B := 6ft + 3in}$ Type 6*ft+3*in:

$\mathbf{Units_B} = 1.91\,\mathbf{m}$ Mathcad's default unit system is set to SI, so Mathcad displays units of length in meters.

$\mathbf{Units_C := 6N + 1kip}$

$\mathbf{Units_C} = 4454.22\,\mathbf{N}$

$\mathbf{Units_D := 1day + 2hr + 25s}$

$\mathbf{Units_D} = 93625.00\,\mathbf{s}$

$\mathbf{Units_E := 5m + 3N}$

$\mathbf{Units_E} = ▮$

Mathcad warns that the fundamental units do not match. The variables are in red, and there is no result shown.

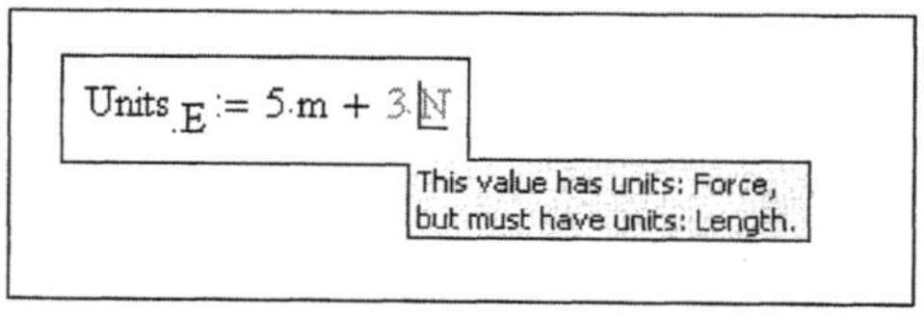

Units.E := 5·m + 3·N
This value has units: Force, but must have units: Length.

FIGURE 4.2 Unit addition

$\mathbf{Units_F := 5.605m}$

If you want to change the displayed results for a number already displayed, then click on the displayed unit. If it is displayed in the default unit system, then a placeholder will appear to the right of the unit. Click on the placeholder, and type the unit you want displayed. You may also double click on the displayed unit and the Insert Unit dialog box will appear.

$\mathbf{Units_F} = 5.61\,\mathbf{m}$

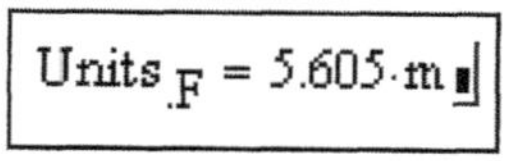

Units.F = 5.605·m ▮

Type the desired display unit in the placeholder

FIGURE 4.3 Unit placeholder

If the result has already had a different unit attached to the result, then the placeholder will not appear. In this case delete the displayed unit, and type a new unit. You may also double click on the displayed unit and select a new unit from the Insert Unit dialog box.

$\mathbf{Units_F} = 18.39 \cdot \mathbf{ft}$ | $\mathbf{Units_F} = 18.39 \cdot \mathbf{ft}$

$\mathbf{Units_F} = 0.003483 \cdot \mathbf{mi}$ | To display a different unit, delete ft and type a new unit.

FIGURE 4.4 Changing displayed units

If you type inconsistent units in the unit placeholder, Mathcad will add units that make the result consistent. See Figure 4.5.

Mathcad can combine units. Look at the examples in Figure 4.6.

> **Tip!** To attach units of area or volume, use the caret (^) symbol to raise the unit to a power. For area type `m^2` or `ft^2`. For volume type `m^3` or `ft^3`.

$\mathbf{Units_G} := 5\mathbf{ft} + 4\mathbf{in}$

$\mathbf{Units_G} = 1.63\,\mathbf{m}$ — Mathcad defaults to m in the SI default unit system.

$\mathbf{Units_G} = 1.63 \frac{\mathbf{m}}{\mathbf{s}} \cdot \mathbf{sec}$ — If you type "sec" in the unit placeholder, Mathcad adds an "s" in the denominator in order for the final units to be units of length.

FIGURE 4.5 Balancing displayed units

$\mathbf{Units_H} := 40\mathbf{mm} \cdot 5\mathbf{mm}$

$\mathbf{Units_H} = 0.0002\,\mathbf{m}^2$

$\mathbf{Units_H} = 200.00 \cdot \mathbf{mm}^2$ — Hint: To get the superscript, use the "^" character above the number 6 key. Type mm^2.

$\mathbf{Units_I} := \frac{10\mathbf{ft}}{2\mathbf{sec}}$

$\mathbf{Units_I} = 1.52 \frac{\mathbf{m}}{\mathbf{s}}$ — Default display

$\mathbf{Units_I} = 5.00 \cdot \frac{\mathbf{ft}}{\mathbf{s}}$ — Displayed as feet per second

$\mathbf{Units_I} = 3.41 \cdot \mathbf{mph}$ — Displayed as miles per hour

FIGURE 4.6 Combining units

DERIVED UNITS

A derived unit dimension is derived from combinations of any of the seven base unit dimensions. A derived unit measures a derived unit dimension. Some examples of derived unit dimensions include acceleration, area, conductance, permeability, permittivity, pressure, and viscosity, among others. Some examples of derived unit names are atmosphere, hectare, farad, joule, newton, and watt, among others.

Mathcad can display derived unit dimensions by the derived unit name or as a combination of the base unit names. The default is to use the derived unit names. To display derived unit dimensions as a combination of base unit names, open the Result Format dialog box by selecting **Result** from the **Format** menu. Then select the Unit Display tab. See Figure 4.7. This tab has three check boxes that control how units are displayed. The "Simplify Units When Possible" box is checked by default. If it is checked, Mathcad displays a derived unit name—if it is available. If a derived unit is not available, Mathcad displays a combination of base unit names. When the "Simplify Units When Possible" box is unchecked, derived unit dimensions are displayed in a combination of base unit names. The "Format units" check box is checked by default. If it is checked, Mathcad displays combinations of base units as mixed fractions with both numerators and denominators. Figure 4.8 illustrates the use of these boxes.

CUSTOM DEFAULT UNIT SYSTEM

When you choose one of the four default unit systems, you get preselected base units and preselected derived units. The addition of the custom default unit system in Mathcad version 13 is a great improvement because it allows you to tell

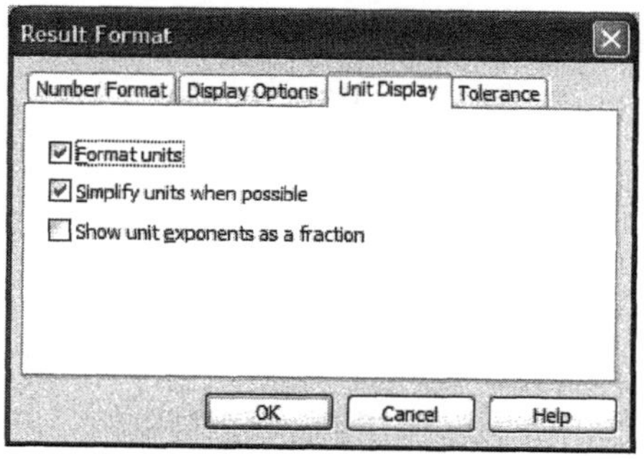

FIGURE 4.7 Unit display

Compare the display of derived units using the check boxes on the "Unit Display" tab of the Result Format dialog box.

Pressure	Energy	Work	Volume
$\mathbf{Press} := 1\mathbf{Pa}$	$\mathbf{Energy} := 1\mathbf{J}$	$\mathbf{Work} := 1\mathbf{J}$	$\mathbf{Vol} := 1\mathbf{m}^3$

"Simplify units when possible" is checked.
"Format units is unchecked." .

$\mathbf{Press} = 1.00\,\mathbf{Pa}$	$\mathbf{Energy} = 1.00\,\mathbf{J}$	$\mathbf{Work} = 1.00\,\mathbf{J}$	$\mathbf{Vol} = 1.00\,\mathbf{m}^{3.00}$

"Simplify units when possible" is checked.
"Format units" is checked."

$\mathbf{Press} = 1.00\,\mathbf{Pa}$	$\mathbf{Energy} = 1.00\,\mathbf{J}$	$\mathbf{Work} = 1.00\,\mathbf{J}$	$\mathbf{Vol} = 1.00\,\mathbf{m}^{3.00}$

"Simplify units when possible" is unchecked.
"Format units" is checked.

$\mathbf{Press} = 1\dfrac{\mathbf{kg}}{\mathbf{m}\cdot\mathbf{s}^2}$	$\mathbf{Energy} = 1\dfrac{\mathbf{m}^2\cdot\mathbf{kg}}{\mathbf{s}^2}$	$\mathbf{Work} = 1\dfrac{\mathbf{m}^2\cdot\mathbf{kg}}{\mathbf{s}^2}$	$\mathbf{Vol} = 1\,\mathbf{m}^3$

"Simplify units when possible" is unchecked.
"Format units" is unchecked.

$\mathbf{Press} = 1\,\mathbf{m}^{-1}\cdot\mathbf{kg}\cdot\mathbf{s}^{-2}$	$\mathbf{Energy} = 1\,\mathbf{m}^2\cdot\mathbf{kg}\cdot\mathbf{s}^{-2}$	$\mathbf{Work} = 1\,\mathbf{m}^2\cdot\mathbf{kg}\cdot\mathbf{s}^{-2}$	$\mathbf{Vol} = 1\,\mathbf{m}^3$

FIGURE 4.8 Compare results of units formatting

Mathcad what base units to use and what derived units to use. For example, if you are using the US default unit system, you can change the base unit of length from feet (ft) to inches (in). If you are using the SI system, you can change the derived unit of force from Pascal (Pa) to kilogram force (kgf).

To create a custom default unit system, open the Worksheet Options dialog box from the **Format** menu. On the Unit Systems tab, select **Custom** and select a Based On unit system. You can now change the base units from any of the seven base dimensions. Next, add or remove any of the derived units. If you see a derived unit that you do not want to use, select it and click **Remove**. For example, the derived unit of volume for SI is Liter (L) and for US it is gallons (gal). If you do not like it, remove it. If you leave the derived unit of volume, when you have a unit of length cubed you will get a display of liters (L) in SI and gallons (gal) in US. If you delete the derived unit of volume, you will get m^3 or ft^3. If you changed the base unit of length from feet to inches, you will get in^3. You can add additional derived units by clicking the **Insert** button. This will give you a choice of all the Mathcad derived units. Select the desired unit, and click **OK**. Unfortunately, you can only choose a built-in Mathcad unit. You cannot choose units you have defined.

> **Tip!** I always use the custom default unit system because it allows me to select the derived units to display. I highly recommend it.

UNITS OF FORCE AND UNITS OF MASS

It is important to understand how Mathcad considers units of force and units of mass. In the US unit system, lbf (pound force) is a unit of force, and lb, lbm, (pound mass), and slug are units of mass. In the SI unit system, N (Newton) and kgf (kilogram force) are units of force, and kg is a unit of mass. Figure 4.9 shows the relationship between various units of force and mass in the US unit system. Figure 4.10 shows the relationship between various units of force and mass in the SI unit system.

> **Tip!** In order to avoid confusion in the US unit system, use the Mathcad unit lbm in lieu of the Mathcad unit lb. This unit was added with Mathcad version 14. You should use lbm as a unit of mass and lbf as a unit of force in the US system.

Note: For this example, the default unit system was changed to US.

$g = 32.17\frac{\mathbf{ft}}{\mathbf{s}^{2.00}}$ g is a built-in Mathcad unit for the acceleration of gravity.

$\mathbf{Units_J} := 1\mathbf{lbm}\cdot\mathbf{g}$ lbf (pound force) = lbm (mass) * acceleration of gravity

$\mathbf{Units_J} = 1.00\,\mathbf{lbf}$

$1\mathbf{lbf} = 32.17\cdot\frac{\mathbf{lbm}\cdot\mathbf{ft}}{\mathbf{sec}^2}$ Shows the relationship between lbf (pound force) and lbm (pound mass).

$1\mathbf{slug} = 32.17\,\mathbf{lb}$ 1 slug = 32.174 pound mass

Note: In order to eliminate confusion as to whether lb mean mass or force, it is suggested to always use lbm in lieu of lb.

FIGURE 4.9 Relationship between various units of mass and force in the US system

$g = 9.81 \dfrac{m}{s^{2.00}}$	g is a built-in Mathcad unit for the acceleration of gravity.
$Units_K := 1kg \cdot g$	mass * acceleration of gravity = Force
$Units_K = 9.81\,N$	This is the way Mathcad defaults in the SI unit system.
$Units_K = 1.00 \cdot kgf$	Force displayed as kgf. 1 kgf =1 kg * g
$1kgf = 9.81\,N$	Shows relationship between kgf and N.
$1N = 0.10 \cdot kgf$	

Note: Sometimes it is easier to display force as kgf rather than N because it eliminates the 9.807 factor.

FIGURE 4.10 Relationship between various units of mass and force in the SI system

CREATING CUSTOM UNITS

Even though Mathcad has over 100 built-in units, you will still need to create your own custom units from time to time. This is very easy to do. You define custom units the same way you define variables. For example, if you want to define a unit of cfs for cubic feet per second, follow the steps in Figure 4.11.

In order to have the custom unit available anywhere in your worksheet, place the definition at the top of your worksheet. You can also use the global definition symbol when creating custom units. The global definition symbol appears as a triple equal sign. All global definitions in the worksheet are scanned by Mathcad prior to scanning the normal definitions. This way, the unit definition does not need to be at the top of your worksheet. To define a global custom unit, type the name of the unit, press the tilde key [~], and then type the definition. See Figure 4.12.

$cfs := \dfrac{ft^3}{sec}$	This defines "cfs" as a custom unit of ft^3 divided by second.
$Units_L := \dfrac{1000ft^3}{1min}$	
$Units_L = 0.47 \dfrac{m^{3.00}}{s}$	Mathcad default in SI units.
$Units_L = 16.67 \cdot cfs$	Type: Units.L=[tab]cfs[enter] The custom unit "cfs" is attached to the result.

FIGURE 4.11 Creating custom units

Create a custom unit for Million Gallons per Day (MGD)

$$\mathbf{mgd} \equiv \frac{1000000}{\mathbf{day}}$$

This is a global definition. The unit will now be available anywhere in the worksheet rather than just below and to the right. Use the tilde key (~) to get the global definition symbol. You may also choose the global definition equal sign from the Evaluation Toolbar.

FIGURE 4.12 Global definition of custom unit

Custom Defined Unit		Arbitrary Usage: Mathcad Default Display (SI)	Arbitrary Usage: Using Custom Unit
$\mathbf{MGD} := \frac{1000000\mathbf{gal}}{\mathbf{day}}$	$\mathbf{Custom_1} := \frac{2\mathbf{m}^3}{45\mathbf{sec}}$	$\mathbf{Custom_1} = 0.04\frac{\mathbf{m}^3}{\mathbf{s}}$	$\mathbf{Custom_1} = 1.01 \cdot \mathbf{MGD}$
$\mathbf{cup} := 8\mathbf{fl\_oz}$	$\mathbf{Custom_2} := 1\mathbf{gal}$	$\mathbf{Custom_2} = 0\,\mathbf{m}^3$	$\mathbf{Custom_2} = 16.00 \cdot \mathbf{cup}$
$\mathbf{tbsp} := \frac{1}{2}\mathbf{fl\_oz}$			$\mathbf{Custom_2} = 256.00 \cdot \mathbf{tbsp}$
$\mathbf{tsp} := \frac{\mathbf{tbsp}}{3}$			$\mathbf{Custom_2} = 768.00 \cdot \mathbf{tsp}$

FIGURE 4.13 Examples of custom units

Tip! Global definitions generally are discouraged because they do not participate in redefinition warnings, and can create confusing redefinition chains if used in the middle of a document. I have had good experience using them for unit definitions. It is still a good idea to include global unit definitions at the top of your worksheet.

Figure 4.13 gives some more examples of custom units. Notice how easily Mathcad deals with the mixing of US and SI units.

UNITS IN EQUATIONS

Now that you understand the concept of units, let's explore the use of units in equations and functions. Units are almost always a part of any engineering equation. In order to take advantage of Mathcad's wonderful unit system, it is

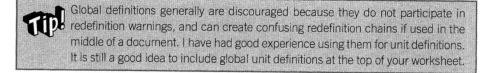

The formula for kinetic energy is: $\frac{\mathbf{Mass} \cdot \mathbf{Velocity}^2}{2}$

$\mathbf{Mass} := 2\mathbf{kg}$ $\mathbf{Velocity} := 200 \frac{\mathbf{cm}}{\mathbf{s}}$

$$\mathbf{KineticEnergy_1} := \frac{\mathbf{Mass} \cdot \mathbf{Velocity}^2}{2}$$

$$\mathbf{KineticEnergy_1} = 4.00\,\mathbf{J}$$

FIGURE 4.14 Example of units in an equation

important to always attach units to your variables. Even for simple calculations, you should get in the habit of using units. There are a few cases where this is not possible. These cases will be noted as they occur.

Figure 4.14 shows the formula for kinetic energy. Notice how the dimension of mass, length, and time combine to form a unit of energy (Joules).

Do Not Redefine Built-in Units

It is important not to redefine built-in Mathcad units. Figure 4.15 shows what will happen if you define m:=2 kg. Note the squiggle line below the m. This means

The formula for kinetic energy is:

$$\frac{\mathbf{m} \cdot \mathbf{v}^2}{2}$$

$\mathbf{v} := 200 \frac{\mathbf{cm}}{\mathbf{s}}$

$\mathbf{m} := 2\mathbf{kg}$

$$\mathbf{KineticEnergy_2} := \frac{\mathbf{m} \cdot \mathbf{v}^2}{2}$$

$$\mathbf{KineticEnergy_2} = 4.00\,\mathbf{J}$$

If the same equation as used in Figure 4.14 is rewritten with only single letter variables, the variable "m" redefines a built-in variable for meter. This is indicated by the squiggle line below "m."

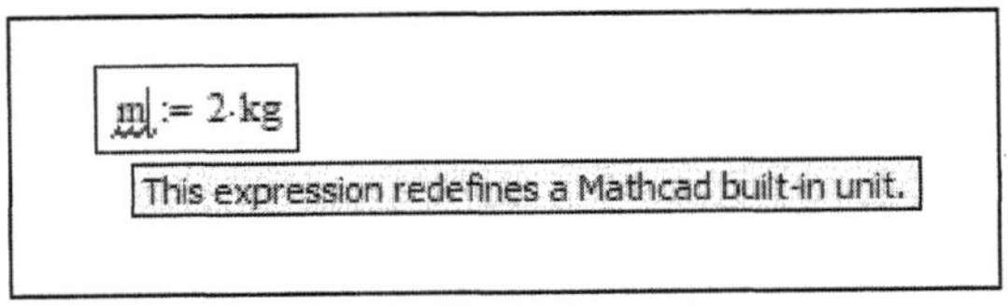
m := 2·kg
This expression redefines a Mathcad built-in unit.

The variable "m" now represents 2 kg instead of 1000mm. Notice what will happen when you try to use the variable "m" to represent a length of 1 meter.

$2\mathbf{ft} = 0.61\,\mathbf{m}$ Meters is still a unit and Mathcad will still display "m" for meters.

$2 \cdot \mathbf{m} = 4.00\,\mathbf{kg}$ If you try to attach units of meter to a number, Mathcad thinks you want to attach 2 kg, because that is how you have now defined "m."

Be very careful. Do not redefine built-in Mathcad units.

FIGURE 4.15 Redefinition warning

Note: For this example, the default unit system was changed to US.

The formula for final velocity based on initial velocity (V_0), acceleration (a), and distance (s) is: $\sqrt{v_0^2 + 2 \cdot a \cdot s}$

$\mathbf{InitialVelocity_1} := 80\frac{ft}{sec}$ $\mathbf{Acceleration_1} := 10\frac{ft}{sec^2}$ $\mathbf{Distance_1} := 300ft$

$\mathbf{FinalVelocity_1} := \sqrt{\mathbf{InitialVelocity_1}^2 + 2 \cdot \mathbf{Acceleration_1} \cdot \mathbf{Distance_1}}$

$\mathbf{FinalVelocity_1} = 111.36\frac{ft}{s}$ Mathcad defaults to ft/s in US default unit system.

$\mathbf{FinalVelocity_1} = 75.92 \cdot mph$ Result after attaching mph to the unit placeholder.

$\mathbf{FinalVelocity_1} = 33.94 \cdot \frac{m}{s}$ Result after attaching m/s to the unit placeholder.

FIGURE 4.16 Example of units in an equation

that the value of m is being redefined. The variable m is a built-in variable for meter. If you use m for meters below this point in your worksheet, Mathcad uses m=2 kg (mass), not m=1 meter.

Figure 4.16 shows the final velocity of an object based on initial velocity (v0), acceleration (a), and distance (s). For this example the default unit system was changed to US.

Let's look at the same equation, but using units from different unit systems. Notice how Mathcad does all the conversions for you. See Figure 4.17.

Note: For this example, the default unit system was changed to US.

The following input quantities are the same as in Figure 4.16. They are just input using different units.

$\mathbf{InitialVelocity_2} := 54.545mph$ $\mathbf{Acceleration_2} := 3.048\frac{m}{sec^2}$ $\mathbf{Distance_2} := 100yd$

$\mathbf{FinalVelocity_2} := \sqrt{\mathbf{InitialVelocity_2}^2 + 2 \cdot \mathbf{Acceleration_2} \cdot \mathbf{Distance_2}}$

$\mathbf{FinalVelocity_2} = 111.35\frac{ft}{s}$ Mathcad defaults to ft/s in US default unit system.

$\mathbf{FinalVelocity_2} = 75.92 \cdot mph$ Result after attaching mph to the unit placeholder.

$\mathbf{FinalVelocity_2} = 33.94 \cdot \frac{m}{s}$ Result after attaching m/s to the unit placeholder.

Notice how the input units of initial velocity do not need to be in ft/sec. The input units of acceleration do not need to be in ft/sec^2, nor does the input distance need to be in feet. They can be in any units of length and time. Mathcad does all the conversion you! The result is exactly the same as in Figure 4.16, even though the input units were all different.

FIGURE 4.17 Same example as Figure 4.16, but using mixed units

UNITS IN USER-DEFINED FUNCTIONS

Using units in user-defined functions is similar to using units in equations, except that you need to include the units in the arguments and not the function. Figure 4.18 uses a function similar to the equation used in Figure 4.16.

UNITS IN EMPIRICAL FORMULAS

Many engineering equations have empirical formulas. There are times when units might not work with the empirical equations. This can occur when the empirical formula raises a number to a power that is not an integer such as $x^{1/2}$ or $x^{2/3}$. If units are attached to x, the units of the result will not be accurate. In order to resolve this problem, divide the variable by the units expected of the equation, and then multiply the results by the same unit. For example, the shear strength of concrete is based on the square root of the concrete strength in psi (lbf/in^2). See Figures 4.19, 4.20, and 4.21 to see how to resolve the use of units in empirical formulas.

Figures 4.19, 4.20, and 4.21 illustrate the following when dealing with units in empirical equations:

- Don't stop using units if they appear to not work with your equation.
- Divide the affected variables by the units in the system expected in the equation.

$$\mathbf{FinalVelocity}(v_0, a, s) := \sqrt{v_0^2 + 2 \cdot a \cdot s}$$

Creates a user-defined function based on initial velocity (v_0), acceleration (a), and distance (s).

$$\mathbf{FinalVelocity}\left(80\frac{ft}{s}, 10\frac{ft}{sec^2}, 300ft\right) = 111.36\frac{ft}{s}$$

Units must be attached to each argument.

$$\mathbf{FinalVelocity}(80, 10, 300) = 111.36$$

In this example, the numbers used for the arguments are the same as above, but no units are attached to the numbers. The numeric result is the same as above, but no units are attached to the result. For engineering calculations you want units attached to all results. Therefore, make sure that units are attached to all the input information.

$$\mathbf{FinalVelocity}\left(54.545\mathbf{mph}, 3.048\frac{m}{sec^2}, 100\mathbf{yd}\right) = 111.35\frac{ft}{s}$$

Input arguments can be mixed units.

$$\mathbf{FinalVelocity}(54.545, 3.048, 100) = 59.87$$

If no units are attached to the arguments, and the units do not match, then the numeric result is incorrect. BE SURE TO ATTACH UNITS TO ALL THE INPUT ARGUMENTS.

FIGURE 4.18 Units in user-defined functions

$\phi := 0.85 \quad f'_c := 4000\,psi$

Input variables: phi and strength of concrete.

$ShearStrength_1 := 2 \cdot \phi \cdot \sqrt{f'_c}$

Empirical Formula for shear strength of concrete based on f'_c.

Result should be in psi.

$ShearStrength_1 = 7318.35 \dfrac{lb^{0.50}}{s \cdot ft^{0.50}}$

Incorrect result because of the units under the square root.

The result needs to be in psi. The empirical formula takes the square root of f'c (in psi) and expects the result in psi, but this did not happen in the above equation.

In order to resolve this problem, divide f'_c by psi to make it unitless, and then multiply the result by psi.

$ShearStrength_2 := 2 \cdot \phi \cdot \sqrt{\dfrac{f'_c}{psi}} \cdot psi \qquad ShearStrength_2 = 107.52\,psi$

Correct result.

FIGURE 4.19 Units in empirical formulas

If an empirical formula expects a number to be in a particular unit (in this case psi), then it is important to divide the number by psi, even if the variable was input in a different unit system. For example, if you are using a US formula, and if f'_c were input in MPa, you would still need to divide by psi, not MPa. The result can be converted to MPa. See below for an example.

$f'_c := 27.579\,MPa$

Change the input to MPa. (This step isn't really necessary. Mathcad already knew that f'_c was 27.579 MPa. It is done only to emphasize that f'c was input as SI.)

$f'_c = 4000.00\,psi$

Even though f'_c was input in metric units, Mathcad knows that it is the same as 4000 psi and the result will be the same.

$ShearStrength_3 := 2 \cdot \phi \cdot \sqrt{\dfrac{f'_c}{psi}} \cdot psi$

Since the equation is meant for US units, divide by psi, not MPa.

$ShearStrength_3 = 107.52\,psi$

The result is the same, even though f'_c was input in metric units.

$ShearStrength_3 = 0.74 \cdot MPa$

The result can be displayed as MPa.

$ShearStrength_4 := 2 \cdot \phi \cdot \sqrt{\dfrac{f'_c}{MPa}} \cdot MPa$

Result is incorrect if you try to divide by MPa. (Because the formula was written for US units.)

$ShearStrength_4 = 1294.85\,psi$

$ShearStrength_4 = 8.93 \cdot MPa$

Incorrect result

FIGURE 4.20 Units in empirical formulas

The same shear strength equation in SI form is: $0.166 \cdot \phi \cdot \sqrt{f'_c}$ where f'_c is in MPa.

$f'_c = 4000.00\ \mathbf{psi}$ $f'_c = 27.58 \cdot \mathbf{MPa}$ These values were input in the previous example.

$\mathbf{SI\_ShearStrength} := 0.166 \cdot \phi \cdot \sqrt{\frac{f'_c}{\mathbf{MPa}}} \cdot \mathbf{MPa}$ This equation was written for SI units; therefore, divide by MPa.

$\mathbf{SI\_ShearStrength} = 107.47\ \mathbf{psi}$

$\mathbf{SI\_ShearStrength} = 0.74 \cdot \mathbf{MPa}$

The results are the same as using the US equation.

When using empirical formulas it is critical to know what units the equation was written for. It is then critical to divide by the units the equation was written for, not what the input units are.
The following illustrates why this is important. Notice how f'c divided by psi = 4000 and f'c divided by MPa = 27.58. These are the numeric numbers expected in the US and SI shear equations respectively.

$\frac{f'_c}{\mathbf{psi}} = 4000.00$ $\frac{f'_c}{\mathbf{MPa}} = 27.58$

FIGURE 4.21 Same example as in Figure 4.20, but in SI form

- It doesn't matter what input units you used as long as you divide by the units in the system expected to be used in the equation.
- After you divide by the units, you might need to multiply again at some point in the equation by the same units.

Figure 4.22 illustrates another empirical equation using units.

SIUnitsOf()

Mathcad has a function called ***SIUnitsOf(x)***. The Mathcad definition of this function is, "Returns the units of x scaled to the default SI unit, regardless of your chosen unit system. If x has no units, returns 1." This means Mathcad takes the unit dimension and returns the SI base unit for that unit dimension, or a number equivalent to the SI base unit if another default unit system is used. The SIUnitsOf function is intended only for the SI unit system. It should never be used with any other unit system including a customized SI unit system. Let's give an example. If you have chosen SI as your default unit system and you define Length:=2 m, then ***SIUnitsOf(Length)*** returns 1 m, because meter is the default unit of length in the SI unit system. If you have chosen US as your default unit system and you define Length:=2 ft, then ***SIUnitsOf(Length)*** returns 3.281 ft. This is 1 m converted to 3.281 ft. If you have a customized default unit system, and have changed the base unit of length to cm, the ***SIUnitsOf(x)*** function will still return 1 m. Remember, the ***SIUnitsOf(x)*** function returns the default SI unit, not customized units. The actual quantity of the dimension returned by the function is the same no matter what default unit system you have chosen; only the displayed unit value changes. See Figure 4.23.

The Hazen-Williams equation for calculating the fluid velocity in a pipe system is Velocity=1.318*C_H*$R^{0.63}$*$S^{0.54}$. Where C_H is a coefficient related to the pipe material, L is length of pipe in feet, R is the hydraulic radius (which for a pipe flowing full is the Area in feet divided by the circumference in feet), and S is the slope of the energy line or the hydraulic gradient h_f/L.

Calculate the velocity given the following:

Hydraulic gradient: 90 feet over 1000 feet

$C_H := 110 \quad D := 24in \quad h_f := 90 \cdot ft \quad Length := 1000ft$

$$Velocity := 1.318 \cdot C_H \cdot \left(\frac{\frac{\pi \cdot D^2}{4}}{\pi \cdot D}\right)^{0.63} \cdot \left(\frac{h_f}{Length}\right)^{0.54}$$

Result is not accurate because of the empirical formula and units.

$Velocity = 25.52\,ft^{0.63}$

$$Velocity := 1.318 \cdot C_H \cdot \left[\frac{\frac{\pi \cdot \left(\frac{D}{ft}\right)^2}{4}}{\pi \cdot \frac{D}{ft}}\right]^{0.63} \cdot \left(\frac{h_f}{Length}\right)^{0.54} \cdot \frac{ft}{sec}$$

$Velocity = 25.52\frac{ft}{s}$

$$1.318 \cdot 110 \cdot \left(\frac{\frac{\pi \cdot 2^2}{4}}{\pi \cdot 2}\right)^{0.63} \cdot \left(\frac{90}{1000}\right)^{0.54} = 25.52$$

With no units attached

FIGURE 4.22 Example of another empirical formula

WARNING: You might see examples where the ***SIUnitsOf(x)*** function is used to create a unitless number. Do not use this function to create a unitless number. It is much better to divide the number by the unit you want the unitless number to represent. For example, if you have a number with length units attached, and you want to display a unitless number representing the length in meters, then divide the number by m. If you want to display a unitless number representing the length in feet, then divide by ft. This way you are sure to get the unitless number you want. Dividing by ***SIUnitsOf()*** can give you a different result than what you want. Let's relook at Figure 4.21, but using the ***SIUnitsOf()*** function. See Figure 4.24.

As seen in Figure 4.24, you receive unexpected results when you use the function ***SIUnitsOf()*** incorrectly. It is best to divide by the expected units instead of using the ***SIUnitsOf()*** function.

The function *SIUnitsOf(x)* returns the SI units of the variable "x".

When the SI is chosen as the default unit system, then the *SIUnitsOf(x)* function always returns the value of 1 times the default base unit. It does not matter what the magnitude of the unit is, the *SIUnitOf(x)* function always returns the value of one times the default base unit.

$\mathbf{Length_1} := 5\mathbf{cm}$

$\mathbf{SIUnitsOf}(\mathbf{Length_1}) = 1.00\,\mathbf{m}$

$\mathbf{Length_2} := 500\mathbf{m}$

$\mathbf{SIUnitsOf}(\mathbf{Length_2}) = 1.00\,\mathbf{m}$

$\mathbf{Torque} := 25\mathbf{N}\cdot\mathbf{m}$

$\mathbf{SIUnitsOf}(\mathbf{Torque}) = 1.00\,\mathbf{J}$

$\mathbf{Pressure} := 25\mathbf{Pa}$

$\mathbf{SIUnitsOf}(\mathbf{Pressure}) = 1.00\,\mathbf{Pa}$

$\mathbf{Power} := 4\frac{\mathbf{N}\cdot\mathbf{m}}{\mathbf{s}}$

$\mathbf{SIUnitsOf}(\mathbf{Power}) = 1.00\,\mathbf{W}$

The quantity of the result of *SIUnitOf(x)* is always the same; however, it may be displayed differently depending on the chosen default unit system. The following values are displayed when the U.S. default unit system is chosen. When converted back to the SI default units, these quantities are exactly the same as shown above.

$\mathbf{SIUnitsOf}(\mathbf{Length_1}) = 3.28\cdot\mathbf{ft}$

$\mathbf{SIUnitsOf}(\mathbf{Length_2}) = 3.28\cdot\mathbf{ft}$

$\mathbf{SIUnitsOf}(\mathbf{Torque}) = 23.73\frac{1}{\mathbf{s}^{2.00}}\cdot\mathbf{ft}^2\mathbf{lbm}$

$\mathbf{SIUnitsOf}(\mathbf{Pressure}) = 0.000145\cdot\mathbf{psi}$

$\mathbf{SIUnitsOf}(\mathbf{Power}) = 1.00\,\mathbf{W}$

FIGURE 4.23 SIUnitsOf(x)

CUSTOM SCALING UNITS

Beginning with Mathcad version 13, functions can be used in the units placeholder. This allows the use of degree Fahrenheit and degree Celsius, which could not be used in previous versions of Mathcad.

Fahrenheit and Celsius

The °***F*** and °***C*** are actually functions, which means that you type the function name and then list its arguments in parentheses. In order to make it easier to input these functions, there is a Custom Character toolbar that contains the °***F*** and °***C*** function icons. If you type **°F(32)=** you get 273.15 K. If you click the units placeholder and insert °*C*, you get the display °*F*(32) = 0.00°C. Figure 4.25 shows some examples of using different units systems.

The *SIUnitsOf()* function is difficult to understand and use. This example illustrates why it should never be used when you have chosen US as your default unit system.

Use the same information as used in Figures 4.21.

$f'_c = 4000.00\,psi \qquad \frac{f'_c}{psi} = 4000.00 \qquad \frac{f'_c}{MPa} = 27.58$

The following example shows the values Mathcad returns when using the *SIUnitsOf()* function.

$$SI\_ShearStrength := 0.166 \cdot \phi \cdot \sqrt{\frac{f'_c}{SIUnitsOf(f'_c)}} \cdot SIUnitsOf(f'_c)$$

$SI\_ShearStrength = 0.11\,psi$ Should be 107 psi

$SI\_ShearStrength = 0.00074 \cdot MPa$ Should be 0.74 MPa

Let's examine why unexpected results were returned.

$SIUnitsOf(f'_c) = 0.000145\,psi$ Value Mathcad returns. (psi is shown because the default unit system is set to US. This is the US value of 1 Pa.)

$SIUnitsOf(f'_c) = 1.00 \cdot Pa$ Pa is the SI default unit for the base dimension of pressure.

$\frac{f'_c}{SIUnitsOf(f'_c)} = 2.76 \times 10^7$ The unitless number is the result of f'c being divided by Pa. For the metric shear equation f'c must be divided by MPa, not Pa. Therefore, the result of using the *SIUnitsOf()* function is inaccurate.

$\frac{f'_c}{Pa} = 2.76 \times 10^7$

Do not use the *SIUnitsOf()* function. In order to avoid getting an unexpected result, divide by the appropriate unit and not *SIUnitsOf()*.

FIGURE 4.24 Example of using the SIUnitsOf() function

°F(32) = 273.15 K	°C(100) = 373.15 K	32°C = ▮
°F(32) = 491.67 · R	°C(100) = 671.67 · R	32°F = ▮
°F(32) = 0.00 · °C	°C(100) = 100.00 · °C	
°F(32) = 32.00 · °F	°C(100) = 212.00 · °F	
250K = 250.00 K	300R = 166.67 K	
250K = 450.00 · R	300R = 300.00 · R	
250K = −23.15 · °C	300R = −106.48 · °C	
250K = −9.67 · °F	300R = −159.67 · °F	

Notice how the degree Fahrenheit and degree Celsius must be entered as functions. You cannot type 32°C or 32°F.

In this example °C and °F are actually functions in the unit placeholder, while K and R are units

FIGURE 4.25 Fahrenheit and Celsius

Use the "xf" icon on the Evaluation toolbar to get the postfix operator.

32 °F = 273.15 **K** °F(32) = 273.15 **K**	This is a function in postfix notation, but it appears as a unit. It functions the same as using a standard function notation. It is not the same as typing "32°F=" or typing "32*" and then inserting "°F" from the Insert Unit dialog box.
100 °C = 373.15 **K** °C(100) = 373.15 **K**	This is the °C function in postfix notation.

FIGURE 4.26 Postfix notation

Even though *°F* and *°C* are functions, there is a way to make it appear as if these are unit assignments. This is by the use of a postfix operator. This operator was discussed in Chapter 3. The postfix operator xf is on the Evaluation toolbar (x meaning argument, and f meaning function). Place the argument in the first placeholder, and place the function in the second placeholder. See Figure 4.26.

Change in Temperature

You can add or subtract Kelvin and Rankine temperatures as you would any units. For example, 400K − 300K = 100K, or 200R + 50R = 250R.

When adding or subtracting Fahrenheit and Celsius temperatures, you must remember that Mathcad converts to Kelvin before doing any addition or subtraction. Thus, 212 °F − 200 °F is the same as taking 373.15 K − 366.48 K = 6.67 K (−447.67 °F) or 671.67R − 659.67R=12R (−447.67 °F). The answer you want to obtain is 12 °F. Mathcad has solved this issue by creating a unit called Δ °F. This simply means the change in temperature in degree Fahrenheit. Therefore, if you do a Fahrenheit subtraction you must display the result in Δ °F, not °F. If you do a Celsius subtraction, you must display the result in Δ °C, not °C. The Δ °F and Δ °C are available on the Insert Unit dialog box. See Figure 4.27.

When adding Fahrenheit and Celsius temperatures, you actually want to add a change in temperature, not add an absolute Kelvin or Rankine temperature. This is illustrated in Figure 4.28.

Figure 4.29 gives an example of using Fahrenheit in engineering calculation.

Degrees Minutes Seconds (DMS)

Another custom scaling function added in Mathcad version 13 is the ***DMS*** function. This function converts degrees, minutes, and seconds to decimal degrees. It will also display radians or decimal degrees as degrees, minutes, and seconds. For the ***DMS*** function, you need a 3 row, 1 column matrix.

The format and use of the ***DMS*** function is illustrated in Figure 4.30.

You may subtract Kelvin and Rankine temperatures as you would any units.The result is an absolute temperature. When subtracting Fahrenheit or Celsius temperatures you usually want to find the difference between the temperatures.

When you calculate the change in temperature you must display the result in terms of the change in degrees Δ°F and not the actual temperature. The Δ°F and Δ°C are available in the Insert Unit dialog box.

$\mathbf{TempChange_1} := \mathbf{°F}(212) - \mathbf{°F}(200)$

$\mathbf{TempChange_1} = 6.67\,\mathbf{K}$

$\mathbf{TempChange_1} = 12.00 \cdot \mathbf{\Delta°F}$

$\mathbf{TempChange_1} = 12.00 \cdot \mathbf{R}$

$\mathbf{TempChange_1} = 6.67 \cdot \mathbf{\Delta°C}$

If you do not display the result in Δ°F or Δ°C, then Mathcad will display the temperature difference as the Rankine or Kelvin temperature.

$\mathbf{TempChange_1} = -447.67 \cdot \mathbf{°F}$ $\quad 12\mathbf{R} = -447.67 \cdot \mathbf{°F}$

$\mathbf{TempChange_1} = -266.48 \cdot \mathbf{°C}$ $\quad 12\mathbf{R} = -266.48 \cdot \mathbf{°C}$

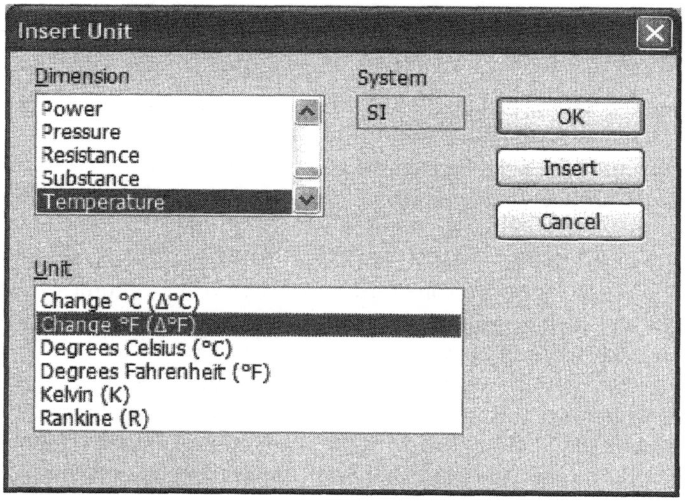

FIGURE 4.27 Change in temperature

Hours Minutes Seconds (hhmmss)

The ***hhmmss*** function converts hours, minutes, and seconds into decimal time—either seconds, minutes, hours, days, and so on. It can also be used in the units placeholder to convert decimal units of time into hours, minutes, and seconds. The input and display of ***hhmmss*** differs from the ***DMS*** function. Instead of

You may add Kelvin and Rankine temperatures just as you would any unit.

$200\mathbf{K} + 50\mathbf{K} = 250.00\,\mathbf{K}$

$200\mathbf{R} + 50\mathbf{R} = 138.89\,\mathbf{K}$ $\quad$ $200\mathbf{R} + 50\mathbf{R} = 250.00 \cdot \mathbf{R}$

If you are adding Fahrenheit or Celsius temperatures, you must use Δ°F and Δ°C

$\mathbf{TempChange_2} := \mathbf{°F}(50) + \mathbf{°F}(30)$

$\mathbf{TempChange_2} = 555.19\,\mathbf{K}$

$\mathbf{TempChange_2} = 539.67 \cdot \mathbf{°F}$

Adding 50°F and 30°F is the same as adding 283.15K and 272.04K, which is equal to 555.19K or 539.67°F.

$\mathbf{°F}(50) = 283.15\,\mathbf{K}$

$\mathbf{°F}(30) = 272.04\,\mathbf{K}$

$283.15\mathbf{K} + 272.039\mathbf{K} = 555.19\,\mathbf{K}$

$\mathbf{TempChange_3} := \mathbf{°F}(50) + \mathbf{30\Delta°F}$

$\mathbf{TempChange_3} = 80.00 \cdot \mathbf{°F}$

$\mathbf{TempChange_3} = 299.82\,\mathbf{K}$

$\mathbf{°F}(80) = 299.82\,\mathbf{K}$

To be accurate, you must take 50°F and add a change of 30°F.

This is the same thing as taking 50°F and adding 30R or 30*(5/9)K.

In other words 1Δ°F=1R and 1Δ°C=1K.

$$283.15 + 30 \cdot \frac{5}{9} = 299.82$$

FIGURE 4.28 Adding temperatures

Calculate the temperature of mixed air from two sources.

$\mathbf{Temp_1} := \mathbf{°F}(85)$ $\quad$ Remember that °F is a function, not a unit.

$\mathbf{Temp_2} := \mathbf{°F}(20)$

$\mathbf{CFM_1} := 200\mathbf{cfm}$

$\mathbf{CFM_2} := 50\mathbf{cfm}$

$$\mathbf{Temp_{Final}} := \frac{\mathbf{Temp_1} \cdot \mathbf{CFM_1} + \mathbf{Temp_2} \cdot \mathbf{CFM_2}}{\mathbf{CFM_1} + \mathbf{CFM_2}}$$

$\mathbf{Temp_{Final}} = 295.37\,\mathbf{K}$

$\mathbf{Temp_{Final}} = 72.00 \cdot \mathbf{°F}$

FIGURE 4.29 Example of using Fahrenheit in engineering calculations

Using *DMS* as a function

DMS(▪) Insert the DMS function from the Insert Unit dialog box.

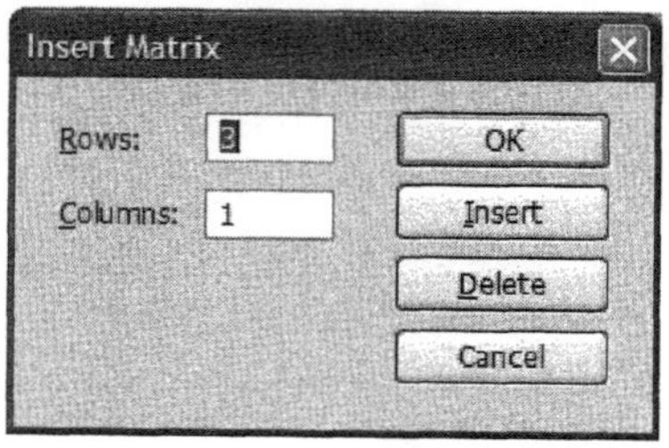

Click on the placeholder and type CTRL+m.
Change the Rows to 3 and Columns to 1.
Click OK.

$$\mathbf{DMS}\left(\begin{pmatrix} \blacksquare \\ \blacksquare \\ \blacksquare \end{pmatrix}\right)$$

Add degrees to the top placeholder.
Add minutes to the middle placeholder.
Add seconds to the bottom placeholder.

$$\mathbf{DMS}\left(\begin{pmatrix} 29 \\ 30 \\ 15 \end{pmatrix}\right) = 0.51$$

The result default display is in radians

$$\mathbf{DMS}\left(\begin{pmatrix} 29 \\ 30 \\ 15 \end{pmatrix}\right) = 29.50 \cdot \mathbf{deg}$$

Click on the units placeholder and type "deg" to get decimal degrees.

Using *DMS* in the units placeholder

$45\mathbf{deg} = 0.79$ The default display is in radians.

$$\frac{\pi}{4} = \begin{pmatrix} 45.00 \\ 0.00 \\ 0.00 \end{pmatrix} \cdot \mathbf{DMS}$$

Type DMS in the units placeholder to get degrees, minutes seconds. The top placeholder is degrees, the middle placeholder is minutes, the bottom placeholder is seconds.

$$29.345321\mathbf{deg} = \begin{pmatrix} 29.00 \\ 20.00 \\ 43.16 \end{pmatrix} \cdot \mathbf{DMS}$$

FIGURE 4.30 Degree minutes seconds

using a vector, the ***hhmmss*** function uses a text string with the hours, minutes, and seconds separated by colons. Thus, 2 hours, 32 minutes, and 14 seconds would be 2:32:14. Figure 4.31 gives some examples.

Feet Inch Fraction (FIF)

The ***FIF*** function is similar to the ***hhmmss*** function. It uses a string as the argument for the function, and when used in the units placeholder it also displays a string. The format uses a string with a single quote for feet and double quote for inches and fractions of inches. A dimension of 2 feet 3½ inches would be input as a text string like this: ***FIF***("2′ 3-1/2" "). Notice the two double quotes

Using *hhmmss* as a function

$\mathbf{Time_1} := \mathbf{hhmmss}(\text{"12:32:15"})$

The hours, minutes, and seconds must be input between quote marks " and must be separated by colons.

$\mathbf{Time_1} = 45135.00\,\mathbf{s}$

$\mathbf{Time_1} = 752.25 \cdot \mathbf{min}$

$\mathbf{Time_1} = 12.54 \cdot \mathbf{hr}$

$\mathbf{Time_1} = 0.52 \cdot \mathbf{day}$

Using *hhmmss* in the units placeholder

$\mathbf{Time_2} := 40000\mathbf{s}$

$\mathbf{Time_2} = \text{"11:6:40"} \cdot \mathbf{hhmmss}$

$54.25\mathbf{min} = \text{"0:54:15"} \cdot \mathbf{hhmmss}$

$4.7525\mathbf{day} = \text{"114:3:36"} \cdot \mathbf{hhmmss}$

$0.125\mathbf{yr} = \text{"1095:43:35.747"} \cdot \mathbf{hhmmss}$

FIGURE 4.31 Hours minutes seconds (hhmmss)

at the end; one indicates inches, the other closes the text string. The keystrokes are `FIF, left parenthesis, double quote, feet, single quote, space, inches, dash, numerator/denominator, double quote, double quote, right parenthesis. FIF ("feet' inches-numerator/denominator" ")`. Figure 4.32 provides examples of its use.

Using *FIF* as a function

FIF_Length := FIF("1'2-1/2"")

The feet, inches, and fractions must be input between quote marks. The format is: FIF, left parenthesis, double quote, feet, single quote, space, inches, dash, numerator/denominator, double quote, double quote, right parenthesis.
FIF("feet' inches-numerator/denominator"")

FIF_Length = 0.37 **m**

FIF_Length = 14.50 · **in**

FIF_Length = 1.21 · **ft**

Using *FIF* in the units placeholder

18.5**ft** = "18' 6"" · **FIF**

25.5**in** = "2' 1-1/2"" · **FIF**

157.75**in** = "13' 1-3/4"" · **FIF**

2.575**ft** = "2' 6-9/10"" · **FIF**

1**m** = "3' 3-47/127"" · **FIF**

FIGURE 4.32 Feet inch fraction (FIF)

Money

Beginning with Mathcad version 14, you can use money as a unit. Mathcad uses a base currency symbol, ¤ , which you can insert from the Insert Unit dialog box (it is listed under **Money**). You need to select a local currency symbol, which will replace the base currency symbol. Select **Worksheet Options** from the **Tools** Menu. This opens the Worksheet Options dialog box. On the Display tab, use the drop-down arrows to select your local currency. Now whenever Mathcad sees the base currency symbol, it displays your local currency. You can set relationships between the local currency and other currencies such as 1*€:=1.50*$.

Mathcad does not maintain any relationship because currency rates constantly change.

The local currency setting is only for the current worksheet. See Chapters 15 and 16 to see how to make the currency setting applicable to all new worksheets.

Creating Your Own Custom Scaling Function

Beginning with Mathcad version 13 you can create your own custom scaling function by defining the function and its inverse function. The inverse function is the same name as the function preceded by a forward slash. You must use the special text mode (CTRL+SHIFT+K) to input the forward slash.

Figure 4.33 illustrates an example given in the Mathcad Help to create a custom scaling unit for decibels.

Create a unit function for decibels.

$dB_{20}(x) := 10^{\frac{x}{20}}$	Create a user-defined function. This is called the "forward function."
$/dB_{20}(x) := 20\log(x)$	Create its "inverse function." The name uses the forward slash in front of the "forward function." To do this, type a letter to begin a math region, then type **CTRL+SHIFT+K**, then arrow to the beginning of the the name and type / and the rest of the variable name. Type **CTRL+SHIFT+K** to close the special text mode.
$Test\_1 := dB_{20}(23)$	
$Test\_2 := 40\,dB_{20}$	You can also use the postfix operator to input the value.
$Test\_1 = 14.13$	Result with no units attached.
$Test\_1 = 23.00 \cdot dB_{20}$	Result with dB_{20} attached to the units placeholder.
$Test\_2 = 100.00$	Result with no units attached.
$Test\_2 = 40.00 \cdot dB_{20}$	Result with dB_{20} attached to the units placeholder.

FIGURE 4.33 Creating your own custom scaling function

DIMENSIONLESS UNITS

There may be times when you want to attach a unit to a number that is not one of the seven basic dimensions of length, mass, time, temperature, luminous intensity, substance, or current or charge. In order to do this, type the name of your unitless dimension and define it as the number 1. This means that you can attach the unit to a number and not affect its value. Once you have defined a dimensionless unit, you can create other units that have relationships with this unit. Figure 4.34 gives two examples of dimensionless units. You can create dimensionless units for just about anything.

Currency Conversion

USD := 1¤	**USD** = 1.00 **$**
EUR := 1.5¤	**EUR** = 1.50 **$**
GBP := 1.8¤	**GBP** = 1.80 **$**

Create a dimensionless unit of USD and assign it to the unit of money. This unit is set to one. Attaching it to a number is the same as multiplying by one. Create additional units and set their relationship to USD.

500**USD** = 500.00 **$**	10**EUR** = 15.00 **$**	100**GBP** = 180.00 **$**
500**USD** = 500.00 · **USD**	10**EUR** = 15.00 · **USD**	100**GBP** = 180.00 · **USD**
500**USD** = 333.33 · **EUR**	10**EUR** = 10.00 · **EUR**	100**GBP** = 120.00 · **EUR**
500**USD** = 277.78 · **GBP**	10**EUR** = 8.33 · **GBP**	100**GBP** = 100.00 · **GBP**

With the exception of the base currency, Mathcad cannot automatically attach custom dimensionless units. You must place a unit in the unit placeholder.

Nonsense Example

Widget := 1	$\mathbf{Product_1}$:= 4**Widget**
Gadget := 2**Widget**	$\mathbf{Product_2}$:= 4**Gadget**
Watzut := 3**Gadget**	$\mathbf{Product_3}$:= 4**Watzut**

Bottles := 1 **Case** := 24 · **Bottles**

5 · **Case** = 120.00 **Bottles**

$\mathbf{Product_1}$ = 4.00	$\mathbf{Product_2}$ = 8.00	$\mathbf{Product_3}$ = 24.00
$\mathbf{Product_1}$ = 4.00 · **Widget**	$\mathbf{Product_2}$ = 8.00 · **Widget**	$\mathbf{Product_3}$ = 24.00 · **Widget**
$\mathbf{Product_1}$ = 2.00 · **Gadget**	$\mathbf{Product_2}$ = 4.00 · **Gadget**	$\mathbf{Product_3}$ = 12.00 · **Gadget**
$\mathbf{Product_1}$ = 0.67 · **Watzut**	$\mathbf{Product_2}$ = 1.33 · **Watzut**	$\mathbf{Product_3}$ = 4.00 · **Watzut**

Be careful to not attach the wrong unit.

$\mathbf{Product_1}$ = 4.00 **Bottles**

$\mathbf{Product_1} = 2.67 \frac{1}{\$} \cdot \mathbf{EUR}$

Mathcad will not warn you if you use inconsistent units. The unit Bottles has a value of 1. Attaching Bottles to "$Product_1$" did not change its value, but an incorrect unit was attached and Mathcad did not warn of an incorrect unit.

FIGURE 4.34 Dimensionless units

Here are a few things to consider when using dimensionless units:

- Mathcad will not automatically attach these types of units; you must attach them yourself.
- Mathcad does not do a consistency check or warn you if you have attached inconsistent units.
- Inconsistent units can create wrong results. See Figure 4.34.

LIMITATION OF UNITS

There are some limitations when using units with the exponent function. When using the exponent function x^y, when x is a dimensioned variable, y must be a constant—it cannot be a variable. In addition, y must be a multiple of 1/60000 (previous versions of Mathcad limited this to 1/60). See Figure 4.35.

SUMMARY

Units are essential for scientific and engineering calculations. If you don't use them, you will be missing out on one of the greatest features of Mathcad.

In Chapter 4 we:

- Showed how to attach units to numbers by multiplying the number by the unit.
- Explained that Mathcad will keep track of all units in variables with similar unit dimensions no matter what units are used.
- Illustrated that units in equations do not need to match as long as the unit dimensions are the same. Mathcad does the conversion for you!
- Discussed how results can be displayed in any unit system.
- Demonstrated how to use the TAB key to move you to the unit placeholder after pressing =.
- Showed that if Mathcad does not have the unit you want, you can define it yourself.
- Encouraged the use of lbm instead of lb to avoid confusion of mass and force in the US unit system.
- Emphasized that user-defined functions need to have units attached to the numbers in the arguments, not the function.
- Learned that when using empirical formulas, divide the numbers (which need to be unitless) by the units expected in the equation, then multiply by the expected units.
- Suggested avoiding the use of the function ***SIUnitOf()*** when creating a unitless number.
- Explained the use of custom scaling functions such as ***°F, °C, DMS, hhmmss***, and ***FIF***.
- Introduced the concept of dimensionless units.

$x := 2m \quad y := 3$

$Sample_1 := x^y$ A dimensioned value cannot be raised to a variable power

$Sample_1 = \blacksquare$

$Sample_2 := x^3$ However, you may raise a dimensioned value to a constant power.

$Sample_2 = 8.00\,m^{3.00}$

$Sample_2 = 8.00 \cdot m^3$

The constant power must be a multiple of 1/60000

$Sample_3 := x^{\frac{1}{70}}$ $Sample_3 = \blacksquare$

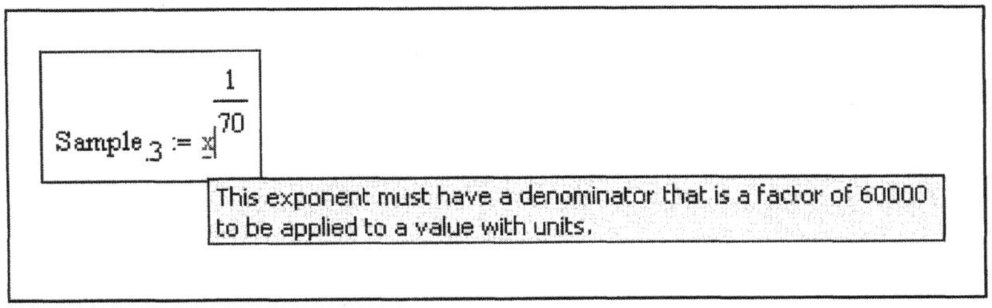

$Sample_4 := x^{\frac{1}{60}}$

$Sample_4 = 1.01\,m^{0.02}$

$Sample_4 = 1.01\,m^{\frac{1}{60}}$

Note:
If you want the exponent to be displayed as a fraction, rather than as a decimal, then double click the result and check "Show unit exponents as a fraction" from the Unit Display tab of the Result Format dialog box.

FIGURE 4.35 Limitation of units

PRACTICE

Additional problems and applications can be found on the companion site: www.elsevierdirect.com/9780123747839.

1. On a Mathcad worksheet type `Unit.1=10 *m`. Now display this variable with the following units: meters, centimeters, millimeters, kilometers, feet, inches, yards, miles, furlongs, and bohr.
2. In order to get a feel for the Mathcad default unit systems, open four different Mathcad worksheets and set each worksheet to a different default unit system

(SI, MKS, CGS, and US). Type the following in each of the four different default unit systems:

a. g=(Gravity)
b. 25ft^2=(Area)
c. 25F=(Capacitance)
d. 25C=(Charge)
e. 25A=(Current)
f. 25BTU=(Energy)
g. 25N*m=(Energy)
h. 25ft^3[spacebar]/min=(Flowrate)
i. 25lbf=(Force)
j. 25N/m=(Force per length)
k. 25km=(Length)
l. 25cd=(Luminous intensity)
m. 25 G=(Magnetic flux density)
n. 25kg=(Mass)
o. 251bf/ft^3=(Mass density)
p. 25V=(Potential)
q. 25kW=(Power)
r. 25N/m^2=(Pressure)
s. 25 K=(Temperature)
t. 25m/s=(Velocity)
u. 25ft^3=(Volume)

3. Go through each of the displayed units from the previous exercise and change the displayed units to a different unit (if it is available).

4. Create a custom unit system and display the results from exercise 2 in the custom unit system.

5. From your field of study (or another field), try to find five (or more) units (or combination of units) that are not defined by Mathcad. Create custom units for these units. Use the global definition (for example, `mgd~1000000*gal [spacebar] /day`).

6. Go back to the user-defined functions you created in Chapter 3 practice exercises. Attach units to the input arguments. Make sure the result is in the unit dimension you expect to see. Then change the displayed units to other units of your choice.

7. Create four dimensionless units that are all related, and perform arithmetic operations using them.

8. John did not have a tape measure to measure a room, so he used his shoes. Using his shoes end to end, the room measured 20 shoes wide by 25.5 shoes long. When he returned, he measured his shoes. They measured 280mm long. Create a new unit "JohnShoes." Define it as 280mm. Define the width and length of the room in units of "JohnShoes." Display the width and length in meters and feet.

PART

Hand Tools for Your Mathcad Toolbox

2

You have now built your Mathcad toolbox, and it is time to start filling it with tools. Part II will introduce some simple Mathcad features. The more complex topics will be discussed in Part III. The goal of Part II is to get you comfortable with Mathcad. You will soon see that Mathcad is an easy-to-learn, yet powerful resource.

The chapters in this part will focus on features essential to understand before the more powerful tools are introduced in Part III. The topics covered in this part include vectors, matrices, simple Mathcad functions, plotting, and simple logic programming.

CHAPTER

Arrays, Vectors, and Matrices 5

An understanding of vectors and matrices can make engineering calculations much more effective. Mathcad will perform many complex vector and matrix operations. This chapter will not delve into all these functions. The purpose of this chapter is to illustrate the benefits that can be had by using simple vectors and matrices in engineering calculations.

Chapter 5 will:

- Review how to define arrays.
- Review the concept of array subscripts.
- Review the ORIGIN variable and recommend changing its default value.
- Review the concept of range variables and tell how to define them and use them.
- Show how to format vector and matrix output in either matrix or table format.
- Illustrate how to attach units to vectors and matrices.
- Describe how to use simple math operators with vectors and matrices.
- Provide several examples of how vectors and matrices can be used in engineering calculations.
- Demonstrate how to use vectors to evaluate the same equation or function for various input values.

REVIEW OF CHAPTER 1

Chapter 1 introduced and defined arrays. It also described how to create and modify arrays. Let's review a few of the important concepts covered in Chapter 1.

To insert a vector or matrix use the Insert Matrix dialog box. This is opened by typing **CTRL+M**, clicking the matrix icon on the Matrix toolbar, or by selecting

Matrix from the **Insert** menu. Use the Insert Matrix dialog box to add or remove rows and columns.

> **Tip!** Be sure to click **Close** to close the box. If you click **OK**, additional rows and/or columns will be added.

The built-in variable ORIGIN tells Mathcad the starting index of your array. This book recommends changing the value of ORIGIN to 1, and examples are given with ORIGIN set at 1. See Figure 1.17.

Chapter 1 distinguished between literal subscripts and array subscripts. Array subscripts are not a part of the variable name (as are literal subscripts). Array subscripts are a means of displaying or defining the value of a particular element in an array. The array subscript is created by using the [key.

RANGE VARIABLES

Range variables were introduced in Chapter 1. Remember that range variables have a beginning value, an incremental value, and an ending value. To define a range variable, type the variable name followed by the colon :. In the placeholder, type the beginning value followed by a comma. In the new placeholder type the incremental value followed by a semicolon ;. This inserts two dots and adds a third placeholder. Now enter the ending value. Remember that you cannot access individual elements of a range variable as you can a vector. It is important to understand the difference between a range variable and a vector. Please review the comparisons in Chapter 1 before proceeding with the next section.

Range variables are best used to increment expressions, iterate calculations, and to set plotting limits. When you use range variables to iterate a calculation, it is important to understand that Mathcad begins at the beginning value and iterates every value in the range. You cannot tell Mathcad to use only part of the range variable. If you use a range variable as an argument for a function, the result is another range variable, which means that the result is displayed, but you cannot access individual elements of the result. You cannot assign the result to a variable. Even though it is possible to use a range variable as an argument for a function, it is best to use a vector, so that each element of the result can be assigned and accessed. Figures 5.1 and 5.2 illustrate this. Later in this chapter we show how to convert a range variable to a vector, so that it can be used as an argument for a function.

This example compares the difference between using a range variable and a vector as input to a function.

1. Using a Range Variable

$o := 0.5, 1 .. 3$ — Create range variable o

$\mathrm{CircleArea}(r) := \pi \cdot r^2$ — Create user-defined function

$\mathrm{CircleArea}(o \cdot cm) =$

0.79
3.14
7.07
12.57
19.63
28.27

$\cdot\ cm^2$

The range variable may be used as the argument of the function. Units may also be added to the function argument. The range variables do not need to be integers when using them in a function.

The results are displayed,but this does not allow you to access or reuse individual results.

Mathcad gives you an error if you try to assign this result to a variable.

$\mathrm{CircleOutput} := \mathrm{CircleArea}(o \cdot cm)$

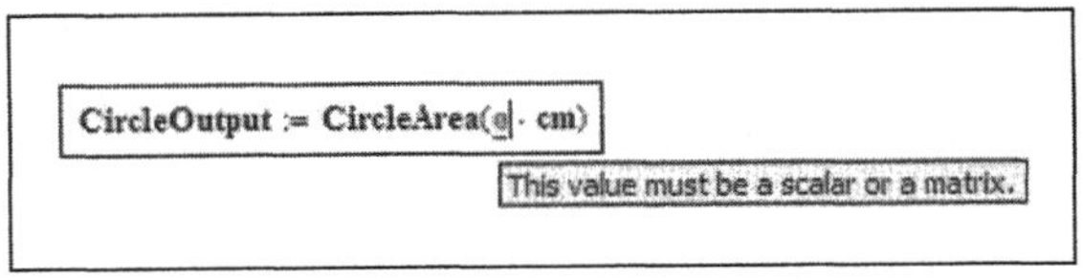
CircleOutput := CircleArea(o · cm)
This value must be a scalar or a matrix.

2. Using a Vector

In order to make each result accessible, you need to assign the results to a vector.

$$v := 1 \begin{pmatrix} 0.5 \\ 1 \\ 1.5 \\ 2 \\ 2.5 \\ 3 \end{pmatrix} \cdot cm$$

Create vector v similar to the range variable. Attach units of cm to the vector. Type CTRL+M to start the vector.

$\mathrm{Area} := \mathrm{CircleArea}(v)$

$$\mathrm{Area} = \begin{pmatrix} 0.79 \\ 3.14 \\ 7.07 \\ 12.57 \\ 19.63 \\ 28.27 \end{pmatrix} \cdot cm^2$$

$\mathrm{Area}_1 = 0.79 \cdot cm^2$

$\mathrm{Area}_6 = 28.27 \cdot cm^2$

By using a vector, each element of the results is now accessible.

FIGURE 5.1 Range variable vs vector in user-defined functions

Use the range variable "o" and the vector "v" from Figure 5.1.

1. Using a range variable

Type the expression π*o² because we are using the previously defined variable "o". The results are displayed, but not accessible, and the expression cannot be assigned to a variable.

$\pi \cdot (o \cdot cm)^2 =$

0.79
3.14
7.07
12.57
19.63
28.27

$\cdot cm^2$

$o =$

0.50
1.00
1.50
2.00
2.50
3.00

Mathcad gives you an error if you try to assign this result to a variable.

$\mathbf{CircleOutput} := \pi \cdot (o \cdot cm)^2$

$\mathbf{CircleOutput} := \pi \cdot (o \cdot cm)^2$

This value must be a scalar or a matrix.

2. Using a Vector

Assign the variable CircleArea_1 to the expressions π*v², where "v" is a previously defined vector variable.

$\mathbf{CircleArea\_1} := \pi \cdot v^2$

$$\mathbf{CircleArea\_1} = \begin{pmatrix} 0.79 \\ 3.14 \\ 7.07 \\ 12.57 \\ 19.63 \\ 28.27 \end{pmatrix} \cdot cm^2 \qquad v = \begin{pmatrix} 0.50 \\ 1.00 \\ 1.50 \\ 2.00 \\ 2.50 \\ 3.00 \end{pmatrix} \cdot cm$$

$\mathbf{CircleArea\_1}_1 = 0.79 \cdot cm^2$

$\mathbf{CircleArea\_1}_4 = 12.57 \cdot cm^2$

FIGURE 5.2 Using multiple range variables in user-defined functions

Using Range Values to Create Arrays

Figures 5.1 and 5.2 showed how Mathcad does not allow range variables to be assigned to a variable. There is an exception to this if you use the same range variable on both sides of the definition, and if the range variable begins with (or is greater than) ORIGIN and uses positive integer increments. The range variable is used as the array subscripts for defining the array, and can also be used to define the value of each element. Figures 5.3 and 5.4 give examples of how you can use range variables to create vectors. In both of these figures, the value of each element created is based on the values of the range variable. Figure 5.5 gives an example of using two range variables to create a matrix.

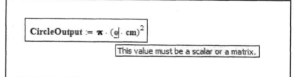

In the following example, the range variable "a" is used to create the vector "Sample." The subscript "a" in the definition is an array subscript created using the [key. The definition tells Mathcad to create a vector called "Sample." Each value of the range variable will be used to create the vector.

$\mathbf{a} := 1, 2 .. 6$

$\mathbf{Sample_a} := \mathbf{a}^2 \cdot 2$

Mathcad uses the procedure listed below to calculate the values of the vector "Sample." It uses every value in the range variable from 1 to 6. The element value is the same as the range value.

Element	Range Variable	Value
1	1	1^2*2=2
2	2	2^2*2=8
3	3	3^2*2=18
4	4	4^2*2=32
5	5	5^2*2=50
6	6	6^2*2=72

There are two ways to display the results of the variable "Sample." Typing, "Sample=" will display the entire vector results in a table. Typing, "Sample[a=" will display all the elements of the variable "Sample" associated with the range variable "a." Thus, if you create a new range variable "b" with values 3, 4, 5, & 6 and type, "Sample[b=", then Mathcad will only display elements 3, 4, 5, & 6 of the vector "Sample."

$$\mathbf{Sample} = \begin{pmatrix} 2.00 \\ 8.00 \\ 18.00 \\ 32.00 \\ 50.00 \\ 72.00 \end{pmatrix}$$

$\mathbf{Sample_a} =$

2.00
8.00
18.00
32.00
50.00
72.00

$\mathbf{b} := 3, 4 .. 6$

$\mathbf{Sample_b} =$

18.00
32.00
50.00
72.00

FIGURE 5.3 Using range variables to create a vector

Using Units in Range Variables

It is possible to add units to the range variable definition, but it is generally discouraged. Range variables with units should be used primarily for plotting.

To add units you simply attach units to the beginning value, second value, and ending value. You are required to input a second value when using units. The second and ending values do not need to use the same units, but the units used must be from the same unit dimension. For example, your beginning value can be in feet, the second value can be in inches, and the ending value in feet. This way the increment will be in inches. See Figure 5.6.

Let's look at the same equation as in Figure 5.3, but using different range variables.

Values less than ORIGIN

$\mathbf{d} := 0, 1 .. 10$

$\mathbf{Sample\_2}_{\mathbf{d}} := \mathbf{d}^2 \cdot 2$

The range variable "d" does not work because ORIGIN is set to 1 and there is not a zero element in the vector.

$\mathbf{Sample\_2}_{\mathbf{d}} := \mathbf{d}^2 \cdot 2$

This array index is invalid for this array.

Non-consecutive numbers

$\mathbf{f} := 2, 4 .. 9$

$\mathbf{Sample\_3}_{\mathbf{f}} := \mathbf{f}^2 \cdot 2$

$$\mathbf{Sample\_3} = \begin{pmatrix} 0.00 \\ 8.00 \\ 0.00 \\ 32.00 \\ 0.00 \\ 72.00 \\ 0.00 \\ 128.00 \end{pmatrix}$$

The range variable "f" has the values 2, 4, 6, 8, and 10. This creates the 2nd, 4th, 6th, and 8th elements of the vector "Sample_3." The elements not specifically defined are assigned the value zero.

Element	Range Variable	Value
1		0
2	2	2^2*2=8
3		0
4	4	4^2*2=32
5		0
6	6	6^2*2=72
7		0
8	8	8^2*2=128
9		0

Non-integers

$\mathbf{h} := 1, 1.5 .. 5$

$\mathbf{Sample\_3}_{\mathbf{h}} := \mathbf{h}^2 \cdot 2$

To create a vector with a range variable, the values in the range variable must be integers to correspond with the element numbers in the vector.

$\mathbf{Sample\_3}_{\mathbf{h}} := \mathbf{h}^2 \cdot 2$

This value must be an integer.

FIGURE 5.4 Using range variables to create vectors

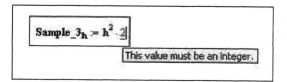

$\mathbf{k} := 1..4$ $\mathbf{n} := 1..3$

$\mathbf{Sample\_4}_{k,n} := \mathbf{k} + 2\mathbf{n}$

$$\mathbf{Sample\_4} = \begin{pmatrix} 3.00 & 5.00 & 7.00 \\ 4.00 & 6.00 & 8.00 \\ 5.00 & 7.00 & 9.00 \\ 6.00 & 8.00 & 10.00 \end{pmatrix}$$

		Column 1 n=1	Column 2 n=2	Column 3 n=3
Row 1	k=1	1+2*1=3	1+2*2=5	1+2*3=7
Row 2	k=2	2+2*1=4	2+2*2=6	2+2*3=8
Row 3	k=3	3+2*1=5	3+2*2=7	3+2*3=9
Row 4	k=4	4+2*1=6	4+2*2=8	4+2*3=10

FIGURE 5.5 Using range variables to create a matrix

Using consistent units

$\mathbf{q} := 1\mathbf{ft}, 2\mathbf{ft}..6\mathbf{ft}$

q = (m)	q = (· ft)
0.30	1.00
0.61	2.00
0.91	3.00
1.22	4.00
1.52	5.00
1.83	6.00

s = 1.00 s

Using units of feet and inches

$\mathbf{r} := 1\mathbf{ft}, 13\mathbf{in}..2\mathbf{ft}$

r = (m)	r = (· in)	r = (· ft)
0.30	12.00	1.00
0.33	13.00	1.08
0.36	14.00	1.17
0.38	15.00	1.25
0.41	16.00	1.33
0.43	17.00	1.42
0.46	18.00	1.50
0.48	19.00	1.58
0.51	20.00	1.67
0.53	21.00	1.75
0.56	22.00	1.83
0.58	23.00	1.92
0.61	24.00	2.00

Using functions

$\mathbf{t} := °\mathbf{F}(0), °\mathbf{F}(10)..°\mathbf{F}(100)$

t = (K)	t = (· °F)	t = (· °C)
255.37	0.00	-17.78
260.93	10.00	-12.22
266.48	20.00	-6.67
272.04	30.00	-1.11
277.59	40.00	4.44
283.15	50.00	10.00
288.71	60.00	15.56
294.26	70.00	21.11
299.82	80.00	26.67
305.37	90.00	32.22
310.93	100.00	37.78

Using mixed units of minutes and seconds

$\mathbf{u} := 1\mathbf{min}, 61\mathbf{s}..2\mathbf{min}$

u = (s)	u = (· min)
60.00	1.00
61.00	1.02
62.00	1.03
63.00	1.05
64.00	1.07
65.00	1.08
66.00	1.10
67.00	1.12
68.00	1.13
69.00	1.15
...	...

FIGURE 5.6 Range variables with units

Calculating Increments from the Beginning and Ending Values

If the increments do not allow the range variable to stop exactly on the ending value, Mathcad will stop the range short of the last value. See Figure 5.7. This could cause some unexpected results in your calculations. If it is difficult to calculate an increment that will stop exactly at the last value, you can enter a formula into the second placeholder. The formula is (Last value − First value)/(Number of increments) + First value. See Figure 5.8.

RangeVariable_D := 0, 0.4 .. 3

RangeVariable_D =

0
0.4
0.8
1.2
1.6
2
2.4
2.8

The last value of RangeVariable_D does not end at the ending value in the definition. This could cause unexpected results in your calculations.

FIGURE 5.7 Range variable where increment does not stop at ending value

RangeVariable_E := 1, 1.9 .. 10

RangeVariable_F := $1, \frac{10-1}{10} + 1 .. 10$

RangeVariable_E =

1
1.9
2.8
3.7
4.6
5.5
6.4
7.3
8.2
9.1
10

RangeVariable_F =

1
1.9
2.8
3.7
4.6
5.5
6.4
7.3
8.2
9.1
10

The values for RangeVariable_E and RangeVariable_F are the same. RangeVariable_F used a formula to get 10 increments between the first and last values. The increment for RangeVariable_E needed to be calculated prior to entering the increment.

FIGURE 5.8 Calculating increments

DISPLAYING ARRAYS

There are two ways to display vectors and matrices. In the previous figures, the arrays were displayed in matrix form. The range variables were displayed in table form. Mathcad uses these two methods of displaying vectors and matrices.

The display of arrays is controlled by the Matrix Display Style in the Result Format dialog box. To access this dialog box, select **Result** from the **Format** menu and click the Display Options tab. See Figure 5.9. The Mathcad default display style is Automatic. When Automatic is selected, matrices smaller than 10 by 10 are displayed in matrix form. When a matrix is 10 by 10 or larger, the default display is in table form. Range variables, by default, are displayed in table form.

If you prefer to have all array output in either matrix form or table form, change the Matrix Display Style from Automatic to either matrix form or table form from the drop-down box adjacent to the Matrix Display Style. You may also change the display of individual results by double-clicking the result, clicking the Display Options tab, and selecting the display form from the Matrix Display Style drop-down box. You may also click the result, and select **Result** from the **Format** menu.

> **Tip!** Remember that when a result is selected, the Result Format dialog box only changes the format of the selected result unless the Set As Default button is clicked after setting your result formats.

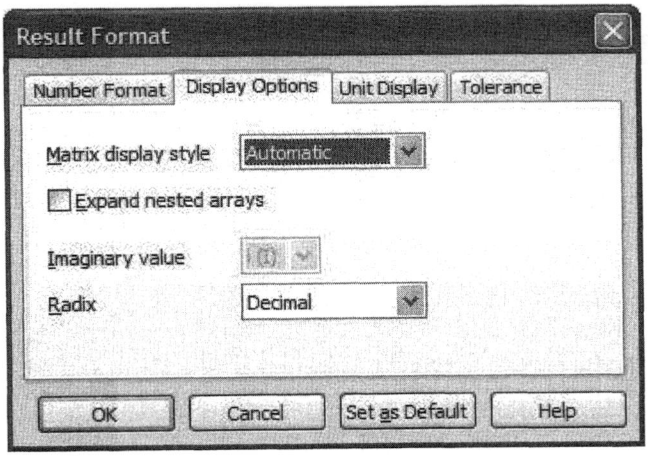

FIGURE 5.9 Result Format dialog box

Table Display Form

When vectors and matrices are displayed in table form, there are several display options to consider. Some of these options include columns and row labels, table size, font size, and the display location of the variable name in relationship to the output table.

Column and Row Labels

Column and row labels appear along the top and left side of an output table. To turn the labels on or off, right-click within the table and select **Properties**. This opens the Component Properties dialog box. To turn the labels on, place a check in the "Show Column/Row Labels" check box. To turn the labels off, clear the check box. The Mathcad default is to have labels turned on for vectors and matrices, and to have labels turned off for range variables. See Figure 5.10 for sample output with and without labels.

Output Table Size

When a vector or matrix becomes too large, Mathcad will display only a portion of the output table. In order to view all the output data, you must click inside the

$$\text{Matrix\_3} := \begin{pmatrix} 1 & 2 & 3 & 4 & 5 & 6 \\ 2 & 3 & 4 & 5 & 6 & 7 \\ 3 & 4 & 5 & 6 & 7 & 8 \\ 4 & 5 & 6 & 7 & 8 & 9 \\ 5 & 6 & 7 & 8 & 9 & 0 \\ 6 & 7 & 8 & 9 & 0 & 1 \end{pmatrix}$$

Matrix output form.

Matrix_3 =

	1	2	3	4	5	6
1	1.00	2.00	3.00	4.00	5.00	6.00
2	2.00	3.00	4.00	5.00	6.00	7.00
3	3.00	4.00	5.00	6.00	7.00	8.00
4	4.00	5.00	6.00	7.00	8.00	9.00
5	5.00	6.00	7.00	8.00	9.00	0.00
6	6.00	7.00	8.00	9.00	0.00	1.00

Output table with row and column labels turned on.

Matrix_3 =

1.00	2.00	3.00	4.00	5.00	6.00
2.00	3.00	4.00	5.00	6.00	7.00
3.00	4.00	5.00	6.00	7.00	8.00
4.00	5.00	6.00	7.00	8.00	9.00
5.00	6.00	7.00	8.00	9.00	0.00
6.00	7.00	8.00	9.00	0.00	1.00

Output table with row and column labels turned off.

FIGURE 5.10 Sample output—matrix form and table form

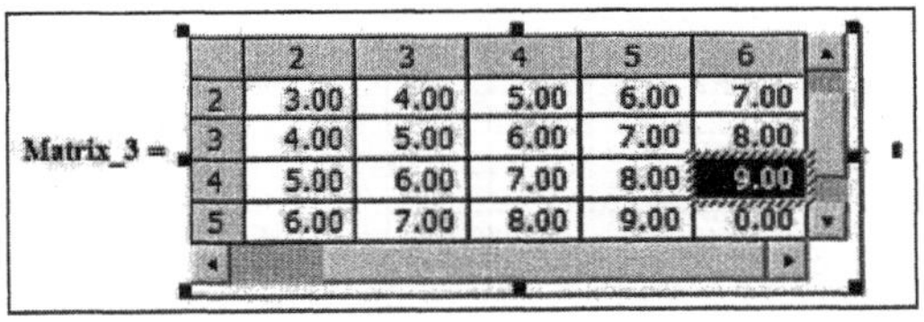

Matrix_3 =

	2	3	4	5	6
2	3.00	4.00	5.00	6.00	7.00
3	4.00	5.00	6.00	7.00	8.00
4	5.00	6.00	7.00	8.00	9.00
5	6.00	7.00	8.00	9.00	0.00

FIGURE 5.11 Output table with scroll bars

table. This will cause scroll bars to appear at the right side and/or bottom of the table. Use the scroll bars to view the remaining output data. In order to display more of the output table, you can make the table larger by clicking and dragging one of the black anchor boxes located in the corners and at midpoints of the table. See Figure 5.11. If the table crosses the right margin, you will not be able to print the entire table if you have Print Single Page Width selected from the Page Setup menu. If this is the case, you can try to make the column widths narrower.

You can only make the columns narrower if the column and row labels are turned on. Move your cursor to the column label and place it on top of a line between columns. The cursor should change to a vertical line with arrows on either side. Click and drag to the left to make the columns narrower. You may also do a similar thing to the rows to make them shorter. See Figure 5.12.

Output Table Font Size and Font Properties

You can also decrease the font size in the output table in order to display more table cells on one page. To change the output table font size, right-click within the table and select **Properties**. This opens the Component Properties dialog box. Click the **Font** button to open the Font dialog box. From this box, you may change the font style, font properties, and font size. If you want a size different from what is listed, then type the font size in the Size box. Remember that the changes you make to this output table will not affect other output tables. See Figure 5.13.

Matrix_3 =

	1	2	3	4	5	6
1	1.00	2.00	3.00	4.00	5.00	6.00
2	2.00	3.00	4.00	5.00	6.00	7.00
3	3.00	4.00	5.00	6.00	7.00	8.00
4	4.00	5.00	6.00	7.00	8.00	9.00
5	5.00	6.00	7.00	8.00	9.00	0.00
6	6.00	7.00	8.00	9.00	0.00	1.00

Matrix_3 with the columns narrowed.

FIGURE 5.12 Output table with columns narrowed

Matrix_3 =

	1	2	3	4	5	6
1	1.00	2.00	3.00	4.00	5.00	6.00
2	2.00	3.00	4.00	5.00	6.00	7.00
3	3.00	4.00	5.00	6.00	7.00	8.00
4	4.00	5.00	6.00	7.00	8.00	9.00
5	5.00	6.00	7.00	8.00	9.00	0.00
6	6.00	7.00	8.00	9.00	0.00	1.00

Matrix_3 with a smaller font.

FIGURE 5.13 Output table with a smaller font

Variable Name Location

You can adjust where the variable name is located in relationship to the output table. To change the location of the variable name, right-click in the table, pass your cursor over **Alignment** and select from the menu of options: Top, Center, Bottom, Above, and Below. Remember that you are selecting the location of the variable name in relationship to the table. The Top, Center, and Bottom alignments have the variable name on the left side of the table. See Figure 5.14.

USING UNITS WITH ARRAYS

Units are just as important with vectors and matrices as they are with other Mathcad variables. Units are essential in engineering calculations. Unfortunately, all data within a vector or matrix must be of the same unit dimension (length, mass, pressure, time, etc.). You cannot have mixed unit dimensions in a vector or matrix.

Matrix_3 =

	1	2	3	4	5	6
1	1.00	2.00	3.00	4.00	5.00	6.00
2	2.00	3.00	4.00	5.00	6.00	7.00
3	3.00	4.00	5.00	6.00	7.00	8.00
4	4.00	5.00	6.00	7.00	8.00	9.00
5	5.00	6.00	7.00	8.00	9.00	0.00
6	6.00	7.00	8.00	9.00	0.00	1.00

	1	2	3	4	5	6
1	1.00	2.00	3.00	4.00	5.00	6.00
2	2.00	3.00	4.00	5.00	6.00	7.00
3	3.00	4.00	5.00	6.00	7.00	8.00
4	4.00	5.00	6.00	7.00	8.00	9.00
5	5.00	6.00	7.00	8.00	9.00	0.00
6	6.00	7.00	8.00	9.00	0.00	1.00

Matrix_3 =

	1	2	3	4	5	6
1	1.00	2.00	3.00	4.00	5.00	6.00
2	2.00	3.00	4.00	5.00	6.00	7.00
3	3.00	4.00	5.00	6.00	7.00	8.00
4	4.00	5.00	6.00	7.00	8.00	9.00
5	5.00	6.00	7.00	8.00	9.00	0.00
6	6.00	7.00	8.00	9.00	0.00	1.00

Matrix_3 =

Matrix_3 =

	1	2	3	4	5	6
1	1.00	2.00	3.00	4.00	5.00	6.00
2	2.00	3.00	4.00	5.00	6.00	7.00
3	3.00	4.00	5.00	6.00	7.00	8.00
4	4.00	5.00	6.00	7.00	8.00	9.00
5	5.00	6.00	7.00	8.00	9.00	0.00
6	6.00	7.00	8.00	9.00	0.00	1.00

FIGURE 5.14 Output table with different variable name locations

$$\text{Matrix\_4} := \begin{pmatrix} 1 & 2 & 3 & 4 \\ 2 & 3 & 4 & 5 \\ 3 & 4 & 5 & 6 \\ 4 & 5 & 6 & 7 \end{pmatrix} \cdot \mathbf{m}$$

If all input values are of the same unit, then multiply the entire matrix by the unit.

$$\text{Matrix\_5} := \begin{pmatrix} 1\mathbf{m} & 2\mathbf{cm} & 3\mathbf{km} & 4\mathbf{mm} \\ 2\mathbf{ft} & 3\mathbf{in} & 4\mathbf{yd} & 5\mathbf{mi} \\ 3\mathbf{furlong} & 4\cdot\mathbf{nmi} & 5\cdot\mathbf{Angstrom} & 6\cdot\mathbf{bohr} \\ 4\cdot\mathbf{cubit} & 5\cdot\mathbf{micron} & 6\cdot\boldsymbol{\mu}\mathbf{m} & 7\cdot\mathbf{mil} \end{pmatrix}$$

If the input values have different units (of the same unit dimension), then multiply each matrix entry by the appropriate unit.

$$\text{Matrix\_6} := \begin{pmatrix} 1\mathbf{m} \\ 2\mathbf{cm} \\ 3\mathrm{mm} \\ 4\mathbf{Pa} \end{pmatrix}$$

Mathcad will not allow vectors or matrices with units from different unit dimensions.

FIGURE 5.15 Using units with arrays

If all the input values have the same units, then you can multiply the entire matrix by the unit. If the input values have different units (of the same unit dimension), then multiply each individual value by the unit. See Figure 5.15. The input may have different units, but the output (in both matrix form and table form) will be in a single unit system. See Figure 5.16.

CALCULATING WITH ARRAYS

Mathcad allows you to use many math operators with arrays just as you do with scalar variables, but you need to make sure that you follow the basic rules for matrix math. For example, you can add and subtract arrays just as you would other variables as long as the arrays are the same size. If they are different size arrays, Mathcad will give you an error warning. You can multiply a scalar and an array. The result will be each element of the array multiplied by the scalar. If you multiply two array variables, Mathcad assumes that you want a matrix dot product and gives a result (assuming that the two arrays are compatible with the dot product rules). There is a way to tell Mathcad that you want the matrix cross product. There is also a way to tell Mathcad that you want to multiply two arrays on an element-by-element basis and return a similar size array. These will be discussed shortly.

Addition and Subtraction

If vectors and matrices are exactly the same size, you may add or subtract them as you would any Mathcad variable. The addition and subtraction is on an element-by-element basis. See Figure 5.17.

$$Matrix\_4 = \begin{pmatrix} 1.00 & 2.00 & 3.00 & 4.00 \\ 2.00 & 3.00 & 4.00 & 5.00 \\ 3.00 & 4.00 & 5.00 & 6.00 \\ 4.00 & 5.00 & 6.00 & 7.00 \end{pmatrix} \mathbf{m}$$

All elements in the output will have the same units.

Matrix_4 =

	1	2	3	4
1	3.28	6.56	9.84	13.12
2	6.56	9.84	13.12	16.40
3	9.84	13.12	16.40	19.69
4	13.12	16.40	19.69	22.97

· ft

Matrix_5 =

	1	2	3	4
1	1.00	0.02	3000.00	0.00
2	0.61	0.08	3.66	8046.72
3	603.50	7408.00	$5.00 \cdot 10^{-10}$	$3.18 \cdot 10^{-10}$
4	1.83	$5.00 \cdot 10^{-6}$	$6.00 \cdot 10^{-6}$	0.00

m

$$Matrix\_5 = \begin{pmatrix} 3.28 & 0.07 & 9842.52 & 0.01 \\ 2.00 & 0.25 & 12.00 & 26400.00 \\ 1980.00 & 24304.46 & 1.64 \times 10^{-9} & 1.04 \times 10^{-9} \\ 6.00 & 0.00 & 0.00 & 0.00 \end{pmatrix} \cdot \mathbf{ft}$$

FIGURE 5.16 All output will have the same units

Multiplication

We now discuss several different ways to multiply arrays.

Scalar Multiplication

You can multiply any vector or matrix by a scalar number. Mathcad multiplies each element in the array by the scalar and returns the results. The scalar can be before or after the array. See Figure 5.18.

Dot Product Multiplication

When you multiply two arrays, Mathcad assumes that you want the matrix dot product of the two arrays. (The dot product is calculated by multiplying each element of the first vector by the corresponding element of the complex conjugate of the second vector, and then summing the result. Refer to a text on matrix math for a discussion of the matrix dot product.) In order for the dot product to work, the number of columns in the first matrix must match the number of rows in the second matrix. In other words, the matrices must be of the size $m^x n$ and $n^x p$. See Figure 5.19.

Vector Cross Product Multiplication

Mathcad can perform a vector cross product on two column vectors. Each vector must have three elements. The result is a vector perpendicular to the plane for

Addition and Subtraction: Arrays must be the same size

$$\mathbf{Matrix\_6} := \begin{pmatrix} 11 & 22 \\ 33 & 44 \end{pmatrix} \quad \mathbf{Matrix\_7} := \begin{pmatrix} 2 & 3 \\ 4 & 5 \end{pmatrix} \quad \mathbf{Matrix\_8} := \begin{pmatrix} 1 & 2 & 3 \\ 2 & 3 & 4 \end{pmatrix}$$

$$\mathbf{AA} := (1 \;\; 2)$$

$$\mathbf{Matrix\_9} := \mathbf{Matrix\_6} + \mathbf{Matrix\_7}$$

$$\mathbf{Matrix\_9} = \begin{pmatrix} 13.00 & 25.00 \\ 37.00 & 49.00 \end{pmatrix}$$

The addition operator adds element by element.

$$\mathbf{Matrix\_10} := \mathbf{Matrix\_6} - \mathbf{Matrix\_7}$$

$$\mathbf{Matrix\_10} = \begin{pmatrix} 9.00 & 19.00 \\ 29.00 & 39.00 \end{pmatrix}$$

The subtraction operator subtracts element by element.

Does not work because the arrays are different size.

$$\mathbf{Matrix\_11} := \mathbf{Matrix\_6} + \mathbf{Matrix\_8}$$

Matrix_11 := Matrix_6 + Matrix_8

These array dimensions do not match.

FIGURE 5.17 Addition and subtraction

You can multiply any vector or matrix by a scalar.

Refer to Figure 5.17 for the definition of $Matrix_6$, $Matrix_7$, and $Matrix_8$.

$$\mathbf{Matrix\_12} := 2 \cdot \mathbf{Matrix\_6} \qquad \mathbf{Matrix\_12} = \begin{pmatrix} 22.00 & 44.00 \\ 66.00 & 88.00 \end{pmatrix}$$

$$\mathbf{Matrix\_13} := 3 \cdot \mathbf{Matrix\_7} \qquad \mathbf{Matrix\_13} = \begin{pmatrix} 6.00 & 9.00 \\ 12.00 & 15.00 \end{pmatrix}$$

$$\mathbf{Matrix\_14} := \mathbf{Matrix\_8} \cdot 4 \qquad \mathbf{Matrix\_14} = \begin{pmatrix} 4.00 & 8.00 & 12.00 \\ 8.00 & 12.00 & 16.00 \end{pmatrix}$$

FIGURE 5.18 Scalar multiplication

the first two vectors. The direction is according to the right-hand rule. (Refer to a text on matrix math for a discussion of the vector cross product.) Use the matrix toolbar to insert the vector cross product operator, or type **CTRL+8**. See Figure 5.20.

The multiplication operator returns the matrix dot product of the arrays. The arrays must be of the size m×n and n×p. The number or rows of the second matrix must match the number of columns in the first matrix.

$$\mathbf{Matrix\_6} = \begin{pmatrix} 11.00 & 22.00 \\ 33.00 & 44.00 \end{pmatrix} \quad \mathbf{Matrix\_7} = \begin{pmatrix} 2.00 & 3.00 \\ 4.00 & 5.00 \end{pmatrix} \quad \mathbf{Matrix\_8} = \begin{pmatrix} 1.00 & 2.00 & 3.00 \\ 2.00 & 3.00 & 4.00 \end{pmatrix}$$

$$\mathbf{Matrix\_15} := \mathbf{Matrix\_6} \cdot \mathbf{Matrix\_7} \qquad \mathbf{Matrix\_15} = \begin{pmatrix} 110.00 & 143.00 \\ 242.00 & 319.00 \end{pmatrix}$$

$$\mathbf{Matrix\_16} := \mathbf{Matrix\_7} \cdot \mathbf{Matrix\_8} \qquad \mathbf{Matrix\_16} = \begin{pmatrix} 8.00 & 13.00 & 18.00 \\ 14.00 & 23.00 & 32.00 \end{pmatrix}$$

The first matrix must have two columns.

$$\mathbf{Matrix\_17} := \begin{pmatrix} 1 \\ 2 \end{pmatrix} \cdot \mathbf{Matrix\_7} \qquad \mathbf{Matrix\_17} := \begin{pmatrix} 1 \\ 2 \end{pmatrix} \cdot \mathbf{Matrix\_7}$$

These array dimensions do not match.

FIGURE 5.19 Array dot product multiplication

$$\mathbf{Vector\_2} := \begin{pmatrix} 1 \\ 1 \\ 50 \end{pmatrix} \qquad \mathbf{Vector\_3} := \begin{pmatrix} 2 \\ 3 \\ 60 \end{pmatrix}$$

$$\mathbf{Vector\_4} := \mathbf{Vector\_2} \times \mathbf{Vector\_3} \qquad \mathbf{Vector\_4} = \begin{pmatrix} -90.00 \\ 40.00 \\ 1.00 \end{pmatrix}$$

FIGURE 5.20 Vector cross product

Element-By-Element Multiplication (Vectorize)

In order to do an element-by-element multiplication, you need to use the vectorize operator. This will tell Mathcad to ignore the normal matrix rules and perform the operation on each element. To vectorize the multiplication operation, select the entire expression, and then type **CTRL+MINUS SIGN**.

Vectorize operator is used to multiply arrays on an element by element basis. Compare this to Figure 5.19 - dot product multiplication.

$$\text{Matrix\_6} = \begin{pmatrix} 11.00 & 22.00 \\ 33.00 & 44.00 \end{pmatrix} \quad \text{Matrix\_7} = \begin{pmatrix} 2.00 & 3.00 \\ 4.00 & 5.00 \end{pmatrix} \quad \text{Matrix\_8} = \begin{pmatrix} 1.00 & 2.00 & 3.00 \\ 2.00 & 3.00 & 4.00 \end{pmatrix}$$

$$\text{Matrix\_18} := \overrightarrow{(\text{Matrix\_6} \cdot \text{Matrix\_7})} \quad \text{Matrix\_18} = \begin{pmatrix} 22.00 & 66.00 \\ 132.00 & 220.00 \end{pmatrix}$$

To apply the vectorize operator, select the expression and type [CTRL]-.

$$\text{Matrix\_19} := \overrightarrow{(\text{Matrix\_7} \cdot \text{Matrix\_8})}$$

The size of the arrays must match in order to use the vectorize operator.

$$\text{Matrix\_19} := \overrightarrow{(\text{Matrix\_7} \cdot \text{Matrix\_8})}$$

These vectors must have the same number of rows.

FIGURE 5.21 Array element-by-element multiplication

This places an arrow above the expression and tells Mathcad to perform the operation on an element-by-element basis. When using a dot product multiplication, the number of columns in the first array must match the number of rows in the second array. When using the vectorize operation, the arrays must be exactly the same size because the multiplication is being done on an element-by-element basis (similar to addition and subtraction). See Figure 5.21 for examples of using the vectorize operator.

Division

For the case of X/Y, the result is dependent on whether X or Y are scalars or arrays. The result is also dependant on whether Y is a square matrix. Here are the rules:

- If Y is a square matrix, the result is the dot product of $X^* Y^{-1}$, where Y^{-1} is the inverse of the square matrix Y.
- If either X or Y is a scalar, then the division is done element by element.
- If both X and Y are arrays, then both arrays must be of the same size. If they are not square matrices, then Mathcad does an element-by-element division. If they are square matrices, then the result is the dot product $X^*Y^{-1.}$
- If both X and Y are square matrices, then in order to do an element-by-element division, you must use the vectorize operator. To do this, select the expression and type `CTRL+MINUS SIGN`.

See Figures 5.22 through 5.25 for examples of array division.

$$\textbf{Matrix\_6} = \begin{pmatrix} 11.00 & 22.00 \\ 33.00 & 44.00 \end{pmatrix} \quad \textbf{Matrix\_7} = \begin{pmatrix} 2.00 & 3.00 \\ 4.00 & 5.00 \end{pmatrix} \quad \textbf{Matrix\_8} = \begin{pmatrix} 1.00 & 2.00 & 3.00 \\ 2.00 & 3.00 & 4.00 \end{pmatrix}$$

$$\textbf{Matrix\_20} := \frac{\textbf{Matrix\_6}}{2} \qquad \textbf{Matrix\_20} = \begin{pmatrix} 5.50 & 11.00 \\ 16.50 & 22.00 \end{pmatrix}$$

Each element of Matrix_6 is divided by the scalar.

$$\textbf{Matrix\_21a} := \frac{2}{\textbf{Matrix\_7}} \qquad \textbf{Matrix\_21a} = \begin{pmatrix} -5.00 & 3.00 \\ 4.00 & -2.00 \end{pmatrix}$$

Because Matrix 7 is a square matrix, the result is equal to 2*Matrix_7^{-1}.

$$\textbf{Matrix\_7}^{-1} = \begin{pmatrix} -2.50 & 1.50 \\ 2.00 & -1.00 \end{pmatrix} \qquad 2 \cdot \textbf{Matrix\_7}^{-1} = \begin{pmatrix} -5.00 & 3.00 \\ 4.00 & -2.00 \end{pmatrix}$$

$$\textbf{Matrix\_21b} := \frac{2}{\textbf{Matrix\_8}} \qquad \textbf{Matrix\_21b} = \begin{pmatrix} 2.00 & 1.00 & 0.67 \\ 1.00 & 0.67 & 0.50 \end{pmatrix}$$

Because Matrix_8 is not a square matrix, the result is element-by-element division.

FIGURE 5.22 Scalar division

$$\textbf{Matrix\_6} = \begin{pmatrix} 11.00 & 22.00 \\ 33.00 & 44.00 \end{pmatrix} \quad \textbf{Matrix\_7} = \begin{pmatrix} 2.00 & 3.00 \\ 4.00 & 5.00 \end{pmatrix} \quad \textbf{Matrix\_8} = \begin{pmatrix} 1.00 & 2.00 & 3.00 \\ 2.00 & 3.00 & 4.00 \end{pmatrix}$$

$$\textbf{Matrix\_22} := \frac{\textbf{Matrix\_6}}{\textbf{Matrix\_7}} \qquad \textbf{Matrix\_22} = \begin{pmatrix} 16.50 & -5.50 \\ 5.50 & 5.50 \end{pmatrix}$$

Because Matrix_7 is a square matrix, the result is equal to Matrix_6 * Matrix_7^{-1}.

$$\textbf{Matrix\_7}^{-1} = \begin{pmatrix} -2.50 & 1.50 \\ 2.00 & -1.00 \end{pmatrix} \qquad \textbf{Matrix\_6} \cdot \textbf{Matrix\_7}^{-1} = \begin{pmatrix} 16.50 & -5.50 \\ 5.50 & 5.50 \end{pmatrix}$$

$$\textbf{Matrix\_23} := \frac{\textbf{Matrix\_8}}{\begin{pmatrix} 2 & 4 & 6 \\ 4 & 6 & 8 \end{pmatrix}} \qquad \textbf{Matrix\_23} = \begin{pmatrix} 0.50 & 0.50 & 0.50 \\ 0.50 & 0.50 & 0.50 \end{pmatrix}$$

Both matrices must be of the same size. Because the bottom matrix is not a square matrix, the result is element-by-element division.

FIGURE 5.23 Array division

$$\mathbf{Matrix\_7} = \begin{pmatrix} 2.00 & 3.00 \\ 4.00 & 5.00 \end{pmatrix} \qquad \mathbf{Matrix\_8} = \begin{pmatrix} 1.00 & 2.00 & 3.00 \\ 2.00 & 3.00 & 4.00 \end{pmatrix}$$

$$\frac{\mathrm{Matrix\_7}}{\mathrm{Matrix\_8}} = \blacksquare$$

This division will not work because the matrices are different sizes.

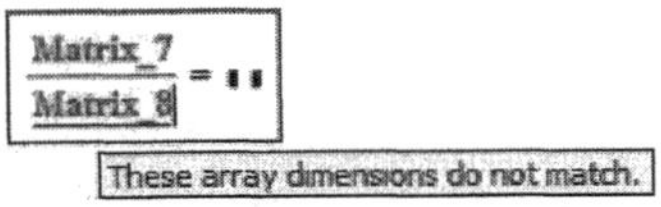

FIGURE 5.24 Array division

Array element-by-element division using the vectorize operator when both matrices are square.

$$\mathbf{Matrix\_6} = \begin{pmatrix} 11.00 & 22.00 \\ 33.00 & 44.00 \end{pmatrix} \qquad \mathbf{Matrix\_7} = \begin{pmatrix} 2.00 & 3.00 \\ 4.00 & 5.00 \end{pmatrix} \qquad \mathbf{Matrix\_7}^{-1} = \begin{pmatrix} -2.50 & 1.50 \\ 2.00 & -1.00 \end{pmatrix}$$

$$\mathbf{Matrix\_24a} := \overrightarrow{\frac{\mathbf{Matrix\_6}}{\mathbf{Matrix\_7}}} \qquad \mathbf{Matrix\_24a} = \begin{pmatrix} 5.50 & 7.33 \\ 8.25 & 8.80 \end{pmatrix}$$

To apply the vectorize operator, select the expression and type CTRL+Minus Sign. This allows an element-by-element division to occur.

This is the same as:

$$\mathbf{Matrix\_24b} := \overrightarrow{\left(\mathbf{Matrix\_6} \cdot \mathbf{Matrix\_7}^{-1}\right)} \qquad \mathbf{Matrix\_24b} = \begin{pmatrix} 5.50 & 7.33 \\ 8.25 & 8.80 \end{pmatrix}$$

FIGURE 5.25 Array division

ARRAY FUNCTIONS

Mathcad has many array functions that add and extract data from arrays. This section will discuss a few of these functions.

Creating Array Functions

augment(A,B,C, . . .): Returns an array formed by placing A, B, C,. . . left to right. A, B, C, . . . are scalars or they are column vector or arrays having the same number of rows.

stack(A,B,C, . . .): Returns an array formed by placing A, B, C,. . . top to bottom. A, B, C, . . . are scalars and column vectors, or they are arrays having the same number of columns.

Size Functions

cols(A): Returns the number of columns in A.
rows(A): Returns the number of rows in A.

Lookup Functions

lookup(z,A,B): Looks in a vector or matrix, A, for a given value, z, and returns the value(s) in the same position(s) (i.e., with the same row and column numbers) in another matrix, B. When multiple values are returned, they appear in a vector.

match(z,A): Looks in a vector or matrix, A, for a given value, z, and returns the index(es) of its positions in A.

hlookup(z,A,r): Looks in the first row of a matrix, A, for a given value, z, and returns the value(s) in the same columns(s) in the row specified, r. When multiple values are returned, they appear in a vector.

Vlookup(z,A,c): Looks in the first column of a matrix, A, for a given value, z, and returns the value(s) in the same row(s) in the column specified, c. When multiple values are returned, they appear in a vector.

Extracting Functions and Operators

max(A,B,C,. . .): Returns the largest value from A, B, C. This function will be discussed in Chapter 6. It works well for extracting the maximum value of a matrix.

min(A,B,C,. . .): Returns the smallest value in A, B, C.

submatrix(A,ir,jr,ic,jc): Returns the submatrix of array A consisting of elements in rows ir through jr and columns ic through jc of A. Remember i is the beginning and j is the ending index of the row r and column c.

Matrix Column operator $M^{<n>}$. This is located on the Matrix toolbar. It is also inserted by typing CTRL+6. Returns the nth column of matrix M.

Sorting Functions

sort(v): Returns a vector with the values from v sorted in ascending order.

reverse (A): Reverses the order of elements in a vector, or the order of rows in a matrix A.

csort (A,n): Returns an array formed by rearranging rows of A until column n is in ascending order.

rsort (A,n): Returns an array formed by rearranging columns of A until row n is in ascending order.

reverse (sort(v)): Sorts in descending order.

Refer to Mathcad Help for additional information. The September 2004 Mathcad Advisor Newsletter has an excellent article on sorting functions.

CALCULATION SUMMARY

It is not the intent of this chapter to provide a complete discussion of all the many different ways that Mathcad can be used to process and manipulate vectors and matrices. Mathcad has some very useful and powerful matrix features such as transpose, inverse, determinant, and statistical functions. An excellent discussion of vectors and matrices can be found in Mathcad Help. Two sections can be referenced. The first section is found as a topic in the Contents. The second section is contained under Operators. This section discusses the many different operators that can be used with vectors and matrices. The Mathcad Tutorial also has an excellent discussion of arrays.

CONVERTING A RANGE VARIABLE TO A VECTOR

Earlier in this chapter we discussed the differences between range variables and vectors. We showed that results obtained using a range variable are only displayed, while results obtained using a vector are accessible and may be used in subsequent calculations. Because of this it is better to use vectors than range variables; however, Mathcad does not have a simple way to create a vector with a range of values. It would be nice to be able to create a vector similar to a range variable or to convert a range variable to a vector. This section will show you how to do this.

This function converts a range variable to a range vector.

$$\mathbf{Range2Vec(Range)} := \left| \begin{array}{l} \mathbf{Count} \leftarrow \mathbf{ORIGIN} \\ \mathbf{for}\ \ \mathbf{i} \in \mathbf{Range} \\ \quad \left| \begin{array}{l} \mathbf{Vec_{Count}} \leftarrow \mathbf{i} \\ \mathbf{Count} \leftarrow \mathbf{Count} + 1 \end{array} \right. \\ \mathbf{Vec} \end{array} \right.$$

See Figure 11.8 for an expaination of this function.

$\mathbf{RV1} := 1, 1.2 .. 2.0$ Create a range variable.

Use the range variable RV1 as the argument for the function.

These two variables look the same, but they behave very differently.

$\mathbf{NewVector\_1} := \mathbf{Range2Vec(RV1)}$

$\mathbf{NewVector\_1_5} = 1.80$

$$\mathbf{RV1} = \begin{pmatrix} 1.00 \\ 1.20 \\ 1.40 \\ 1.60 \\ 1.80 \\ 2.00 \end{pmatrix} \qquad \mathbf{NewVector\_1} = \begin{pmatrix} 1.00 \\ 1.20 \\ 1.40 \\ 1.60 \\ 1.80 \\ 2.00 \end{pmatrix}$$

FIGURE 5.26 Range2Vec function for converting a range variable into a vector

Figure 5.26 provides a user-defined function called ***Range2Vec***. In order convert a range variable to a vector, you need to know some advanced programming. This will not be discussed until Chapter 11. This function is provided so that you can begin using vectors instead of range variables for your multiple calculations. Refer to Figure 11.8 to see how this function is created.

ENGINEERING EXAMPLES

We have spent several pages introducing vectors and matrices. Let's now give some examples of how to use arrays in your engineering equations.

Engineering Example 5.1 calculates fluid pressure and shows how vectors can be used in a user-defined function.

Engineering Example 5.2 shows how vectors can be used in expressions for many different input cases.

Engineering Example 5.3 shows the use of element-by-element multiplication when using matrices as input values.

Engineering Example 5.4 calculates the charge on a capacitor and compares the use of a range variable and a vector for multiple results.

Engineering Example 5.1: Using Vectors in a User-Defined Function

The density of glycerine is 1259 kg/m^3. Calculate the pressure in a tank of gycerine at depths of 2, 4, 6, 8, 10 and 12 meters.

$$\textbf{Density} := 1259 \frac{\textbf{kg}}{\textbf{m}^3}$$

Input variable for the user-defined function

$$\textbf{NetPressure(d)} := \textbf{Density} \cdot \textbf{g} \cdot \textbf{d}$$

Define user-defined function to calculate the pressure based on depth.

$$\textbf{g} = 9.81 \frac{\textbf{m}}{\textbf{s}^{2.00}}$$

g is acceleration of gravity - a Built-in constant.

$$\textbf{i} := 2,\, 4.. \ 12$$

$$\textbf{Depths} := \textbf{Range2Vec(i)} \cdot \textbf{m}$$

Create a vector of input depth values.
Attach units of meters.

$$\textbf{Depths} = \begin{pmatrix} 2.00 \\ 4.00 \\ 6.00 \\ 8.00 \\ 10.00 \\ 12.00 \end{pmatrix} \textbf{m}$$

$$\textbf{Pressures} := \textbf{NetPressure(Depths)}$$

The variable "Pressures" is defined using the user defined function "NetPressure". The argument for the user-defined function is the vector "Depths".

$$\textbf{Pressures} = \begin{array}{|c|c|} \hline & 1 \\ \hline 1 & 24.693 \\ \hline 2 & 49.386 \\ \hline 3 & 74.079 \\ \hline 4 & 98.773 \\ \hline 5 & 123.466 \\ \hline 6 & 148.159 \\ \hline \end{array} \cdot \textbf{kPa}$$

Note: The array output was changed from matrix form to an output table to show the vector indeces.

$$\textbf{Pressures}_2 = 49386.29 \ \textbf{Pa}$$

$$\textbf{Pressures}_6 = 148158.87 \ \textbf{Pa}$$

Compare to just using the range variable "i"

$$\textbf{NetPressures} \ (\textbf{i} \cdot \textbf{m}) =$$

$$\textbf{Depths} = \begin{array}{|c|} \hline 2.00 \\ \hline 4.00 \\ \hline 6.00 \\ \hline 8.00 \\ \hline 10.00 \\ \hline 12.00 \\ \hline \end{array} \textbf{m} \qquad \textbf{Pressures} = \begin{array}{|c|} \hline 24.69 \\ \hline 49.39 \\ \hline 74.08 \\ \hline 98.77 \\ \hline 123.47 \\ \hline 148.16 \\ \hline \end{array} \cdot \textbf{kPa} \qquad \begin{array}{|c|} \hline 24.69 \\ \hline 49.39 \\ \hline 74.08 \\ \hline 98.77 \\ \hline 123.47 \\ \hline 148.16 \\ \hline \end{array} \cdot \textbf{kPa}$$

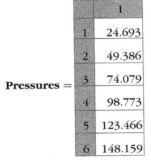

Engineering Example 5.2: Using Vectors in Expressions

Find the force caused by four different mass elements of 5 kg, 3 gm, 6 lbm, and 2 oz. and four different accelerations of 3 m/s^2, 2 m/s^2, 4 ft/s^2, and 5 ft/s^2.

Using assigned scalar values:

$\mathbf{Mass} := 5\mathbf{kg}$ $\qquad \mathbf{Acceleration} := 3\dfrac{\mathbf{m}}{\mathbf{sec}^2}$

$\mathbf{Force} := \mathbf{Mass}\cdot\mathbf{Acceleration}$ $\qquad \mathbf{Force} = 15.00\ \mathbf{N}$

$\mathbf{Mass} := 3\mathbf{gm}$ $\qquad \mathbf{Accelaration} := 2\dfrac{\mathbf{m}}{\mathbf{sec}^2}$

$\mathbf{Force} := \mathbf{Mass}\cdot\mathbf{Acceleration}$ $\qquad \mathbf{Force} = 0.01\cdot\mathbf{N}$

An easier way is to create a vector for the mass values and a vector for the acceleration values. The same equation, now gives output for the four input conditions.

$$\mathbf{Mass} := \begin{pmatrix} 5\cdot\mathbf{kg} \\ 3\cdot\mathbf{gm} \\ 6\cdot\mathbf{lbm} \\ 2\cdot\mathbf{oz} \end{pmatrix} \qquad \mathbf{Acceleration} := \begin{pmatrix} 3\dfrac{\mathbf{m}}{\mathbf{sec}^2} \\ 2\dfrac{\mathbf{m}}{\mathbf{sec}^2} \\ 4\dfrac{\mathbf{ft}}{\mathbf{sec}^2} \\ 5\dfrac{\mathbf{ft}}{\mathbf{sec}^2} \end{pmatrix}$$

Note the use of mixed input units, but same derived unit dimensions.

$\mathbf{Force_1} := \mathbf{Mass}\cdot\mathbf{Acceleration}$ $\qquad \mathbf{Force_2} := \overrightarrow{(\mathbf{Mass}\cdot\mathbf{Acceleration})}$

Use the vectorize operator to cause an element-by-element multiplication.

$\mathbf{Force_1} = 18.41\ \mathbf{N}$

The above did not perform an element by element multiplication.

$$\mathbf{Force_2} = \begin{pmatrix} 15.00 \\ 0.01 \\ 3.32 \\ 0.09 \end{pmatrix}\mathbf{N}$$

$\mathbf{Force_{2_2}} = 0.01\,\mathbf{N}$

Note: In this example, "Force$_2$" is the name of the variable using a literal subscript. The additional subscript is an array subscript.

Engineering Example 5.3: Using Element by Element Multiplication (Vectorize Operator)

Use a 2×2 mass matrix and a 2×2 acceleration matrix to calculate the force on various elements.

$$\mathbf{MassInput} := \begin{pmatrix} 5 & 7 \\ 6 & 8 \end{pmatrix} \cdot \mathbf{kg} \qquad \mathbf{AccelerationInput} := \begin{pmatrix} 3 & 7 \\ 5 & 9 \end{pmatrix} \frac{\mathbf{m}}{\mathbf{sec^2}}$$

1) **Use a function:**

$\mathbf{ForceFunction(m, a) := m \cdot a}$ — Define a function

$\mathbf{ForceOutput_1 := ForceFunction(MassInput, AccelerationInput)}$

$$\mathbf{ForceOutput_1} = \begin{pmatrix} 50.00 & 98.00 \\ 58.00 & 114.00 \end{pmatrix} \mathbf{N}$$

Not accurate.

$$\mathbf{ForceOutput_2} := \overrightarrow{\mathbf{ForceFunction(MassInput,\ AccelerationInput)}}$$

$$\mathbf{ForceOutput_2} = \begin{pmatrix} 15.00 & 49.00 \\ 30.00 & 72.00 \end{pmatrix} \mathbf{N}$$

You must use the vectorize operator to get an element by element multiplication.

2) **Use an expression:**

$\mathbf{ForceOutput_3 = MassInput \cdot AccelerationInput}$ — Not accurate.

$$\mathbf{ForceOutput_3} = \begin{pmatrix} 50.00 & 98.00 \\ 58.00 & 114.00 \end{pmatrix} \mathbf{N}$$

$$\mathbf{ForceOutput_4} := \overrightarrow{\mathbf{(MassInput . AccelerationInput)}}$$

$$\mathbf{ForceOutput_4} = \begin{pmatrix} 15.00 & 49.00 \\ 30.00 & 72.00 \end{pmatrix} \mathbf{N}$$

You must use the vectorize operator to get an element by element multiplication.

Engineering Example 5.4: Charge on a capacitor

A charge Q (Coulomb) builds up on a capacitor when a resistor, capacitor, and battery are connected in series. The formula to define the charge is:

$$\mathbf{Q(t)} = \mathbf{C}\cdot\mathbf{V}\cdot\left(1 - \mathbf{e}^{\frac{-\mathbf{t}}{\mathbf{R}\cdot\mathbf{C}}}\right).$$

Calculate the results for the first 4 seconds at half-second intervals, given R=4 ohm, C=1 F, and V=9 Volts.

Define equation: $\mathbf{Q(t, R, C, V)} := \mathbf{C}\cdot\mathbf{V}\cdot\left(1 - \mathbf{e}^{\frac{-\mathbf{t}}{\mathbf{R}\cdot\mathbf{C}}}\right)$

Assign constants **Resistance** :=4Ω **Capacitor** := 1F **Voltage** :=9 · V

1) Use a range variable Create a range variable from 0 to 5 **t** := 0s, 0.5s.. 4s

Charge_1 := **Q(t, Resistance, Capacitor, Voltage)**

Charge_1 := Q(t,Resistance,Capacitor,Voltage)
This value must be a scalar or a matrix.

When using range variables the results are displayed, but the results cannot be assigned to a variable.

Q(t, Resistance, Capacitor, Voltage) =

	1
1	0.00
2	1.06
3	1.99
4	2.81
5	3.54
6	4.18
7	4.75
8	5.25
9	5.69

C

2) Use a vector

Convert "t" to a vector. Option 2 is a much better option because results can be assigned and accessible.

Time := **Range2Vec(t)** **Charge_2** := **Q(Time, Resistance, Capacitor, Voltage)**

Time =

0.00
0.50
1.00
1.50
2.00
2.50
3.00
3.50
4.00

s

Charge_2 =

	1
1	0.00
2	1.06
3	1.99
4	2.81
5	3.54
6	4.18
7	4.75
8	5.25
9	5.69

C

Individual results are now accessible.

$\mathbf{Charge\_2_4} = 2.81\ \mathbf{C}$

SUMMARY

Arrays are a very useful tool in engineering calculations. Mathcad can do very advanced matrix computations. This chapter focused mostly on using arrays to perform multiple iterations of engineering expressions.

In Chapter 5 we:

- Reviewed how to create and modify the size of vectors and matrices.
- Set the ORIGIN built-in variable to 1.
- Learned how to attach units to variables.
- Demonstrated different ways to have array output data displayed.
- Learned how to make output tables smaller so that more information can be displayed on a single page.
- Learned how to add, subtract, multiply, and divide arrays.
- Introduced the user-defined fuction Range2Vec.
- Illustrated how arrays can be used in engineering calculations.

PRACTICE

Additional problems and applications can be found on the companion site: www.elsevierdirect.com/9780123747839.

1. Open the Insert Matrix dialog box. Create a 4×5 matrix. Fill in the matrix with some numbers.
2. In the matrix just created, insert a new column between the third and fourth columns. Insert a new row between the second and third rows. Fill in new data.
3. In the matrix, delete the third row. Delete the fifth column.
4. Create 10 different arrays. Practice inserting and deleting rows and columns in the different arrays. Have at least two arrays larger than 20×20.
5. Assign variable names to the arrays. Display each array in matrix form, in an output table with row and column labels, and in an output table without row and column labels. If the entire matrix does not display in table form, then enlarge the output table. Practice changing the column widths, row heights, and table font. In the output tables, use all five different alignment options for locating the variable name.
6. Select 20 elements from the variables, and use the subscript operator to display the elements.
7. Create 10 range variables. Use various increments, some using fractions, decimals, and negative numbers. Use some formulas in the range variables.

8. Create three arrays of the following sizes: 1×3, 2×2, 2×3, and 3×3. Attach units of length to each matrix. Assign each array a variable name.

9. Using the arrays, create expressions to perform the following calculations. Practice using different forms for displaying results.

   a. Six addition expressions
   b. Six subtraction expressions
   c. Six dot product expressions
   d. Six element-by-element multiplication expressions
   e. Twelve division expressions; for square matrices, create a dot product solution and an element-by-element solution

10. Define a simple function with only one argument (see Figure 9.11). Create a range variable that has at least six elements. Create an expression that uses the function with the range variable as the argument. Display the result in matrix form and in table form. Use the subscript operator to display at least three individual elements of the output vector.

11. Define another simple function with two arguments. Create two input vectors. Create an expression that uses the function and the input vectors (see Engineering Examples 5.1 and 5.2). Display the result in matrix form and in table form. Use the subscript operator to display at least three individual elements of the output vector.

12. Use the function and expression created earlier, but instead of using input vectors, use input matrices.

13. Verify that the results in the preceding three practice exercises are correct. Do you need to use the vectorize operator to get the correct results?

CHAPTER

6 Selected Mathcad Functions

By now, you should be familiar with many Mathcad functions. There are hundreds of Mathcad functions. The following is a partial list of the different categories that have functions: Bessel Functions, Complex Numbers, Curve Fitting, Data Analysis, Differential Equation Solving, Finance, Fourier Transforms, Graphing, Hyperbolic Functions, Image Processing, Interpolation, Logs, Number Theory, Probability, Solving, Sorting, Statistics, Trigonometry, Vectors, and Waves.

There are many functions many of us have never heard of, and will never use. Several books have been written discussing the many Mathcad functions. As explained earlier, the purpose of this book is to teach essential Mathcad skills and the application of Mathcad to scientific and engineering calculations; therefore, a great deal of time will not be spent discussing specific Mathcad functions. The *Mathcad Users Guide* is an excellent resource to learn the details of specific functions. The Mathcad Help is also useful to learn about the many Mathcad functions.

Of all the many Mathcad functions, which ones are most important for scientific and engineering calculations? The answer to this question depends on your own perspective. A function that is important for one person may not be important for another.

With that said, this chapter will introduce and discuss a few functions that (in the author's opinion) will be beneficial to many scientists and engineers doing technical calculations. These functions were chosen because they are easy to learn, and add power to technical calculations.

Chapter 6 will:

- Review the basic concept of built-in functions.
- Discuss the calculation toolbar.
- Introduce the following functions:
  - ***max***
  - ***min***
  - ***mean***
  - ***median***

- ***floor*** and ***Floor***
- ***ceil*** and ***Ceil***
- ***trunc*** and ***Trunc***
- ***round*** and ***Round***
- ***Vector Sum***
- ***Summation***
- ***Range Sum***
- ***if***
- ***linterp***

■ Discuss various categories of functions:
- Curve fitting functions
- String functions
- Picture functions
- Mapping functions
- Polar notation
- Angle functions
- Reading and writing functions

REVIEW OF BUILT-IN FUNCTIONS

In Chapters 1 and 3, we learned that every built-in Mathcad function is set up in a similar way. The name of the function is given, followed by a pair of parentheses. The information required within the parentheses is called the argument. Mathcad processes the argument(s) based on rules that are defined for the specific function.

To insert a function into a worksheet, use the Insert Function dialog box. To open this dialog box, select **Function** from the **Insert** menu. You can also type CTRL+E, or click the measuring cup icon on the Standard toolbar.

Refer to Chapter 3 for a more detailed description of functions.

TOOLBARS

There are many useful functions and operators located as icons on many of the Mathcad toolbars. You can access many of these toolbars from the Math toolbar. To open the Math toolbar, hover your mouse over **Toolbars** on the **View** menu and click **Math**.

In addition to math operators, the Calculator toolbar is a good place to look for icons, which insert simple math operators and functions. This toolbar contains simple trigonometry functions, log functions, factorial, and absolute value.

SELECTED FUNCTIONS

max and *min* Functions

The ***max*** and ***min*** functions are useful to select the maximum or minimum values from a list of values. The ***max*** function takes the form ***max***(A,B,C, . . .). Mathcad returns the largest value from A, B, C, and so on. The ***min*** function takes the same form, ***min***(A,B,C, . . .). In its simplest form, you can just type the list of values in the function. See Figure 6.1.

A, B, C, . . . can also be variable names. See Figure 6.2. A, B, C, . . . can also be a list of arrays. See Figure 6.3.

The *max* function selects the maximum value from a list of arguments. The *min* function selects the minimum value.

$\mathbf{Test\_max_1} := \mathbf{max}(1,3,5,7,9,8,6,4,2,-4)$ $\mathbf{Test\_max_1} = 9.00$

$\mathbf{Test\_min_1} := \mathbf{min}(1,3,5,7,9,8,6,4,2,-4)$ $\mathbf{Test\_min_1} = -4.00$

FIGURE 6.1 *max* and *min* functions with a list

The *max* and *min* functions can include a list of variable names.

$\mathbf{Var_1} := 3$ $\mathbf{Var_2} := 5$ $\mathbf{Var_3} := 7$

$\mathbf{max}(\mathbf{Var_1}, \mathbf{Var_2}, \mathbf{Var_3}) = 7.00$

$\mathbf{min}(\mathbf{Var_1}, \mathbf{Var_2}, \mathbf{Var_3}) = 3.00$

$\mathbf{Text\_max_2} := \mathbf{max}(\mathbf{Var_1}, \mathbf{Var_2}, \mathbf{Var_3})$ $\mathbf{Text\_max_2} = 7.00$

$\mathbf{Test\_min_2} := \mathbf{min}(\mathbf{Var_1}, \mathbf{Var_2}, \mathbf{Var_3})$ $\mathbf{Test\_min_2} = 3.00$

FIGURE 6.2 *max* and *min* functions with variables

The *max* and *min* function can also be used to select the maximum and minimum values from vectors and matrices.

$$\mathbf{Matrix_{6.2}} := \begin{pmatrix} 1 & 2 & 3 \\ 4 & 5 & 6 \\ 7 & 8 & 9 \end{pmatrix} \qquad \mathbf{Vector_{6.2}} := \begin{pmatrix} 13 \\ 16 \\ 19 \end{pmatrix}$$

$\mathbf{Test\_max_3} := \mathbf{max}(\mathbf{Matrix_{6.2}})$ $\qquad$ $\mathbf{Test\_max_4} := \mathbf{max}(\mathbf{Vector_{6.2}})$

$\mathbf{Test\_max_3} = 9.00$ $\qquad$ $\mathbf{Test\_max_4} = 19.00$

$\mathbf{Test\_min_3} := \mathbf{min}(\mathbf{Matrix_{6.2}})$ $\qquad$ $\mathbf{Test\_min_4} := \mathbf{min}(\mathbf{Vector_{6.2}})$

$\mathbf{Test\_min_3} = 1.00$ $\qquad$ $\mathbf{Test\_min_4} = 13.00$

max and *min* can also select from multiple arrays.

$\mathbf{Test\_max_5} := \mathbf{max}(\mathbf{Matrix_{6.2}}, \mathbf{Vector_{6.2}})$ $\qquad$ $\mathbf{Test\_min_5} := \mathbf{min}(\mathbf{Matrix_{6.2}}, \mathbf{Vector_{6.2}})$

$\mathbf{Test\_max_5} = 19.00$ $\qquad$ $\mathbf{Test\_min_5} = 1.00$

FIGURE 6.3 ***max*** and ***min*** functions with arrays

Units can also be attached to A, B, C, . . . as long as they are of the same unit dimension. See Figure 6.4.

If A, B, C, . . . include a complex number, the ***max*** function returns the largest real part of any value, and i times the largest imaginary part of any value. For the ***min*** function, Mathcad returns the smallest real part of any value, and i times the smallest imaginary part of any value. See Figure 6.5.

A, B, C, . . . can even be strings. For strings, z is larger than a, thus the string "cat" is larger than the string "alligator." You cannot mix strings and numbers. See Figure 6.6.

mean and *median* Functions

Mathcad has many statistical and data analysis functions. We will discuss only two of these functions. The ***mean*** function is useful for calculating averages of a list of values. The ***median*** function returns the value above and below which there are an equal number of values.

The *max* and *min* functions may also include units from the same unit dimension. Mathcad converts all units to consistent SI units, and then selects the maximum or minimum values. The result can be displayed in any unit.

$\mathbf{Test\_max_6 := max(1m, 3m, 5m, 7m, 9cm, 8m, 6m, 4m, 25ft)}$ $\mathbf{Test\_max_6} = 8.00\,\mathbf{m}$

$\mathbf{Test\_min_6 := min(1m, 3m, 5m, 7m, 9cm, 8m, 6m, 4m, 2m)}$ $\mathbf{Test\_min_6} = 0.09\,\mathbf{m}$

$\mathbf{Test\_max_7 := max(1min, 30sec, 0.5day, 5hr)}$ $\mathbf{Test\_max_7} = 12.00 \cdot \mathbf{hr}$

$\mathbf{Test\_min_7 := min(1min, 30sec, 0.5day, 5hr)}$ $\mathbf{Test\_min_7} = 30.00\,\mathbf{s}$

$$\mathbf{Matrix_{6.4}} := \begin{pmatrix} 3\mathbf{ft} & 5\mathbf{in} & 10\mathbf{mm} \\ 1\mathbf{m} & 56\mathbf{cm} & 1\mathbf{yd} \\ 0.0005\mathbf{mile} & 5000\mathbf{mil} & 0.0005\mathbf{furlong} \end{pmatrix}$$

$$\mathbf{Matrix_{6.4}} = \begin{pmatrix} 0.91 & 0.13 & 0.01 \\ 1.00 & 0.56 & 0.91 \\ 0.80 & 0.13 & 0.10 \end{pmatrix} \mathbf{m}$$

$\mathbf{Test\_max_8 := max(Matrix_{6.4})}$ $\mathbf{Test\_max_8} = 1.00\,\mathbf{m}$

$\mathbf{Test\_min_8 := min(Matrix_{6.4})}$ $\mathbf{Test\_min_8} = 10.00 \cdot \mathbf{mm}$

FIGURE 6.4 ***max*** and ***min*** functions with units

The ***mean*** function takes the form ***mean***(A,B,C,. . .). The arguments A, B, C, . . . can be scalars, arrays, or complex numbers. The arguments can also have units attached. The ***mean*** function sums all the elements in the argument list and divides by the number of elements. If one or all of the arguments are arrays, Mathcad sums all the elements in the arrays and counts all the elements in the arrays. If any of the arguments are complex numbers, Mathcad returns i times the sum of the imaginary parts divided by the total number of all elements (not just the complex numbers). When using units, Mathcad converts all units to SI units before taking the average. It then displays the average in the desired unit system. See Figure 6.7.

If the argument list for the *max* and *min* functions include complex numbers, then Mathcad selects the largest or smallest real values and the largest or smallest imaginary part, even though they may be from different elements.

$$\mathbf{Vector_{6.5a}} := \begin{pmatrix} 1+9i \\ 8 \\ 3+3i \\ 4+4i \end{pmatrix} \qquad \mathbf{Vector_{6.5b}} := \begin{pmatrix} 1+9i \\ 8 \\ 3-3i \\ 4+4i \end{pmatrix}$$

$\mathbf{Test\_max_9} := \mathbf{max}(\mathbf{Vector_{6.5a}})$ $\mathbf{Test\_max_9} = 8.00 + 9.00i$

$\mathbf{Test\_min_9} := \mathbf{min}(\mathbf{Vector_{6.5a}})$ $\mathbf{Test\_min_9} = 1.00$ In this case, 0*i is the smallest imaginary part.

$\mathbf{Test\_max_{10}} := \mathbf{max}(\mathbf{Vector_{6.5b}})$ $\mathbf{Test\_max_{10}} = 8.00 + 9.00i$

$\mathbf{Test\_min_{10}} := \mathbf{min}(\mathbf{Vector_{6.5b}})$ $\mathbf{Test\_min_{10}} = 1.00 - 3.00i$

FIGURE 6.5 ***max*** and ***min*** functions with complex numbers

The *max* and *min* functions can also select from a list of string variables.

Alpha_max := **max**("cat" , "alligator") **Alpha_max** = "cat" For string variables, letter "b" > letter "a".

Alpha_min := **min**("cat" , "alligator") **Alpha_min** = "alligator"

FIGURE 6.6 ***max*** and ***min*** functions with string variables

The *mean* function calculates the average of a list of variables.

$$\mathbf{Matrix_{6.7a}} := \begin{pmatrix} 3 & 5 & 7 \\ 9 & 11 & 13 \\ 15 & 17 & 19 \end{pmatrix} \qquad \mathbf{Vector_{6.7a}} := \begin{pmatrix} 5 \\ 15 \\ 30 \end{pmatrix}$$

$$\mathbf{Test\_mean_1} := \mathbf{mean}(\mathbf{Matrix_{6.7a}}) \qquad \mathbf{Test\_mean_1} = 11.00$$

$$\mathbf{Test\_mean_2} := \mathbf{mean}(\mathbf{Vector_{6.7a}}) \qquad \mathbf{Test\_mean_2} = 16.67$$

$$\mathbf{Test\_mean_3} := \mathbf{mean}(\mathbf{Matrix_{6.7a}}, \mathbf{Vector_{6.7a}}) \qquad \mathbf{Test\_mean_3} = 12.42$$

The arguments may include units. Mathcad converts all units to consistent SI units, and then takes the mean.

$$\mathbf{Matrix_{6.7b}} := \begin{pmatrix} 3\mathbf{ft} & 5\mathbf{in} & 10\mathbf{mm} \\ 1\mathbf{m} & 56\mathbf{cm} & 1\mathbf{yd} \\ 0.0005\mathbf{mile} & 5000\mathbf{mil} & 0.0005\mathbf{furlong} \end{pmatrix} \qquad \mathbf{Matrix_{6.7b}} = \begin{pmatrix} 0.91 & 0.13 & 0.01 \\ 1.00 & 0.56 & 0.91 \\ 0.80 & 0.13 & 0.10 \end{pmatrix} \mathbf{m}$$

$$\mathbf{Test\_mean_4} := \mathbf{mean}(\mathbf{Matrix_{6.7b}}) \qquad \mathbf{Test\_mean_4} = 0.51\,\mathbf{m}$$

If any of the arguments are complex numbers, Mathcad takes the average of the real parts and the average of the imaginary parts (including the numbers with 0i).

$$\mathbf{Vector_{6.7c}} := \begin{pmatrix} 1 + 9i \\ 8 \\ 3 + 3i \\ 4 + 4i \end{pmatrix} \qquad \mathbf{Vector_{6.7d}} := \begin{pmatrix} 5 + 9i \\ 8 \\ 3 - 3i \\ 4 + 4i \end{pmatrix}$$

$$\mathbf{Test\_mean_5} := \mathbf{mean}(\mathbf{Vector_{6.7c}}) \qquad \mathbf{Test\_mean_5} = 4.00 + 4.00i$$

$$\mathbf{Test\_mean_6} := \mathbf{mean}(\mathbf{Vector_{6.7d}}) \qquad \mathbf{Test\_mean_6} = 5.00 + 2.50i$$

$$\mathbf{Test\_mean_7} := \mathbf{mean}(\mathbf{Vector_{6.7c}}, \mathbf{Vector_{6.7d}}) \qquad \mathbf{Test\_mean_7} = 4.50 + 3.25i$$

FIGURE 6.7 ***mean*** function

The ***median*** function returns the median of all the elements in the argument list. The median is the value above and below which there are an equal number of values. If there are an even number of values, the median is the arithmetic mean of the two central values. The arguments A, B, C, ... can be scalars or arrays, but not complex numbers. The arguments can also have units attached. Mathcad first sorts the arguments from lowest to highest prior to taking the median. See Figure 6.8.

The *median* function returns the value above and below which there are an equal number of values.

If the lists of values used to the left are rearranged, it is easier to see how Mathcad selects the median.

$\mathbf{Test\_median_1} := \mathbf{median}(3,5,7,9,8,6,4)$ $\quad \mathbf{Test\_median_1} = 6.00$ $\quad \mathbf{median}(3,4,5,6,7,8,9) = 6.00$

$\mathbf{Test\_median_2} := \mathbf{median}(5\mathbf{m},8\mathbf{m},6\mathbf{m},9\mathbf{m})$ $\quad \mathbf{Test\_median_2} = 7.00\,\mathbf{m}$ $\quad \mathbf{median}(5,6,8,9) = 7.00$

$\mathbf{Test\_median_3} := \mathbf{median}(5\mathbf{m},8\mathbf{m},6\mathbf{cm},9\mathbf{mm})$ $\quad \mathbf{Test\_median_3} = 2.53\,\mathbf{m}$ $\quad \mathbf{median}(0.009,0.06,5,8) = 2.53$

$$\frac{0.06+5}{2} = 2.53$$

If there are an even number of values, then Mathcad takes the arithmetic mean of the two central values.

$$\mathbf{Matrix_{6.7a}} = \begin{pmatrix} 3.00 & 5.00 & 7.00 \\ 9.00 & 11.00 & 13.00 \\ 15.00 & 17.00 & 19.00 \end{pmatrix} \qquad \mathbf{Vector_{6.7a}} = \begin{pmatrix} 5.00 \\ 15.00 \\ 30.00 \end{pmatrix}$$

$\mathbf{Test\_median_4} := \mathbf{median}(\mathbf{Matrix_{6.7a}})$ $\quad \mathbf{Test\_median_4} = 11.00$

$\mathbf{Test\_median_5} := \mathbf{median}(\mathbf{Vector_{6.7a}})$ $\quad \mathbf{Test\_median_5} = 15.00$

$\mathbf{Test\_median_6} := \mathbf{median}(\mathbf{Matrix_{6.7a}}, \mathbf{Vector_{6.7a}})$ $\quad \mathbf{Test\_median_6} = 12.00$

Check

$\mathbf{median}(3,5,5,7,9,11,13,15,15,17,19,30) = 12.00$

$$\mathbf{Matrix_{6.7b}} = \begin{pmatrix} 0.91 & 0.13 & 0.01 \\ 1.00 & 0.56 & 0.91 \\ 0.80 & 0.13 & 0.10 \end{pmatrix} \mathbf{m}$$

$\mathbf{Test\_median_7} := \mathbf{median}(\mathbf{Matrix_{6.7b}})$ $\quad \mathbf{Test\_median_7} = 0.56\,\mathbf{m}$

Check

$\mathbf{median}(0.01,0.10,0.13,0.13,0.56,0.80,0.91,0.91,1.00) = 0.56$

FIGURE 6.8 ***median*** function

Truncation and Rounding Functions

We will discuss four truncation and rounding functions (***floor***, ***ceil***, ***trunc***, and ***round***). Each of these functions has two forms, lowercase and uppercase. We will discuss the lowercase forms first.

The function ***floor***(z) returns the greatest integer less than z.
The function ***ceil***(z) returns the smallest integer greater than z.
The function ***trunc***(z) returns the integer part of z by removing the fractional part. If z is greater than zero, this function is identical to ***floor***(z). If z is less then zero, this function is identical to ***ceil***(z).
The function ***round***(z,[n]) returns z rounded to n decimal places. The argument n must be an integer. If n is omitted (or equal to zero), it returns z rounded to the nearest integer. If n is less than zero, it returns z rounded to n places to the left of the decimal point.

The argument z can be a real or complex scalar or vector. It cannot be an array. See Figures 6.9 and 6.10 for examples of the lowercase truncate and round functions. The argument z in these lowercase functions cannot have units attached, but there is still a way to use units with these functions. If your list of values has units attached, divide the list of values by the unit you want to use, then multiply the function by the same unit. See Figure 6.11 for an example of using units with the lowercase truncate and round functions.

The uppercase forms of these equations are a bit more complicated. The uppercase functions introduce an additional argument y. This argument must be a real, nonzero scalar or vector. The argument y tells Mathcad to truncate or round to a multiple of y. The lowercase functions are equivalent to the uppercase

Use the following four values to examine the results of the following four functions: *floor, ceil, trunc,* and *round.*

$Var_{6.9a} := 3.49$	$Var_{6.9b} := 3.51$	$Var_{6.9c} := -3.49$	$Var_{6.9d} := -3.51$
$Test\_floor_1 := floor(Var_{6.9a})$	$Test\_floor_2 := floor(Var_{6.9b})$	$Test\_floor_3 := floor(Var_{6.9c})$	$Test\_floor_4 := floor(Var_{6.9d})$
$Test\_floor_1 = 3.00$	$Test\_floor_2 = 3.00$	$Test\_floor_3 = -4.00$	$Test\_floor_4 = -4.00$
$Test\_ceil_1 := ceil(Var_{6.9a})$	$Test\_ceil_2 := ceil(Var_{6.9b})$	$Test\_ceil_3 := ceil(Var_{6.9c})$	$Test\_ceil_4 := ceil(Var_{6.9d})$
$Test\_ceil_1 = 4.00$	$Test\_ceil_2 = 4.00$	$Test\_ceil_3 = -3.00$	$Test\_ceil_4 = -3.00$
$Test\_trunc_1 := trunc(Var_{6.9a})$	$Test\_trunc_2 := trunc(Var_{6.9b})$	$Test\_trunc_3 := trunc(Var_{6.9c})$	$Test\_trunc_4 := trunc(Var_{6.9d})$
$Test\_trunc_1 = 3.00$	$Test\_trunc_2 = 3.00$	$Test\_trunc_3 = -3.00$	$Test\_trunc_4 = -3.00$
$Test\_round_1 := round(Var_{6.9a})$	$Test\_round_2 := round(Var_{6.9b})$	$Test\_round_3 := round(Var_{6.9c})$	$Test\_round_4 := round(Var_{6.9d})$
$Test\_round_1 = 3.00$	$Test\_round_2 = 4.00$	$Test\_round_3 = -3.00$	$Test\_round_4 = -4.00$

FIGURE 6.9 Lowercase truncate and round functions

The lower case truncate and round functions may have vectors as arguments, but not matrices.

$$\mathbf{Var_{6.10a}} := \begin{pmatrix} 4.49 \\ 49.50 \\ -150.01 \\ -1499.99 \\ 5000.01 \end{pmatrix}$$

$\mathbf{Test\_floor_5} := \mathbf{floor}(\mathbf{Var_{6.10a}})$ $\quad$ $\mathbf{Test\_ceil_5} := \mathbf{ceil}(\mathbf{Var_{6.10a}})$ $\quad$ $\mathbf{Test\_trunc_5} := \mathbf{trunc}(\mathbf{Var_{6.10a}})$ $\quad$ $\mathbf{Test\_round_5} := \mathbf{round}(\mathbf{Var_{6.10a}})$

$$\mathbf{Test\_floor_5} = \begin{pmatrix} 4.00 \\ 49.00 \\ -151.00 \\ -1500.00 \\ 5000.00 \end{pmatrix} \quad \mathbf{Test\_ceil_5} = \begin{pmatrix} 5.00 \\ 50.00 \\ -150.00 \\ -1499.00 \\ 5001.00 \end{pmatrix} \quad \mathbf{Test\_trunc_5} = \begin{pmatrix} 4.00 \\ 49.00 \\ -150.00 \\ -1499.00 \\ 5000.00 \end{pmatrix} \quad \mathbf{Test\_round_5} = \begin{pmatrix} 4.00 \\ 50.00 \\ -150.00 \\ -1500.00 \\ 5000.00 \end{pmatrix}$$

The round function has the form round(z,[n]), where n is an integer telling Mathcad at which decimal place to round. If n is less than zero, Mathcad rounds to n places to the left of the decimal point. If n is left blank (zero), Mathcad rounds to an integer.

n=1 rounds to 1 decimal place $\quad$ n=-1 rounds to the 10's $\quad$ n=-2 rounds to the 100's

$\mathbf{Test\_round_6} := \mathbf{round}(\mathbf{Var_{6.10a}}, 1)$ $\quad$ $\mathbf{Test\_round_7} := \mathbf{round}(\mathbf{Var_{6.10a}}, -1)$ $\quad$ $\mathbf{Test\_round_8} := \mathbf{round}(\mathbf{Var_{6.10a}}, -2)$

$$\mathbf{Test\_round_6} = \begin{pmatrix} 4.50 \\ 49.50 \\ -150.00 \\ -1500.00 \\ 5000.00 \end{pmatrix} \quad \mathbf{Test\_round_7} = \begin{pmatrix} 0.00 \\ 50.00 \\ -150.00 \\ -1500.00 \\ 5000.00 \end{pmatrix} \quad \mathbf{Test\_round_8} = \begin{pmatrix} 0.00 \\ 0.00 \\ -200.00 \\ -1500.00 \\ 5000.00 \end{pmatrix}$$

FIGURE 6.10 Lowercase truncate and round functions with vectors

functions with y equal to one. (The n in round(z,[n]) being zero.) If y is equal to three, Mathcad truncates or rounds the argument z to a multiple of three. The uppercase functions also allow the use of units. Let's look at a few examples. See Figures 6.12, 6.13 (without units), and 6.14 (with units).

Figure 6.11 is the same as figure 6.10 except that units are used in 6.11.

Even though the lower case truncate and round functions do not allow the use of units, you can still use units with these functions. To do this, divide the list of units by the unit you want to truncate or round to, then multiply the function by the same unit.

$$\mathbf{Var_{6.11}} := \begin{pmatrix} 4.49 \\ 49.50 \\ -150.01 \\ -1499.99 \\ 5000.01 \end{pmatrix} \cdot \mathbf{m} \qquad \mathbf{Var_{6.11}} = \begin{pmatrix} 4.49 \\ 49.50 \\ -150.01 \\ -1499.99 \\ 5000.01 \end{pmatrix} \mathbf{m} \qquad \mathbf{Var_{6.11}} = \begin{pmatrix} 14.73 \\ 162.40 \\ -492.16 \\ -4921.23 \\ 1.64 \times 10^4 \end{pmatrix} \cdot \mathbf{ft}$$

In order to fit all examples in this figure, the results are not attached to a variable. This is not good engineering practice, because the results will not be available for later use.

$$\mathbf{floor}\left(\frac{\mathbf{Var_{6.11}}}{\mathbf{m}}\right) \cdot \mathbf{m} = \begin{pmatrix} 4.00 \\ 49.00 \\ -151.00 \\ -1500.00 \\ 5000.00 \end{pmatrix} \mathbf{m} \qquad \mathbf{trunc}\left(\frac{\mathbf{Var_{6.11}}}{\mathbf{m}}\right) \cdot \mathbf{m} = \begin{pmatrix} 4.00 \\ 49.00 \\ -150.00 \\ -1499.00 \\ 5000.00 \end{pmatrix} \mathbf{m}$$

$$\mathbf{ceil}\left(\frac{\mathbf{Var_{6.11}}}{\mathbf{m}}\right) \cdot \mathbf{m} = \begin{pmatrix} 5.00 \\ 50.00 \\ -150.00 \\ -1499.00 \\ 5001.00 \end{pmatrix} \mathbf{m} \qquad \mathbf{round}\left(\frac{\mathbf{Var_{6.11}}}{\mathbf{m}}\right) \cdot \mathbf{m} = \begin{pmatrix} 4.00 \\ 50.00 \\ -150.00 \\ -1500.00 \\ 5000.00 \end{pmatrix} \mathbf{m}$$

$$\mathbf{floor}\left(\frac{\mathbf{Var_{6.11}}}{\mathbf{ft}}\right) \cdot \mathbf{ft} = \begin{pmatrix} 14.000 \\ 162.000 \\ -493.000 \\ -4922.000 \\ 1.640 \times 10^4 \end{pmatrix} \cdot \mathbf{ft} \qquad \mathbf{round}\left(\frac{\mathbf{Var_{6.11}}}{\mathbf{ft}}\right) \cdot \mathbf{ft} = \begin{pmatrix} 15.00 \\ 162.00 \\ -492.00 \\ -4921.00 \\ 1.64 \times 10^4 \end{pmatrix} \cdot \mathbf{ft}$$

The rounding unit may be different than the input units. You can use feet, even though the input was in meters.

The round function has the form round(z,[n]), where n is an integer telling Mathcad at which decimal place to round. If n is less than zero, Mathcad rounds to n places to the left of the decimal point. If n is left blank (zero), Mathcad rounds to an integer.

n=1 rounds to 1 decimal place — n=-1 rounds to the 10's — n=-2 rounds to the 100's

$$\mathbf{round}\left(\frac{\mathbf{Var_{6.11}}}{\mathbf{m}}, 1\right) \cdot \mathbf{m} = \begin{pmatrix} 4.50 \\ 49.50 \\ -150.00 \\ -1500.00 \\ 5000.00 \end{pmatrix} \mathbf{m} \qquad \mathbf{round}\left(\frac{\mathbf{Var_{6.11}}}{\mathbf{m}}, -1\right) \cdot \mathbf{m} = \begin{pmatrix} 0.00 \\ 50.00 \\ -150.00 \\ -1500.00 \\ 5000.00 \end{pmatrix} \mathbf{m} \qquad \mathbf{round}\left(\frac{\mathbf{Var_{6.11}}}{\mathbf{ft}}, -2\right) \cdot \mathbf{ft} = \begin{pmatrix} 0.00 \\ 200.00 \\ -500.00 \\ -4900.00 \\ 1.64 \times 10^4 \end{pmatrix} \cdot \mathbf{ft}$$

FIGURE 6.11 Lowercase truncate and round functions with units

The upper case truncate and round functions introduce a second argument y. The functions have the form Floor(z,y). The argument y tells Mathcad to truncate or round to a multiple of y. Let's look at the same variables as in Figure 6.9 and look at two values of y.

$\mathbf{Var_{6.9a}} = 3.49$

$\mathbf{Var_{6.9b}} = 3.51$

$\mathbf{Var_{6.9c}} = -3.49$

Using y=2 means that all results will be a multiple of 2 (or an even number.)

$\mathbf{Test\_Floor_9} := \mathbf{Floor}(\mathbf{Var_{6.9a}}, 2)$

$\mathbf{Test\_Floor_9} = 2.00$

$\mathbf{Test\_Ceil_9} := \mathbf{Ceil}(\mathbf{Var_{6.9a}}, 2)$

$\mathbf{Test\_Ceil_9} = 4.00$

$\mathbf{Test\_Trunc_9} := \mathbf{Trunc}(\mathbf{Var_{6.9a}}, 2)$

$\mathbf{Test\_Trunc_9} = 2.00$

$\mathbf{Test\_Round_9} := \mathbf{Round}(\mathbf{Var_{6.9a}}, 2)$

$\mathbf{Test\_Round_9} = 4.00$

$\mathbf{Test\_Floor_{10}} := \mathbf{Floor}(\mathbf{Var_{6.9b}}, 2)$

$\mathbf{Test\_Floor_{10}} = 2.00$

$\mathbf{Test\_Ceil_{10}} := \mathbf{Ceil}(\mathbf{Var_{6.9b}}, 2)$

$\mathbf{Test\_Ceil_{10}} = 4.00$

$\mathbf{Test\_Trunc_{10}} := \mathbf{Trunc}(\mathbf{Var_{6.9b}}, 2)$

$\mathbf{Test\_Trunc_{10}} = 2.00$

$\mathbf{Test\_Round_{10}} := \mathbf{Round}(\mathbf{Var_{6.9b}}, 2)$

$\mathbf{Test\_Round_{10}} = 4.00$

$\mathbf{Test\_Floor_{11}} := \mathbf{Floor}(\mathbf{Var_{6.9c}}, 2)$

$\mathbf{Test\_Floor_{11}} = -4.00$

$\mathbf{Test\_Ceil_{11}} := \mathbf{Ceil}(\mathbf{Var_{6.9c}}, 2)$

$\mathbf{Test\_Ceil_{11}} = -2.00$

$\mathbf{Test\_Trunc_{11}} := \mathbf{Trunc}(\mathbf{Var_{6.9c}}, 2)$

$\mathbf{Test\_Trunc_{11}} = -2.00$

$\mathbf{Test\_Round_{11}} := \mathbf{Round}(\mathbf{Var_{6.9c}}, 2)$

$\mathbf{Test\_Round_{11}} = -4.00$

Using y=0.1 means that all results will be a multiple of 0.1, or to the first decimal place.

$\mathbf{Test\_Floor_{12}} := \mathbf{Floor}(\mathbf{Var_{6.9a}}, 0.1)$

$\mathbf{Test\_Floor_{12}} = 3.40$

$\mathbf{Test\_Ceil_{12}} := \mathbf{Ceil}(\mathbf{Var_{6.9a}}, 0.1)$

$\mathbf{Test\_Ceil_{12}} = 3.50$

$\mathbf{Test\_Trunc_{12}} := \mathbf{Trunc}(\mathbf{Var_{6.9a}}, 0.1)$

$\mathbf{Test\_Trunc_{12}} = 3.40$

$\mathbf{Test\_Round_{12}} := \mathbf{Round}(\mathbf{Var_{6.9a}}, 0.1)$

$\mathbf{Test\_Round_{12}} = 3.50$

$\mathbf{Test\_Floor_{13}} := \mathbf{Floor}(\mathbf{Var_{6.9b}}, 0.1)$

$\mathbf{Test\_Floor_{13}} = 3.50$

$\mathbf{Test\_Ceil_{13}} := \mathbf{Ceil}(\mathbf{Var_{6.9b}}, 0.1)$

$\mathbf{Test\_Ceil_{13}} = 3.60$

$\mathbf{Test\_Trunc_{13}} := \mathbf{Trunc}(\mathbf{Var_{6.9b}}, 0.1)$

$\mathbf{Test\_Trunc_{13}} = 3.50$

$\mathbf{Test\_Round_{13}} := \mathbf{Round}(\mathbf{Var_{6.9b}}, 0.1)$

$\mathbf{Test\_Round_{13}} = 3.50$

$\mathbf{Test\_Floor_{14}} := \mathbf{Floor}(\mathbf{Var_{6.9c}}, 0.2)$

$\mathbf{Test\_Floor_{14}} = -3.60$

$\mathbf{Test\_Ceil_{14}} := \mathbf{Ceil}(\mathbf{Var_{6.9c}}, 0.1)$

$\mathbf{Test\_Ceil_{14}} = -3.40$

$\mathbf{Test\_Trunc_{14}} := \mathbf{Trunc}(\mathbf{Var_{6.9c}}, 0.1)$

$\mathbf{Test\_Trunc_{14}} = -3.40$

$\mathbf{Test\_Round_{14}} := \mathbf{Round}(\mathbf{Var_{6.9c}}, 0.1)$

$\mathbf{Test\_Round_{14}} = -3.50$

FIGURE 6.12 Uppercase truncate and round functions

The upper case truncate and round functions may have vectors as arguments, but not matrices. Let's look at the same variables from Figure 6.10 and use 3 values of y.

$$\mathbf{Var_{6.10a}} = \begin{pmatrix} 4.49 \\ 49.50 \\ -150.01 \\ -1499.99 \\ 5000.01 \end{pmatrix}$$

In order to fit all examples in this figure, the results are not attached to a variable. This is not good engineering practice, because the results will not be available for later use.

y=2 y=0.1 y=0.01

$$\mathbf{Floor}(\mathbf{Var_{6.10a}}, 2) = \begin{pmatrix} 4.00 \\ 48.00 \\ -152.00 \\ -1500.00 \\ 5000.00 \end{pmatrix} \quad \mathbf{Floor}(\mathbf{Var_{6.10a}}, 0.1) = \begin{pmatrix} 4.40 \\ 49.50 \\ -150.10 \\ -1500.00 \\ 5000.00 \end{pmatrix} \quad \mathbf{Floor}(\mathbf{Var_{6.10a}}, 0.01) = \begin{pmatrix} 4.49 \\ 49.50 \\ -150.01 \\ -1499.99 \\ 5000.01 \end{pmatrix}$$

$$\mathbf{Ceil}(\mathbf{Var_{6.10a}}, 2) = \begin{pmatrix} 6.00 \\ 50.00 \\ -150.00 \\ -1498.00 \\ 5002.00 \end{pmatrix} \quad \mathbf{Ceil}(\mathbf{Var_{6.10a}}, 0.1) = \begin{pmatrix} 4.50 \\ 49.50 \\ -150.00 \\ -1499.90 \\ 5000.10 \end{pmatrix} \quad \mathbf{Ceil}(\mathbf{Var_{6.10a}}, 0.01) = \begin{pmatrix} 4.49 \\ 49.50 \\ -150.01 \\ -1499.99 \\ 5000.01 \end{pmatrix}$$

$$\mathbf{Trunc}(\mathbf{Var_{6.10a}}, 2) = \begin{pmatrix} 4.00 \\ 48.00 \\ -150.00 \\ -1498.00 \\ 5000.00 \end{pmatrix} \quad \mathbf{Trunc}(\mathbf{Var_{6.10a}}, 0.1) = \begin{pmatrix} 4.40 \\ 49.50 \\ -150.00 \\ -1499.90 \\ 5000.00 \end{pmatrix} \quad \mathbf{Trunc}(\mathbf{Var_{6.10a}}, 0.01) = \begin{pmatrix} 4.49 \\ 49.50 \\ -150.01 \\ -1499.99 \\ 5000.01 \end{pmatrix}$$

$$\mathbf{Round}(\mathbf{Var_{6.10a}}, 2) = \begin{pmatrix} 4.00 \\ 50.00 \\ -150.00 \\ -1500.00 \\ 5000.00 \end{pmatrix} \quad \mathbf{Round}(\mathbf{Var_{6.10a}}, 0.1) = \begin{pmatrix} 4.50 \\ 49.50 \\ -150.00 \\ -1500.00 \\ 5000.00 \end{pmatrix} \quad \mathbf{Round}(\mathbf{Var_{6.10a}}, 0.01) = \begin{pmatrix} 4.49 \\ 49.50 \\ -150.01 \\ -1499.99 \\ 5000.01 \end{pmatrix}$$

FIGURE 6.13 Uppercase truncate and round functions with vectors

Summation Operators

Mathcad has three ways of summing data: the ***Vector Sum*** operator, the ***Summation*** operator, and the ***Range Sum*** operator. These are technically not functions, but are useful in engineering calculations, so they are included in this chapter.

The upper case truncate and round functions allow the use of units. You do not need to divide out the units as you do with the lower case functions.

$$\mathbf{Var_{6.14}} := \begin{pmatrix} 4.49 \\ 49.50 \\ 150.01 \\ 1499.99 \\ 5000.01 \end{pmatrix} \mathbf{m} \qquad \mathbf{Var_{6.14}} = \begin{pmatrix} 14.73 \\ 162.40 \\ 492.16 \\ 4921.23 \\ 1.64 \times 10^{4} \end{pmatrix} \cdot \mathbf{ft}$$

$$\mathbf{Test\_Floor_{15}} := \mathbf{Floor}(\mathbf{Var_{6.14}}, 1\mathbf{m}) \qquad \mathbf{Test\_Round_{15}} := \mathbf{Round}(\mathbf{Var_{6.14}}, 1\mathbf{m})$$

$$\mathbf{Test\_Floor_{15}} = \begin{pmatrix} 4.00 \\ 49.00 \\ 150.00 \\ 1499.00 \\ 5000.00 \end{pmatrix} \mathbf{m} \qquad \mathbf{Test\_Round_{15}} = \begin{pmatrix} 4.00 \\ 50.00 \\ 150.00 \\ 1500.00 \\ 5000.00 \end{pmatrix} \mathbf{m}$$

Floor and Round to the nearest meter

$$\mathbf{Test\_Floor_{16}} := \mathbf{Floor}(\mathbf{Var_{6.14}}, 1\mathbf{ft}) \qquad \mathbf{Test\_Round_{16}} := \mathbf{Round}(\mathbf{Var_{6.14}}, 1\mathbf{ft})$$

$$\mathbf{Test\_Floor_{16}} = \begin{pmatrix} 4.27 \\ 49.38 \\ 149.96 \\ 1499.92 \\ 4999.94 \end{pmatrix} \mathbf{m} \qquad \mathbf{Test\_Round_{16}} = \begin{pmatrix} 4.57 \\ 49.38 \\ 149.96 \\ 1499.92 \\ 4999.94 \end{pmatrix} \mathbf{m}$$

Floor and Round to the nearest foot.
Displayed in meters.

$$\mathbf{Test\_Floor_{16}} = \begin{pmatrix} 14.00 \\ 162.00 \\ 492.00 \\ 4921.00 \\ 1.64 \times 10^{4} \end{pmatrix} \cdot \mathbf{ft} \qquad \mathbf{Test\_Round_{16}} = \begin{pmatrix} 15.00 \\ 162.00 \\ 492.00 \\ 4921.00 \\ 1.64 \times 10^{4} \end{pmatrix} \cdot \mathbf{ft}$$

Floor and Round to the nearest foot.
Displayed in feet.

FIGURE 6.14 Uppercase truncate and round functions with units

The simplest summation operator to use is the ***Vector Sum***. This operator adds all the elements in a vector. This operator is useful when you want to add a variable series of numbers. If you include all the numbers you want to add in a vector, this function will give the sum of all the elements. To insert the ***Vector Sum*** operator, type CTRL+4 or use the summation icon on the Matrix toolbar. See Figure 6.15 for some examples.

The ***Summation*** operator allows you to sum an expression or a function over a range of values. The operator has four placeholders. The placeholder below the

$\sum$ ▪ Type CTRL+4 or use the Matrix toolbar to get the Vector Sum operator. Type the name of the vector in the placeholder.

$$\mathbf{Vector_{6.15}} := \begin{pmatrix} 5 \\ 4 \\ 3 \\ 2 \\ 1 \end{pmatrix} \qquad \sum \mathbf{Vector_{6.15}} = 15.00$$

Suppose you want to input various data, and then sum them.

$\mathbf{Input_1} := 50\mathbf{Pa}$

$\mathbf{Input_2} := 100\mathbf{Pa}$

$\mathbf{Input_3} := 40\mathbf{Pa}$

$\mathbf{Input_4} := 120\mathbf{Pa}$

Note that these variables are not literal subscripts (typed with the period key). They are array subscripts typed with the [key.

$$\mathbf{Input} = \begin{pmatrix} 50.00 \\ 100.00 \\ 40.00 \\ 120.00 \end{pmatrix} \mathbf{Pa}$$

Give the sum of the input variables.

$$\sum \mathbf{Input} = 310.00\,\mathbf{Pa}$$

FIGURE 6.15 ***Vector Sum*** operator

sigma and to the left of the equal sign contains the index of the summation. This index is independent of any variable name outside of the operator. It can be any variable name, but since it is independent to the operator, it is best to keep it a single letter. The placeholder to the right of the sigma is the expression that is to be summed. The expression usually contains the index, but it is not necessary. The remaining two placeholders give the beginning and ending limits of the index. Let's look at some examples. See Figures 6.16 and 6.17.

$\sum_{\blacksquare=\blacksquare}^{\blacksquare} \blacksquare$ Type CTRL+SHIFT+4 or use the Calculus toolbar.

Define the index, the beginning limit, the ending limit, and the expression.

$\sum_{i=1}^{4} 2^i = 30.00$ This is equivalent to: $2^1 + 2^2 + 2^3 + 2^4 = 30.00$

$f(x) := 2^x$ Define a function to use in the summation.

BeginningLimit := 1 The beginning limit and ending limit may be variables.

EndingLimit := 4

$\sum_{j=\text{BeginningLimit}}^{\text{EndingLimit}} f(j) = 30.00$ This is the same as above, except the function and limits were defined previously.

The summation expression does not need to contain the index, as shown below.

$\sum_{k=1}^{5} 2 = 10.00$ This is equivalent to: $2 + 2 + 2 + 2 + 2 = 10.00$

FIGURE 6.16 ***Summation*** operator

You can also use vectors and arrays with the Summation operator.

$$\text{Vector}_{6.17} := \begin{pmatrix} 5 \\ 4 \\ 3 \\ 2 \\ 1 \end{pmatrix} \qquad \text{Matrix}_{6.17} := \begin{pmatrix} 1 & 2 \\ 3 & 4 \end{pmatrix}$$

$$\text{Summation\_1} := \sum_{l=1}^{3} \left(\text{Vector}_{6.17} \cdot l\right)$$

This is equivalent to:

$$\text{Summation\_1} = \begin{pmatrix} 30.00 \\ 24.00 \\ 18.00 \\ 12.00 \\ 6.00 \end{pmatrix} \qquad \begin{pmatrix} 5 \cdot 1 + 5 \cdot 2 + 5 \cdot 3 \\ 4 \cdot 1 + 4 \cdot 2 + 4 \cdot 3 \\ 3 \cdot 1 + 3 \cdot 2 + 3 \cdot 3 \\ 2 \cdot 1 + 2 \cdot 2 + 2 \cdot 3 \\ 1 \cdot 1 + 1 \cdot 2 + 1 \cdot 3 \end{pmatrix} = \begin{pmatrix} 30.00 \\ 24.00 \\ 18.00 \\ 12.00 \\ 6.00 \end{pmatrix}$$

$$\text{Summation\_2} := \sum_{m=1}^{3} \left(\text{Matrix}_{6.17} \cdot m\right)$$

This is equivalent to:

$$\text{Summation\_2} = \begin{pmatrix} 6.00 & 12.00 \\ 18.00 & 24.00 \end{pmatrix} \qquad \begin{pmatrix} 1 \cdot 1 + 1 \cdot 2 + 1 \cdot 3 & 2 \cdot 1 + 2 \cdot 2 + 2 \cdot 3 \\ 3 \cdot 1 + 3 \cdot 2 + 3 \cdot 3 & 4 \cdot 1 + 4 \cdot 2 + 4 \cdot 3 \end{pmatrix} = \begin{pmatrix} 6.00 & 12.00 \\ 18.00 & 24.00 \end{pmatrix}$$

FIGURE 6.17 ***Summation*** operator with vectors

The ***Range Sum*** operator is similar to the ***Summation*** operator, except you need to have a range variable defined before using the ***Range Sum*** operator. This operator has only two placeholders. The placeholder below the sigma is for the name of a previously defined range variable. The placeholder to the right of the sigma is the expression that is to be summed. The expression usually contains the range variable, but it is not necessary. Let's look at some examples. See Figures 6.18 and 6.19.

Figure 6.20 shows the use of the ***Range Sum*** and ***Summation*** operator in an engineering example.

if Function

The ***if*** function allows Mathcad to make a determination between two or more choices. The ***if*** function is similar to the ***if*** function in Microsoft Excel. It takes the form ***if***(Cond,x,y). Cond is an expression, typically involving a logical or Boolean operator. The function returns x if Cond is true, and y otherwise.

Let's look at some engineering examples. See Figures 6.21 and 6.22.

$\sum_{\blacksquare} \blacksquare$ Type $ or use the Calculus toolbar

These examples are exactly the same as in Figure 6.16, except the Range Sum operator is used.

In order to use the Range Sum operator, the range must be defined previous to using the operator.

$i := 1, 2 .. 4$ Define a range variable

$\sum_{i} 2^{i} = 30.00$ This is the same as in Figure 6.16, except the limits are defined by the range variable i.

$gg(x) := 2^{x}$ Define a function to use in the summation.

$BeginningCounter := 1$ Define limits for the range variable

$EndingCounter := 4$

$j := BeginningCounter .. EndingCounter$ Define a range variable using previously defined variables.

$\sum_{j} gg(j) = 30.00$ This is the same as above, except the function was defined previously.

The summation expression does not need to contain the range variable.

$k := 1 .. 5$

$\sum_{k} 2 = 10.00$ This is equivalent to: $2 + 2 + 2 + 2 + 2 = 10.00$

FIGURE 6.18 ***Range Sum*** operator

These examples are exactly the same as in Figure 6.17, except the Range Sum operator is used

You can also use vectors and arrays with the Range Sum operator.

$$\mathbf{Vector_{6.17}} = \begin{pmatrix} 5.00 \\ 4.00 \\ 3.00 \\ 2.00 \\ 1.00 \end{pmatrix} \qquad \mathbf{Matrix_{6.17}} = \begin{pmatrix} 1.00 & 2.00 \\ 3.00 & 4.00 \end{pmatrix}$$

$\mathbf{I} := 1..3$

$$\mathbf{Summation\_3} := \sum_{\mathbf{I}} \left(\mathbf{Vector_{6.17}} \cdot \mathbf{I}\right)$$

This is equivalent to:

$$\mathbf{Summation\_3} = \begin{pmatrix} 30.00 \\ 24.00 \\ 18.00 \\ 12.00 \\ 6.00 \end{pmatrix} \qquad \begin{pmatrix} 5 \cdot 1 + 5 \cdot 2 + 5 \cdot 3 \\ 4 \cdot 1 + 4 \cdot 2 + 4 \cdot 3 \\ 3 \cdot 1 + 3 \cdot 2 + 3 \cdot 3 \\ 2 \cdot 1 + 2 \cdot 2 + 2 \cdot 3 \\ 1 \cdot 1 + 1 \cdot 2 + 1 \cdot 3 \end{pmatrix} = \begin{pmatrix} 30.00 \\ 24.00 \\ 18.00 \\ 12.00 \\ 6.00 \end{pmatrix}$$

$\mathbf{M} := 1..3$

$$\mathbf{Summation\_4} := \sum_{\mathbf{M}} \left(\mathbf{Matrix_{6.17}} \cdot \mathbf{M}\right)$$

This is equivalent to:

$$\mathbf{Summation\_4} = \begin{pmatrix} 6.00 & 12.00 \\ 18.00 & 24.00 \end{pmatrix} \qquad \begin{pmatrix} 1 \cdot 1 + 1 \cdot 2 + 1 \cdot 3 & 2 \cdot 1 + 2 \cdot 2 + 2 \cdot 3 \\ 3 \cdot 1 + 3 \cdot 2 + 3 \cdot 3 & 4 \cdot 1 + 4 \cdot 2 + 4 \cdot 3 \end{pmatrix} = \begin{pmatrix} 6.00 & 12.00 \\ 18.00 & 24.00 \end{pmatrix}$$

FIGURE 6.19 ***Range Sum*** operator with vectors

linterp Function

Mathcad has several interpolation and regression functions. The ***linterp*** function allows straight-line interpolation between points. It is a straight-line interpolation, and is the easiest to use. You might have a specific need to use some of the more advanced functions, but for our discussion, we will use the liner interpolation function.

The ***linterp*** function has the form ***linterp***(vx,vy,x). The value vx is a vector of real data values in ascending order. The value vy is a vector of real data values having the same number of elements as vector vx. The value x is the value of the independent variable at which to interpolate the value. It is best if the value x

Use the summation operator to calculate the total floor mass of a structure.

$\mathbf{TopFloor} := 4$ — Define the top floor

$\mathbf{Counter} := 1..\mathbf{TopFloor}$ — Create a range variable for each floor

$$\mathbf{Area} := \begin{pmatrix} 1400 \\ 1200 \\ 1200 \\ 1000 \end{pmatrix} \cdot \mathbf{m}^2 \qquad \mathbf{Mass} := \begin{pmatrix} 700 \\ 700 \\ 700 \\ 500 \end{pmatrix} \cdot \frac{\mathbf{kg}}{\mathbf{m}^2}$$

Create two vectors with the area and mass/m^2 of each floor.

$$\mathbf{TotalMass} := \sum_{\mathbf{Counter}} \left(\mathbf{Area}_{\mathbf{Counter}} \cdot \mathbf{Mass}_{\mathbf{Counter}}\right)$$

This equation takes the first element of Area and the first element of Mass and multiplies them. The counter then increments and the 2nd elements are multiplied. The process continues for the 3rd and 4th elements. It then takes the sum of the results.

$\mathbf{TotalMass} = 3.16 \times 10^6\,\mathbf{kg}$

Check

$$\mathbf{test} := \overrightarrow{\left[\begin{pmatrix} 1400 \\ 1200 \\ 1200 \\ 1000 \end{pmatrix} \mathbf{m}^2 \cdot \begin{pmatrix} 700 \\ 700 \\ 700 \\ 500 \end{pmatrix} \cdot \frac{\mathbf{kg}}{\mathbf{m}^2}\right]} \qquad \mathbf{test} = \begin{pmatrix} 9.80 \times 10^5 \\ 8.40 \times 10^5 \\ 8.40 \times 10^5 \\ 5.00 \times 10^5 \end{pmatrix} \mathbf{kg}$$

Note: Use CTRL+MINUS SIGN to get the vectorize operator.

$\sum \mathbf{test} = 3.16 \times 10^6\,\mathbf{kg}$ **OK. Same result**

FIGURE 6.20 ***Range Sum*** operator—engineering example

is contained within the data range of vx. If x is below the first value of vx, then Mathcad extrapolates a straight line between the first two data points. If x is above the last value of vx, Mathcad extrapolates a straight line between the last two data points.

The ***linterp*** function draws a straight line between each data point and uses straight-line interpolation between the pairs of points. The linterp function is very useful if you have a table or graph of data and need to interpolate between data points. If your data is scattered, you might want to consider using a regression function instead.

Let's first look at a simple example. See Figure 6.23.

In this next example, there is a longer list of data values. Mathcad uses linear interpolation between each pair of data points. See Figure 6.24.

The *if* function takes the form *if*(cond,x,y). If cond is true then x. If cond is false then y.

Assume that some Mathcad expressions returned two results.

$\mathbf{Result_1} := 10 \cdot \mathbf{Hz}$

$\mathbf{Result_2} := 0 \cdot \mathbf{Hz}$

IfTrue := "Result1 is greater than Result2" Create two strings to use in the if function.

IfFalse := "Result1 is less than Result 2"

Is Result_1 greater than Result_2?

$\mathbf{Result_3} := \mathbf{if}(\mathbf{Result_1} > \mathbf{Result_2}, \mathbf{IfTrue}, \mathbf{IfFalse})$

$\mathbf{Result_3}$ = "Result1 is greater than Result2"

$$\mathbf{Result_4} := \mathbf{if}\left(\mathbf{Result_2} = 0, \text{"Division by Zero"}, \frac{\mathbf{Result_1}}{\mathbf{Result_2}}\right)$$

$\mathbf{Result_4}$ = "Division by Zero"

FIGURE 6.21 *if* function

In this example, there is an input length. A specific formula needs to use the input length, but the length cannot be less than 25 ft.

InputLength := 20ft

Length_1 := **if(InputLength** < 25ft, 25ft, **InputLength)** **Length_1** = 25.00 · **ft**

InputFunction(L) := **if(L** < 25ft, 25ft, **InputLength)**

Length_2 := **InputFunction(InputLength)** **Length_2** = 25.00 **ft**

FIGURE 6.22 *if* function

The *linterp* function takes the form linterp(vx,vy,x), where vx and vy are vectors in ascending order, and x is the independent variable.

$$\mathbf{vx} := \begin{pmatrix} 1 \\ 2 \end{pmatrix} \qquad \mathbf{vy} := \begin{pmatrix} 4 \\ 8 \end{pmatrix}$$

The vectors vx and vy are the same length and have corresponding data.

linterp(vx, vy, 1.5) = 6.00

Mathcad calculates the value of y for x=1.5.

FIGURE 6.23 ***linterp*** function

Time Velocity

$$\textbf{Time} := \begin{pmatrix} 1 \\ 2 \\ 3 \\ 4 \\ 5 \end{pmatrix} \cdot \textbf{sec} \qquad \textbf{Velocity} := \begin{pmatrix} 2.1 \\ 3.9 \\ 5.9 \\ 7.8 \\ 10.1 \end{pmatrix} \cdot \frac{\mathbf{m}}{\mathbf{s}}$$

The vectors Time and Velocity must be the same length and have corresponding data. Mathcad uses linear interpolation between each data point.

$$\textbf{linterp}(\textbf{Time}, \textbf{Velocity}, 0.5\textbf{s}) = 1.20\frac{\mathbf{m}}{\mathbf{s}}$$

$$\textbf{linterp}(\textbf{Time}, \textbf{Velocity}, 1.5\textbf{s}) = 3.00\frac{\mathbf{m}}{\mathbf{s}}$$

$$\textbf{linterp}(\textbf{Time}, \textbf{Velocity}, 2.2\textbf{s}) = 4.30\frac{\mathbf{m}}{\mathbf{s}}$$

$$\textbf{linterp}(\textbf{Time}, \textbf{Velocity}, 6\textbf{s}) = 12.40\frac{\mathbf{m}}{\mathbf{s}}$$

Use the *linterp* function to interpolate the velocity at different moments in time.

Assign a vector of values to the x argument to calculate all values at the same time.

$$\textbf{xValues} := \begin{pmatrix} 0.5 \\ 1.5 \\ 2.2 \\ 6 \end{pmatrix} \cdot \textbf{s} \qquad \textbf{linterp}(\textbf{Time}, \textbf{Velocity}, \textbf{xValues}) = \begin{pmatrix} 1.20 \\ 3.00 \\ 4.30 \\ 12.40 \end{pmatrix} \frac{\mathbf{m}}{\mathbf{s}}$$

In order to reuse the interpolated values, assign the output to a variable.

$$\textbf{InterpolatedResults} := \textbf{linterp}(\textbf{Time}, \textbf{Velocity}, \textbf{xValues})$$

$$\textbf{InterpolatedResults} = \begin{pmatrix} 1.20 \\ 3.00 \\ 4.30 \\ 12.40 \end{pmatrix} \frac{\mathbf{m}}{\mathbf{s}} \qquad \textbf{xValues} = \begin{pmatrix} 0.50 \\ 1.50 \\ 2.20 \\ 6.00 \end{pmatrix} \mathbf{s}$$

You can now access individual results from the vector "Interpolated Results."

$$\textbf{InterpolatedResults}_1 = 1.20\frac{\mathbf{m}}{\mathbf{s}}$$

$$\textbf{InterpolatedResults}_2 = 3.00\frac{\mathbf{m}}{\mathbf{s}}$$

FIGURE 6.24 *linterp* function

Figure 6.25 is a bit more complicated engineering example. It uses interpolation to calculate pressures at different heights above the ground. It then uses some of the array information discussed in Chapter 9 to calculate force and overturning moments.

This example uses linear interpolation to calculate the pressure on areas located at various heights above the ground. It uses pressure data taken from a table that lists pressures at various heights. The vectors Ht_{Data} and $Pressure_{Data}$ are the same length and have corresponding data.

Ht_{Data} and $Pressure_{Data}$ are taken from tables.

Area is the surface area of an object. Ht is the height above the ground to the centroid of the corresponding area.

$$\mathbf{Ht_{Data}} := \begin{pmatrix} 0 \\ 10 \\ 20 \\ 30 \\ 40 \\ 50 \\ 60 \\ 70 \end{pmatrix} \cdot \mathbf{ft} \qquad \mathbf{Pressure_{Data}} := \begin{pmatrix} 24 \\ 24 \\ 25 \\ 27 \\ 30 \\ 34 \\ 39 \\ 45 \end{pmatrix} \cdot \mathbf{psf} \qquad \mathbf{Area} := \begin{pmatrix} 10 \\ 12 \\ 9 \\ 11 \\ 20 \end{pmatrix} \cdot \mathbf{ft}^2 \qquad \mathbf{Ht} := \begin{pmatrix} 12 \\ 22 \\ 27 \\ 33 \\ 45 \end{pmatrix} \cdot \mathbf{ft}$$

$$\mathbf{Pressure} := \mathbf{linterp}(\mathbf{Ht_{Data}}, \mathbf{Pressure_{Data}}, \mathbf{Ht})$$

Each value of Ht is interpolated between values of Ht_{Data} and $Pressure_{Data}$.

$$\mathbf{Pressure} = \begin{pmatrix} 24.20 \\ 25.40 \\ 26.40 \\ 27.90 \\ 32.00 \end{pmatrix} \cdot \mathbf{psf}$$

Pressure is a vector of the interpolated pressures at the heights in the vector Ht.

The following part of this example uses vector information discussed in Chapter 5.

Calculate the force on each area, which is the area times the pressure

$$\mathbf{Force} := \overrightarrow{(\mathbf{Area} \cdot \mathbf{Pressure})} \qquad \mathbf{Force} = \begin{pmatrix} 242.00 \\ 304.80 \\ 237.60 \\ 306.90 \\ 640.00 \end{pmatrix} \cdot \mathbf{lbf}$$

Use the vectorize operator to ensure that there is an element-by element multiplication.

Calculate the overturning moment, which is the force times the height.

$$\mathbf{OTM} := \overrightarrow{(\mathbf{Force} \cdot \mathbf{Ht})} \qquad \mathbf{OTM} = \begin{pmatrix} 2904.00 \\ 6705.60 \\ 6415.20 \\ 10127.70 \\ 28800.00 \end{pmatrix} \cdot \mathbf{ft} \cdot \mathbf{lbf}$$

Calculate the sum of the overturning moments. Use the Vector Sum operator contained on the matrix toolbar.

$$\mathbf{SumOTM} := \sum \mathbf{OTM} \qquad \mathbf{SumOTM} = 54952.50 \cdot \mathbf{ft} \cdot \mathbf{lbf}$$

FIGURE 6.25 *linterp* function—engineering example

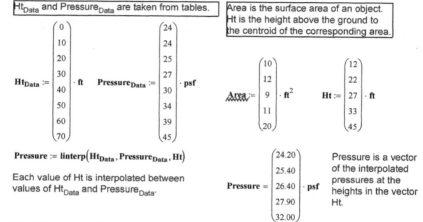

MISCELLANEOUS CATEGORIES OF FUNCTIONS

Curve Fitting, Regression, and Data Analysis

Mathcad has numerous curve fitting and data analysis functions. It has functions for linear regression, generalized regression, polynomial regression, and specialized regression. There is even a Data Analysis Extension Pack that can be purchased, which adds additional capacity to analyze data. The topic of data analysis can fill chapters and is beyond the scope of this book. For additional information, we refer you to Mathcad Help, QuickSheets, and Tutorials. There are also many resources available at www.ptc.com such as the Mathcad Advisory Newsletter.

Namespace Operator

The namespace operator was introduced in Mathcad version 12.

If you redefine a unit or a built-in variable, the new variable definition is used by Mathcad and the old variable definition is no longer available. The namespace operator allows you to reference the original value.

To use the namespace operator, type the name of the variable and then type `CTRL+SHIFT+N`. This places bracketed subscripts below the variable name. Four module names can be used within the brackets:

- mc: For any built-in Mathcad function or built-in dimensionless constant such as e or π
- unit: For any built-in Mathcad unit or dimensioned constant
- doc: Referring to the most recent previous definition in the document
- user: For any function in a UserDLL (Dynamic Link Library)

For example, if you use the variable name "V" for a vector, the unit V (for Volt) is redefined. By using the namespace operator it is possible to still use the original unit V = Volt. You can also use the namespace operator for a built-in Mathcad function that is redefined. See Figure 6.26 for examples.

Even though this operator is available, it is still best to follow the practice of not overwriting built-in Mathcad units, functions, or constants.

Error Function

The ***error***() function allows you to create your own custom error messages. This is useful when you are writing user-defined functions, or if you are creating Mathcad programs. See Figure 6.27 for examples.

String Functions

Mathcad has many string functions that can add to or manipulate strings. These will not be described in any detail, but are presented to inform you of Mathcad's

Define a vector, V

Assign a voltage:

$V := \begin{pmatrix} 3 \\ 5 \\ 8 \end{pmatrix}$ This was done to overwrite the unit V.

$\mathbf{Voltage_1} := 3\mathbf{V}$ We try to assign a voltage to a number

$\mathbf{Voltage_1} = \begin{pmatrix} 9.00 \\ 15.00 \\ 24.00 \end{pmatrix}$ This gives an incorrect result because "V" was redefined to be a vector.

In order to assign a voltage, we must now use the namespace operator.

$\mathbf{Voltage_2} := 3\mathbf{V}_{[\mathbf{unit}]}$

$\mathbf{Voltage_2} = 3.00\,\mathbf{V}$

$m := 50\mathbf{kg}$ If the mass of an object is defined as a variable "m", it redefines the built-in variable "m" for meters.

$\mathbf{Length_1} := 50\mathbf{m}$ $\mathbf{Length_1} = 2500.00\,\mathbf{kg}$ The variable "Length" is now assigned a mass.

$\mathbf{Length_2} := 50\mathbf{m}_{[\mathbf{unit}]}$ $\mathbf{Length_2} = 50.00\,\mathbf{m}$ The namespace operator can be used to tell Mathcad to use meter if the built-in unit "m" has been redefined.

Using the namespace operator for a built-in Mathcad function.

$\mathbf{sin}(\mathbf{x}) := 4 \cdot \mathbf{x}$ The function sin is redefined

$\mathbf{sin}\left(\frac{\pi}{4}\right) = 3.14$ The original Mathcad function is no longer available.

$\mathbf{sin}_{[\mathbf{mc}]}\left(\frac{\pi}{4}\right) = 0.71$ Using namespace operator CTRL+SHIFT+N, you can use the original Mathcad function for sin.

$\mathbf{sin}(\mathbf{x}) := \mathbf{sin}_{[\mathbf{mc}]}(\mathbf{x})$ In order to reassign the function sin to its original function, use the namespace operator with the mc designation.

$\mathbf{sin}\left(\frac{\pi}{4}\right) = 0.71$ The original sin function is now reassigned.

FIGURE 6.26 Namespace operator CTRL+SHIFT+N

Using the *Error* function within a function

$$\mathbf{ErrorTest(a)} := \mathbf{if}\left(\mathbf{a} > 0, \frac{1}{\mathbf{a}}, \mathbf{error}(\text{"Argument must be greater than zero"})\right)$$

$\mathbf{ErrorTest}(2) = 0.50$

A custom error message appears when you use a negative number as an argument

$\mathbf{ErrorTest}(3) = 0.33$

$\mathbf{ErrorTest}(-3) = \blacksquare$

ErrorTest(-3) = ■ ■

Argument must be greater than zero

FIGURE 6.27 *Error* function

capability to work with strings. For further information, refer to Mathcad Help. The following can be done with strings:

- Add strings together to form one string using ***concat***()
- Return only a portion of a string using ***substr***()
- Tell where a certain phrase begins in a string using ***search***()
- Return how many characters are in a string using ***strlen***()
- Convert a scalar into a string using ***num2str***()
- Convert a number to a string using ***str2num***()
- Convert a string to a vector of ANSI codes using ***str2vec***()
- Convert a vector of integer ANSI codes to a string using ***vec2str***()
- Ask Mathcad if a variable is a string using ***IsString***()

The November 2002 Mathcad Advisor Newsletter has an excellent article on the use of the string functions.

Picture Functions and Image Processing

You can insert graphic images into Mathcad. Mathcad supports the following graphic types: BMP, GIF, JPG, PCX, and TGA.

Mathcad currently does not support some digital camera images. If Mathcad does not load a digital camera image, open the file in an editing program and resave the file in JFIF, TIF, or BMP format. PTC is aware of the problem, and is working on a solution.

To insert a picture into Mathcad, click **Picture** from the **Insert** menu, or type CTRL+T. This inserts a square region with a placeholder in the lower-left corner. Enter a string containing the path and filename of the picture you want to insert. Unfortunately, there is not a browse command with this feature. The string with the path and filename can be assigned to a variable. This variable can then be used in the picture placeholder.

Once the picture is inserted, if you click the picture, Mathcad opens the Picture toolbar. From this toolbar you can modify the orientation, brightness, contrast, magnification, and grayscale/color mapping. You can hide the path and filename by right-clicking the image and selecting Hide Arguments from the pop-up menu.

Many other Mathcad functions can be used with picture images. Refer to Mathcad help for additional information.

Complex Numbers, Polar Coordinates, and Mapping Functions

Mathcad recognizes either i or j to represent the imaginary portion of a complex number. When entering complex numbers, you must use a number in front of the i or j. For example, type 1 + 1i or 1+1j, do not type 1+i or 1+j. Also, do not type 1+1 *i or 1+1 *j. If you type either i or j by itself, or if you use a multiplication, Mathcad will look for a variable i or j. Once you type 1i or 1j, the number 1 will disappear and show only an i or j. You can choose to display complex results with either i or j. This is controlled by the Display Options tab in the Result Format dialog box, found on the **Format** menu.

Mathcad has several functions for working with complex numbers. The easy-to-remember ones are ***Re***(Z), which returns the real part of Z, and ***Im***(Z), which returns the imaginary part of Z.

Mapping Functions

Mathcad has several mapping functions that are used in 2D and 3D plotting. These functions convert from rectangular coordinates to polar coordinates [***xy2pol***(x,y)], spherical coordinates [***xyz2sph***(x,y,z)], and cylindrical coordinates [***xyz2cyl***(x,y,z)]. These functions need unitless numbers, and the results are returned as a vector with unitless radius and angles in radians. The inverse to these functions are ***pol2xy***(r,theta), ***sph2xy***(r,theta,phi), and ***cyl2xy***(r,q,f).

Polar Notation

It is possible to use complex numbers to convert between rectangular and polar coordinates, and to add rectangular coordinates. The following procedure was featured in the February 2004 Mathcad Advisor Newsletter.

Let the real part of the complex number represent the x-coordinate and the imaginary part represent the y-coordinate. You can now add or subtract a series of x- and y-coordinates by adding and subtracting the complex numbers.

Now let's see how to convert to polar coordinates. Let the complex number be represented by "z." The radius for polar coordinates is the absolute value of "z." The angle is calculated using the function arg(z). The angle is measured from π to — π. The result can also be displayed in degrees, if you add deg to the unit

Let the x-coordinate be represented by the real part and the y-coordinate be represented by the imaginary part.

$\mathbf{PolarExample_1} := 3 + 3i$

$\mathbf{Radius_1} := |\mathbf{PolarExample_1}|$ $\quad \mathbf{Radius_1} = 4.24$

$\mathbf{Angle_1} := \mathbf{arg}(\mathbf{PolarExample_1})$ $\quad \mathbf{Angle_1} = 0.79$ $\quad \mathbf{Angle_1} = 45.00 \cdot \mathbf{deg}$

$\mathbf{PolarExample_2} := -3 - 3i$

$\mathbf{Radius_2} := |\mathbf{PolarExample_2}|$ $\quad \mathbf{Radius_2} = 4.24$

$\mathbf{Angle_2} := \mathbf{arg}(\mathbf{PolarExample_2})$ $\quad \mathbf{Angle_2} = -2.36$ $\quad \mathbf{Angle_2} = -135.00 \cdot \mathbf{deg}$

This polar notation allows you to add x and y coordinates and then get a final radius and angle.

$\mathbf{PolarExample_3} := (3 + 3i) + (4 + 12i) + (3 - 5i)$ $\quad \mathbf{PolarExample_3} = 10.00 + 10.00i$

$\mathbf{Radius_3} := |\mathbf{PolarExample_3}|$ $\quad \mathbf{Radius_3} = 14.14$

$\mathbf{Angle_3} := \mathbf{arg}(\mathbf{PolarExample_3})$ $\quad \mathbf{Angle_3} = 0.79$ $\quad \mathbf{Angle_3} = 45.00 \cdot \mathbf{deg}$

Create a user defined function to convert from polar coordinates to a complex number.

$i := i$ Reset i from the range variable used in Figure 13.1

$\mathbf{P2i(mag, angle)} := |\mathbf{mag}| \cdot (\mathbf{cos(angle \cdot deg)} + \mathbf{sin(angle \cdot deg)}\ \mathbf{i})$

$\mathbf{P2i}(2, 45) = 1.41 + 1.41i$

$\mathbf{P2i}(10\sqrt{2}, -135) = -10.00 - 10.00i$

Assign a variable so that results can be reused.

$\mathbf{PolarExample_4} := \mathbf{P2i}(20, 145)$ $\quad \mathbf{PolarExample_4} = -16.38 + 11.47i$

FIGURE 6.28 Polar notation

placeholder. You can create a user-defined function to convert from polar coordinates back to polar notation. See Figure 6.28 for the procedure.

Angle Functions

It can be useful to compare the results of Mathcad's various angle functions: ***angle***(x,y), ***atan***(y/x), ***atan2***(x,y), and ***arg***(x+yi). These functions return the angle from the x-axis to a line going through the origin and the point (x,y).

angle(x,y): Returns the angle in radians between 0 and 2π, excluding 2π.
atan(y/x): Returns the angle in radians between $\frac{-\pi}{2}$ and $\frac{\pi}{2}$.
atan2(x,y): Returns the angle in radians between $-\pi$ and π, excluding $-\pi$.
arg(x+yi): Returns the angle in radians between $-\pi$ and π, excluding $-\pi$. The only difference between this function and atan2(x,y) is the way the arguments are input.

Reading From and Writing To Files

Mathcad has numerous read and write functions. These are located in the Function Category File Access from the Insert Function dialog box. These functions are executed each time the worksheet is recalculated. Some key functions include:

READPRN(file): Returns a matrix formed from a structured data file. The argument is a text string with the path and filename to the data file. If a full path is not given, then the file is relative to the current working directory. The current working directory can be shown by typing CWD=. The file should be ASCII text only, with data arranged in rows and columns separated by spaces or tabs. A text header is allowed; however, once READPRN encounters a number, it assumes data has begun, so headers should contain no numbers.

READFILE(file, type, [[colwidths], [rows], [cols] emptyfill]]): Returns an array from the contents of a file of specified type (delimited, fixed-width, or Excel). This function is very similar to the Data Import Wizard discussed in Chapter 19. The arguments are as follows:

- file: A text string with the path and filename of the data file. See earlier for relative path.
- type: A text string with one of the following words: "delimited," "fixed," or "Excel."
- colwidths: (This is required for a Fixed type, and is omitted for other types.) It is a vector. Each row specifies the number of characters in each fixed-width column in the data file.
- rows: (This argument is optional. If omitted, Mathcad reads every row of the data file.) This argument can be a scalar or a two-row vector. A scalar tells Mathcad at which row to start. A two-row vector tells Mathcad a beginning row and an ending row. Row numbering begins with 1. ORIGIN does not affect the numbering.
- cols: This argument is optional; however, if used, rows also should be used. If omitted, Mathcad reads every column of the data file. This argument can be a scalar or a two-row vector. A scalar tells Mathcad at which column to start. A two-row vector tells Mathcad a beginning column and an ending column. Column numbering begins with 1. ORIGIN does not affect the numbering.
- emptyfill. This argument tells Mathcad what to do with missing entries in the data file. It can be a text string or a scalar. The default is NaN. The built-in Mathcad constant NaN represents a missing value. It stands for "not a number." Once input into a matrix it can be detected by the ***IsNaN*** function.

WRITEPRN(file,[M]): The arguments are as follows:

- file: A text string with the path and filename of the data file. See earlier for relative path.

- M: The ***WRITEPRN*** function will write the contents of this variable. It can be an array or scalar. The argument is only needed if the function is used on the right of the definition operator. For example, two ways to use this function are: ***WRITEPRN***("TestFile"):=M, or NewVariable:=***WRITEPRN***("TestFile",M). In the second case, the variable "NewVariable" is assigned the contents of the file.

APPENDPRN(file,[M]): This is similar to ***WRITEPRN*** function, except that it adds the contents of the array M to the end of the file. The number of columns in the array M must match the number of columns in the existing file.

The Mathcad default for the functions ***WRITEPRN*** and ***APPENDPRN*** is to append four significant digits in columns eight digits wide. You can change the defaults by choosing **Worksheet Options** from the **Tools** menu. Change the PRN file setting on the Built-in Variables tab.

SUMMARY

We have reviewed only a handful of the hundreds of Mathcad functions. Hopefully the functions discussed in this chapter will be helpful to your technical calculations. As you become more familiar with Mathcad, you will add other functions to your Mathcad toolbox. We will discuss additional functions in later chapters. In the meantime, open the Insert Function dialog box and look for functions that appear interesting to you. Use the Mathcad Help to learn about these functions.

In Chapter 6 we:

- Reviewed the basics of Mathcad functions.
- Learned about the following functions:
  - ***max*** and ***min*** functions
  - ***mean*** and ***median*** functions
  - truncation and rounding functions
  - ***summation*** operators
  - ***if*** function
  - interpolation functions
- Introduced the following categories of functions:
  - Curve fitting
  - String functions
  - Picture functions
  - Mapping functions
  - Polar notation
  - Angle functions
  - Reading and writing functions
- Encouraged you to search for and learn about additional functions you can add to your Mathcad toolbox.

PRACTICE

Additional problems and applications can be found on the companion site: www.elsevierdirect.com/9780123747839.

1. Use the ***max*** and ***min*** functions to find the maximum and minimum values of the following:

   a. 10 m, 1 km, 3000 mm, 1 mi, 2000 yd, 10 furlong, 1 nmi
   b. 100 s, 1.25 min, 0.15 day, 0.05 week, 0.005 year
   c. 1000 Pa, 1 psi, 1 atm, 10 torr, 0.002 ksi, 0.002 MPa, 1in_Hg
   d. 1 bhp, 10 ehp, 10.1 kW, 19 mhp, 10 hpUK, 10 hhp, 10000 W

2. Use the ***mean*** and ***median*** functions to determine the mean and median values from the list of values used in practice exercise 1.
3. Place the list of values from practice exercise 1 into four different vectors.
4. Use the ***max*** and ***min*** functions and the mean and median functions with the vectors created in practice exercise 3.
5. Use the ***floor ceil***, and ***trunc*** functions with the vectors created in practice exercise 3. Select two different units to use for each vector. For example, truncate to units of meters and feet for the first vector.
6. Use the ***round*** function with the vectors created in practice exercise 3. Round to the following: 2 decimal places, 1 decimal place, 0 decimal places, 10's, and 100's. Select two units for each case. For example, round to units of meters and feet for the first vector.
7. Use the Vector Sum operator to calculate the sum of the vectors created in practice exercise 3. Display the result in three different units.
8. Use the Summation operator and the Range Sum operator to calculate the total area of 10 squares with sides incrementing from 1 m to 10 m.
9. Use the following variables for this practice exercise: R1 = 3Amp, R2 = 4Amp, R3= 10Volt, R4 = 20Volt. Create four variables using the ***if*** function to meet the conditions in the following table:

Condition	First	Operator	Second	If true	If false
1	R1	<	R2	R3	R4
2	R1	=	R2	R3	R4
3	R2	=	0	"Division by zero"	R1/R2
4	R3	>=	R4	R1	R2

   a. Vary the values of the variables, to see how the results are affected.

10. Use the ***linterp*** function to interpolate the following values:

Time	Distance	Interpolate the distance for this time
0 sec	0.0 m	0.5 sec
1 sec	2.5 m	1.2 sec
2 sec	10.0 m	2.4 sec
3 sec	22.5 m	3.3 sec
4 sec	40.0 m	4.6 sec
5 sec	62.5 m	5.1 sec

11. Create an MS Excel file with data filling 10 rows and 10 columns. Save this Excel file on your hard drive. Use the ***READFILE*** function to assign this data to the variable "ExcelData."

12. Use the ***WRITEPRN*** function to save the data from the above problem to a file named "PracticeData."

CHAPTER

Plotting

7

Plots are an important part of engineering calculations because they allow visualization of data and equations. They are an important part of equation solving because they can help you select an initial guess for a solution. Plots also allow you to visualize trends in engineering data. This chapter will focus on using plots as a tool for the visualization and solving of equations.

Chapter 7 will:

- Review how to create simple 2D X-Y plots and Polar plots.
- Show how to set plot ranges.
- Instruct how to graph multiple functions in the same plot.
- Discuss the use of range variables to control plots.
- Tell how to plot data points.
- Describe the steps necessary to format plots, including the use of log scale, grid lines, scaling, numbering, and setting defaults.
- Discuss the use of titles and labels.
- Show how to get numeric readout of plotted coordinates.
- Show how the use of plots can help find the solutions to various engineering problems.
- Discuss plotting over a log scale.
- Introduce 3D plotting.
- Use engineering examples to illustrate the concepts.

CREATING A SIMPLE X-Y QUICKPLOT

In Chapter 1 we showed how to create a simple X-Y QuickPlot by typing [@], or by hovering the mouse over **Graph** on the **Insert** menu and clicking **X-Y Plot**. Remember that the x-axis variable in the bottom middle placeholder must be a previously undefined variable. If you use a previously defined scalar variable, Mathcad will plot only a single point instead of a range of points. The middle left placeholder is where you place an expression or function using the variable on the x-axis.

CREATING A SIMPLE POLAR PLOT

Creating a simple Polar QuickPlot is similar to creating a simple X-Y QuickPlot. Open the Polar plot operator by typing CTRL+7, or hover the mouse over **Graph** on the **Insert** menu, and select **Polar Plot**.

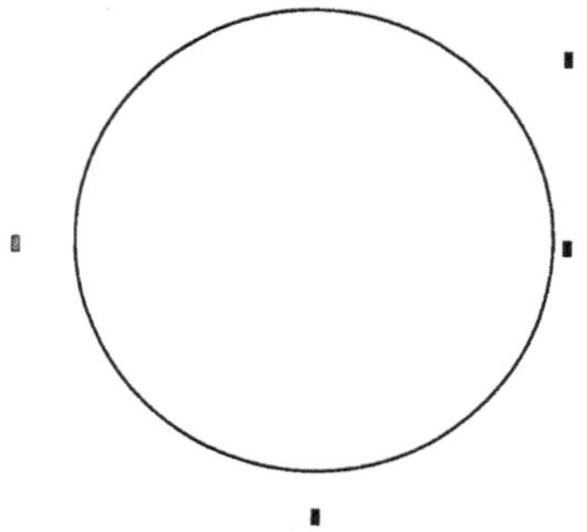

Click the bottom middle placeholder. This is where you type the angular variable. Unless you specify otherwise, Mathcad assumes the variable to be in radians. Type the name of a previously undefined variable. The variable can be any Mathcad variable name. Next, click the middle left placeholder and type an expression using the angular variable defined on the x-axis. This sets the properties of the radial axis. Click outside of the operator to view the plot. For every angle from 0 to 2π, Mathcad plots a radial value. Mathcad automatically selects the radial range. See Figure 7.1.

You can also use a previously defined function in a simple Polar QuickPlot. Open the Polar plot operator by typing CTRL+7. Type the name of the angular variable in the bottom placeholder. Type the name of the function on the left placeholder using the angular variable name as the argument of the function. See Figure 7.2.

USING RANGE VARIABLES

Range variables are used to tell Mathcad what range of values to use when graphing an expression or function. When you create simple X-Y QuickPlots and Polar QuickPlots, Mathcad sets the range of values. By using range variables, you can control the range of values. To graph using a range variable, create the range

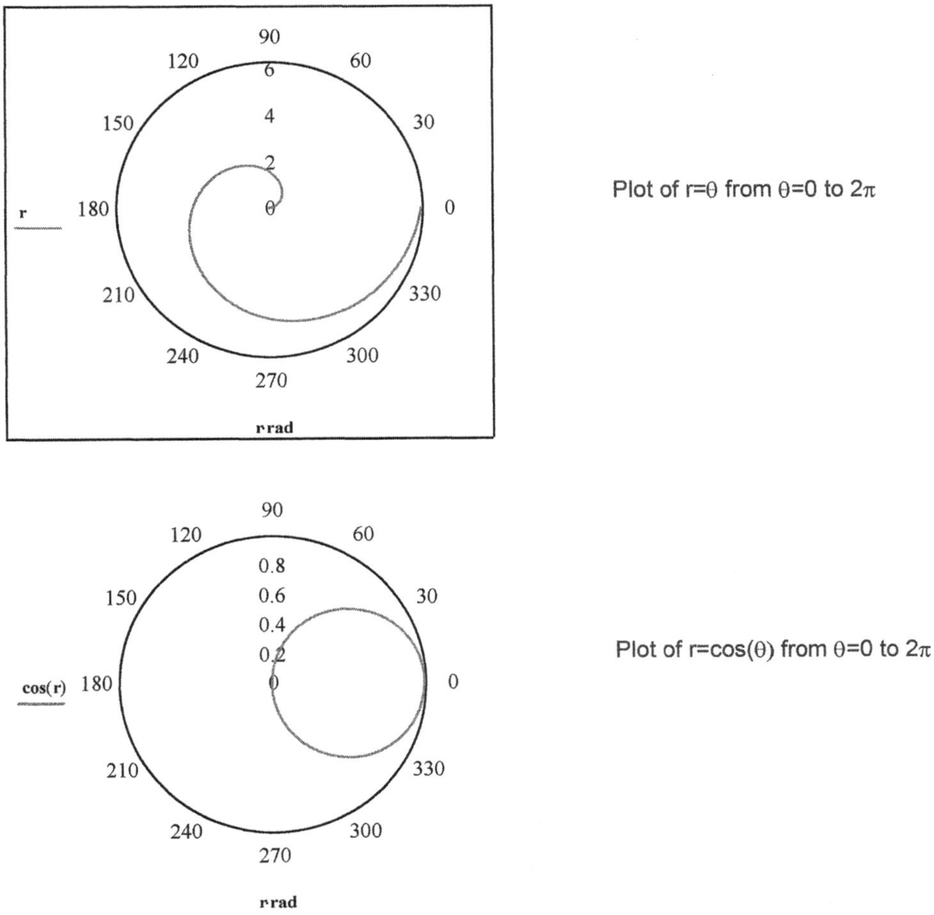

FIGURE 7.1 Polar QuickPlot of functions

variable before using the plot operator. Next, open the plot operator, and type the name of the range variable in the placeholder on the x-axis. In the left placeholder, type the name of a function or expression using the range variable. See Figures 7.3 and 7.4. When using range variables, you are actually telling Mathcad to plot each point in the range variable and to draw a line between the points. This will be illustrated later.

ii(r) := r   jj(r) := cos(6r)   Define functions

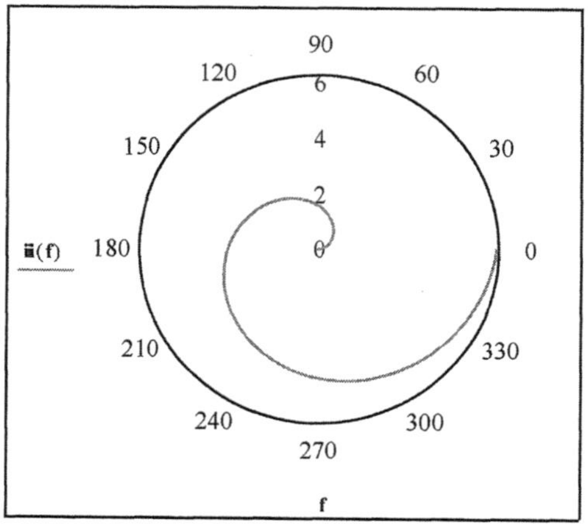

Plot of ii(r)=r from r=0 to 2π

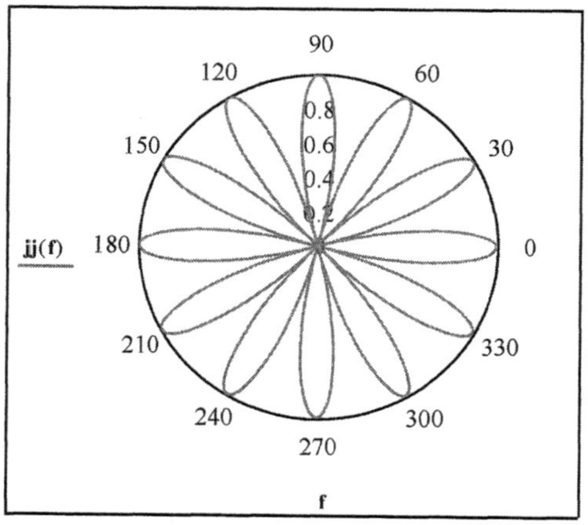

Plot of jj(r)=cos(6r) from r=0 to 2π

FIGURE 7.2 Polar QuickPlot of functions

SETTING PLOTTING RANGES

To change the plotting ranges for an X-Y plot change the placeholders in the lower left and lower right and on the left side upper and left side lower. For a Polar plot, the limits of the radial axis are set by the two placeholders on the right.

For a Polar plot you can experiment with changing the lower placeholder, but it seems to work best when this placeholder is left at zero. See Figures 7.5 and 7.6.

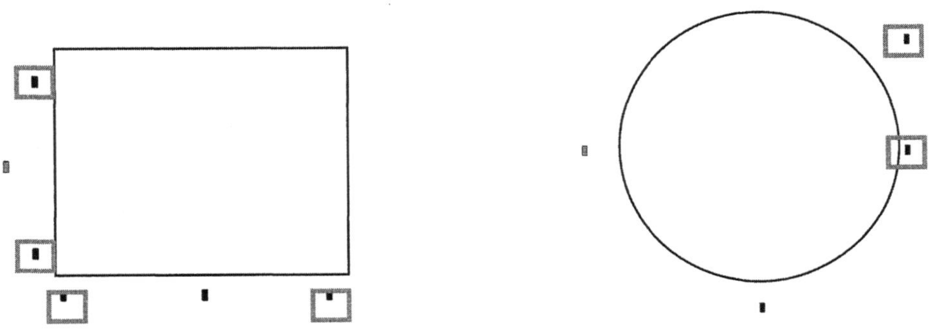

$D(t) := \frac{1}{2} \cdot 9.81 \cdot t^2$ Define function

$t_1 := 0, 1 .. 21$ Define range variables

$t_2 := 0, 1 .. 100$

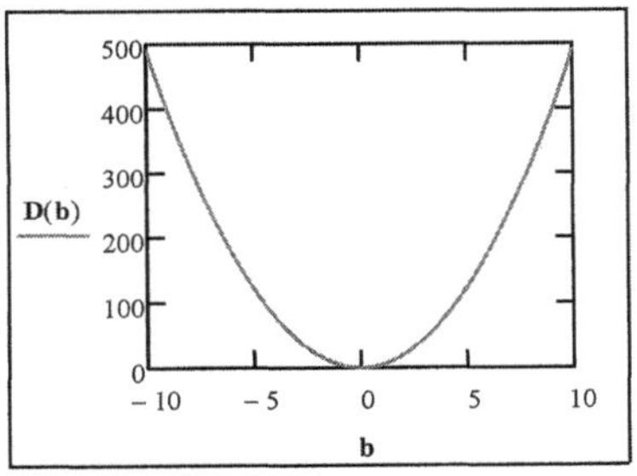

Using a simple X-Y quick plot with Mathcad default ranges.

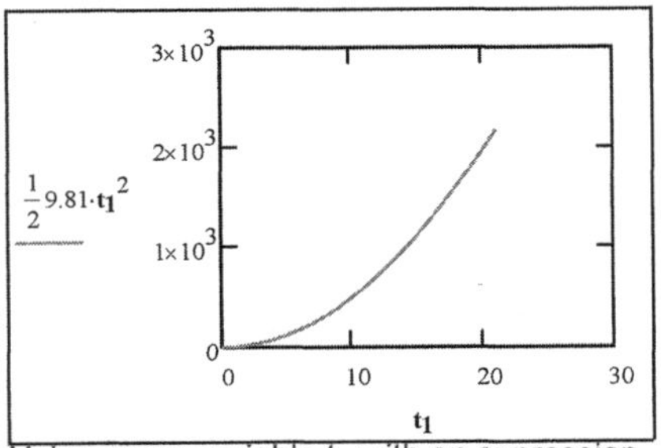

Using range variable t_1 with an expression.

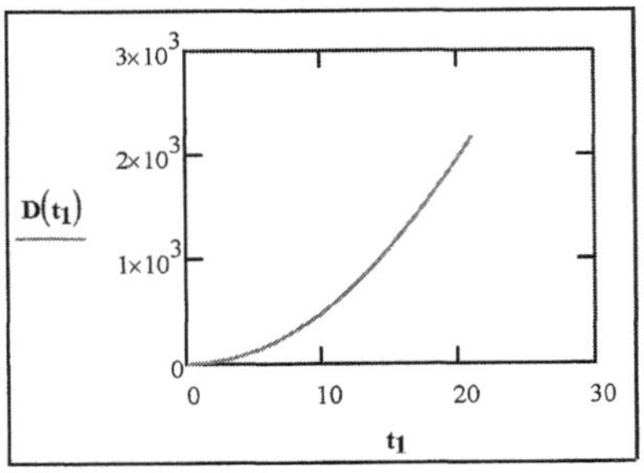

Function "D(t)" using the range variable t_1.

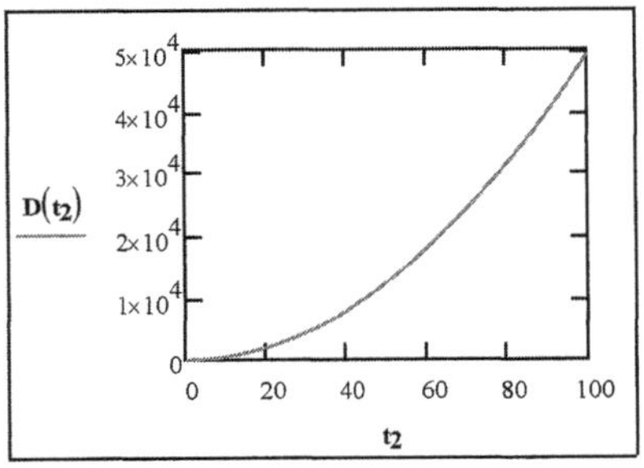

Function "D(t)" using the range variable t_2.

FIGURE 7.3 Using range variables to set plot range

Create range variables and functions.

$\mathbf{Angle_1} := 0, \frac{2\pi}{100} .. 1.8\pi$ $\mathbf{Function_1(\alpha)} := \mathbf{sin(\alpha)^2}$

$\mathbf{Angle_2} := 0, \frac{\pi}{50} .. \pi$ $\mathbf{Function_2(\alpha)} := \mathbf{sin(\alpha)^2 + cos(\alpha)}$

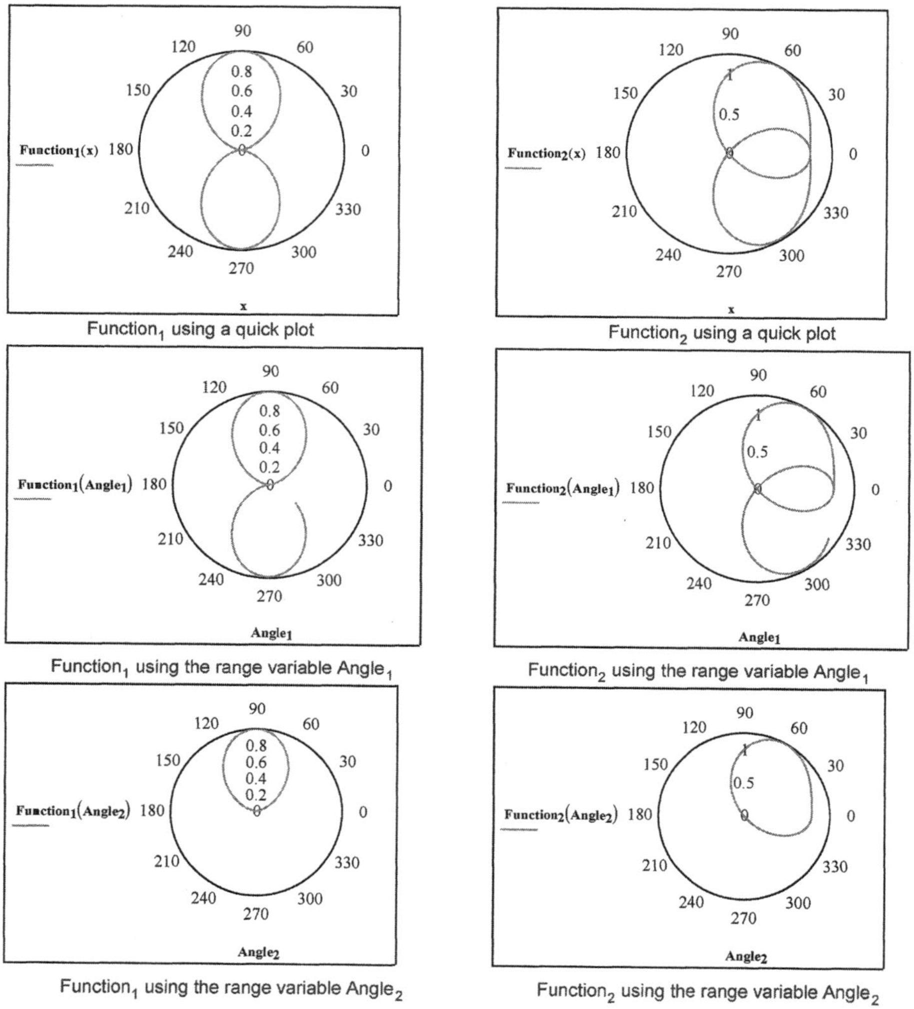

FIGURE 7.4 Using range variables to set plot range

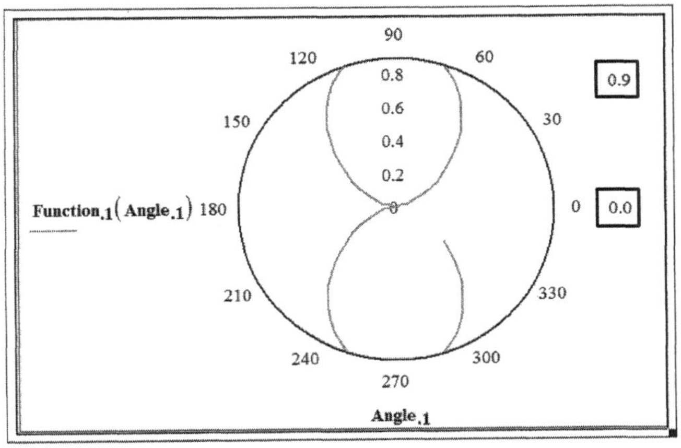

This plot is the same as in Figure 7.4, but in this Figure, the radial limits are 0 to 0.9. The range variable $Angle_1$ is still used.

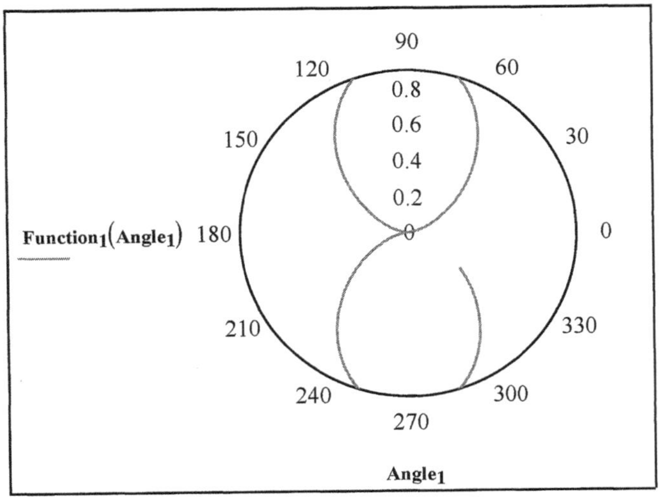

FIGURE 7.5 Setting plot range

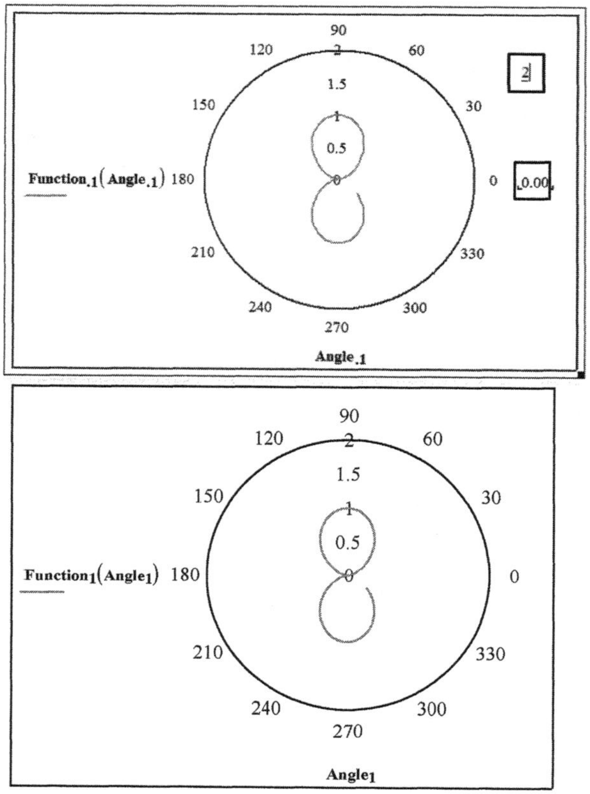

This plot is the same as in Figure 7.4, but in this Figure, the radial limits are 0 to 2.0. The range variable $Angle_1$ is still used.

FIGURE 7.6 Setting plot range

GRAPHING WITH UNITS

When you plot functions and data with units attached, the numbers displayed on the axis correspond to the base unit or the derived unit for the unit dimension. If you are plotting distance, the numbers on the axis will correspond to meters (if SI is the default unit system). If you are plotting pressure, the numbers on the axis will correspond to Pa (if SI is the default unit system). In Figure 7.7 we want to plot pressure in psf, but the worksheet default unit system is SI. This causes the depth to be plotted in meters and the pressure to be plotted as Pa. How can the pressure be plotted in psf, psi, or other pressure units? How can the depth be plotted in feet?

The density of water is about 62.24 pound force per cubic foot.
Graph the water pressure (in psf) at various depths (in feet).

Density := 62.24**pcf**

Pressure(**Depth**) := **Density** · **Depth** **Pressure**(10**ft**) = 622.40 · **psf**

i := 0**ft**, 1**ft**.. 10**ft** Set plot limits using a range variable.

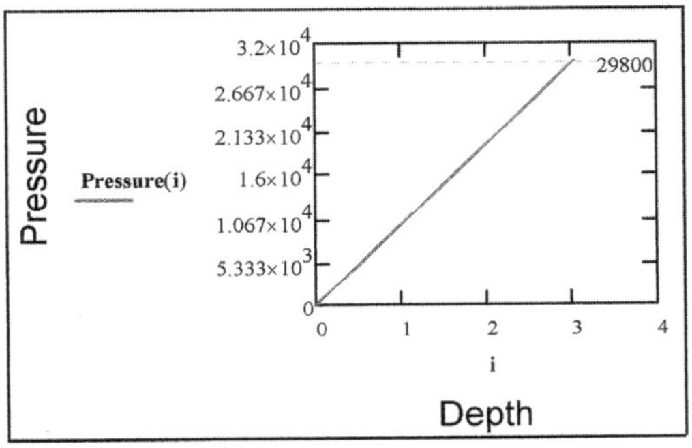

Note: The dashed line is a "marker" set at 29800. See Figure 7.12 for additional examples of markers.

Note that the x-axis is not plotting the values of feet and the y-axis is not plotting the values of psf. At 10 feet, the value should be 622.

The x-axis represents meters, and the y-axis represents Pa. These are the base units and derived units in the SI unit system. How can you plot the depth in feet and the pressure in psf? See Figure 7.8.

Pressure(10**ft**) = 29800.67 **Pa** 10**ft** = 3.05 · **m**

FIGURE 7.7 Using units with plots

In Chapter 4, we discussed using units in empirical equations. Plots are similar. To plot functions and data with units attached, divide the function or data by the units you want plotted. This creates unitless data with the values you want displayed.

In Figures 7.7 and 7.8, the range variable i has units attached to it. If the range variable did not have units attached, the function argument would need to have units added. The function would then need to be divided by the pressure units to be plotted. The y-axis function would be Pressure(i*ft)/psf.

GRAPHING MULTIPLE FUNCTIONS

You can graph up to 16 multiple functions or expressions on the same plot. To graph multiple expressions using the same x-axis variable, type a comma after entering the function or expression in the left middle placeholder. This places a new placeholder below the original placeholder. You can now type a new

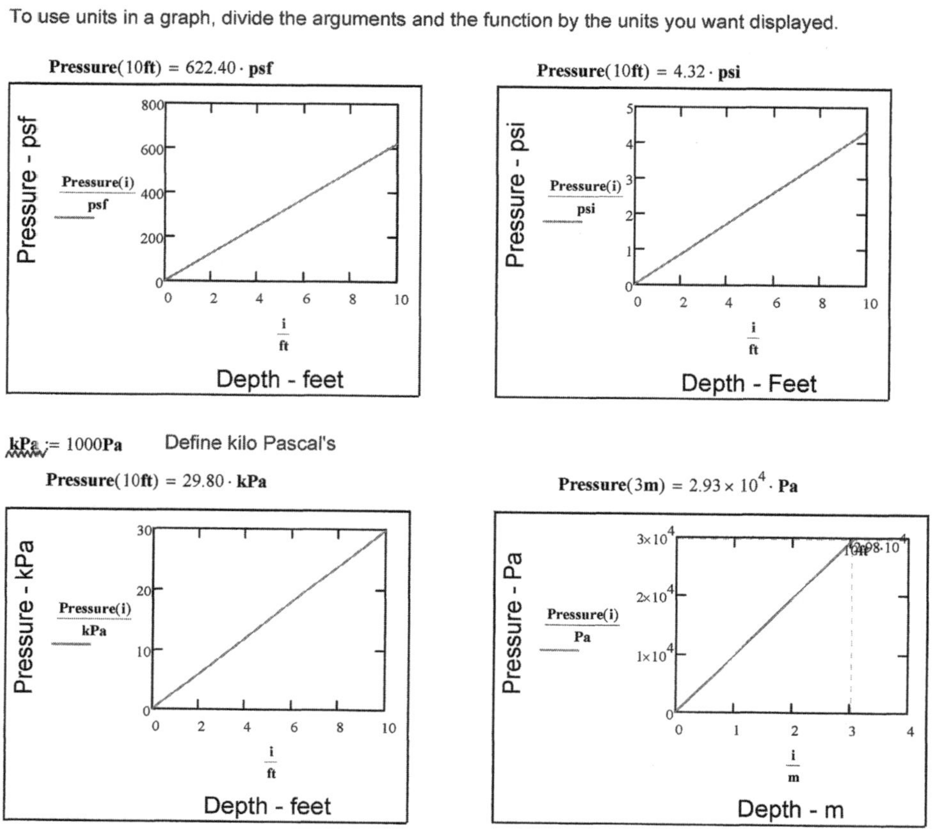

FIGURE 7.8 Using units with plots

function or expression. You can repeat the process until you have up to 16 functions or expressions. Each plot is called a trace. See Figure 7.9.

You can also use multiple variables on the x-axis and then plot corresponding expressions on the y-axis. To do this, type a comma after entering the variable name on the x-axis. This places a new placeholder adjacent to the original placeholder. On the y-axis, create new placeholders in the same way, and use the corresponding x-axis variable name in your expression. See Figure 7.10.

Beginning with Mathcad 12, you can create a secondary y-axis on the right-hand vertical axis of an X-Y plot. This secondary y-axis can be used to graph

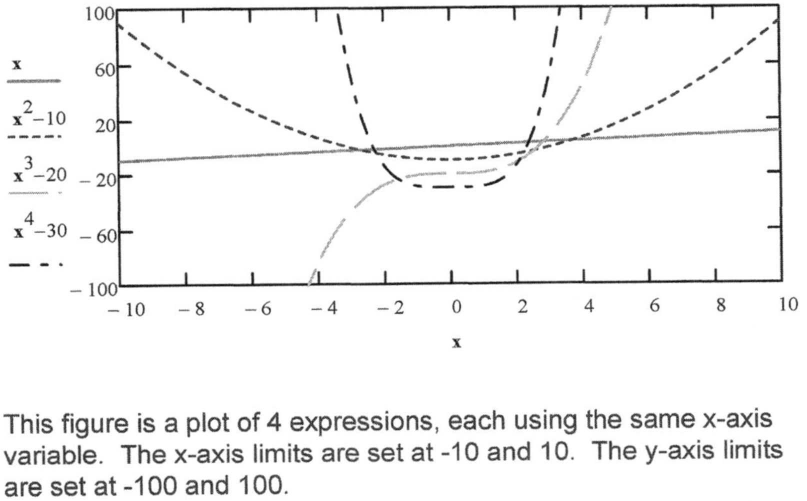

FIGURE 7.9 Multiple plots

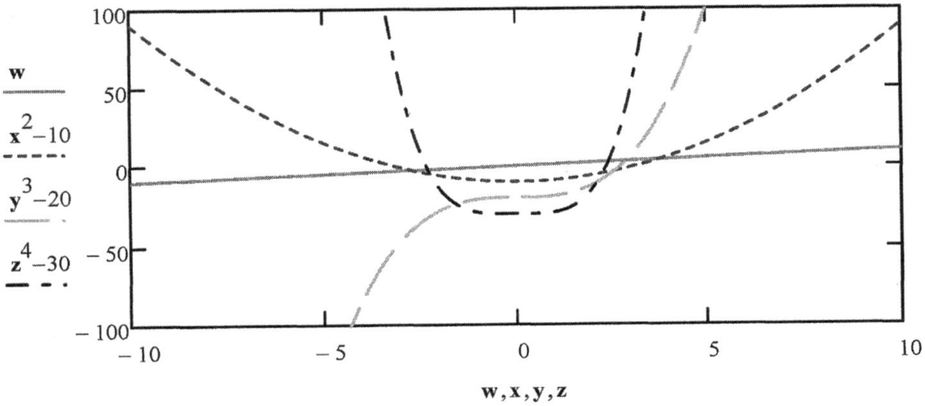

FIGURE 7.10 Multiple plots with multiple variables

additional traces at a different scale than the primary y-axis. To create a secondary y-axis, double-click on the plot. This opens a plot formatting dialog box. Place a check in the box Enable Secondary Y-Axis. This will add new placeholders on the right side of the plot. You can now type expressions, functions, and range limits in the new placeholders. See Figure 7.11.

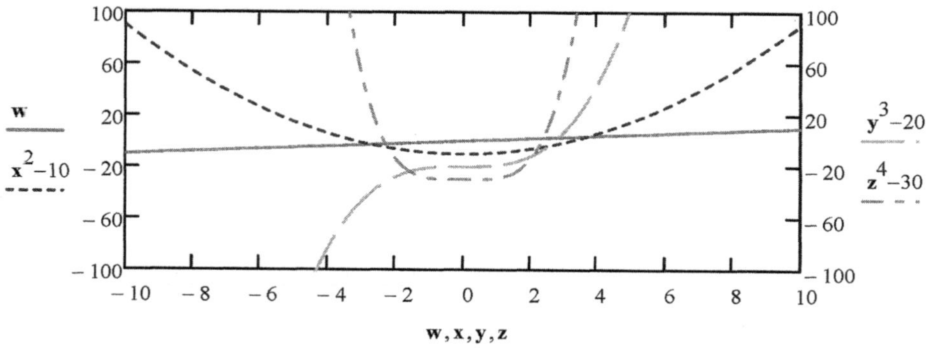

FIGURE 7.11 Multiple plots using Secondary Y-Axis

FORMATTING PLOTS

Mathcad allows you to customize many aspects of your plots. You can add grid lines; change the spacing of the grid lines; plot with equal x- and y-axes; change the color, line weight, or line type of each trace; or add symbols to the trace. You can even plot in log scale.

To make customizations, double-click the plot. This opens a plot formatting dialog box. The features in this box will depend on whether you are working with an X-Y plot or a Polar plot. Each version of Mathcad seems to have a slightly different dialog box, but they all control similar features.

Axes Tab

This tab will be either the X-Y Axes tab, or the Polar Axes tab, depending on the type of plot you have open. This tab controls the appearance of the axes and grids, and allows you to:

- Change an axis to log scale.
- Add grid lines to an axis.
- Display or not display numbers on the axes.
- Tell Mathcad whether the axis limits are set at the data limits or whether the axis limits are set to the next major tick mark beyond the end of the data (Auto scale).
- Add horizontal, vertical, or radial marker lines at values that you set. The marker lines are dashed horizontal or vertical lines that can be set at specified locations. Each plot axis may have two marker locations.
- Change the spacing of the grid lines.
- Change from axes on the sides of the plot to axes in the center of the plot.

Mathcad Help has an excellent description of each of these various features. To access these descriptions click the **Help** button within the Formatting dialog box. See Figures 7.12 and 7.13 for examples of these different features.

$$f(x) := x^2 - 8 \qquad h(x) := -x^2 + 8$$

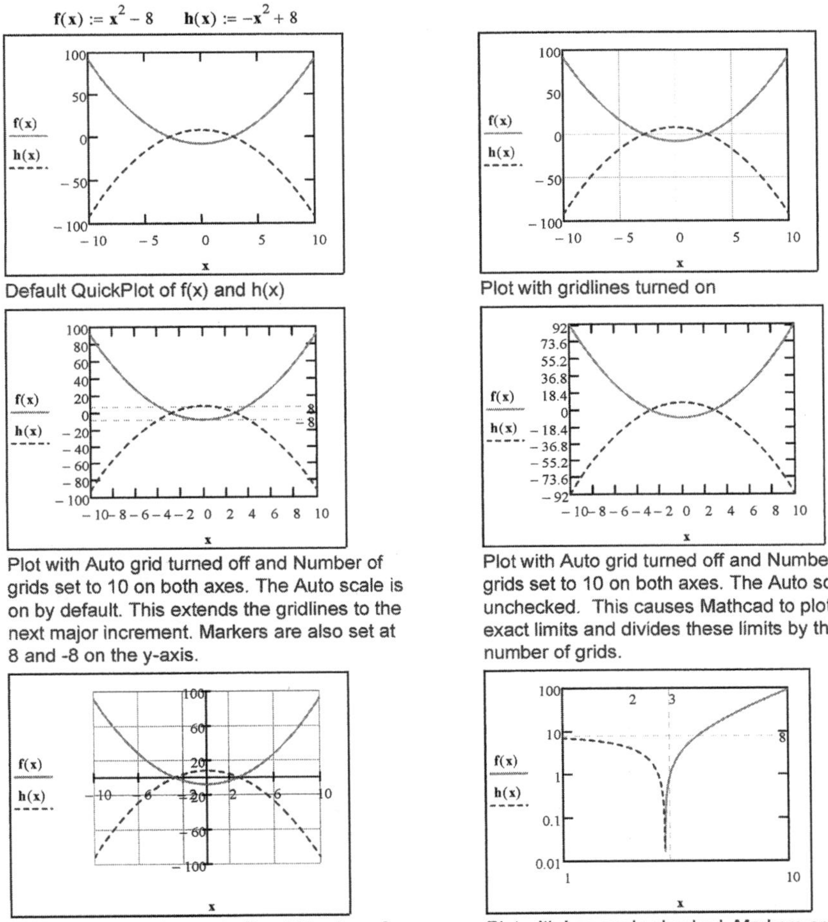

Default QuickPlot of f(x) and h(x)

Plot with gridlines turned on

Plot with Auto grid turned off and Number of grids set to 10 on both axes. The Auto scale is on by default. This extends the gridlines to the next major increment. Markers are also set at 8 and -8 on the y-axis.

Plot with Auto grid turned off and Number of grids set to 10 on both axes. The Auto scale is unchecked. This causes Mathcad to plot to the exact limits and divides these limits by the number of grids.

Plot with Auto grid turned off and Number of grids set to 5 on both axes. Grid lines is checked. The Axis Style is set to Crossed.

Plot with Log scale checked. Markers are also set at 2 and 3 on the x-axis and at 8 on the y-axis. Note that with log scale turned on, only positive values are plotted.

FIGURE 7.12 Formatting examples

Traces Tab

This tab allows you to change how each trace appears. You can change line type, line weight, and line color. You can also change the type of plot from a line plot to various forms of bar graphs, or change to just plotting points. You can also add a legend giving titles to the different traces. You can also add different symbols for the data points. Data points will be discussed later in the chapter.

See Figures 7.14 through 7.16 for a display of these various features. The best way to learn these features is to try them. See the practice exercises at the end of the chapter.

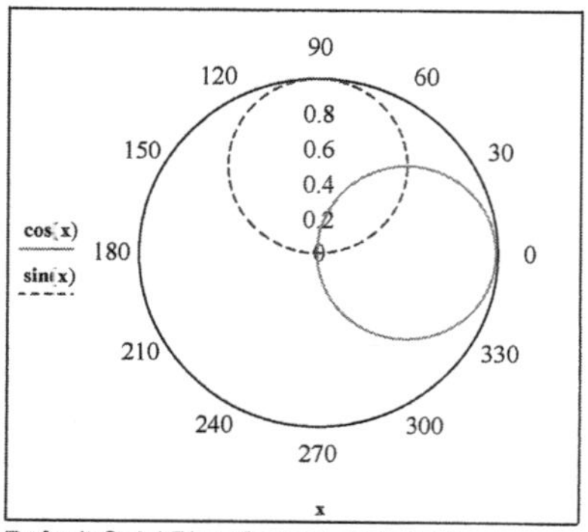

Default QuickPlot of cos(x) and sin(x)

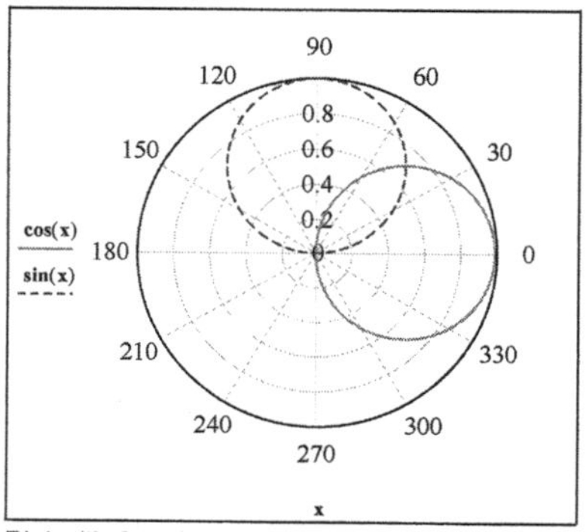

Plot with Angular and Radial grid lines turned on

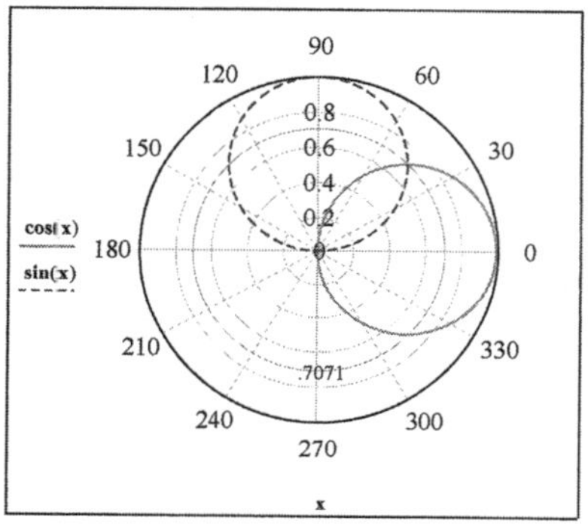

Plot with Angular and Radial grid lines turned on.
Radial marker set at 0.7.

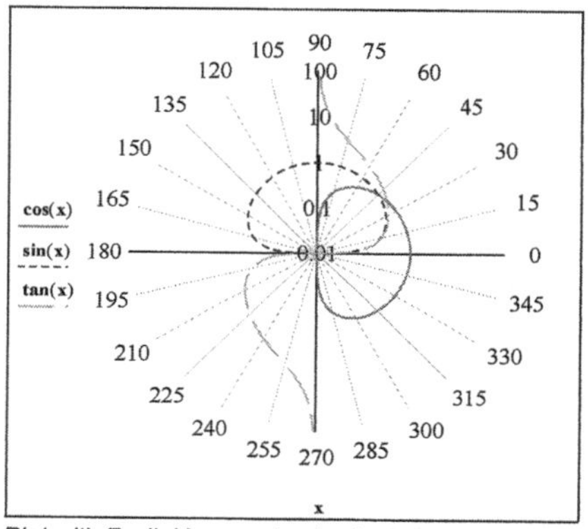

Plot with Radial Log scale checked. Radial limits set at 0.01 and 100. Axis style is set to Crossed. Angular Number of grids changed to 24. Angular grid lines turned on. Tangent function also added to plot.

FIGURE 7.13 Formatting examples

Number Format Tab

The Number Format tab was added with Mathcad version 14. This tab allows you to set the accuracy of tick mark labels displayed on the x- and y-axes. Prior to this dialog box, Mathcad would draw the plots accurately, but round the displayed axis tick mark labels to two decimal points. This could cause some plots to appear to be inaccurate, but it was only due to the numbers displayed on the axes. This new dialog box allows plot tick mark labels to be displayed with more accuracy.

Labels Tab

The Labels tab allows you to apply titles to your plot and to each of the axes. In order for your titles to be visible, the check box associated with each title must be checked. See Figure 7.17 for an example.

Use the same functions as in Figure 7.12.

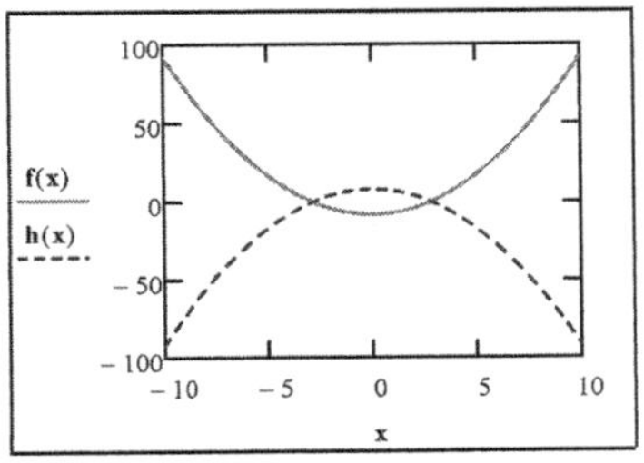

QuickPlot with defaults

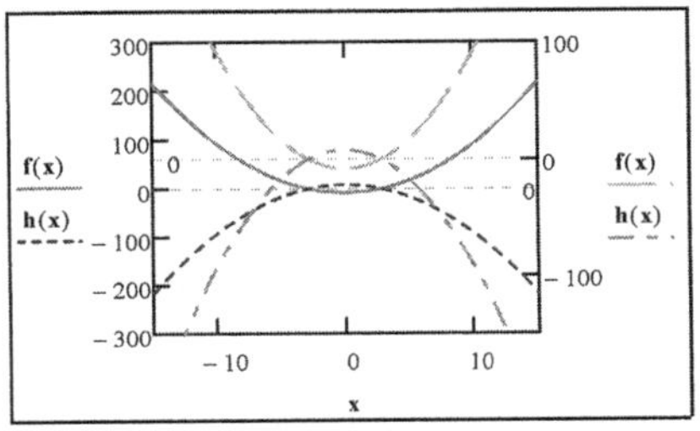

Plot with secondary y axis using same functions, but with different y-axis limits.

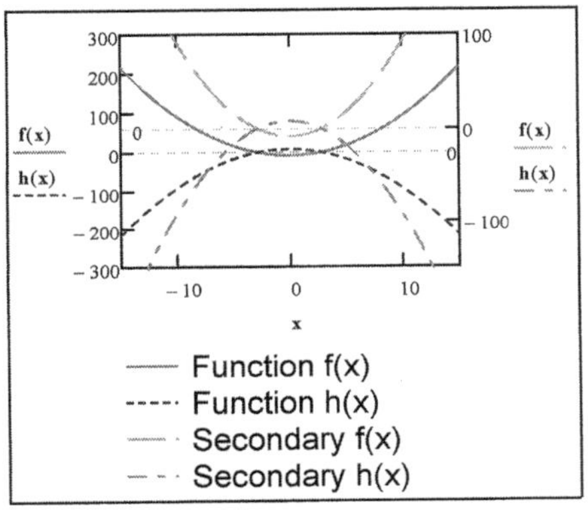

Legend turned on. Positioned below plot.

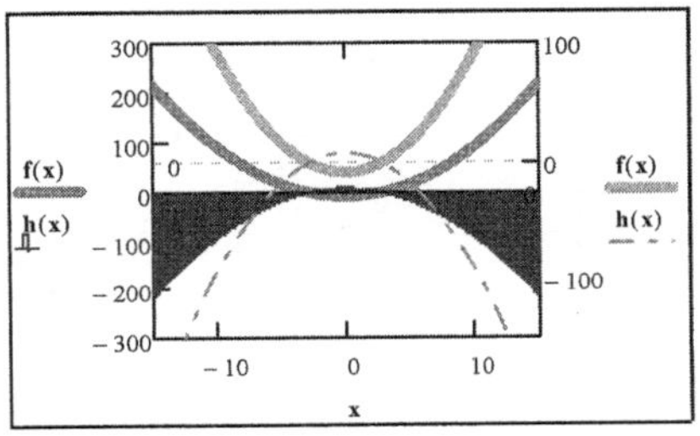

Trace 2 changed to a bar plot. Line weight of trace 1 and trace 3 changed from 1 to 4.

FIGURE 7.14 Formatting examples

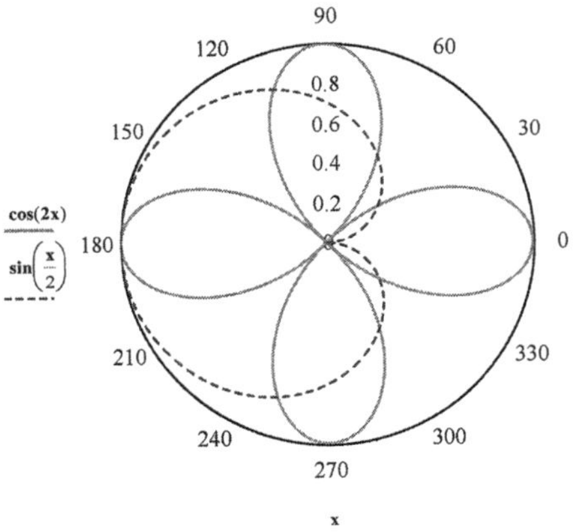

Default QuickPlot of cos(2x) and sin(x/2)

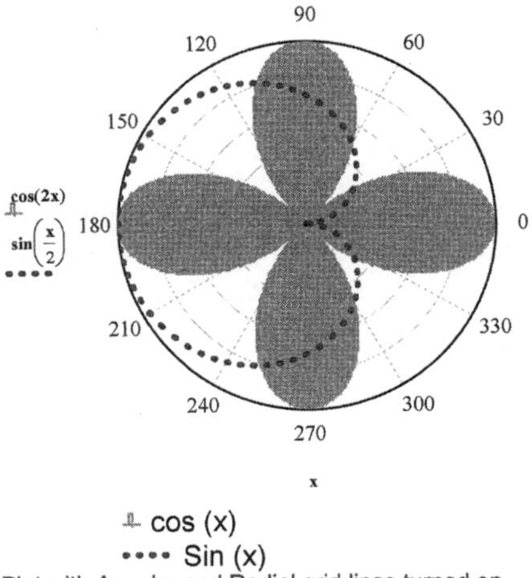

Plot with Angular and Radial grid lines turned on. Trace 1 changed to bar plot. Line weight on trace 2 changed to 3.

FIGURE 7.15 Formatting examples

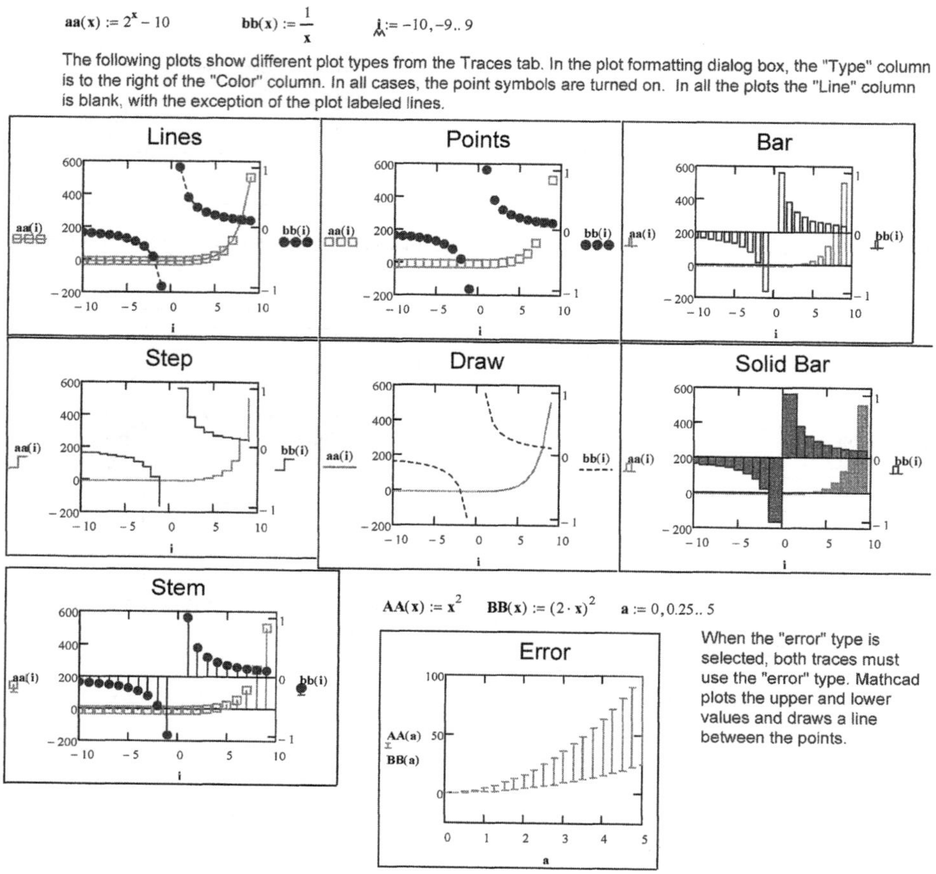
aa(x) := 2^x − 10
bb(x) := 1/x
i := −10, −9 .. 9
The following plots show different plot types from the Traces tab. In the plot formatting dialog box, the "Type" column is to the right of the "Color" column. In all cases, the point symbols are turned on. In all the plots the "Line" column is blank, with the exception of the plot labeled lines.
Lines
Points
Bar
Step
Draw
Solid Bar
Stem
aa(i)
bb(i)
i
AA(x) := x^2
BB(x) := (2 · x)^2
a := 0, 0.25 .. 5
Error
AA(a)
BB(a)
a
When the "error" type is selected, both traces must use the "error" type. Mathcad plots the upper and lower values and draws a line between the points.

FIGURE 7.16 Plot types

The font used in the plot title and axis labels comes from the Math Text Font variable style. To change the font, size, color, or style of your title and labels, change the Math Text Font style. To do this, click **Equation** from the **Format** menu. Refer to Chapters 14 and 15 for additional information on customizing styles.

Defaults Tab

This tab allows you to reset your plot to the Mathcad defaults. It also allows you to use the current plot settings as the default settings for the current document.

If you have customized the plot settings and want to reuse the settings for future documents, save the document as a template. You can even save the plot settings to your customized normal.xmct file so that they will be available for all new documents. See Chapter 16 for additional information about templates.

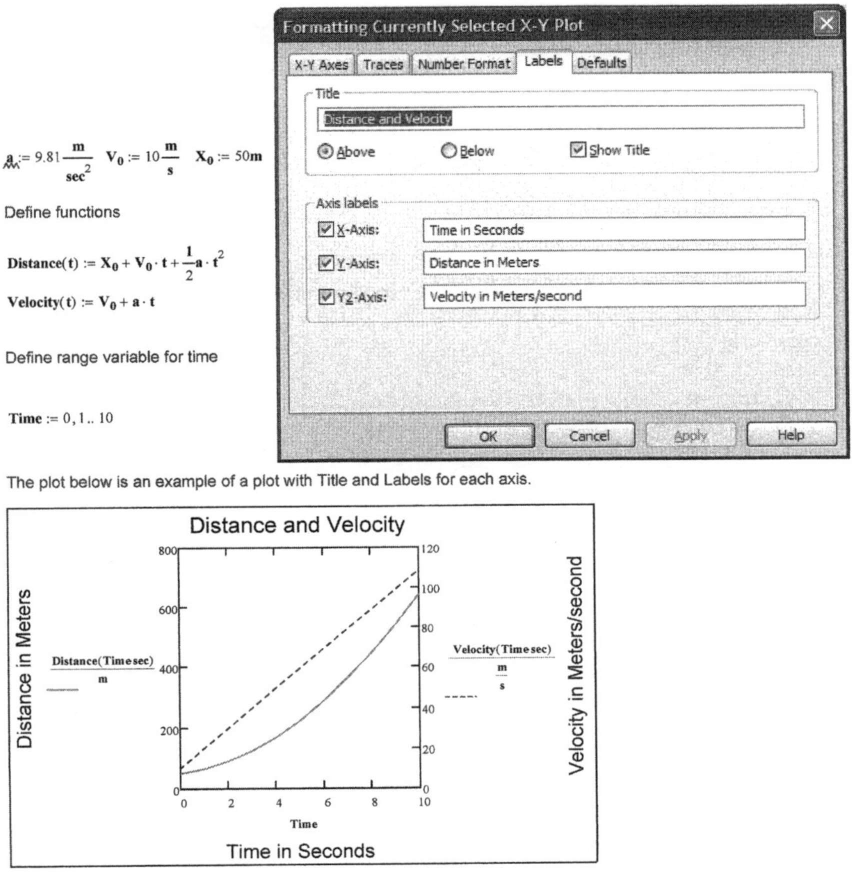

FIGURE 7.17 Using titles

ZOOMING

There are times when you might want to zoom in on a plot. Perhaps you want to see where a plot crosses the x-axis, or perhaps you want to see where two plots intersect. To zoom in on a plot, right-click within the plot and select **Zoom**. You can also click on the plot to select it, and then click on the **Zoom** icon from the **Graph** toolbar. You can also select **Zoom** from the **Graph** option on the **Format** menu. This opens a Zoom dialog box. See Figure 7.18. Once the dialog box is open, click your mouse at one corner of the plot region you want to zoom. Then drag the mouse to include the area you want to zoom. A dashed selection outline will show you what area of the plot will be enlarged. The coordinates of the

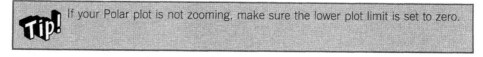

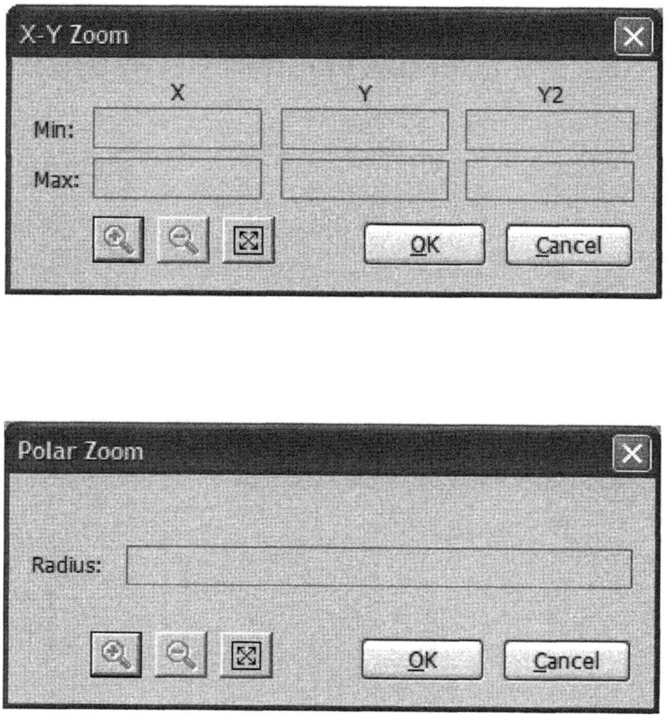

FIGURE 7.18 Zoom dialog boxes

selection outline will also appear in the Zoom dialog box. Once the selection outline encloses the area of the plot you want to zoom, let go of the mouse button. You can drag the selection outline to fine-tune its location on the plot.

Once you are satisfied with the location of the selection outline, click the plus icon in the Zoom dialog box. This temporarily sets the axis limits to the coordinates specified in the Zoom dialog box. You can now zoom in again using the same procedure, or you can zoom back out by clicking the minus icon button in the Zoom dialog box. If you have zoomed in several times, you can step out by repeatedly clicking the minus icon, or you can zoom out to the original view by clicking the right icon button in the Zoom dialog box. If you want to make the zoomed-in region the permanent axis limits, then click OK.

> **Tip!** If your Polar plot is not zooming, make sure the lower plot limit is set to zero.

Another easy way to zoom a plot is to make the plot larger. You can do this by clicking and dragging the bottom right corner grip.

PLOTTING DATA POINTS

Up until now, we have focused on graphing functions and expressions. These plots are easily represented with lines. Mathcad also allows you to plot data points. These data points can be created by using range variables, or they can be from a vector or matrix.

Range Variables

When we discussed using range variables earlier, we were actually graphing data points. When we used a range variable on the x-axis, Mathcad created a data point for each value in the range variable, and then plotted the corresponding value on the y-axis. Mathcad drew a line between all the points.

Let's look at a few examples of using range variables to plot data points. See Figures 7.19 and 7.20.

$RV_1 := -20, -19 .. 20$ $RV_2 := -20, -10 .. 20$

Range variables actually create data points that are connected by straight lines. The plots used earlier in the chapter did not have the data points displayed.

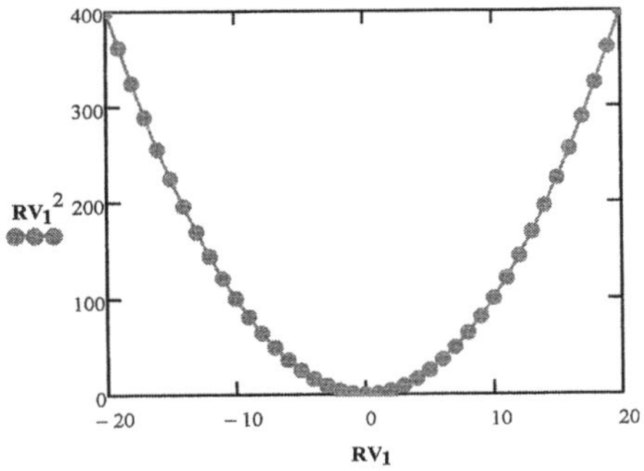

This plot uses a range variable with an increment of 1. Notice how smooth the graph is.

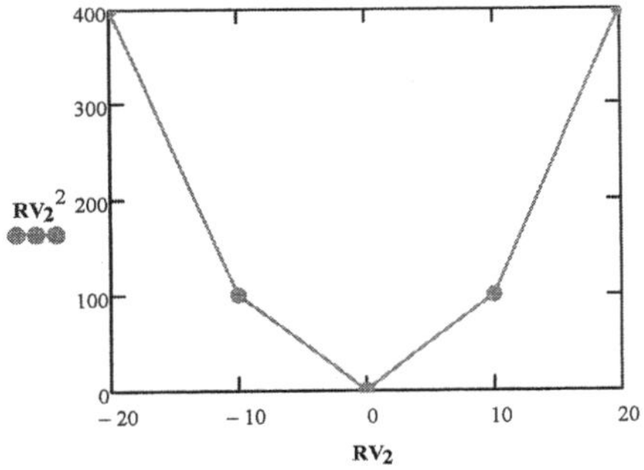

This plot uses a range variable with an increment of 10. Notice how it is not as smooth as the above plot.

FIGURE 7.19 Plotting data points

$$RV_3 := \frac{-\pi}{2}, \frac{-\pi}{2} + \frac{\pi}{24} .. \frac{\pi}{2} \qquad RV_4 := \frac{-\pi}{2}, \frac{-\pi}{2} + \frac{\pi}{8} .. \frac{\pi}{2}$$

Range variables actually create data points that are connected by straight lines.

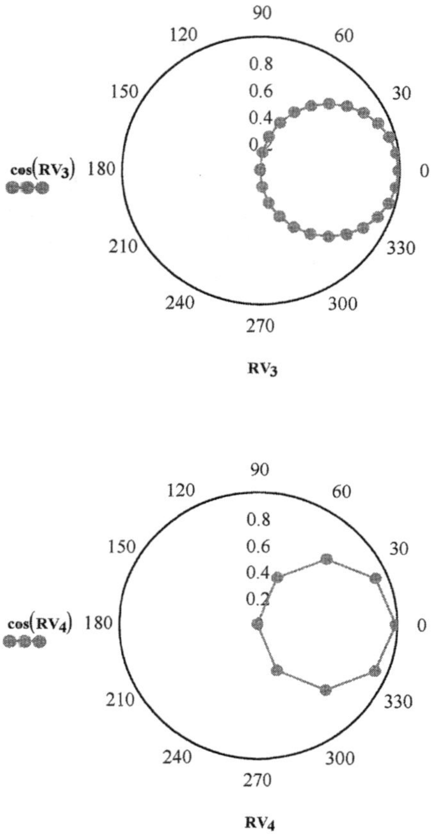

This plot uses a range variable with an increment of π/24. Notice how smooth the graph is.

This plot uses a range variable with an increment of π/8. Notice how it is not as smooth as the above plot.

FIGURE 7.20 Plotting data points

Data Vectors

See Figure 7.21 for an example of plotting a vector of data points. See Figure 7.22 for an example of plotting matrix data points.

NUMERIC DISPLAY OF PLOTTED POINTS (TRACE)

You can get numeric display of plotted points by using the Trace dialog box. To open the Trace dialog box, right-click within the plot and choose **Trace**.

$$\mathbf{DataVector} := \begin{pmatrix} 0.9 \\ 1.9 \\ 3.1 \\ 4.2 \\ 5.1 \\ 5.9 \\ 7.0 \\ 7.9 \\ 9.1 \\ 10.0 \end{pmatrix}$$

To plot a data vector, create a range variable to provide the x-axis value, and to tell Mathcad which data points to plot. The range variable must be in integer increments.

Use the *last*(v) function, which returns the element number of the last element of the vector v.

$\mathbf{last(DataVector)} = 10.00$

$i := 1, 2 .. \mathbf{last(DataVector)}$

With this range variable, we are telling Mathcad to plot the first through the last elements of the vector "DataVector".

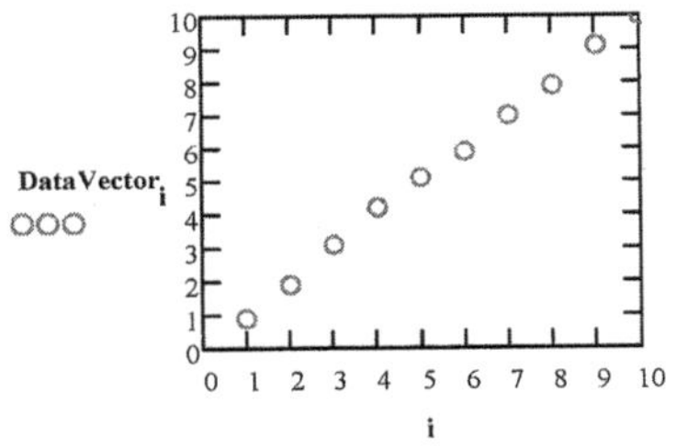

Type the name of the range variable on the x-axis. On the y-axis, type the name of the data vector and use the vector operator (vector subscript) with the range variable as the subscript. (Type [to get the vector operator.)

For this plot, the trace type is changed to "Points". The Symbol type is set to an open circle.

FIGURE 7.21 Plotting vector data points

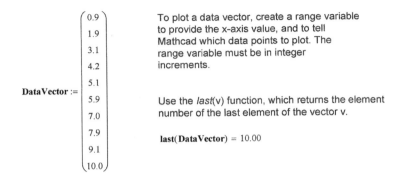

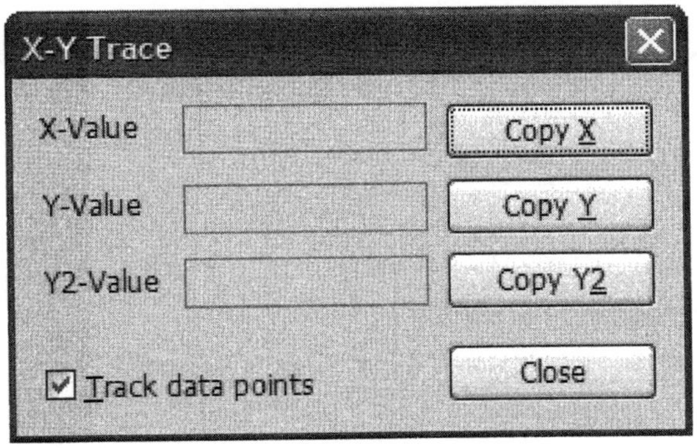

Place a check in the "Track Data Points" box. Click and drag your mouse along the trace whose coordinates you want to see. You should see a dotted crosshair move from top point along the trace. The coordinates of each data

$$\mathbf{DataMatrix} := \begin{pmatrix} 0 & 0.1 \\ 1 & 0.9 \\ 2 & 1.9 \\ 3 & 3.1 \\ 4 & 4.2 \\ 5 & 5.1 \\ 6 & 4.3 \\ 7 & 4.0 \\ 10 & 2.9 \\ 11 & 2.1 \\ 12 & 0.8 \end{pmatrix}$$

To plot data from a matrix, assign each column in the matrix to a variable. To do this, use the matrix column operator "M<>". The column operator returns a vector of the matrix column numbered between the brackets. This operator is located on the Matrix toolbar, or press CTRL+6. (The ***submatrix*** function also could have been used.)

$\mathbf{X} := \mathbf{DataMatrix}^{\langle 1 \rangle}$ Note: Type CTRL+6 to get the brackets.

$\mathbf{Y} := \mathbf{DataMatrix}^{\langle 2 \rangle}$

X =

	1
1	0.00
2	1.00
3	2.00
4	3.00
5	4.00
6	5.00
7	6.00
8	7.00
9	10.00
10	11.00
11	12.00

Y =

	1
1	0.10
2	0.90
3	1.90
4	3.10
5	4.20
6	5.10
7	4.30
8	4.00
9	2.90
10	2.10
11	0.80

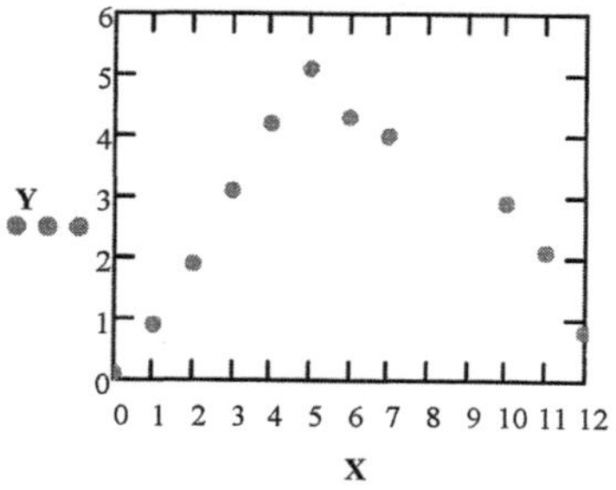

For this plot, the trace type is changed to "Points". The data points are: (X_1,Y_1), (X_2,Y_2), etc.

FIGURE 7.22 Plotting matrix data points

point will be displayed in the dialog box. If you release the mouse button, you can use the left and right arrows to move to the previous and next data points, respectively.

USING PLOTS FOR FINDING SOLUTIONS TO PROBLEMS

One great use of plots is to find the intersection of two functions. In Chapters 9 and 10, we will be discussing the Mathcad solving functions. You can use a plot to get a quick guess for the required input in the ***find*** function. Using the trace feature just described, you can also get a quick approximation of the solution. See Figure 7.23 for an example.

PARAMETRIC PLOTTING

A parametric plot is one in which a function or expression is plotted against another function or expression that uses the same independent variable. See Figure 7.24 for an example.

$y_1(x) := x^2 + 2$ $\qquad$ $y_2(x) := -x + 15$

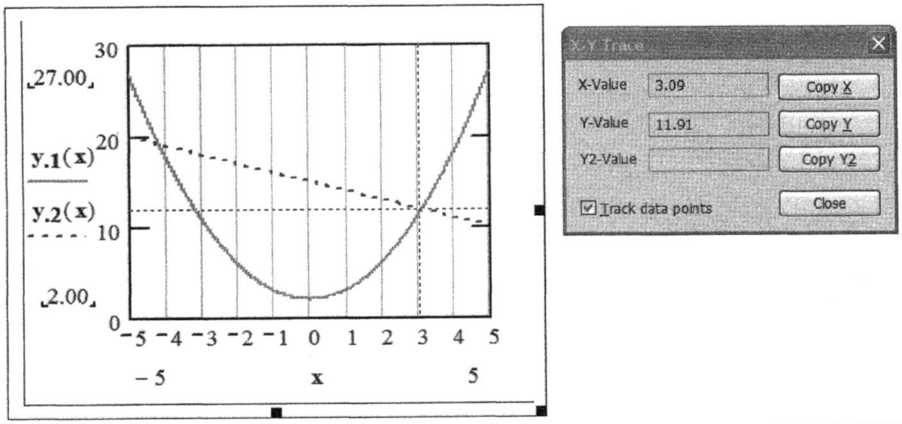

Using gridlines you can see that one solution is just greater than x=3. The other solution is just less than x=-4.

Using the Trace dialog box, you can see that one solution is very close to x=3.09.

FIGURE 7.23 Using trace to find approximate solutions

$\mathbf{iii} := 1, 2 .. 10$

$\mathbf{ZZ}(\mathbf{x}) := \frac{1}{\mathbf{x}^2}$

$\mathbf{YY}(\mathbf{x}) := \mathbf{x}^2$

$\mathbf{ZZ}(\mathbf{iii}) =$

	1
1	1.000
2	0.250
3	0.111
4	0.063
5	0.040
6	0.028
7	0.020
8	0.016
9	0.012
10	0.010

$\mathbf{YY}(\mathbf{iii}) =$

	1
1	1.00
2	4.00
3	9.00
4	16.00
5	25.00
6	36.00
7	49.00
8	64.00
9	81.00
10	100.00

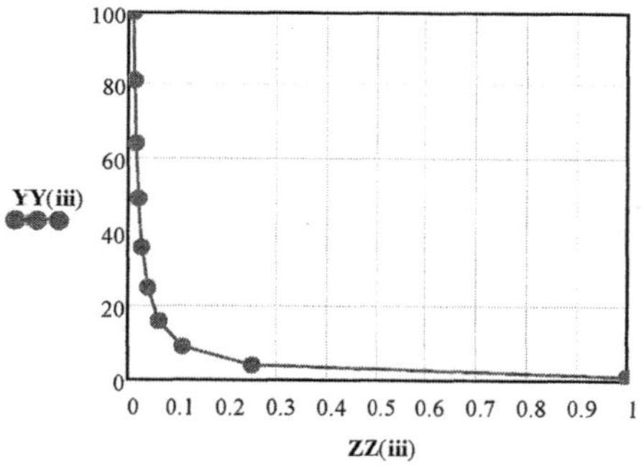

Parametric plot of ZZ(x) on the x-axis and YY(x) on the y-axis.

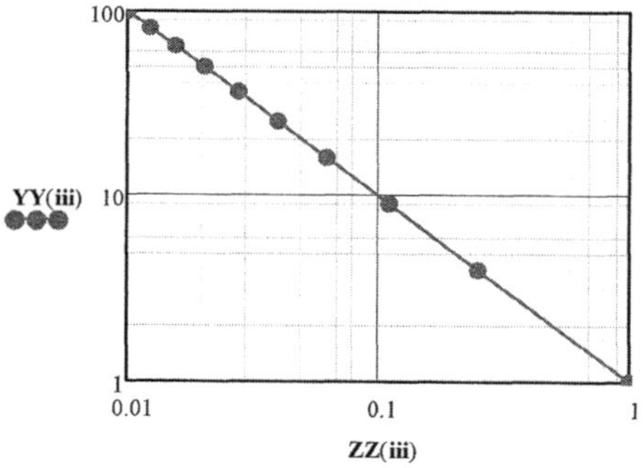

Parametric plot of ZZ(x) on the x-axis and YY(x) on the y-axis using Log scale on both axes.

FIGURE 7.24 Example of parametric plot

PLOTTING OVER A LOG SCALE

If you are plotting a series of points using a log scale, you will want to have your points closer together the closer you get to zero, and further apart the further you get away from zero. For example, if you are plotting from 0.001 to 10,000, you need to have a very small increment in order to see the points near the beginning of the plot. However, if you select a small increment for a range variable, you will be plotting millions of unnecessary points as you move toward 10,000.

Most plot ranges occur over a uniform increment. This example uses a variable increment.

For this example, plot 21 points from a uniformly sloping line. The minimum value is 0.001. The maximum value is 10,000. Use a log scale on both axes.

Create a vector of values to be plotted using the Mathcad function ***logspace***. This function returns a vector of n points logarithmically spaced from the minimum to the maximum value. The function takes the form logspace(min,max,npts), where npts is the numberf of points. Notice that as the vector gets larger, so does the distance between each point.

$\mathbf{X} := \mathbf{logspace}(0.001, 10000, 21)$

$\mathbf{X} =$

	1
1	0.00100
2	0.00224
3	0.00501
4	0.01122
5	0.02512
6	0.05623
7	0.12589
8	0.28184
9	0.63096
10	1.41254
11	3.16228
12	7.07946
13	15.84893
14	35.48134
15	79.43282
16	...

FIGURE 7.25 Plotting using a variable range

The solution to this is to use a variable plotting range using the Mathcad function ***logspace***. This function creates a vector of n logarithmically spaced points. See Figures 7.25 and 7.26 for the example.

Plotting Conics

The January 2002 issue of the Mathcad Advisor Newsletter has an excellent discussion on the graphing of circles, ellipses, and hyperbolas. Please refer to this article for information on this topic.

Plotting a Family of Curves

The following is discussed in the November 2001 Mathcad Advisor Newsletter.

It is possible to plot two parameters on a 2D graph. For example, you can plot F(a,b): = a2*cos(b) over the ranges a = 0,1..4 and b = 0,0.1.. 15, with b plotted on the x-axis. This is actually a single trace. Mathcad will plot all values of b for a

Continued from Figure 7.25

Function(a) := **a**

$s := 0.001, \frac{10000 - 0.001}{20} .. 10000$

(Creates a range variable "s" with 21 uniformly spaced points and with 20 uniformly spaced intervals)

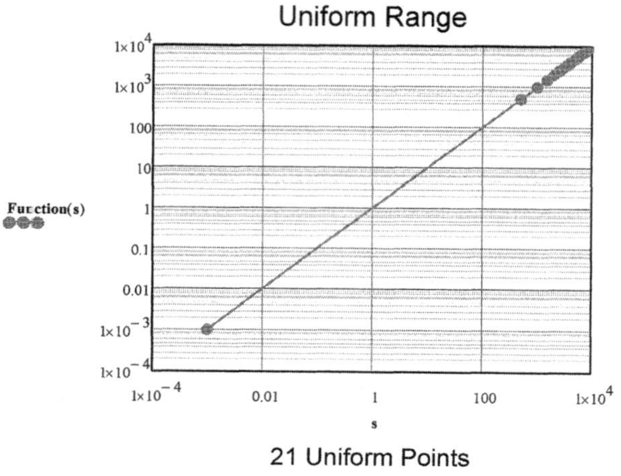

s =

	1
1	0.0
2	500.0
3	1000.0
4	1500.0
5	2000.0
6	2500.0
7	3000.0
8	3500.0
9	4000.0
10	4500.0
11	5000.0
12	5500.0
13	6000.0
14	6500.0
15	7000.0
16	...

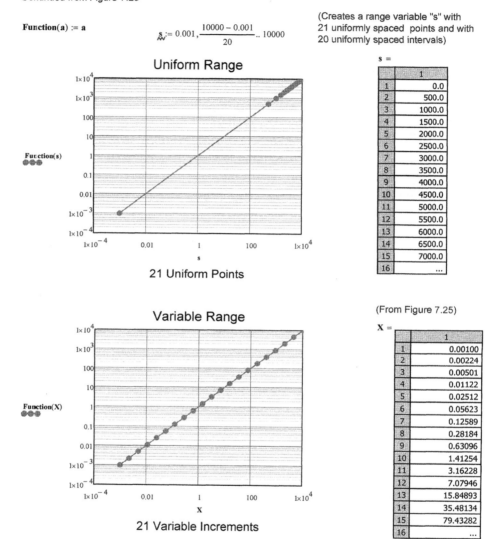

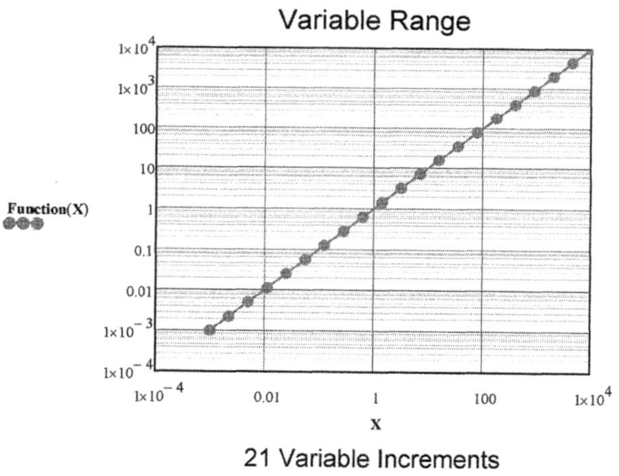

(From Figure 7.25)

X =

	1
1	0.00100
2	0.00224
3	0.00501
4	0.01122
5	0.02512
6	0.05623
7	0.12589
8	0.28184
9	0.63096
10	1.41254
11	3.16228
12	7.07946
13	15.84893
14	35.48134
15	79.43282
16	...

FIGURE 7.26 Plotting uniform versus variable ranges on a log plot

single value of a, and then the line doubles back to zero to plot the next value of a. In order to prevent this double-back line from plotting, double-click the plot and select **Draw** from the trace type on the Traces tab. Be sure that the upper limit on your x-axis does not exceed the maximum value in the range over which you are plotting. See Figure 7.27.

$\mathbf{Family(a,b) := a^2 \cdot cos(b)}$ $\quad a := 0, 1 .. 4 \quad$ $\mathbf{b} := 0, 0.1 .. 15$

Since this is actually only one trace, the lines double back for each value of "a" plotted.

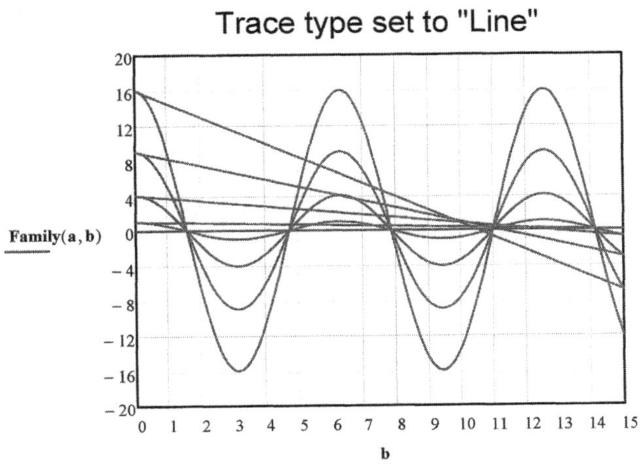

In order to prevent the extra lines, select "Draw" from the trace type.

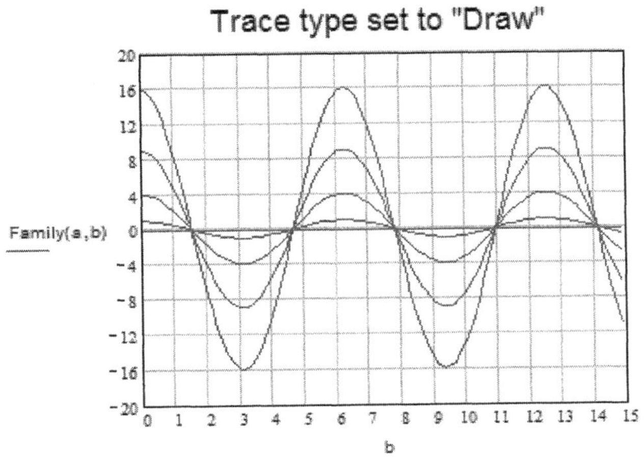

Note: Mathcad 14.0 version M020, has a bug, which prevents the "Draw" trace type from plotting.

FIGURE 7.27 Plotting a family of curves

3D PLOTTING

Mathcad can create many types of three-dimensional plots such as a surface plot, contour plot, 3D bar plot, 3D scatter plot, 3D bar chart, and vector field plot. These plot types are all accessible from the Graph tool bar, or you can choose the type of plot after hovering your mouse over **Graph** on the **Insert** menu. The primary difference between these plot types is how the plots are initially formatted. After a plot type is selected, you can change the plot formatting characteristics to display the data in other types of plots.

Three-dimensional plots can be created from a matrix, a function, or a set of vectors. The matrix row represents the x-axis, the matrix column represents the y-axis, and the matrix element represents the height above the x-y plane. A function uses the ranges of x and y to generate values for height above the x-y plane. You can also have three vectors representing the x, y, and z coordinates.

Once you have selected your plot type you should have a blank 3D plot operator on your worksheet.

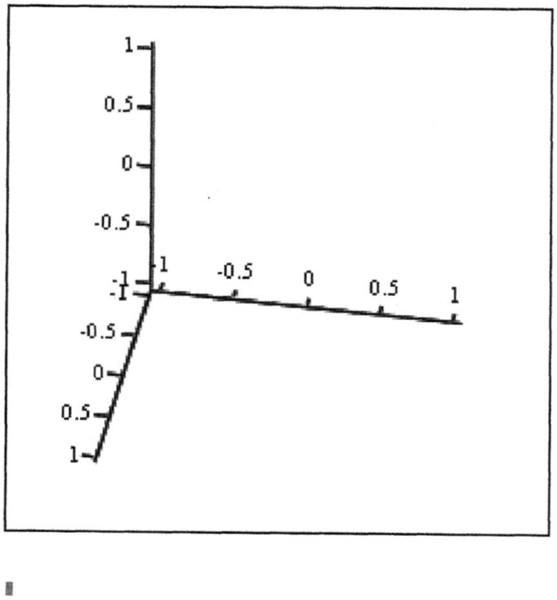

Notice the placeholder located in the lower left corner of the 3D plot operator. This is where you place the name of your data matrix, your function, or three data vectors enclosed in parentheses and separated by commas. You can plot multiple 3D plots by separating the data by commas. See Figure 7.28 for an example of four different plot types.

Create three data vectors.

$$X := \begin{pmatrix} 1 \\ 1 \\ 1 \\ 2 \\ 2 \\ 2 \\ 3 \\ 3 \\ 3 \end{pmatrix} \quad Y := \begin{pmatrix} 1 \\ 2 \\ 3 \\ 1 \\ 2 \\ 3 \\ 1 \\ 2 \\ 3 \end{pmatrix} \quad Z := \begin{pmatrix} 2 \\ 2.5 \\ 2 \\ 3 \\ 4 \\ 3 \\ 2 \\ 1.5 \\ 2 \end{pmatrix}$$

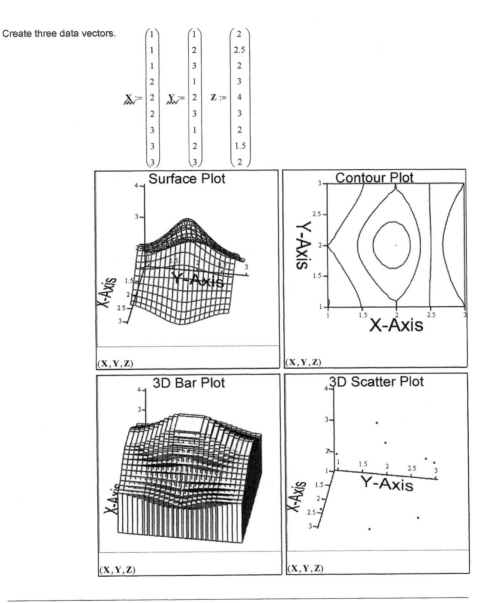

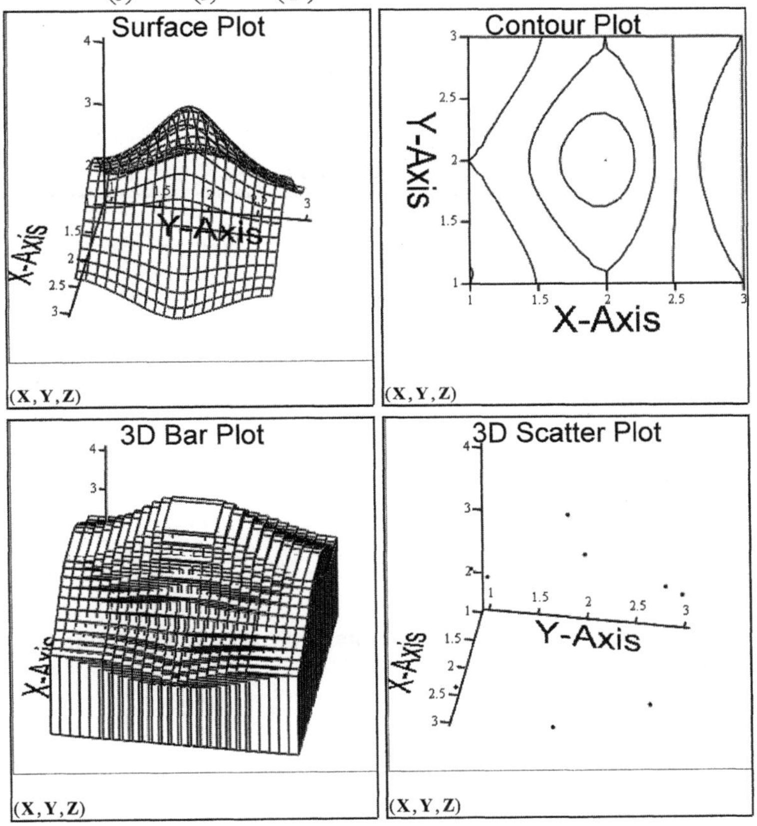

FIGURE 7.28 3D Plot types

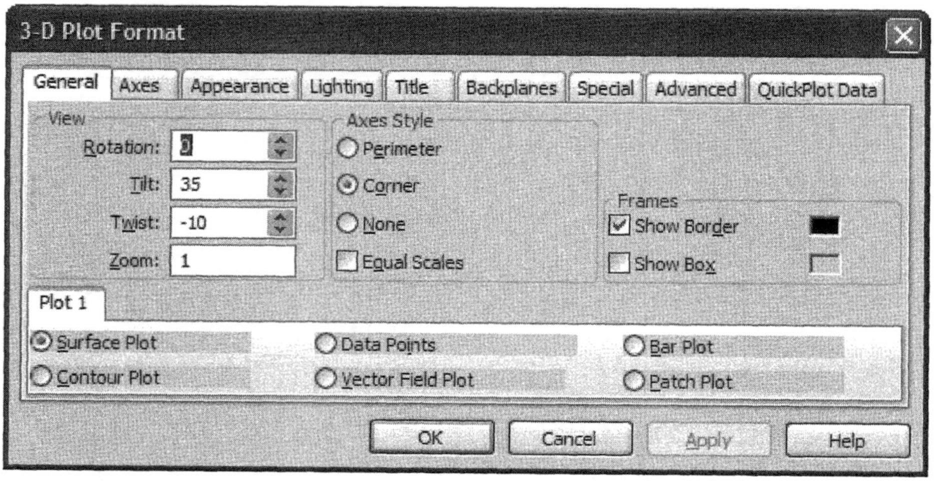

FIGURE 7.29 3-D Plot Format dialog box

Three-dimensional plots are controlled by the 3-D Plot Format dialog box. See Figure 7.29. You access this dialog box by double-clicking the plot operator. From this dialog box you can format such things as:

- Plot type
- Axis style
- Plot limits
- Axis labels
- Grid and axis colors
- Wireframe and contour options, such as number of contours and mesh points
- Color options and shading
- Lighting options
- Plot title and location
- Axis grids
- Transparency, perspective, fog, and printing quality

Figure 7.30 shows the same plots as from Figure 7.28, but with adjustments made to the plot formats from the 3-D Plot Format dialog box. To change the viewing angle, left-click and hold the left mouse button while moving the mouse. The plot rotation will change depending on where you first click.

The PTC Web site has an electronic book that shows how to create amazing 3D images. Go to http://www.ptc.com/appserver/mkt/products/resource/mathcad/index.jsp and click on the "Creating Amazing Images with Mathcad 14" link.

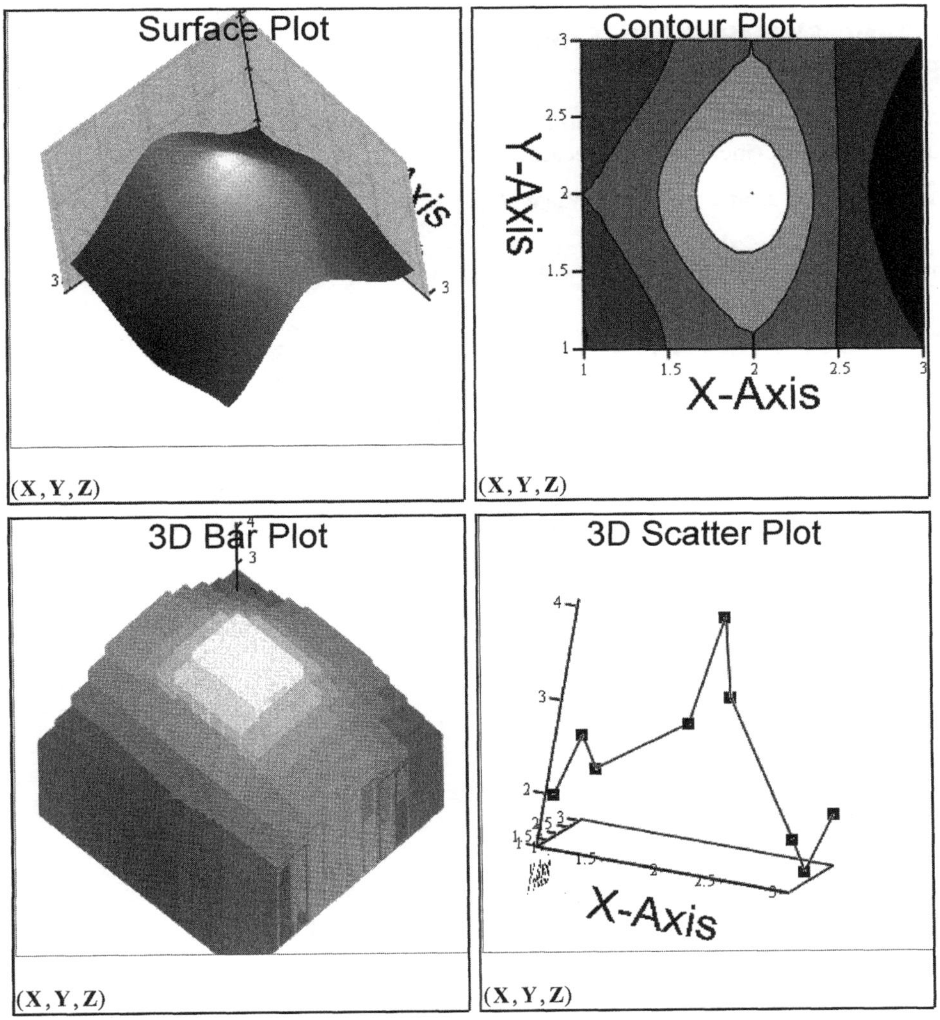

FIGURE 7.30 3D Plot types after formatting modifications

ENGINEERING EXAMPLE 7.1: SHEAR, MOMENT, AND DEFLECTION DIAGRAMS

For a simple span beam with uniform load (w) and length (L), plot the shear (kips), moment (ft*kips), and deflection (inches) of the beam.

The formula for shear at any point x is $\mathbf{V_x}(\mathbf{w},\mathbf{L},\mathbf{x}) := \mathbf{w}\cdot\left(\frac{\mathbf{L}}{2}-\mathbf{x}\right)$.

The formula for moment at any point x is $\mathbf{M_x}(\mathbf{w},\mathbf{L},\mathbf{x}) := \frac{\mathbf{w}\cdot\mathbf{x}}{2}(\mathbf{L}-\mathbf{x})$.

The formula for deflection at any point x is $\mathbf{\Delta_x}(\mathbf{w},\mathbf{L},\mathbf{x},\mathbf{E},\mathbf{I}) := \frac{\mathbf{w}\cdot\mathbf{x}}{24\cdot\mathbf{E}\cdot\mathbf{I}}\cdot(\mathbf{L}^3-2\cdot\mathbf{L}\cdot\mathbf{x}^2+\mathbf{x}^3)$.

$\mathbf{Length} := 20\mathbf{ft}$ $\mathbf{Load} := 2500\frac{\mathbf{lbf}}{\mathbf{ft}}$ $\mathbf{E} := 29000\mathbf{ksi}$ $\mathbf{I} := 428\cdot\mathbf{in}^4$

$\mathbf{x} := 0\mathbf{ft}, 1\mathbf{ft}\,..\,\mathbf{Length}$ Use a range variable to set the plot points.

Because the values of deflection are much smaller than the values for shear and moment, plot the deflection values on the right-hand side as a secondary y-axis.

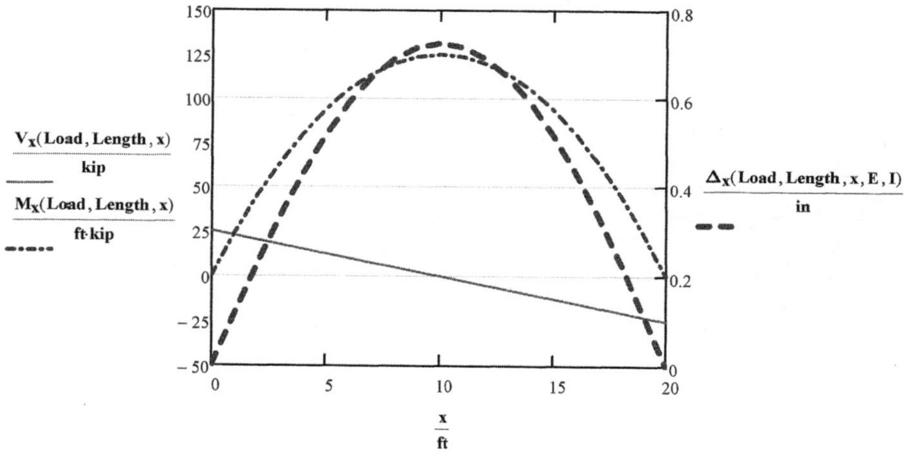

ENGINEERING EXAMPLE 7.2: DETERMINING THE FLOW PROPERTIES OF A CIRCULAR PIPE FLOWING PARTIALLY FULL

Mannings equation $\mathbf{Q} = \frac{1.49}{\mathbf{n}}\cdot\mathbf{A_f}\cdot\mathbf{R}^{\frac{2}{3}}\cdot\mathbf{S}^{\frac{1}{2}}$ is commonly used for computing the discharge in circular pipes flowing full or partially full. However, a difficulty in using it is computing the area of flow and wetted perimeter when the pipe is NOT either full or 1/2 full. This is because the geometric relationship between depth of flow and area of flow is not simple. There are several trigonometric formulas where the angle α can be used to compute the various characteristics required for flow problems. However, α is not generally known. We can get around this problem by computing all partial characteristics as a function of α and then plotting them against each other directly, as shown below.

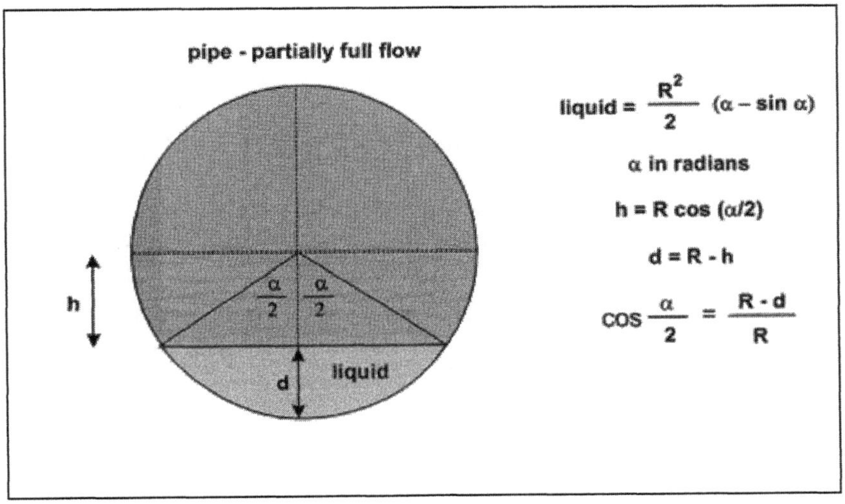

Create range variable of angle increments to be used in plots $\alpha := 0 \cdot \mathbf{rad}, .1 \cdot \mathbf{rad} .. 2 \cdot \pi \cdot \mathbf{rad}$

The trigonometric equations given above allow calculation of the area of flow and the depth of flow for any angle α from 0 to 2π radians. We can use these to compute the values needed for Manning's Equation:

Area of Flow, A_f

Wetted Perimeter, P_w

Hydraulic Radius, R_h (Af/Pw)

Write all indepent variables, A_f, R_h, d in terms of the angle, α

Area of flow $$A_f(\alpha, D) := \frac{\left(\frac{D}{2}\right)^2}{2} \cdot (\alpha - \sin(\alpha))$$

Wetted perimeter $$P_w(\alpha, D) := \left(\frac{D}{2} \cdot \alpha\right)$$

Hydraulic radius $$R_h(\alpha, D) := \frac{A_f(\alpha, D)}{P_w(\alpha, D)}$$ Hydraulic radius is the ratio of area of flow divided by the wetted perimeter.

$$h = \frac{D}{2} \cdot \cos\left(\frac{\alpha}{2}\right)$$

Depth of flow is radius minus h. $\text{Depth}(\alpha, D) := \left(\frac{D}{2} - \frac{D}{2} \cdot \cos\left(\frac{\alpha}{2}\right)\right)$

The equations below allow calculation of the flow rate and velocity for any pipe diameter and angle α.

Flow rate based on Manning's equation $Q(\alpha, D, n, S) := \frac{1.49}{n} \cdot A_f(\alpha, D) \cdot R_h(\alpha, D)^{\frac{2}{3}} \cdot S^{\frac{1}{2}}$

Velocity of flow based on Manning's equation $V(\alpha, D, n, S) := \frac{1.49}{n} \cdot R_h(\alpha, D)^{\frac{2}{3}} \cdot S^{\frac{1}{2}}$

Since all flow characteristics are in terms of the angle α, · use a parametric plot using α as the independent variable. In the graph below we can get the area of flow, wetted perimeter, and velocity as a function of the depth of flow in the pipe.

Plot values for n = 0.013 and S = 0.007 $n := 0.063$ $S := 0.007$ $D := 1\ \text{ft}$

Note: D can be any number.

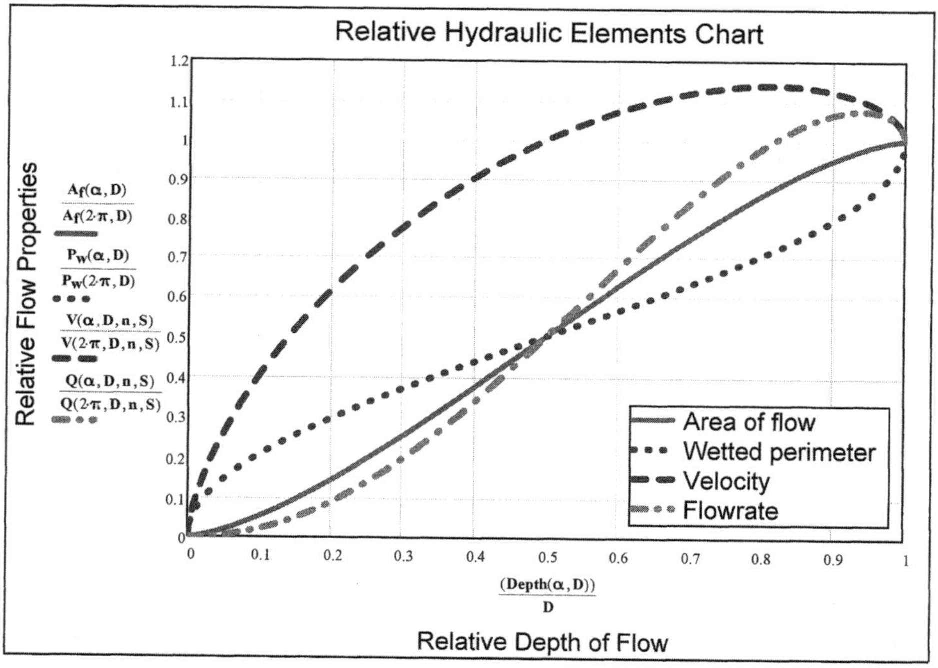

Now lets examine the actual flow velocity as a function of depth in two pipes, one 10" and one 24" in diameter.

$D_1 := 10 \cdot in$ $D_2 := 24 \cdot in$ $n = 0.0130$ $S = 0.0070$

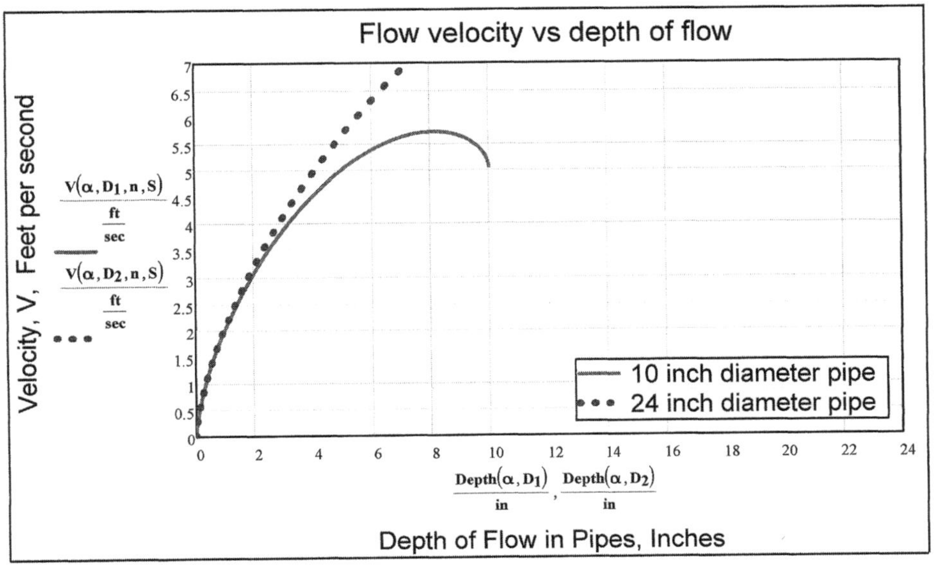

Notice that the maximum flow rate in a circular pipe does NOT occur when the pipe flows full. Note also that the velocity in the pipe actually drops as the depth of flow approaches the crown (top) of the pipe.

Using the above information, determine the diameter of pipe needed to carry 300 cfs when flowing 60% full at a slope of 0.001, with n = 0.013.

Input Parameters

Pipe slope	$S := 0.001$
Manning's coefficient	$n := 0.013$
Pipe diameter	$D_3 := 9.4 \cdot ft$
Depth when flowing 0.6 full	$.6D_3 = 5.64 \cdot ft$

In Chapter 10 we introduce Mathcad's solving functions, which can solve this solution directly. For this solution, we plot the flow and set plot markers at 300 cfs and 0.6*D. We then vary the pipe diameter until the line passes through the markers.

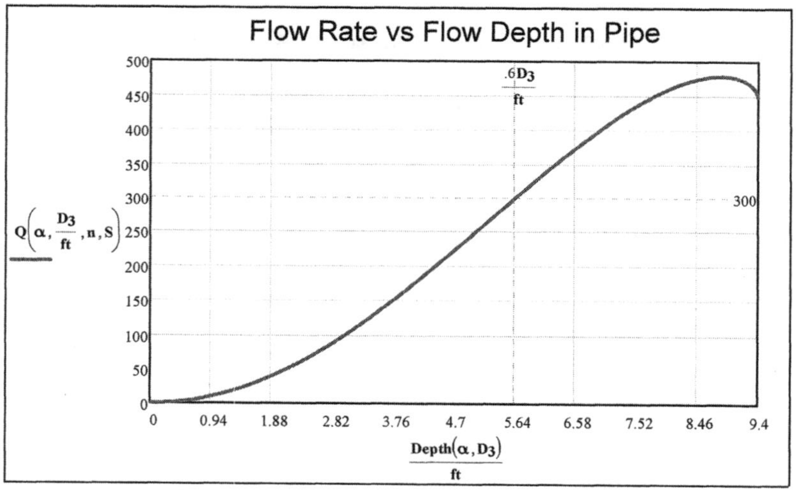

As can be seen from the plot above a $\mathbf{D_3} = 9.40 \cdot \mathbf{ft}$ diameter pipe delivers 300 cfs when the depth of flow is $0.6 \cdot \mathbf{D_3} = 5.64 \cdot \mathbf{ft}$

Verify:
For depth of 0.6 depth, the α is 3.551.

$$\mathbf{Depth}(3.551, \mathbf{D_3}) = 5.66 \cdot \mathbf{ft} \qquad \frac{5.66}{.6} = 9.43 \qquad \mathbf{Q}\left(3.551, \frac{\mathbf{D_3}}{\mathbf{ft}}, \mathbf{n}, \mathbf{S}\right) = 299.95$$

SUMMARY

Plots are useful, easy-to-use tools. They can help visualize equations or functions. They can also be used to help solve equations.

In Chapter 7 we:

- Showed how to create simple X-Y QuickPlots and Polar QuickPlots.
- Showed how to plot functions and expressions.
- Discussed using range variables to set range limits.
- Showed how to plot multiple functions on the same plot.
- Showed how to plot functions on the y-axis and on a secondary y-axis.
- Showed how to plot using units, and how to change the displayed units.

- Explained how to format plots with grid lines, numbers, labels, and titles.
- Explained how to change the trace colors, line type, and line weight.
- Discussed plotting data information from range variables, data vectors, and data matrices.
- Demonstrated how to get numeric readout of plotted traces.
- Explained how plots can be used to help solve systems of equations.
- Used the ***logspace*** function to create logarithmically spaced points.
- Introduced 3D plotting.

PRACTICE

Additional problems and applications can be found on the companion site: www.elsevierdirect.com/9780123747839.

1. Create separate simple X-Y QuickPlots of the following equations. Use expressions rather than functions on the y-axis.

   a. x^2
   b. $x^3 + 2x^2 + 3x - 10$
   c. sin(x)

2. Create separate Polar QuickPlots of the following equations. Use expressions rather than functions on the radial axis.

   a. x/2
   b. cos(6x)
   c. tan(x)

3. Create functions for the expressions in Exercises 1 and 2 and plot these functions.

4. Create two range variables for each of the previous functions. One range variable should have a small increment; the other should have a large increment. Plot these functions using each range variable. Use the Format dialog box and the Traces tab to make each plot look different from the others.

5. The formula to calculate the bending moment at point x in a beam (with uniform loading) is M = (1\2*w*x)*(L - x), where w is in force/length, L is total length of beam, and x is distance from one end of the beam. Create a plot with distance x on the x-axis (from zero to L), and moment on the y-axis. Use w = 2 N/m and L = 10 m. Moment should be displayed as N*m. Provide a title and axis labels.

6. Plot the following data points. Use a range variable for the x-axis. Use a solid box as the symbol. Connect the data points with a dashed line.

1	19.1
2	29.5
3	40.3
4	52.4
5	59.3
6	70.5

7. Plot the following data points. Use a blue solid circle as the symbol. Do not connect the data points.

1.2	2.4
2.3	3.3
4.5	5.3
5.2	6.3
4.5	4.6
5.5	6.4

8. Plot the following equations and use the Trace dialog box to find approximate solutions where the plots intersect: $y_1(x) = 2x^2 + 3x - 10$, $y_2(x) = -x^2 + 2x + 20$.

9. Write a function to describe the vertical motion and a function to describe the horizontal motion of a projectile fired at 700 ft/s with a 35-degree inclination from the horizontal. Each function should be a function of time. Create a parametric plot with the following:

   a. Use a range variable to set the range of the plot. Use a range from 0 to 20 with an increment of 1.

   b. Create a parametric plot with horizontal motion on the x-axis and the vertical motion on the y-axis. Use the range variable for the argument of both functions.

   c. Use units in the functions and the plots.
   Hint: Remember to multiply the function argument by seconds.

10. Copy the plot from Exercise 8 and plot in terms of meters instead of feet.

CHAPTER

Simple Logic Programming

8

Scientific and engineering calculations must have a way to logically reach conclusions based on data calculated. Mathcad programming allows you to write logic programs. Mathcad calls this "programming," but it is essentially a better way to use the *if* function. These "programs" allow Mathcad to choose a result based on specific parameters. Programs can be very complex, and can be used for many things. This chapter focuses solely on simple logic programming. The more complex programs are discussed in Chapter 11, "Advanced Programming."

Chapter 8 will:

- Provide several simple Mathcad examples to illustrate the concept of logic programming.
- List the steps necessary to create a logic program.
- Describe the logic Mathcad uses to arrive at a conclusion.
- Warn the user about violating this logic, which could cause Mathcad to make an inaccurate conclusion.
- Reveal new ways of creating logic programs that are not provided in the Mathcad documentation.
- Show how to use a logic program to draw and display conclusions.

INTRODUCTION TO THE PROGRAMMING TOOLBAR

The purpose of this chapter is to get you comfortable with the concept of Mathcad programming. We will use simple examples. You will soon see that you do not need to be a computer programmer to use the Mathcad programming features. You do not need to learn complex programming commands. All the operators you need to use are contained on the Programming toolbar.

To open the Programming toolbar, hover your mouse over **Toolbars** on the **View** menu and click **Programming**. You may also click the Programming toolbar icon on the Math toolbar. The Programming toolbar contains the operators you will use when writing a Mathcad program.

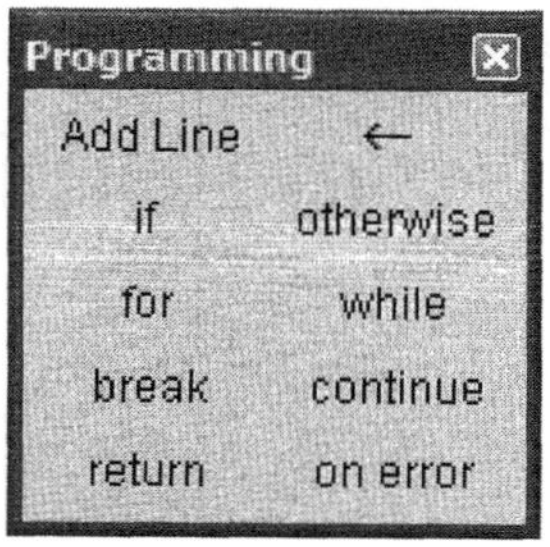

This chapter will focus on the following operators: ***Add Line, if, otherwise***, and ***return***. The remaining operators will be discussed in Chapter 11, "Advanced Programming."

One important thing to remember when using Mathcad programming is that all the programming operators must be inserted using the Programming toolbar, or by using keyboard shortcuts. You cannot type "if," "otherwise," or "return." They are operators and must be inserted.

CREATING A SIMPLE PROGRAM

You begin a Mathcad program by clicking ***Add Line***. This inserts a vertical line with two placeholders.

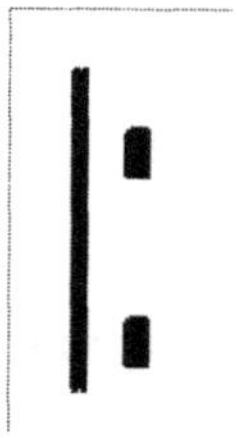

This is called the Programming operator. We will place a conditional statement in the top placeholder. In the bottom placeholder, we will place a statement about what to do if the conditional statement is false.

The following example is similar to Figure 6.22, which used the *if* function.

InputLength := 20**ft**

DesignLength := | 25**ft** **if** **InputLength** < 25**ft**
| **InputLength** **otherwise**

DesignLength = 25.00 · **ft**

To create this program follow these steps:

1. Open the Programming toolbar.
2. Type the variable name followed by the colon.
3. Click the **Add Line** button on the Programming toolbar.
4. Select the top placeholder and click the **if** button on the Programming toolbar.
5. Select the bottom placeholder and click the **otherwise** button on the Programming toolbar.
6. Fill-in the placeholders.

DesignLength := | ▪
| ▪

DesignLength := | ▪ **if** ▪
| ▪ **otherwise**

Remember. . . When creating a program, typing "if" is not the same as using the **if** operator on the programming toolbar.

FIGURE 8.1 Simple program

Let's look at a two simple examples. See Figures 8.1 and 8.2.

The logic used by the program in Figure 8.1 is, "Do this if this statement is true, otherwise do this." The ***if*** function just as easily could have been used in place of the program. The benefit of using the program operator comes when there are multiple ***if*** statements.

When using multiple ***if*** statements in a program, it is important to understand how Mathcad treats the ***if*** statements. Mathcad evaluates every ***if*** statement. If a statement is false, the statement is not executed, but Mathcad proceeds to the next line. If the statement is true, the statement is executed, and Mathcad proceeds to the next line. The ***otherwise*** statement is executed only if all previous ***if*** statements are false. If there are more than one true ***if*** statements, Mathcad returns the last true ***if*** statement. This is important to understand, or you may get incorrect results. Let's look at a few examples. See Figures 8.3 and 8.4.

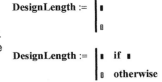

Use three methods to check the program for different lengths.

Refer to Chapter 5 for a discussion of using arrays with equations and functions.

Method 1 -- Use and expression with a range variable and vector to get multiple results.

$$i := 1, 2 .. 3$$

$$\text{InputLength} := \begin{pmatrix} 5\text{ft} \\ 25\text{ft} \\ 50\text{ft} \end{pmatrix}$$

Create and input vector

$$\text{DesignLength}_i := \begin{vmatrix} 25\text{ft} \ \text{if} \ \text{InputLength}_i < 25\text{ft} \\ \text{InputLength}_i \ \ \text{otherwise} \end{vmatrix}$$

Define the expression.
Note how both sides of the assignment operator must have the range variable i, and that i starts at ORIGIN and increments by one.

$$\text{DesignLength} = \begin{pmatrix} 25.00 \\ 25.00 \\ 50.00 \end{pmatrix} \cdot \text{ft}$$

OK
OK
OK

Method 2 -- Use a function. Use different arguments in the function.

$$\text{NewLength}(x) := \begin{vmatrix} 25\text{ft} \ \text{if} \ x < 25\text{ft} \\ x \ \ \text{otherwise} \end{vmatrix}$$

$$\text{NewLength}(5\text{ft}) = 25.00 \cdot \text{ft}$$

$$\text{NewLength}(25\text{ft}) = 25.00 \cdot \text{ft}$$

$$\text{NewLength}(50\text{ft}) = 50.00 \cdot \text{ft}$$

Method 3 -- Use a function with a vector as the argument.

$$\text{ResultLength} := \overrightarrow{\text{NewLength}(\text{InputLength})}$$

Note that the vectorize operator CTRL+MINUS needs to be added to the function so that Mathcad will recognize the vector with the function.

$$\text{ResultLength} = \begin{pmatrix} 25.00 \\ 25.00 \\ 50.00 \end{pmatrix} \cdot \text{ft}$$

FIGURE 8.2 Check the program in Figure 8.1 for different lengths

RETURN OPERATOR

The ***return*** operator is used in conjunction with an ***if*** statement. When the ***if*** statement is true, the ***return*** operator tells Mathcad to stop the program and return the value, rather than proceed to the next line in the program. Figure 8.5 shows how to use the ***return*** statement to make the examples from Figure 8.4 more intuitive.

Assume a previous calculation returned a calculated factor. The minimum value of the factor should be 0.5. The maximum value should be 2.0. Let's write a program as a user-defined function (using the calculated factor as an argument) so that Mathcad will choose a final factor between 0.5 and 2.0.

$0.5 < Factor < 2.0$

Let's look at two examples of a program.
This first example creates inaccurate results.

$$\mathbf{Factor(x)} := \left| \begin{array}{l} 2.0 \;\; \mathbf{if} \;\; \mathbf{x} > 2.0 \\ \mathbf{x} \;\; \mathbf{if} \;\; \mathbf{x} > 0.5 \\ 0.5 \;\; \mathbf{otherwise} \end{array} \right.$$

$\mathbf{F_1} := \mathbf{Factor}(0.25)$ $\quad \mathbf{F_1} = 0.50$ Correct

$\mathbf{F_2} := \mathbf{Factor}(1.0)$ $\quad \mathbf{F_2} = 1.00$ Correct

$\mathbf{F_3} := \mathbf{Factor}(3.0)$ $\quad \mathbf{F_3} = 3.00$ Incorrect (The correct result should be 2.0)

Why did F_3 above example give incorrect results?
It returned incorrect results because Mathcad evaluates every **if** statement, and executes it if it is true. The final true **if** statement is returned. For F_3, the final true statement that is encountered is "if x is greater than 0.5 then return x." Since 3>0.5, Mathcad returned 3.

Rewrite the program so that it creates correct results. Do this by ensuring that the first two statements cannot both be true.

$$\mathbf{RevisedFactor(x)} := \left| \begin{array}{l} 0.5 \;\; \mathbf{if} \;\; \mathbf{x} < 0.5 \\ 2.0 \;\; \mathbf{if} \;\; \mathbf{x} > 2.0 \\ \mathbf{x} \;\; \mathbf{otherwise} \end{array} \right.$$

$\mathbf{F_4} := \mathbf{RevisedFactor}(0.25)$ $\quad \mathbf{F_4} = 0.50$ Correct

$\mathbf{F_5} := \mathbf{RevisedFactor}(1.0)$ $\quad \mathbf{F_5} = 1.00$ Correct

$\mathbf{F_6} := \mathbf{RevisedFactor}(3.0)$ $\quad \mathbf{F_6} = 2.00$ Correct

FIGURE 8.3 Using multiple *if* statements

In wood design there is a depth factor that needs to be applied to different depths of wood joists. Let's look at two examples of a conditional program to determine the proper depth factor.
This first example creates incorrect results.

$$\mathbf{Factor_1(d)} := \left|\begin{array}{l} 1.5 \ \ \mathbf{if}\ \ \mathbf{d} \le 3.5\mathbf{in} \\ 1.4 \ \ \mathbf{if}\ \ \mathbf{d} \le 4.5\mathbf{in} \\ 1.3 \ \ \mathbf{if}\ \ \mathbf{d} \le 5.5\mathbf{in} \\ 1.2 \ \ \mathbf{if}\ \ \mathbf{d} \le 7.25\mathbf{in} \\ 1.1 \ \ \mathbf{if}\ \ \mathbf{d} \le 9.25\mathbf{in} \\ 1.0 \ \ \mathbf{if}\ \ \mathbf{d} \le 11.25\mathbf{in} \\ 0.9 \ \ \mathbf{otherwise} \end{array}\right.$$

$\mathbf{Factor_1}(2.5\mathbf{in}) = 1.00$	Incorrect (The result should be 1.5)
$\mathbf{Factor_1}(3.5\mathbf{in}) = 1.00$	Incorrect (The result should be 1.5)
$\mathbf{Factor_1}(4.5\mathbf{in}) = 1.00$	Incorrect (The result should be 1.4)
$\mathbf{Factor_1}(5.5\mathbf{in}) = 1.00$	Incorrect (The result should be 1.3)
$\mathbf{Factor_1}(7.25\mathbf{in}) = 1.00$	Incorrect (The result should be 1.2)
$\mathbf{Factor_1}(9.25\mathbf{in}) = 1.00$	Incorrect (The result should be 1.1)
$\mathbf{Factor_1}(11.25\mathbf{in}) = 1.00$	Correct
$\mathbf{Factor_1}(13.25\mathbf{in}) = 0.90$	Correct

Why did the above example give so many incorrect results?
It returned incorrect results because Mathcad evaluates every **if** statement and executes it if it is true. The final result is the last true statement that is encountered. The last true statement that is encountered is d <=11.25 in. Most of the depths were less than 11.25 in., so 1.0 is the value assigned to $Factor_1$.

If we reorganize the program, so that false statements are encountered as the depth decreases, we get Correct results.

$$\mathbf{RevisedFactor_2(d)} := \left|\begin{array}{l} 1.0 \ \ \mathbf{if}\ \ \mathbf{d} \le 11.25\mathbf{in} \\ 1.1 \ \ \mathbf{if}\ \ \mathbf{d} \le 9.25\mathbf{in} \\ 1.2 \ \ \mathbf{if}\ \ \mathbf{d} \le 7.25\mathbf{in} \\ 1.3 \ \ \mathbf{if}\ \ \mathbf{d} \le 5.5\mathbf{in} \\ 1.4 \ \ \mathbf{if}\ \ \mathbf{d} \le 4.5\mathbf{in} \\ 1.5 \ \ \mathbf{if}\ \ \mathbf{d} \le 3.5\mathbf{in} \\ 0.9 \ \ \mathbf{otherwise} \end{array}\right.$$

$\mathbf{RevisedFactor_2}(2.5\mathbf{in}) = 1.50$	Correct
$\mathbf{RevisedFactor_2}(3.5\mathbf{in}) = 1.50$	Correct
$\mathbf{RevisedFactor_2}(4.5\mathbf{in}) = 1.40$	Correct
$\mathbf{RevisedFactor_2}(5.5\mathbf{in}) = 1.30$	Correct
$\mathbf{RevisedFactor_2}(7.25\mathbf{in}) = 1.20$	Correct
$\mathbf{RevisedFactor_2}(9.25\mathbf{in}) = 1.10$	Correct
$\mathbf{RevisedFactor_2}(11.25\mathbf{in}) = 1.00$	Correct
$\mathbf{RevisedFactor_2}(13.25\mathbf{in}) = 0.90$	Correct

FIGURE 8.4 Multiple *if* statements

BOOLEAN OPERATORS

Boolean operators are very useful when writing logical Mathcad programs. The Boolean operators are located on the Boolean toolbar. To open the Boolean toolbar, hover your mouse over **Toolbars** on the **Format** menu and select **Boolean**. You may also click the button on the Math toolbar.

The return operator stops the program. In conjunction with an **if** statement, it can be used to stop the program if the **if** statement is true. If the **if** statement is false, then Mathcad proceeds to the next line.

The first example from Figure 8.4 could be written in the following manner using the return operator:

$$\mathbf{Factor_2(d)} := \left| \begin{array}{l} \mathbf{return}\ 1.5\ \ \mathbf{if}\ \ \mathbf{d} \le 3.5\mathbf{in} \\ \mathbf{return}\ 1.4\ \ \mathbf{if}\ \ \mathbf{d} \le 4.5\mathbf{in} \\ \mathbf{return}\ 1.3\ \ \mathbf{if}\ \ \mathbf{d} \le 5.5\mathbf{in} \\ \mathbf{return}\ 1.2\ \ \mathbf{if}\ \ \mathbf{d} \le 7.25\mathbf{in} \\ \mathbf{return}\ 1.1\ \ \mathbf{if}\ \ \mathbf{d} \le 9.25\mathbf{in} \\ \mathbf{return}\ 1.0\ \ \mathbf{if}\ \ \mathbf{d} \le 11.25\mathbf{in} \\ 0.9\ \ \mathbf{otherwise} \end{array} \right.$$

$\mathbf{Factor_2}(2.5\mathbf{in}) = 1.50$	Correct
$\mathbf{Factor_2}(3.5\mathbf{in}) = 1.50$	Correct
$\mathbf{Factor_2}(4.5\mathbf{in}) = 1.40$	Correct
$\mathbf{Factor_2}(5.5\mathbf{in}) = 1.30$	Correct
$\mathbf{Factor_2}(7.25\mathbf{in}) = 1.20$	Correct
$\mathbf{Factor_2}(9.25\mathbf{in}) = 1.10$	Correct
$\mathbf{Factor_2}(11.25\mathbf{in}) = 1.00$	Correct
$\mathbf{Factor_2}(13.25\mathbf{in}) = 0.90$	Correct

Note: A return could have been placed in front of the last line to make it read more consistent, but it is not required.

FIGURE 8.5 Return operator

To use the Boolean operators you need to use the buttons on the Boolean toolbar

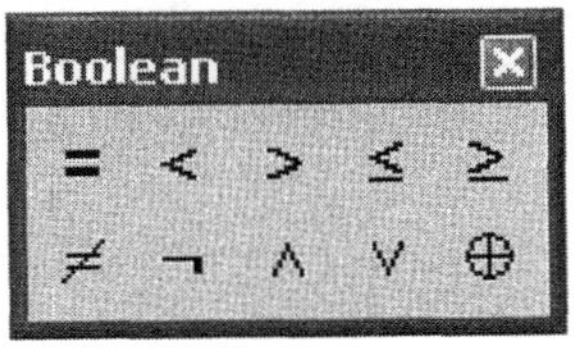

or use keyboard shortcuts. The exception is the ***Greater Than*** and ***Less Than*** operators, which can be entered from the keyboard.

The first six icons on the toolbar are obvious. The last four are not as obvious.

The ¬ button is Boolean ***Not***. This is true if x is zero, and false if x is nonzero. See Figure 8.6 for an example.

The ∧ button is Boolean ***And***. This is true if both statements are true. See Figure 8.7 for an example.

The ∨ button is Boolean ***Or***. This is true if either statement is true.

Boolean **Not** operator returns true (1) if the value is zero. It returns false (0) if the value is non-zero

$\neg 1 = 0.00 \qquad \neg 0 = 1.00$

Example using the **Not** operator.

$$\mathbf{AA}(\mathbf{x}) := \left| \begin{array}{ll} \text{"X is zero"} & \mathbf{if}\ \neg\mathbf{x} \\ \text{"X is not zero"} & \mathbf{otherwise} \end{array} \right.$$

$\mathbf{AA}(2) = \text{"X is not zero"}$ $\qquad$ $\mathbf{AA}(0) = \text{"X is zero"}$

This is the same result as using **if** x=0.

$$\mathbf{AB}(\mathbf{x}) := \left| \begin{array}{ll} \text{"X is zero"} & \mathbf{if}\ \mathbf{x} = 0 \\ \text{"X is not zero"} & \mathbf{otherwise} \end{array} \right.$$

$\mathbf{AB}(2) = \text{"X is not zero"}$ $\qquad$ $\mathbf{AB}(0) = \text{"X is zero"}$

FIGURE 8.6 Boolean *Not*

Boolean **And** operator returns true if both statements are true.

$$\mathbf{BB}(\mathbf{x},\mathbf{y}) := \left| \begin{array}{ll} \text{"This And statement is TRUE"} & \mathbf{if}\ \mathbf{x} > 0 \wedge \mathbf{y} > 0 \\ \text{"This And statement is false"} & \mathbf{otherwise} \end{array} \right.$$

$\mathbf{BB}(5,4) = \text{"This And statement is TRUE"}$ $\qquad$ $\mathbf{BB}(5,0) = \text{"This And statement is false"}$

$\mathbf{BB}(0,3) = \text{"This And statement is false"}$ $\qquad$ $\mathbf{BB}(0,0) = \text{"This And statement is false"}$

FIGURE 8.7 Boolean *And*

Warning: If the first statement is true, then the second statement is not evaluated. Therefore, the second statement may contain an error that is not initially detected because Mathcad stopped evaluating after the first statement. See Figure 8.8 for examples.

The ⊕ button is Boolean ***Xor*** (exclusive Or). This is true if either the first or the second statement is nonzero, but not both. Thus if both statements are nonzero, the result is false. See Figure 8.9 for an example.

Boolean **OR** operator returns true if either statement is true.

$$\mathbf{CC(x,y)} := \left| \begin{array}{ll} \text{"This Or statement is TRUE"} & \textbf{if } \mathbf{x} > 0 \vee \mathbf{y} > 0 \\ \text{"This Or statement is false"} & \textbf{otherwise} \end{array} \right.$$

CC(5,4) = "This Or statement is TRUE"   CC(0,4) = "This Or statement is TRUE"

CC(5,0) = "This Or statement is TRUE"   CC(0,0) = "This Or statement is false"

When using the **OR** operator, if the first statement is true, the second statement will not be checked. This could prevent an error from being detected.

$$\mathbf{DD(x,y)} := \left| \begin{array}{ll} \text{"This Or statement is TRUE"} & \textbf{if } \mathbf{x} > 0 \vee \frac{\mathbf{y}}{0} > 0 \\ \text{"This Or statement is false"} & \textbf{otherwise} \end{array} \right.$$

This divide by zero error is not caught, unless the first statement is false.

DD(5,4) = "This Or statement is TRUE"   DD(0,4) = ▪

DD(0,4) = ▪▪

Divide by zero.

DD(5,0) = "This Or statement is TRUE"   DD(0,0) = ▪

FIGURE 8.8 Boolean *Or*

Boolean **Xor** operator returns true if one statement is true, but not both.

$$\mathbf{EE(x,y)} := \left| \begin{array}{ll} \text{"The xor statement is TRUE"} & \textbf{if } \mathbf{x} > 0 \oplus \mathbf{y} > 0 \\ \text{"The xor statement is false"} & \textbf{otherwise} \end{array} \right.$$

EE(5,6) = "The xor statement is false"   False because one of the statements is not false.

EE(0,6) = "The xor statement is TRUE"   True because one, but not both statements is true.

EE(5,0) = "The xor statement is TRUE"   True because one, but not both statements is true.

EE(0,0) = "The xor statement is false"   False because one of the statements is not false.

FIGURE 8.9 Boolean *Xor*

ADDING LINES TO A PROGRAM

It is possible to add nested programs inside a program. Creative use of nested programs sometimes makes it easier to create conditional programs.

If you are inside a program, Mathcad will add a new programming line or a new placeholder when you click the Add Line button. The location of the new line or placeholder depends where the vertical editing line is located and what information is selected. It is much easier to illustrate the behavior than try to describe it. In the following examples, notice what is selected and where the vertical editing line is located when the Add Line button is clicked.

1.

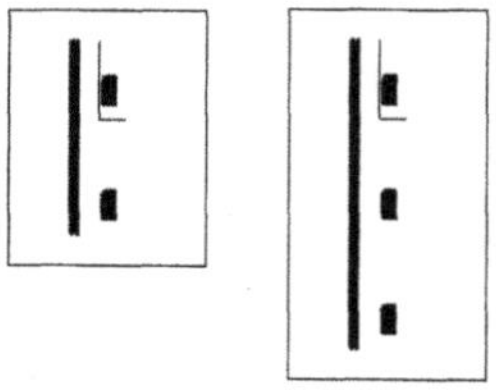

2.

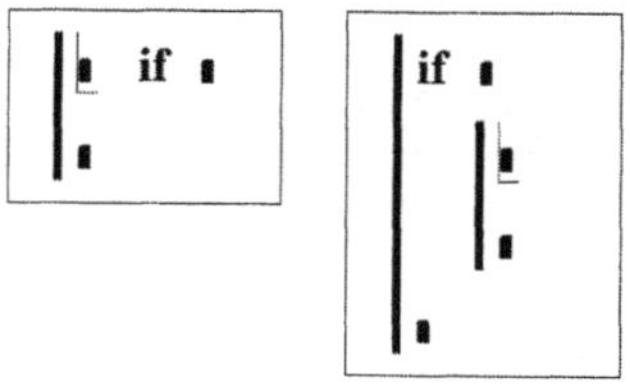

3.

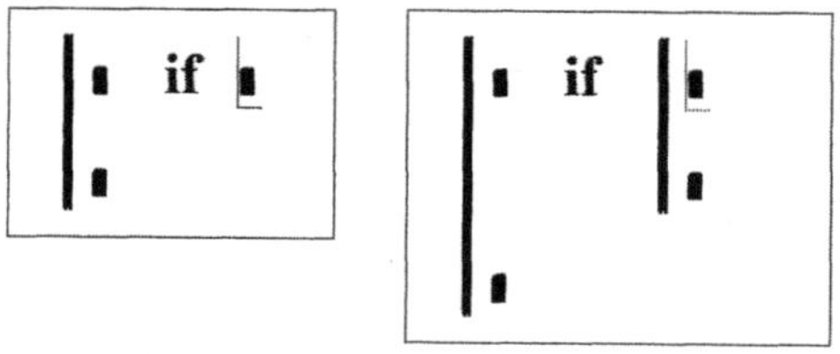

4.

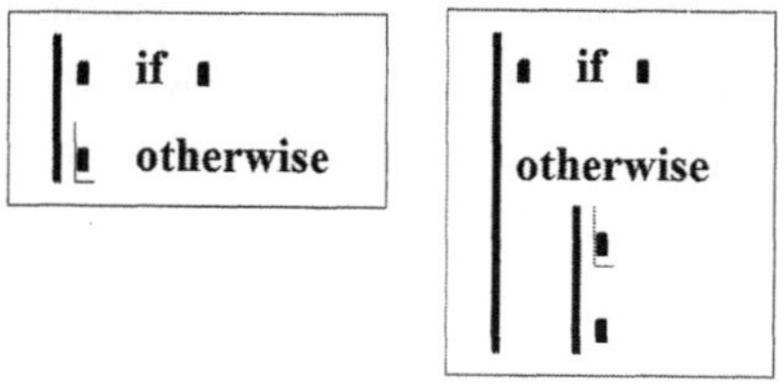

5.

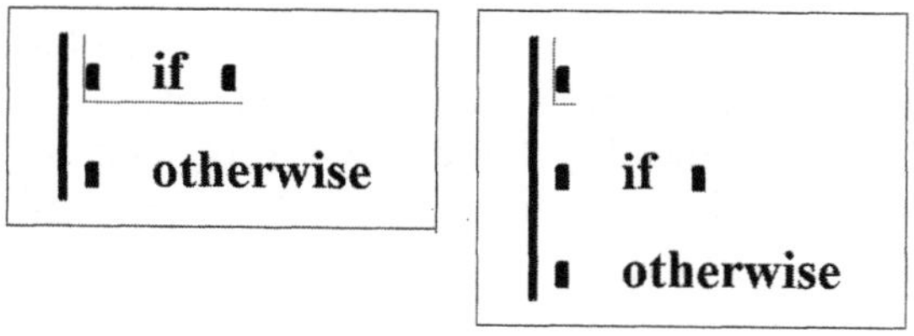

6. 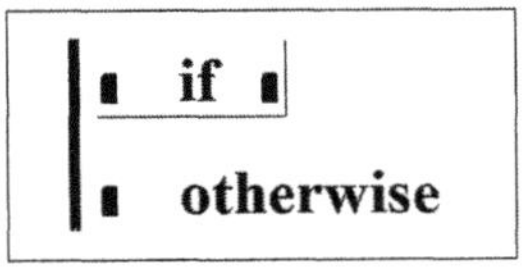 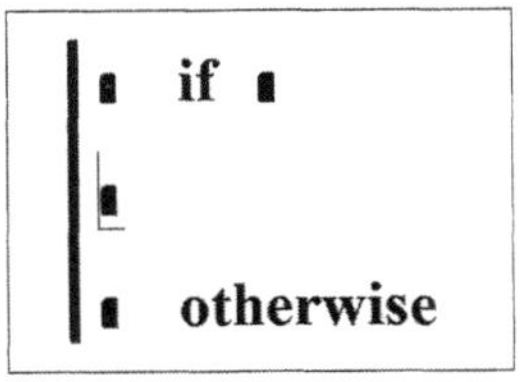

7. 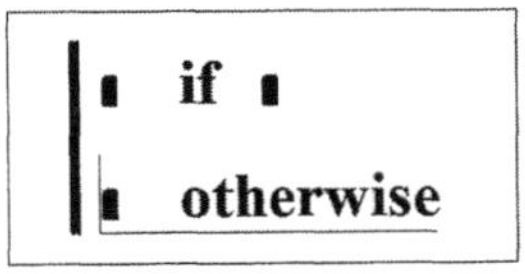 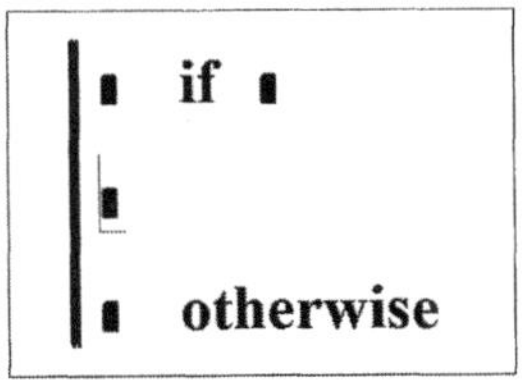

8. 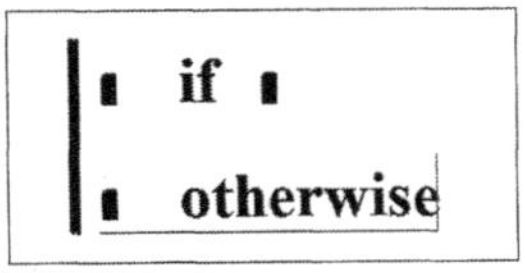 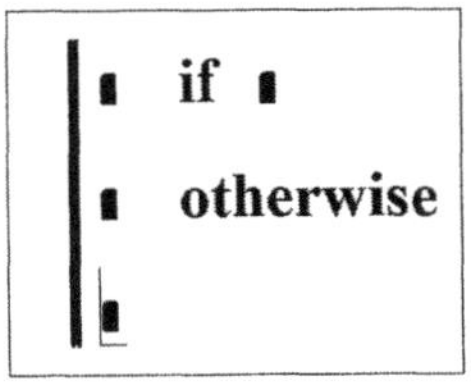

Now, let's illustrate how these different forms of the program work. In Figures 8.10 through 8.13 we try to arrive at consistent results using different forms of the programming lines. Notice how each program is a little different from the

---

$i := 1, 2 .. 4$

$$\mathbf{x} := \begin{pmatrix} 3 \\ 4 \\ -7 \\ -10 \end{pmatrix} \qquad \mathbf{y} := \begin{pmatrix} 2 \\ -1 \\ 3 \\ -2 \end{pmatrix}$$

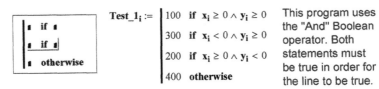

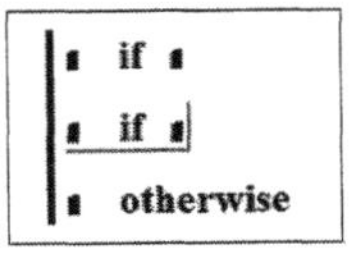

$$\mathbf{Test\_1_i} := \begin{vmatrix} 100 & \textbf{if } \mathbf{x_i} \ge 0 \wedge \mathbf{y_i} \ge 0 \\ 300 & \textbf{if } \mathbf{x_i} < 0 \wedge \mathbf{y_i} \ge 0 \\ 200 & \textbf{if } \mathbf{x_i} \ge 0 \wedge \mathbf{y_i} < 0 \\ 400 & \textbf{otherwise} \end{vmatrix}$$

This program uses the "And" Boolean operator. Both statements must be true in order for the line to be true.

Test_1$_i$ =

100.00
200.00
300.00
400.00

FIGURE 8.10 Programming Form 1

$$\mathbf{x} = \begin{pmatrix} 3.00 \\ 4.00 \\ -7.00 \\ -10.00 \end{pmatrix} \qquad \mathbf{y} = \begin{pmatrix} 2.00 \\ -1.00 \\ 3.00 \\ -2.00 \end{pmatrix}$$

$$\left|\begin{array}{l} \blacksquare \ \textbf{if} \ \blacksquare \\ \blacksquare \ \textbf{otherwise} \end{array}\right.$$

$$\text{Test\_2}_i := \left|\begin{array}{l} \textbf{if } x_i \ge 0 \\ \quad \left|\begin{array}{l} 100 \ \textbf{ if } y_i \ge 0 \\ 200 \ \textbf{ otherwise} \end{array}\right. \\ \textbf{if } x_i < 0 \\ \quad \left|\begin{array}{l} 300 \ \textbf{ if } y_i \ge 0 \\ 400 \ \textbf{ otherwise} \end{array}\right. \end{array}\right.$$

This form of the program allows you to have an additional conditional statement branching off of the first conditional statement.

$\text{Test\_2}_i =$	$\text{Test\_1}_i =$
100.00	100.00
200.00	200.00
300.00	300.00
400.00	400.00

FIGURE 8.11 Programming Form 2

$$\mathbf{x} = \begin{pmatrix} 3.00 \\ 4.00 \\ -7.00 \\ -10.00 \end{pmatrix} \qquad \mathbf{y} = \begin{pmatrix} 2.00 \\ -1.00 \\ 3.00 \\ -2.00 \end{pmatrix}$$

$$\left|\begin{array}{l} \blacksquare \ \textbf{if} \ \blacksquare \\ \blacksquare \ \textbf{otherwise} \end{array}\right.$$

$$\text{Test\_3a}_i := \left|\begin{array}{l} 100 \ \textbf{ if } \left|\begin{array}{l} \text{"This line is a comment line"} \\ x_i \ge 0 \wedge y_i \ge 0 \end{array}\right. \\ 200 \ \textbf{ if } \left|\begin{array}{l} \text{"Use this line for text"} \\ x_i \ge 0 \wedge y_i < 0 \end{array}\right. \\ 300 \ \textbf{ if } \left|\begin{array}{l} \text{"Add info about your program here"} \\ x_i < 0 \wedge y_i \ge 0 \end{array}\right. \\ 400 \ \textbf{ otherwise} \end{array}\right.$$

This form of the program is tricky. Essentially the **if** statement has two condition lines. Regardless of the first answer, the second condition will always be evaluated. If the second statement is true, then it doesn't matter whether or not the first statement is true. There is not an And function that will tie the two statements together.

For this reason, use a string in the first line as a comment line.

$\text{Test\_3a}_i =$	$\text{Test\_1}_i =$
100.00	100.00
200.00	200.00
300.00	300.00
400.00	400.00

FIGURE 8.12 Programming Form 3

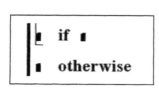

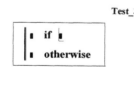

$$x = \begin{pmatrix} 3.00 \\ 4.00 \\ -7.00 \\ -10.00 \end{pmatrix} \qquad y = \begin{pmatrix} 2.00 \\ -1.00 \\ 3.00 \\ -2.00 \end{pmatrix}$$

$$\left|\begin{array}{l} \textbf{if } x_i \ge 0 \\ \quad \left|\begin{array}{l} 100 \ \textbf{ if } \ y_i \ge 0 \\ 200 \ \textbf{ otherwise} \end{array}\right. \\ \blacksquare \ \textbf{ otherwise} \end{array}\right.$$

$$Test\_4_i := \left|\begin{array}{l} \textbf{if } x_i \ge 0 \\ \quad \left|\begin{array}{l} 100 \ \textbf{ if } \ y_i \ge 0 \\ 200 \ \textbf{ otherwise} \end{array}\right. \\ \textbf{otherwise} \\ \quad \left|\begin{array}{l} 300 \ \textbf{ if } \ y_i \ge 0 \\ 400 \ \textbf{ otherwise} \end{array}\right. \end{array}\right.$$

$Test\_4_i =$
100.00
200.00
300.00
400.00

$Test\_1_i =$
100.00
200.00
300.00
400.00

FIGURE 8.13 Programming Form 4

others, but still achieves the same result. In these examples, we use a range variable and vector math.

USING CONDITIONAL PROGRAMS TO MAKE AND DISPLAY CONCLUSIONS

Having Mathcad display a statement at the conclusion of a problem is a great feature. You can have Mathcad display statements such as "Passes" or "Fails," depending on whether or not the calculation worked. You can create other string variables that Mathcad can display.

Let's look at an example. See Figure 8.14.

Create three string variables.

Passes := "Passes - Solution Works"

Fails := "Fails - Retry"

$i := 1, 2 .. 3$

Zero := "Result is Zero"

$$\mathbf{Result} := \begin{pmatrix} -5 \\ 0 \\ 5 \end{pmatrix} \qquad \mathbf{Test}_i := \left| \begin{array}{l} \mathbf{Passes} \ \ \mathbf{if} \ \ \mathbf{Result}_i > 0 \\ \mathbf{Fails} \ \ \mathbf{if} \ \ \mathbf{Result}_i < 0 \\ \mathbf{Zero} \ \ \mathbf{otherwise} \end{array} \right.$$

$\mathbf{Test}_1$ = "Fails - Retry"

$\mathbf{Test}_2$ = "Result is Zero"

$\mathbf{Test}_3$ = "Passes - Solution Works"

FIGURE 8.14 Using conditional programming to display conclusions

ENGINEERING EXAMPLES

Engineering Example 8.1 demonstrates how to use a program to calculate the flow in a pipe or channel. Engineering Example 8.2 demonstrates how to use a program to calculate a progressive income tax.

Engineering Example 8.1

Manning's Equation is an empirical relationship commonly used to predict flow in open channels. It has the form:

$$\mathbf{Q} = \frac{1.49}{\mathbf{n}} \cdot \mathbf{A} \cdot \mathbf{R}^{\frac{2}{3}} \cdot \mathbf{S}^{\frac{1}{2}} \ \ \textbf{flow (Q) in } \frac{\mathbf{ft^3}}{\mathbf{sec}}, \ \textbf{dimensions in feet}$$

Create a worksheet that will calculate the flow of water based on the Manning equation and four types of channels.

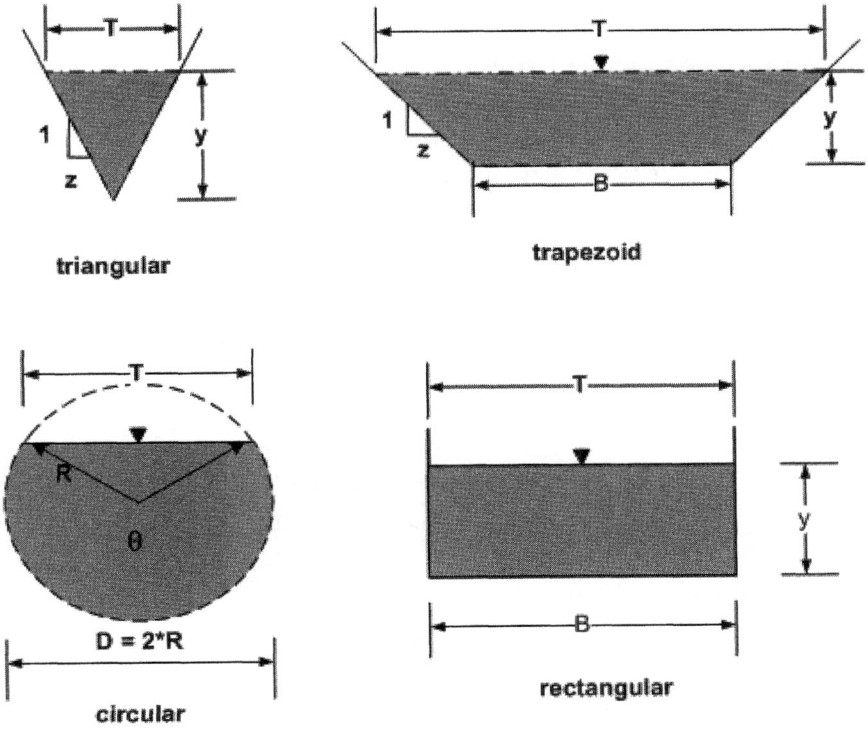

Inputs:

Bottom width of channel **B** := 5**ft**

Depth of flow in channel (Not needed for pipe) **v** := 3**ft**

Diameter of pipe **Diam** := 5**ft**

Side slope for trapezoidal channels (run/rise) **z** := 4

Slope of channel bottom=Energy slope for uniform flow, ft/ft or m/m

Manning roughness cofficient n **n** := 0.015 **S** := 0.005

θ angle of flow in circular pipe (attach units of degrees or radian)

θ := 150**deg**

Shape :=
- ○ Triangular
- ○ Trapezoidal
- ◉ Rectangular
- ○ Circular

Shape= "Rectangular"

Note: The radio box at the left is a web control. See Chapter 21 for a discussion about adding web controls.

The program functions below compute the flow area, wetted perimeter, and hydraulic radius.

$$\mathbf{Flow\_Area} := \left| \begin{array}{l} \text{"Compute flow area based on channel shape"} \\ \textbf{return } \mathbf{z \cdot y^2} \textbf{ if Shape} = \text{"Triangular"} \\ (\mathbf{B + z \cdot y}) \cdot \mathbf{y} \textbf{ if Shape} = \text{"Trapezoidal"} \\ \mathbf{B \cdot y} \textbf{ if Shape} = \text{"Rectangular"} \\ \frac{1}{2} \cdot \left(\frac{\mathbf{Diam}}{2} \right)^2 \cdot (\boldsymbol{\theta} - \mathbf{sin}(\boldsymbol{\theta})) \textbf{ if Shape} = \text{"Circular"} \end{array} \right.$$

$$\mathbf{Flow\_Area} = 15 \cdot \mathbf{ft^2}$$

$$\mathbf{Wetted\_Perimeter} := \left| \begin{array}{l} \text{"Compute wetted perimeter based on channel shape"} \\ 2 \cdot \mathbf{y} \cdot \sqrt{1 + \mathbf{z}^2} \textbf{ if Shape} = \text{"Triangular"} \\ \mathbf{B} + 2 \cdot \mathbf{y} \cdot \sqrt{1 + \mathbf{z}^2} \textbf{ if Shape} = \text{"Trapezoidal"} \\ \mathbf{B} + 2 \cdot \mathbf{y} \textbf{ if Shape} = \text{"Rectangular"} \\ \frac{1}{2} \cdot \frac{\boldsymbol{\theta}}{\mathbf{rad}} \cdot \mathbf{Diam} \textbf{ if Shape} = \text{"Circular"} \end{array} \right.$$

$$\mathbf{Wetted\_Perimeter} = 11.00 \cdot \mathbf{ft}$$

R in the Manning equation is the hydraulic radius, which is the cross sectional area divided by the wetted perimeter.

$$\mathbf{Hyd\_Rad} := \frac{\mathbf{Flow\_Area}}{\mathbf{Wetted\_Perimeter}} \qquad \mathbf{Hyd\_Rad} = 1.36 \cdot \mathbf{ft}$$

Calculate Flow Rate from the Manning equation. This is an empirical equation. Refer to Chapter 4 - Units! The equation is based on English units, so divide by Hyd_Rad by feet. The value of

$$\mathbf{Flow\_Rate} := \frac{1.49}{\mathbf{n}} \cdot \mathbf{Flow\_Area} \cdot \left(\frac{\mathbf{Hyd\_Rad}}{\mathbf{ft}} \right)^{\frac{2}{3}} \cdot \mathbf{S}^{\frac{1}{2}} \cdot \frac{\mathbf{ft}}{\mathbf{sec}}$$

$\mathbf{Flow\_Rate} = 1.16 \frac{\mathbf{m^3}}{\mathbf{s}}$ Flow Rate displayed in metric units

$\mathbf{Flow\_Rate} = 40.97 \cdot \frac{\mathbf{ft^3}}{\mathbf{s}}$ Flow Rate displayed in English units

Engineering Example 8.2

The income tax rate for a state are as follows:
0.023 for income \$0 to \$2,000
0.033 for income \$2,001 to \$4000
0.042 for income \$4001 to \$6000
0.052 for income \$6,001 to 8000
0.060 for income \$8,001 to 11,000
0.0698 for income over \$11,000

Write a program to calculate the taxes owed for various income levels. Write the program so that tax rates and brackets can be easily changed.

Input the tax rates and bracket levels, and calculate maximum taxes paid in each tax bracket.

$\mathbf{Rate_1} := 0.023$ $\mathbf{Max_1} := 2000 \cdot \$$ $\mathbf{Tax_1} := \mathbf{Max_1} \cdot \mathbf{Rate_1}$ $\mathbf{Tax_1} = 46.00\ \$$

$\mathbf{Rate_2} := 0.033$ $\mathbf{Max_2} := 4000 \cdot \$$ $\mathbf{Tax_2} := (\mathbf{Max_2} - \mathbf{Max_1}) \cdot \mathbf{Rate_2}$ $\mathbf{Tax_2} = 66.00\ \$$

$\mathbf{Rate_3} := 0.042$ $\mathbf{Max_3} := 6000 \cdot \$$ $\mathbf{Tax_3} := (\mathbf{Max_3} - \mathbf{Max_2}) \cdot \mathbf{Rate_3}$ $\mathbf{Tax_3} = 84.00\ \$$

$\mathbf{Rate_4} := 0.052$ $\mathbf{Max_4} := 8000 \cdot \$$ $\mathbf{Tax_4} := (\mathbf{Max_4} - \mathbf{Max_3}) \cdot \mathbf{Rate_4}$ $\mathbf{Tax_4} = 104.00\ \$$

$\mathbf{Rate_5} := 0.06$ $\mathbf{Max_5} := 11000 \cdot \$$ $\mathbf{Tax_5} := (\mathbf{Max_5} - \mathbf{Max_4}) \cdot \mathbf{Rate_5}$ $\mathbf{Tax_5} = 180.00\ \$$

$\mathbf{Rate_6} := 0\ .0698$

$$\mathbf{Taxes(Income)} := \left| \begin{array}{l} \mathbf{return}[(\mathbf{Tax_1} + \mathbf{Tax_2} + \mathbf{Tax_3} + \mathbf{Tax_4} + \mathbf{Tax_5}) + (\mathbf{Income} - \mathbf{Max_5}) \cdot \mathbf{Rate_6}] \ \mathbf{if\ Income} > \mathbf{Max_5} \\ \mathbf{return}[(\mathbf{Tax_1} + \mathbf{Tax_2} + \mathbf{Tax_3} + \mathbf{Tax_4}) + (\mathbf{Income} - \mathbf{Max_4}) \cdot \mathbf{Rate_5}] \ \mathbf{if\ Income} > \mathbf{Max_4} \\ \mathbf{return}[(\mathbf{Tax_1} + \mathbf{Tax_2} + \mathbf{Tax_3}) + (\mathbf{Income} - \mathbf{Max_3}) \cdot \mathbf{Rate_4}] \ \mathbf{if\ Income} > \mathbf{Max_3} \\ \mathbf{return}[(\mathbf{Tax_1} + \mathbf{Tax_2}) + (\mathbf{Income} - \mathbf{Max_2}) \cdot \mathbf{Rate_3}] \ \mathbf{if\ Income} > \mathbf{Max_2} \\ \mathbf{return}[(\mathbf{Tax_1}) + (\mathbf{Income} - \mathbf{Max_1}) \cdot \mathbf{Rate_2}] \ \mathbf{if\ Income} > \mathbf{Max_1} \\ \mathbf{Income} \cdot \mathbf{Rate_1}\ \mathbf{otherwise} \end{array} \right.$$

$$\mathbf{IncomeInput} := \begin{pmatrix} 2000 \\ 4000 \\ 6000 \\ 8000 \\ 11000 \\ 1000 \\ 3000 \\ 5000 \\ 7000 \\ 10000 \\ 20000 \end{pmatrix} \cdot \$$$

Use vectorize.
See Ch. 5.

$$\mathbf{Tax} := \overrightarrow{\mathbf{Taxes(IncomeInput)}}$$

IncomeInput =

	1
1	2000.00
2	4000.00
3	6000.00
4	8000.00
5	11000.00
6	1000.00
7	3000.00
8	5000.00
9	7000.00
10	10000.00
11	20000.00

\$

Tax =

	1
1	46.00
2	112.00
3	196.00
4	300.00
5	480.00
6	23.00
7	79.00
8	154.00
9	248.00
10	420.00
11	1108.20

\$

SUMMARY

Logical programs are used in place of the *if* function. They are easier to visualize than the *if* function, and they are easier to write and check. Chapter 11 will discuss programming in much more detail.

In Chapter 8 we:

- Explained how to create and add lines to a program.
- Emphasized the need to use the Programming toolbar to insert the programming operators. Typing the operators will not work.
- Warned about checking the logic in the program so that a subsequent true statement does not change the result.
- Learned about Boolean operators.
- Showed how to insert the *if* operator into different locations within the program.
- Demonstrated how to use logic programs to draw and display conclusions.

PRACTICE

Additional problems and applications can be found on the companion site: www.elsevierdirect.com/9780123747839.

1. Use the *if* function for the following logic:
   - **a.** Create variable x = 3.
   - **b.** If x = 3 the variable "Result" = 40.
   - **c.** If x is not equal to 3 then "Result" = 100.
2. Create a Mathcad program to achieve the same result.
3. Rewrite the program to place the *if* operator in a different location.
4. Write three Mathcad programs that use the Boolean operator ***And***.
5. Write three Mathcad programs that use the Boolean operator ***Or***.
6. Create and solve five problems from your field of study. At the conclusion of each problem, write a logical program to have Mathcad display a concluding statement depending on the result of the problem. Change input variables to ensure that the display is accurate for all conditions.

PART

3

Power Tools for Your Mathcad Toolbox

You have just filled your Mathcad toolbox with many simple yet useful tools. Part III will now add some very powerful Mathcad tools.

The purpose of Part II was to introduce features and functions that were not too confusing. It also gave simple examples. The features and functions introduced in Part III tend to be more complex and require more complex examples. We postponed the discussion of these topics until after you had a thorough understanding of Mathcad basics.

Part III discusses the powerful topics of symbolic calculations, root finding, solve blocks, and advanced Mathcad programming.

CHAPTER

9 Introduction to Symbolic Calculations

Symbolic calculations return algebraic results rather than numeric results. Mathcad has a sophisticated symbolic processor that can solve very complex problems algebraically. The symbolic processor is different from the numeric processor.

The intent of this chapter is to whet your appetite for symbolic calculations. We will cover only a few of the topics in symbolic calculations. The topics we discuss are useful to help solve some basic engineering problems. We will not get into solving complex mathematical equations.

As always, the Mathcad Help and Tutorials provide excellent examples of topics not covered in this chapter.

Chapter 9 will:

- Introduce symbolic calculations.
- Begin by showing how to solve a polynomial expression.
- Show how to solve polynomial expressions for algebraic and numeric solutions.
- Demonstrate how to get quick static symbolic results by using the Symbolics menu commands.
- Discuss live symbolics and use of the symbolic equal sign.
- Tell how to get numeric rather than algebraic results.
- Show how to use the "explicit" keyword to display the values of variables used in your expressions.
- Demonstrate how to string a series of symbolic keywords.

GETTING STARTED WITH SYMBOLIC CALCULATIONS

Let's get started by showing how simple it is to solve for the roots of a polynomial expression. Figure 9.1 shows how to solve for symbolic results with variables as coefficients. You can also solve equations with numbers as coefficients, as shown in Figure 9.2.

The results from using the symbolic solve with variable coefficients can become very complex, as we see in Figure 9.3.

To solve for the following expression, select either one of the instances of variable "x" in the below equation. From the **Symbolics** menu, hover your mouse over **Variable**, and click **Solve**.

$$a \cdot x^2 + b \cdot x + c \qquad \begin{pmatrix} -\dfrac{\dfrac{b}{2} + \dfrac{\sqrt{b^2 - 4 \cdot a \cdot c}}{2}}{a} \\ -\dfrac{\dfrac{b}{2} - \dfrac{\sqrt{b^2 - 4 \cdot a \cdot c}}{2}}{a} \end{pmatrix}$$

Mathcad returns both algebraic solutions to the expression. You will recognize the solutions as both solutions to the quadratic equation.

The below examples use the same procedure, but each example selects a different variable to solve for.

Solving for "a"	Solving for "b"	Solving for "c"
$a \cdot x^2 + b \cdot x + c$	$a \cdot x^2 + b \cdot x + c$	$a \cdot x^2 + b \cdot x + c$
$-\dfrac{c + b \cdot x}{x^2}$	$-\dfrac{a \cdot x^2 + c}{x}$	$-a \cdot x^2 - b \cdot x$

Tip! Note that with the new symbolic solver in Mathcad 14, the negative value in front of many solutions is overlooked because it looks like an extension of the division line. For example, note that there is a negative sign in front of the quadratic equation solutions and in the first and second solutions above.

FIGURE 9.1 Symbolically solving for a variable

Use the same procedure as described in Figure 9.1. The solutions can be real or complex. The symbolic processor does not default to decimal answers.

$$x^2 + 5 \cdot x + 6 \qquad x^2 + x + 3 \qquad 6 \cdot x + 2 \qquad x^3 + 1$$

$$\begin{pmatrix} -2 \\ -3 \end{pmatrix} \qquad \begin{pmatrix} -\dfrac{1}{2} + \dfrac{\sqrt{11} \cdot i}{2} \\ -\dfrac{1}{2} - \dfrac{\sqrt{11} \cdot i}{2} \end{pmatrix} \qquad -\dfrac{1}{3} \qquad \begin{pmatrix} -1 \\ \dfrac{1}{2} + \dfrac{\sqrt{3} \cdot i}{2} \\ \dfrac{1}{2} - \dfrac{\sqrt{3} \cdot i}{2} \end{pmatrix}$$

Note the negative value in front of the second and third solutions.

FIGURE 9.2 Using the symbolic processor to solve for numeric results

$$x^3 + x + 1$$

$$\begin{bmatrix} \dfrac{\left(\dfrac{29}{54}-\dfrac{\sqrt{93}}{18}\right)^{\frac{1}{3}}-\dfrac{1}{3}}{\left(\dfrac{\sqrt{93}}{18}-\dfrac{1}{2}\right)^{\frac{1}{3}}} \\[2ex] \dfrac{1-3\cdot\left(\dfrac{29}{54}-\dfrac{\sqrt{93}}{18}\right)^{\frac{1}{3}}+\sqrt{3}\cdot i+3i\cdot\sqrt{3}\cdot\left(\dfrac{29}{54}-\dfrac{\sqrt{93}}{18}\right)^{\frac{1}{3}}}{6\cdot\left(\dfrac{\sqrt{93}}{18}-\dfrac{1}{2}\right)^{\frac{1}{3}}} \\[2ex] \dfrac{3\cdot\left(\dfrac{29}{54}-\dfrac{\sqrt{93}}{18}\right)^{\frac{1}{3}}-1+\sqrt{3}\cdot i+3i\cdot\sqrt{3}\cdot\left(\dfrac{29}{54}-\dfrac{\sqrt{93}}{18}\right)^{\frac{1}{3}}}{6\cdot\left(\dfrac{\sqrt{93}}{18}-\dfrac{1}{2}\right)^{\frac{1}{3}}} \end{bmatrix}$$

Note the negative sign in front of the third solution.

FIGURE 9.3 Solving for a variable

These examples show how easily you can solve for any variable in your equation. There is a drawback to this approach, however. As you can see, the results are static. If you change the equation, the result does not change. This can cause errors in your calculations. It would be much nicer to have Mathcad provide a dynamic result.

Mathcad does provide this capability! It is called the symbolic equal sign or live symbolics. You can get similar results as we did earlier, but with the advantage of being updated automatically whenever there is a change to the equations.

The symbolic equal sign is an arrow pointing to the right →. The symbolic equal sign is located on the Symbolic toolbar. You can open this toolbar from the **View** menu, or you can click the icon on the Math toolbar. In order to solve an equation with the symbolic equal sign, you need to include a keyword with it. The keyword tells Mathcad what you want to do symbolically. In our case, the keyword is "solve." The keyword is located in a placeholder located just before the symbolic equal sign. To insert the symbolic equal sign with the keyword placeholder, click the icon on the Symbolic toolbar or type **CTRL+SHIFT+PERIOD**. You also need to tell Mathcad which variable to solve for. To do this, type a comma after the keyword, followed by the name of the variable to solve for.

Let's look at how you can use the symbolic equal sign to solve for the same equations as in Figures 9.1 through 9.3. See Figures 9.4 through 9.6.

The previous examples used only equations. You can also create user-defined functions using symbolic math. You do this by naming the user-defined function and listing its arguments. You then enter the equation, the keyword "solve," and the variable you want to solve for. You can now add numeric arguments in the function and the function will return numeric results. See Figure 9.7.

EVALUATE

Whenever you use the symbolic equal sign without any keywords, Mathcad evaluates the expression to the left of the symbolic equal sign. If the expression cannot be simplified further, Mathcad returns the original expression. If variables are not defined prior to using the symbolic equal sign, Mathcad just returns the same expression. When using the symbolic equal sign with numeric input, Mathcad does not express numbers in decimal format. It uses fractions instead of decimals. Let's look at a few examples comparing the symbolic equal sign with the numeric equal sign. See Figures 9.8 and 9.9.

FLOAT

In the previous examples, Mathcad returned the symbolic results as fractions without reducing the numbers to decimals. This is because Mathcad's symbolic processor does not reduce terms. In Figure 9.9, we showed how to have Mathcad display the numeric results by using the numeric equal sign at the end of the expression. There is another way to have the symbolic processor return numeric results. You do this by using the keyword "float." This is called floating point calculation. Using this keyword skips the display of the symbolic result. When using floating point calculations, Mathcad uses a default level of precision equal to 20. You can control the level of precision by typing a comma following the "float" keyword and typing an integer, which defines the desired precision. When you specify a precision, Mathcad returns a result based on that level of precision. It does not calculate to a higher level of precision and then display the result to the requested number of decimal places. The precision can be between 1 and 250 for live symbolics. If you use the **Evaluate > Floating Point** command from the **Symbolics** menu, you can specify a precision as high as 4000. See Figure 9.10.

These examples are the same as used in Figure 9.1, but using live symbolics. If you change the function then the symbolic result will also change.

Follow these steps to solve using the symbolic equal sign:

1. Type the equation
2. Click the symbolic equal sign with a placeholder, or type CTRL+SHIFT+PERIOD.
3. Click in the placeholder
4. Type "solve,x"
5. Click outside the region

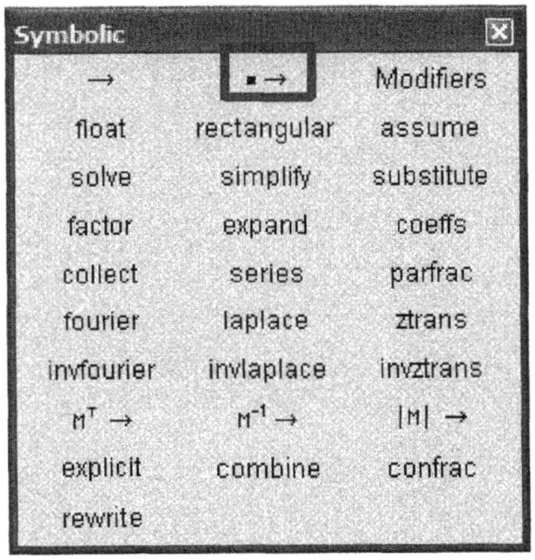

$$\mathbf{a \cdot x^2 + b \cdot x + c\ solve, x} \rightarrow \begin{pmatrix} \dfrac{\dfrac{b}{2} + \dfrac{\sqrt{b^2 - 4 \cdot a \cdot c}}{2}}{a} \\ \dfrac{\dfrac{b}{2} - \dfrac{\sqrt{b^2 - 4 \cdot a \cdot c}}{2}}{a} \end{pmatrix}$$

Solving for "a"

$$\mathbf{a \cdot x^2 + b \cdot x + c\ solve, a} \rightarrow \frac{c + b \cdot x}{x^2}$$

Solving for "b"

$$\mathbf{a \cdot x^2 + b \cdot x + c\ solve, b} \rightarrow \frac{a \cdot x^2 + c}{x}$$

Solving for "c"

$$\mathbf{a \cdot x^2 + b \cdot x + c\ solve, c} \rightarrow -a \cdot x^2 - b \cdot x$$

FIGURE 9.4 Live symbolics

These equations are the same as in Figure 9.2, but they are now dynamic. If you change the equation, then the result will also change.

$$\mathbf{x}^2 + 5\mathbf{x} + 6 \ \mathbf{solve,x} \rightarrow \begin{pmatrix} -2 \\ -3 \end{pmatrix} \qquad 6\mathbf{x} + 2 \ \mathbf{solve,x} \rightarrow -\frac{1}{3}$$

$$\mathbf{x}^2 + \mathbf{x} + 3 \ \mathbf{solve,x} \rightarrow \begin{pmatrix} -\frac{1}{2} + \frac{\sqrt{11}\cdot i}{2} \\ -\frac{1}{2} - \frac{\sqrt{11}\cdot i}{2} \end{pmatrix} \qquad \mathbf{x}^3 + 1 \ \mathbf{solve,x} \rightarrow \begin{pmatrix} -1 \\ \frac{1}{2} + \frac{\sqrt{3}\cdot i}{2} \\ \frac{1}{2} - \frac{\sqrt{3}\cdot i}{2} \end{pmatrix}$$

FIGURE 9.5 Live symbolics

This is the same equation as in Figure 9.3, except that it is dynamic. If you change the equation, then the result will also change.

$$\mathbf{x}^3 + \mathbf{x} + 1 \ \mathbf{solve,x} \rightarrow \begin{bmatrix} \dfrac{\left(\frac{29}{54} - \frac{\sqrt{93}}{18}\right)^{\frac{1}{3}} - \frac{1}{3}}{\left(\frac{\sqrt{93}}{18} - \frac{1}{2}\right)^{\frac{1}{3}}} \\ \dfrac{1 - 3\cdot\left(\frac{29}{54} - \frac{\sqrt{93}}{18}\right)^{\frac{1}{3}} + \sqrt{3}\cdot i + 3i\cdot\sqrt{3}\cdot\left(\frac{29}{54} - \frac{\sqrt{93}}{18}\right)^{\frac{1}{3}}}{6\cdot\left(\frac{\sqrt{93}}{18} - \frac{1}{2}\right)^{\frac{1}{3}}} \\ \dfrac{3\cdot\left(\frac{29}{54} - \frac{\sqrt{93}}{18}\right)^{\frac{1}{3}} - 1 + \sqrt{3}\cdot i + 3i\cdot\sqrt{3}\cdot\left(\frac{29}{54} - \frac{\sqrt{93}}{18}\right)^{\frac{1}{3}}}{6\cdot\left(\frac{\sqrt{93}}{18} - \frac{1}{2}\right)^{\frac{1}{3}}} \end{bmatrix}$$

FIGURE 9.6 Live symbolics

You can assign a user-defined function in conjunction with symbolic calculations.

The function below solves for both roots of a quadratic equation.

$$\mathbf{Function_1(a,b,c)} := \mathbf{a \cdot x^2 + b \cdot x + c \text{ solve}, x} \rightarrow \begin{pmatrix} \dfrac{\dfrac{b}{2} + \dfrac{\sqrt{b^2 - 4 \cdot a \cdot c}}{2}}{a} \\ \dfrac{\dfrac{b}{2} - \dfrac{\sqrt{b^2 - 4 \cdot a \cdot c}}{2}}{a} \end{pmatrix}$$

$$\mathbf{Function_1}(1,5,6) = \begin{pmatrix} -3.00 \\ -2.00 \end{pmatrix}$$

$$\mathbf{Function_1}(2,3,4) = \begin{pmatrix} -0.75 - 1.20i \\ -0.75 + 1.20i \end{pmatrix}$$

You can also create functions from the other equations used in Figure 9.4.

$$\mathbf{Function_2(b,c,x)} := \mathbf{a \cdot x^2 + b \cdot x + c \text{ solve}, a} \rightarrow \frac{c + b \cdot x}{x^2}$$

$$\mathbf{Function_2}(5,6,1) = -11.00$$

$$\mathbf{Function_3(a,c,x)} := \mathbf{a \cdot x^2 + b \cdot x + c \text{ solve}, b} \rightarrow \frac{a \cdot x^2 + c}{x}$$

$$\mathbf{Function_3}(1,6,1) = -7.00$$

$$\mathbf{Function_4(a,b,x)} := \mathbf{a \cdot x^2 + b \cdot x + c \text{ solve}, c} \rightarrow -a \cdot x^2 - b \cdot x$$

$$\mathbf{Function_4}(1,5,1) = -6.00$$

FIGURE 9.7 Creating user-defined functions using symbolic math

Type CTRL+PERIOD to get the symbolic equal sign.

Symbolic

When using the symbolic equal sign, if A_1 and A_2 are not defined, then the symbolic equal sign returns the original expressions.

$A_1 + A_2 \rightarrow A_1 + A_2$

$A_1 \cdot A_2 \rightarrow A_1 \cdot A_2$

$A_1 := 2ft \quad A_2 := 4sec$

When using the symbolic equal sign, if the units of the equation are not consistent, then Mathcad simplifies the equation as much as possible, but does not return an error message. Units are not converted to the default unit system.

$A_1 + A_2 \rightarrow 2 \cdot ft + 4 \cdot sec$

$A_1 \cdot A_2 \rightarrow 8 \cdot ft \cdot sec$

Numeric

When using the numeric equal sign, if A_1 and A_2 are not defined, Mathcad gives an error stating that the variable is not defined.

$A_1 + A_2 = \blacksquare$

$A_1 \cdot A_2 = 2.44\,m \cdot s$

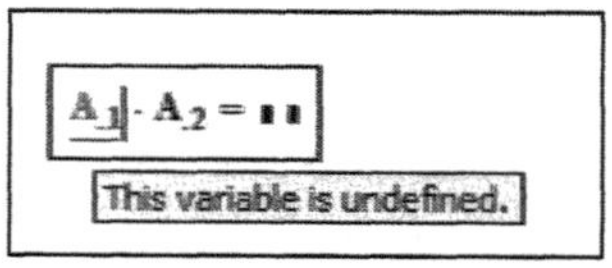

A1 · A2 = ▪▪
This variable is undefined.

When using the numeric equal sign, Mathcad returns an error if it tries to add inconsistent units. The multiplication returns units in the default unit system.

$A_1 + A_2 = \blacksquare$

$A_1 \cdot A_2 = 2.44\,m \cdot s$

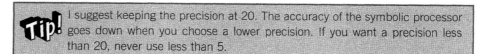

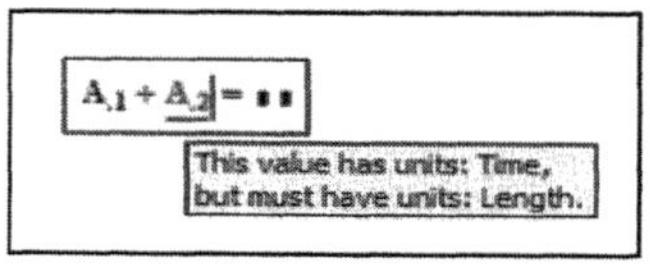

A1 + A2 = ▪▪
This value has units: Time,
but must have units: Length.

FIGURE 9.8 Comparison of how the symbolic equal sign works compared to the numeric equal sign

> **Tip!** I suggest keeping the precision at 20. The accuracy of the symbolic processor goes down when you choose a lower precision. If you want a precision less than 20, never use less than 5.

You can also force Mathcad to use floating point calculations by including a decimal point in the input numbers. See Figure 9.11.

Symbolic evaluation		Numeric evaluation	
$\cos(x_1 + x_2) \rightarrow \cos(x_1 + x_2)$	If x_1 and x_2 are undefined, returns the original expression.	$\cos(x_1 + x_2) = \blacksquare$	If x_1 and x_2 are undefined, gives error for undefined variable
$x_1 := \frac{\pi}{4}$ $\quad x_2 := \frac{4\pi}{3}$	$x_1 \rightarrow \frac{\pi}{4}$	$x_1 = 0.79$	
$\cos(x_1) \rightarrow \frac{\sqrt{2}}{2}$	Symbolic results do not use decimals. They are expressed as fractions.	$\cos(x_1) = 0.71$	Numeric results
$\cos(x_2) \rightarrow -\frac{1}{2}$		$\cos(x_2) = -0.50$	
$\cos(x_1 + x_2) \rightarrow \frac{\sqrt{6}}{4} - \frac{\sqrt{2}}{4}$	Check	$\cos(x_1 + x_2) = 0.26$ $\frac{1}{4}\cdot\sqrt{6}\cdot\left(1 - \frac{1}{3}\cdot\sqrt{3}\right) = 0.26$	
$x_1 + x_2 \rightarrow \frac{19\cdot\pi}{12}$		$x_1 + x_2 = 4.97$	

You can cause Mathcad to return a numeric evaluation of a live symbolics by adding a numeric equal sign at the end of the expression.

$$\cos(x_1 + x_2) \rightarrow \frac{\sqrt{6}}{4} - \frac{\sqrt{2}}{4} = 0.26$$

$$x_1 + x_2 \rightarrow \frac{19\cdot\pi}{12} = 4.97$$

$$x_1 \rightarrow \frac{\pi}{4} = 0.79$$

FIGURE 9.9 Additional examples of symbolic evaluation compared to numeric evaluation

EXPAND, SIMPLIFY, AND FACTOR

You can also use the symbolic menu to expand, simplify, and factor equations.

When Mathcad expands an equation, it multiplies the various elements and expands the exponents. Let's look at how Mathcad can expand algebraic equations. See Figures 9.12 and 9.13.

$y_1 := \frac{\pi}{8}$ $\quad$ $y_2 := \frac{\pi}{4}$ $\quad$ $y_3 := \pi$

You can force numeric evaluation of live symbolic calculations by using the "float" keyword. The default precision is 20.

Using standard evaluation with numeric equal sign	Using keyword "float"
$y_1 \rightarrow \frac{\pi}{8} = 0.392699$	y_1 **float** $\rightarrow 0.39269908169872415481$
$y_2 \rightarrow \frac{\pi}{4} = 0.785398$	y_2 **float** $\rightarrow 0.78539816339744830962$
$y_1 + y_2 \rightarrow \frac{3 \cdot \pi}{8} = 1.178097$	$y_1 + y_2$ **float** $\rightarrow 1.1780972450961724644$

Use a comma and a number after the keyword to control the desired precision.

y_1 **float**, 5 $\rightarrow 0.3927$ $\quad$ y_2 **float**, 5 $\rightarrow 0.7854$ $\quad$ $y_1 + y_2$ **float**, 5 $\rightarrow 1.1781$

$\cos(y_1 + y_2) \rightarrow \frac{\sqrt{2 - \sqrt{2}}}{2} = 0.38268$ $\quad$ $\cos(y_1 + y_2)$ **float**, 4 $\rightarrow 0.3827$

The floating point calculations only provide numbers to the level of precision you specify. Note the lower accuracy when the precision is set low. It is best to keep the default precision. Compare the numbers below with the numeric results. The precision can be as high as 250 for live symbolics.

AA := y_1 **float**, 4 $\rightarrow 0.3927$ $\quad$ **BB** := y_2 **float**, 2 $\rightarrow 0.79$

AA = 0.392700 $\quad$ **BB** = 0.790000

$y_1 = 0.39269908$ $\quad$ $y_2 = 0.78539816$

CC := $\cos(y_1 + y_2)$ **float**, 6 $\rightarrow 0.382683$ $\quad$ **CC** = 0.38268300

$\cos(y_1 + y_2) = 0.38268343$

$y_3 = 3.14159265358979300$ $\quad$ Numeric evaluation

y_3 **float**, 50 $\rightarrow 3.1415926535897932384626433832795028841971693993751$

e **float**, 50 $\rightarrow 2.7182818284590452353602874713526624977572470937$

FIGURE 9.10 Using floating point calculations

You can force Mathcad to use floating point calculations by including a decimal point in your numbers.

$\mathbf{e}^3 \to \mathbf{e}^3 = 20.08553692318766800$ Normal live symbolic result with numeric equal sign.

$\mathbf{e}^3\ \mathbf{float}\ \to 20.085536923187667741$ Live symbolic with floating point calculation and default level of precision = 20.

$\mathbf{e}^{3.0} \to 20.085536923187667741$ Forced floating point calculation (uses default precision of 20) by using decimal number.

$\mathbf{e}^3 = 20.08553692318766400$ Numeric calculation with number of decimal places set to 17. Note slight differences between symbolic and numeric calculation.

See below for additional examples of forced floating point calculations:

$\mathbf{ln}(2) \to \mathbf{ln}(2)$ $\mathbf{cos}(1) \to \mathbf{cos}(1)$

$\mathbf{ln}(2.0) \to 0.69314718055994530942$ $\mathbf{cos}(1.0) \to 0.54030230586813971740$

$5\frac{3}{4} \to 5\frac{3}{4}$ $5\frac{3.0}{4} \to 5\frac{3.0}{4}$ (Note: Use the mixed number icon on the Calculator toolbar get the mixed fraction.)

Note that once a forced floating point symbolic evaluation is made (by using a decimal), you cannot use a precision greater than 20.

$\mathbf{ln}(3) \to \mathbf{ln}(3)$

$\mathbf{ln}(3.0) \to 1.0986122886681096914\ \mathbf{float}, 40\ \to 1.0986122886681096914$

The forced floating point calculation above set the precision at 20. You cannot then use a greater precision by typing "float, 40". Compare above and below examples.

$\mathbf{ln}(3)\ \mathbf{float}, 40\ \to 1.098612288668109691395245236922525704647$

FIGURE 9.11 Forcing floating point calculations

When Mathcad simplifies an equation, it tries to reduce the equation to the simplest form. Figure 9.14 shows the results of some simplifications.

When you factor an expression, Mathcad breaks the equation into all its parts. Figure 9.15 gives examples of what happens when you use "factor." The best way to learn what each of these does is to experiment and practice.

To expand the following expression, select the entire equation. Click on **Expand** from the **Symbolics** menu.

$$(x+1)^3 + (x+3)^2 + x - 3$$

$$\mathbf{x^3 + 4\cdot x^2 + 10\cdot x + 7}$$

You can also expand just a portion of the equation. To do this, just select the portion you want to expand and click **Expand** from the **Symbolics** menu.

$$(x+1)^3 + (x+3)^2 + x - 3$$

$$x^3 + 3\cdot x^2 + 3\cdot x + 1 + (x+3)^2 + x - 3$$

For this case, just the (x+1)^3 was selected before expanding.

$$(x+1)^3 + \left[\left[(x+3)^2\right]\right] + x - 3$$

$$(x+1)^3 + \left(\mathbf{x^2 + 6\cdot x + 9}\right) + x - 3$$

For this case, just the (x+3)^2 was selected before expanding.

Use the symbolic equal sign to expand the above equations. Use the keyword "expand". The entire equation is selected when you use the symbolic equal sign.

$$(x+1)^3 + (x+3)^2 + x - 3 \ \mathbf{expand} \ \rightarrow x^3 + 4\cdot x^2 + 10\cdot x + 7$$

FIGURE 9.12 Expanding expressions

Below are more examples of expanding algebraic equations using **Expand** from the Symbolics menu.

$$(\cot(x) + 1)^2 + (\tan(x) + 2)^2$$

$$4\cdot \tan(x) + \frac{2}{\tan(x)} + \frac{1}{\tan(x)^2} + \tan(x)^2 + 5$$

$$\frac{1}{2} + \frac{1}{3} + \frac{1}{4}$$

$$\frac{13}{12}$$

$$(a\cdot x - b\cdot y)\cdot(c\cdot z + d\cdot w)$$

$$\mathbf{a\cdot d\cdot w\cdot x + a\cdot c\cdot x\cdot z - b\cdot d\cdot w\cdot y - b\cdot c\cdot y\cdot z}$$

Use the symbolic equal sign to expand the above equations.

$$(\cot(x) + 1)^2 + (\tan(x) + 2)^2 \ \mathbf{expand} \ \rightarrow 4\cdot \tan(x) + \frac{2}{\tan(x)} + \frac{1}{\tan(x)^2} + \tan(x)^2 + 5$$

$$(a\cdot x - b\cdot y)\cdot(c\cdot z + d\cdot w) \ \mathbf{expand} \ \rightarrow a\cdot d\cdot w\cdot x + a\cdot c\cdot x\cdot z - b\cdot d\cdot w\cdot y - b\cdot c\cdot y\cdot z$$

$$\frac{1}{2} + \frac{1}{3} + \frac{1}{4} \ \mathbf{expand} \ \rightarrow \frac{13}{12} = 1.08$$

FIGURE 9.13 Expanding expressions

Simplify returns a simplified expression, which cancels common factor, and returns trigonometric identities. Select the entire equation and click **Simplify** from the **Symbolics** menu.

$$(\cot(x)+1)^2+(\tan(x)+2)^2$$

$$\frac{3\cdot\cos(x)^4+2\cdot\cos(x)^3\cdot\sin(x)-4\cdot\sin(x)\cdot\cos(x)-3\cdot\cos(x)^2-1}{\cos(x)^2\cdot\left[(-1)+\cos(x)^2\right]}$$

$$\frac{x^2+5x+6}{x+3} \qquad 1-\sin(x)^2$$

$$x+2 \qquad \cos(x)^2$$

Use the symbolic equal sign to simplify the above equations.

$$(\cot(x)+1)^2+(\tan(x)+2)^2 \ \mathbf{simplify} \ \rightarrow \cot(x)^2+2\cdot\cot(x)+\tan(x)^2+4\cdot\tan(x)+5$$

$$\frac{x^2+5x+6}{x+3} \ \mathbf{simplify} \ \rightarrow x+2$$

$$1-\sin(x)^2 \ \mathbf{simplify} \ \rightarrow \cos(x)^2$$

FIGURE 9.14 Simplifying expressions

Factoring returns the entire expression as a product of all its parts. It is the opposite of expanding. Select the entire equation and click **Factor** from the **Symbolics** menu

$$21 \qquad 462 \qquad x^2-5x+6 \qquad x^3-x^2-17\cdot x-15 \qquad x^4-4\cdot x^3-7\cdot x^2+22\cdot x+24$$

$$3\cdot 7 \qquad 2\cdot 3\cdot 7\cdot 11 \qquad (x-2)\cdot(x-3) \qquad (x+1)\cdot(x-5)\cdot(x+3) \qquad (x+1)\cdot(x+2)\cdot(x-3)\cdot(x-4)$$

Use the symbolic equal sign to factor the above equations.

$$21 \ \mathbf{factor} \ \rightarrow 3\cdot 7$$

$$462 \ \mathbf{factor} \ \rightarrow 2\cdot 3\cdot 7\cdot 11$$

$$x^2+5x+6 \ \mathbf{factor} \ \rightarrow (x+3)\cdot(x+2)$$

$$x^3-x^2-17\cdot x-15 \ \mathbf{factor} \ \rightarrow (x-5)\cdot(x+3)\cdot(x+1)$$

$$x^4-4\cdot x^3-7\cdot x^2+22\cdot x+24 \ \mathbf{factor} \ \rightarrow (x-3)\cdot(x-4)\cdot(x+2)\cdot(x+1)$$

FIGURE 9.15 Factoring equations

EXPLICIT

Mathcad version 13 added the keyword "explicit." This is a very exciting feature because it allows you to display the values of the variables used in your expressions. To use this feature, type "explicit" as a keyword followed by a comma and a list of the variables you want to display. See Figures 9.16 through 9.19 and Figure 3.21.

The explicit keyword allows you to show intermediate results in your calculations.

Given the following input variable definitions:

$C_1 := 5.1N \qquad C_2 := 4.5N \qquad D_1 := 3.2m \qquad D_2 := 4.8m$

A normal equation would look like this:

$M_1 := C_1 \cdot D_1 + C_2 \cdot D_2 \qquad M_1 = 37.92 \cdot N \cdot m$

Using explicit calculations, you are able to show the values that Mathcad uses to calculate the results:

$M_2 := C_1 \cdot D_1 + C_2 \cdot D_2 \text{ explicit}, C_1, D_1, C_2, D_2 \rightarrow 5.1 \cdot N \cdot 3.2 \cdot m + 4.5 \cdot N \cdot 4.8 \cdot m \rightarrow 16.32 \cdot N \cdot m + 21.6 \cdot N \cdot m$

$M_2 = 37.92 \cdot N \cdot m$

$LineLoad_1 := \frac{C_1}{D_1} + \frac{C_2}{D_2} \qquad LineLoad_1 = 2.53 \frac{kg}{s^{2.00}} \qquad LineLoad_1 = 2.53 \cdot \frac{N}{m}$

$LineLoad_2 := \frac{C_1}{D_1} + \frac{C_2}{D_2} \text{ explicit}, C_1, C_2, D_1, D_2 \rightarrow \frac{5.1 \cdot N}{3.2 \cdot m} + \frac{4.5 \cdot N}{4.8 \cdot m} \rightarrow \frac{0.9375 \cdot N}{m} + \frac{1.59375 \cdot N}{m}$

$LineLoad_2 = 2.53 \cdot \frac{N}{m}$

FIGURE 9.16 Using the keyword "explicit"

$\mathbf{Time} := 3\mathbf{sec}$ $\quad \mathbf{Velocity} := 5\frac{\mathbf{m}}{\mathbf{sec}}$ $\quad \mathbf{InitialDist} := 5.2\mathbf{m}$

A normal equation would look like this

$\mathbf{Distance_1} := \mathbf{InitialDist} + \mathbf{Time} \cdot \mathbf{Velocity}$ $\qquad \mathbf{Distance_1} = 20.20\,\mathbf{m}$

In the above example, if the definitions for InitialDist, Time, and Velocity were on a different page, you would not know what values Mathcad is using for the three variables. With the keyword "explicit," you can have Mathcad display the input variables.

$\mathbf{Distance_2} := \mathbf{InitialDist} + \mathbf{Time} \cdot \mathbf{Velocity}\ \mathbf{explicit, InitialDist, Time, Velocity} \rightarrow 5.2 \cdot \mathbf{m} + 3 \cdot \mathbf{sec} \cdot 5 \cdot \frac{\mathbf{m}}{\mathbf{sec}} \rightarrow 15 \cdot \mathbf{m} + 5.2 \cdot \mathbf{m}$

$\mathbf{Distance_2} = 20.20\,\mathbf{m}$

Mathcad allows you to hide the keyword and modifiers to simplify the variable display. To do this, right click on the keyword and select "Hide keywords." This will only show the definition as seen below. If you click on the region, the full definition will appear just as above.

$\mathbf{Distance_2} := \mathbf{InitialDist} + \mathbf{Time} \cdot \mathbf{Velocity} \rightarrow 5.2 \cdot \mathbf{m} + 3 \cdot \mathbf{sec} \cdot 5 \cdot \frac{\mathbf{m}}{\mathbf{sec}} \rightarrow 15 \cdot \mathbf{m} + 5.2 \cdot \mathbf{m}$

You can also change the symbolic equal sign to display as a normal equal sign. To do this, right click in the expression, hover your mouse over "View Evaluation As," and select "Equal sign."

$\mathbf{Distance_2} := \mathbf{InitialDist} + \mathbf{Time} \cdot \mathbf{Velocity} = 5.2 \cdot \mathbf{m} + 3 \cdot \mathbf{sec} \cdot 5 \cdot \frac{\mathbf{m}}{\mathbf{sec}} = 15 \cdot \mathbf{m} + 5.2 \cdot \mathbf{m}$

$\mathbf{Distance_2} = 20.20\,\mathbf{m}$

FIGURE 9.17 Using the keyword "explicit"

USING MORE THAN ONE KEYWORD

Mathcad allows you to string a series of symbolic calculations together. To do this, type the symbolic expression as you normally would. Click outside the region to display the expression. Then click the result and insert another symbolic equal sign (`CTRL+SHIFT+PERIOD`). Enter the keyword and any modifiers. See Figures 9.20 and 9.21.

UNITS WITH SYMBOLIC CALCULATIONS

The symbolic process does not recognize units. It passes units through as undefined variables. Thus, you can use units with your expressions; however, they will not be truly understood. For instance, 10 ft/s * 5 sec is returned 50 ft, but you

You can also include previous results in the list following "explicit." The results vary depending on what variables are included in the modify list. See the examples below.

$E_1 := 3.0m \qquad E_2 := 4.2m \qquad E_3 := 5.5m \qquad E_4 := 6.2m$

$F_1 := \frac{E_1}{E_2} \qquad F_1 = 0.71 \qquad F_2 := \frac{E_3}{E_4} \qquad F_2 = 0.89$

$G_1 := \frac{F_1}{F_2}\ \text{explicit}, F_1, F_2 \rightarrow \frac{\frac{E1}{E2}}{\frac{E3}{E4}} \qquad G_1 = 0.8052$

By listing F_1 and F_2, Mathcad substitutes the definitions of F_1 and F_2.

$G_2 := \frac{F_1}{F_2}\ \text{explicit}, E_1, E_2, E_3, E_4 \rightarrow \frac{F_1}{F_2} \qquad G_2 = 0.8052$

Listing only E_1, E_2, E_3, and E_4, does not have any effect, because they do not appear in the current definition.

By listing F_1 and F_2 in addition to E_1, E_2, E_3, and E_4, Mathcad substitutes all values used.

$G_3 := \frac{F_1}{F_2}\ \text{explicit}, E_1, E_2, E_3, E_4, F_1, F_2 \rightarrow \frac{\frac{3.0 \cdot m}{4.2 \cdot m}}{\frac{5.5 \cdot m}{6.2 \cdot m}} \rightarrow 0.80519480519480519481$

$G_3 = 0.8052$

Repeat the above example with the keyword hidden and symbolic equal sign changed to numeric equal sign.

$G_4 := \frac{F_1}{F_2} = \frac{\frac{3.0 \cdot m}{4.2 \cdot m}}{\frac{5.5 \cdot m}{6.2 \cdot m}} = 0.80519480519480519481$

$G_4 = 0.8052$

FIGURE 9.18 Using the keyword "explicit"

cannot convert the 50 ft to meters. Another example is adding 10 ft + 5 sec. The symbolic processor returns 10 ft + 5 sec, not recognizing the error of adding units from different unit dimensions. In order to process the units, you need to do a numeric evaluation of the result. Do this by assigning a variable to the symbolic calculation and then numerically evaluating the variable. See Figure 9.22 for examples.

This Figure is very similar to Figure 9.18. In this example, the value of F_3 and F_4 are evaluated in addition to being defined.

$$\mathbf{F_3} := \frac{\mathbf{E_1}}{\mathbf{E_2}}\ \mathbf{explicit, E_1, E_2} \rightarrow \frac{3.0\cdot\mathbf{m}}{4.2\cdot\mathbf{m}} \rightarrow 0.71428571428571428571$$

Define and evaluate F_3.

$$\mathbf{F_4} := \frac{\mathbf{E_3}}{\mathbf{E_4}}\ \mathbf{explicit, E_3, E_4} \rightarrow \frac{5.5\cdot\mathbf{m}}{6.2\cdot\mathbf{m}} \rightarrow 0.88709677419354838710$$

Define and evaluate F_4.

Define and evaluate G_5. Because F_3 and F_4 have already been evaluated, the numeric results of F_3 and F_4 are displayed.

$$\mathbf{G_5} := \frac{\mathbf{F_3}}{\mathbf{F_4}}\ \mathbf{explicit, F_3, F_4} \rightarrow \frac{0.71428571428571428571}{0.88709677419354838710} \rightarrow 0.80519480519480519479$$

Repeat the above example with the keyword hidden and symbolic equal sign changed to numeric equal sign.

$$\mathbf{G_6} := \frac{\mathbf{F_3}}{\mathbf{F_4}} = \frac{0.71428571428571428571}{0.88709677419354838710} = 0.805194805194805194805$$

FIGURE 9.19 Using the keyword "explicit"

You can use several keywords in succession.

Input the first symbolic expression.

$$(4x + 5y - 3x - 3y)^3\ \mathbf{simplify} \rightarrow (x + 2\cdot y)^3$$

Next, click on the result and insert another symbolic equal sign.

$$(4x + 5y - 3x - 3y)^3\ \mathbf{simplify} \rightarrow (x + 2\cdot y)^3\ \mathbf{expand} \rightarrow x^3 + 6\cdot x^2\cdot y + 12\cdot x\cdot y^2 + 8\cdot y^3$$

You can also use several keywords in succession, but only display the final result. To do this, select the first keyword with the vertical selection bar on the right side, then press CTRL+SHIFT+PERIOD. This inserts a vertical bar with a placeholder. Type the next keyword in the placeholder. Mathcad only displays the result from the last keyword.

$a := 1 \quad b := 10 \quad c := 12$

$$a\cdot x^2 + b\cdot x + c\ \mathbf{solve, x} \rightarrow \begin{pmatrix} \sqrt{13} - 5 \\ -\sqrt{13} - 5 \end{pmatrix}\ \mathbf{float} \rightarrow \begin{pmatrix} -1.3944487245360107069 \\ -8.6055512754639892931 \end{pmatrix}$$

$$a\cdot x^2 + b\cdot x + c \left|\begin{array}{l} \mathbf{solve, x} \\ \mathbf{float} \end{array}\right. \rightarrow \begin{pmatrix} -1.3944487245360107069 \\ -8.6055512754639892931 \end{pmatrix}$$

FIGURE 9.20 Using keywords in succession

Find a numeric solution to the equation used in Figure 9.6. Use the solve keyword, and then the float keyword.

If you try to display both results, the solution extends well beyond the right margin. For this example, we will only display the final result.

$$\text{Solution} := x^3 + x + 1 \begin{vmatrix} \text{solve, x} \\ \text{float} \end{vmatrix} \rightarrow \begin{pmatrix} -0.68232780382801932737 \\ 0.34116390191400966368 + 1.1615413999972519361i \\ 0.34116390191400966368 - 1.1615413999972519361i \end{pmatrix}$$

$\text{Solution}_1 = -0.68232780382801930$

$\text{Solution}_2 = 0.34116390191400964 + 1.16154139999725200i$

$\text{Solution}_3 = 0.34116390191400964 - 1.16154139999725200i$

FIGURE 9.21 Using keywords in succession

You may attach units to symbolic calculations.

$\text{SampleUnits}_1 := 1\text{ft} + 5\text{in} + 2\text{m}$

$\text{SampleUnits}_1 \rightarrow \text{ft} + 5 \cdot \text{in} + 2 \cdot \text{m}$

The symbolic processor just passes units through. It does not evaluate them.

$\text{SampleUnits}_1 = 2.43\,\text{m}$

You must use the numeric processor to evaluate the units.

$10\text{ft} + 5\text{sec} \rightarrow 10 \cdot \text{ft} + 5 \cdot \text{sec}$

$\text{SampleUnits}_2 := 10\text{ft} + 5\text{sec} \rightarrow 10 \cdot \text{ft} + 5 \cdot \text{sec}$

The symbolic processor did not recognize the error of adding units in different unit dimensions. Once a variable is assigned to the expression, the numeric processor recognizes the error.

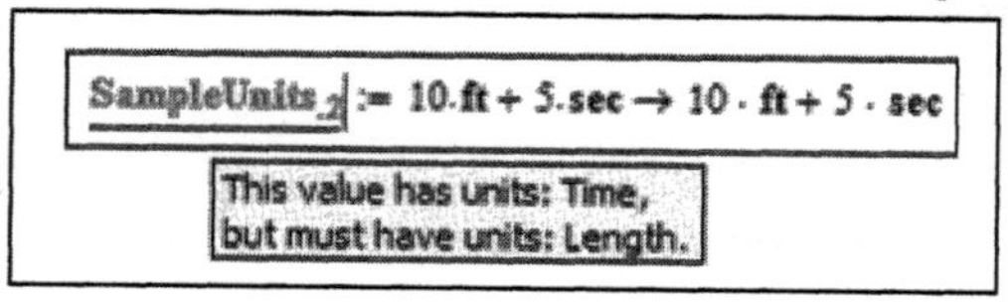

$\boxed{\text{SampleUnits}_2 := 10 \cdot \text{ft} + 5 \cdot \text{sec} \rightarrow 10 \cdot \text{ft} + 5 \cdot \text{sec}}$
This value has units: Time, but must have units: Length.

$10\dfrac{\text{ft}}{\text{s}} \cdot 20\text{s} + \dfrac{1}{2} 50 \dfrac{\text{m}}{\text{s}^2} \cdot 20^2\text{s}^2 \rightarrow 200 \cdot \text{ft} + 10000 \cdot \text{m}$

The symbolic processor cancels common units ((ft/s)*(s)=ft), but it does not convert ft to meters.

$\text{Distance}(v_0, a, t) := v_0 \cdot t + \dfrac{1}{2} \cdot a \cdot t^2$

Assign a user-defined function

$\text{Distance}\left(10\dfrac{\text{ft}}{\text{s}}, 50\dfrac{\text{m}}{\text{s}^2}, 20\text{s}\right) \rightarrow 200 \cdot \text{ft} + 10000 \cdot \text{m}$

The symbolic processor does not convert units.

$\text{Distance}\left(10\dfrac{\text{ft}}{\text{s}}, 50\dfrac{\text{m}}{\text{s}^2}, 20\text{s}\right) = 10060.96\,\text{m}$

The numeric processor does convert units.

$\text{Distance}\left(10\dfrac{\text{m}}{\text{s}}, 50\dfrac{\text{m}}{\text{s}^2}, 20\text{s}\right) \rightarrow 10200 \cdot \text{m}$

Because all the units were consistent, the symbolic processor returned meters.

$\text{Distance}\left(10\dfrac{\text{m}}{\text{s}}, 50\dfrac{\text{m}}{\text{s}^2}, 20\text{s}\right) = 33464.57 \cdot \text{ft}$

The symbolic processor does not have a units placeholder to convert results to other units.

FIGURE 9.22 Using units with symbolic calculations

ADDITIONAL TOPICS TO STUDY

The topic of symbolic calculation could fill an entire book. In this chapter, we have only introduced you to a few symbolic calculation concepts. Even though the discussion of these concepts has been brief, these concepts will be a valuable addition to your Mathcad toolbox.

Some topics we did not discuss in this chapter that may be of interest to you for further study include:

- Partial fractions
- Calculus (see Chapter 12)
- Symbolic transformations
- Series expansion
- Complex input
- Assumptions
- Special functions
- Symbolic optimization

SUMMARY

Mathcad provides two methods to symbolically solve engineering equations. The first method is to use the commands from the **Symbolics** menu. The second method is to use the symbolic equal sign. The results of the first method are static and do not change. The results of the second method are dynamic. Thus, if you change the input equation, the results are updated automatically.

The symbolic equal sign needs to use keywords. These keywords are necessary to tell Mathcad what operation is desired. In addition to the keyword, it is sometimes necessary to tell Mathcad what variable to use with the operation.

In Chapter 9 we:

- Explained how to use the commands from the Symbolics toolbar.
- Showed that the results from using the Symbolics toolbar are static.
- Introduced the symbolic equal sign.
- Illustrated how to use keywords with the symbolic equal sign.
- Emphasized that results are dynamic when using live symbolics.
- Learned how to get numeric results when using the symbolic equal sign.
- Discussed using the "explicit" keyword to display values of variables used in your calculations.
- Recommended studying the Mathcad help and Tutorials to learn more advanced features of symbolic calculations.

PRACTICE

Additional problems and applications can be found on the companion site: www.elsevierdirect.com/9780123747839.

1. Use the following commands from the Symbolics menu on each of the equations: Evaluate (Symbolically), Solve (for each of the variables), Simplify, Expand, and Factor.

   a. $y = x^2 - 3$
   b. $y = (2x + 4)*(x - 3)$
   c. $x^2 + y^2 = 1$
   d. $y = x^3 - 2x^2 - 5x + 6$
   e. $y = a*x^3 + b*x^2 + c*x + d$
   f. $y = (x - 1)(x - 2)(2x - 6)$
   g. $2x^2 - 3x + 1/(2x - 1)$

2. Write symbolic equations to solve for x with the following functions. Assign each to a user-defined function. (See Figures 9.7 and 9.22 for examples of user-defined functions.)

   a. x3 + 6x2 - x - 30
   b. y = x2(x - 1)2
   c. 3x2 + 4y2 = 12

3. Write symbolic equations to evaluate the following functions. Assign each to a user-defined function.

   a. $\cos(\pi/3) + \tan(2\pi/3)$
   b. $[(21/11) + (5/3)]/3$
   c. $\pi + \pi/4 + \ln(2)$

4. Write symbolic equations to expand the following functions. Assign each to a user-defined function.

   a. (x + 2)(x - 3)
   b. (x - 1)(x + 2)(x - 5)
   c. (x + 3 * i)(x - 3 * i) (Note: Use j if that is your default for imaginary value.)

5. Write symbolic equations to simplify the following functions. Assign each to a user-defined function.

   a. $\sin(\pi/2 - x)$
   b. sec(x)
   c. (x4 - 8x3 - 7x2 + 122x - 168)/(x - 2)

6. Write symbolic equations to factor the following functions. Assign each to a user-defined function.

   a. x3 + 5x2 + 6x
   b. x * y + y * x2
   c. 2x2 + llx - 21

7. Use the "float" keyword to evaluate the following functions. Use various levels of precision.

   a. $\cos(\pi/6)$
   b. $\tan(3\pi/4) + \mathrm{COS}(\pi\ /7) - \sin(\pi/5)$
   c. $(\pi + e3)(\tan(3\pi/8))$

8. Use the equations from exercise 7 and show intermediate results before using the "float" keyword.

9. From your field of study, create three expressions using assigned variables. Use the "explicit" keyword to show intermediate results.

10. Open the Mathcad Help and review additional topics on live symbolics from the Contents tab.

CHAPTER

10

Solving Engineering Equations

Once you begin using the solving features of Mathcad, you will wonder how you ever got along without them. Solving engineering equations is one of Mathcad's most useful power tools. It ranks high with the use of units as one of Mathcad's best features. In Chapter 9, we introduced the keyword "solve" with live symbolics. In this chapter, we will add more solving tools to your Mathcad toolbox. The intent of this chapter is to illustrate how engineering problems can be solved using the Mathcad functions ***root, polyroots,*** and ***Find***. Because these are some of the most useful functions for engineers, this chapter will use many examples to illustrate their use. When trying to solve for an equation, it is very useful to plot the equations. This helps to visualize the solution before using Mathcad to solve the equations. Some of the functions discussed require initial guess values, and a plot is useful to help select the initial guess.

Chapter 10 will:

- Introduce the ***root*** function.
- Discuss the two different forms it takes: unbracketed and bracketed.
- Give examples of each form.
- Show how a plot is useful to determine the initial guess.
- Discuss the ***polyroots*** function.
- Give examples of the ***polyroots*** function.
- Show how to use a Solve Block using ***Given***, ***Find***, ***Maximize***, and ***Minimize*** to solve multiple equations with multiple unknowns.
- Note when to use ***Minerr*** instead of ***Find***.
- Illustrate how to use units in solving equations.
- Provide several engineering examples, which illustrate the use of each method of solving equations and illustrate how each would be used.

ROOT FUNCTION

The ***root*** function is used to find a single solution to a single function with a single unknown. In later sections, we will discuss finding all the solutions to a polynomial function. We will also discuss solving multiple equations with multiple unknowns. For now, we will focus on using the ***root*** function.

If a function has several solutions, the solution Mathcad finds is based on the initial guess you give Mathcad. Because of this, it is helpful to plot the function prior to giving Mathcad the initial guess.

The ***root*** function takes the form ***root***(f(var), var, [a, b]). It returns the value of var to make the function f equal to zero. The real numbers a and b are optional. If they are specified (bracketed), ***root*** finds var on this interval. The values of a and b must meet these requirements: $a < b$ and f(a) and f(b) must be of opposite signs. This is because the function must cross the x-axis in this interval. If you do not specify the numbers a and b (unbracketed), then var must be defined with an initial guess prior to using the root function.

> **Tip!** When plotting the function, use a different variable name on the x-axis than the variable you define for your initial guess. If you do not use a different variable name, the plot will not work because Mathcad will plot the value var only on the x-axis. The variable used on the x-axis needs to be a previously undefined variable.

Let's look at some simple examples of using ***root***. See Figures 10.1 and 10.2.

If you do not use a plot to determine your initial guesses, Mathcad might not arrive at the solution you want. If your initial guesses are close to a maximum or minimum point or have multiple solutions between the bracketed initial guesses, then Mathcad might not arrive at a solution, or might arrive at a different solution than you wanted. If Mathcad is not arriving at a solution, you can refer to Mathcad Help for ideas on how to help resolve the issue.

Complex numbers can be used as an initial guess to arrive at a complex solution.

POLYROOTS FUNCTION

The ***polyroots*** function is used to solve for all solutions to a polynomial equation at the same time. The solution is returned in a vector containing the roots of the polynomial.

In order to use this function, you need to create a vector of coefficients of the polynomial. Include all coefficients in the vector even if they are zero. The coefficients of the polynomial $f(x) = 4x^3 - 2x^2 + 3x, - 5$, are (4, −2, 3, −5). For the Mathcad vector, however, you begin with the constant term. The Mathcad vector looks like this:

$$\mathbf{Vector} := \begin{pmatrix} -5 \\ 3 \\ -2 \\ 4 \end{pmatrix}$$

$\textbf{SampleFunction}(\mathbf{x}) := \mathbf{x}^2 + 5\mathbf{x} - 6$ — Define the function.

Find the root closest to x=0

The first form of *root (bracketed)* needs two initial guesses such that a<b, and SampleFunction(a)<0 and SampleFunction(b)>0.

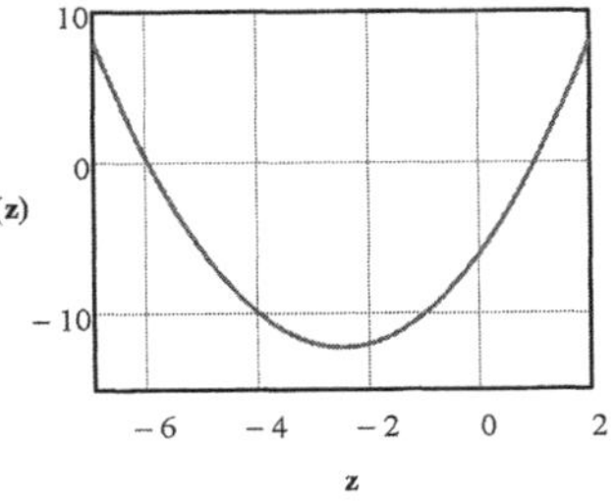

$\textbf{Root}_{1a} := \textbf{root}(\textbf{SampleFunction}(\mathbf{x}), \mathbf{x}, 0, 2)$

$\textbf{Root}_{1a} = 1.00$

The second form of *root* (unbracketed) does not need the two values a and b, but you must define a guess value for x prior to using the *root* function.

$\mathbf{x} := 0$ Guess value for x

$\textbf{Root}_{1b} := \textbf{root}(\textbf{SampleFunction}(\mathbf{x}), \mathbf{x})$ $\textbf{Root}_{1b} = 1.00$

Find the root closest to x= - 5 Bracketed

$\textbf{Root}_{1c} := \textbf{root}(\textbf{SampleFunction}(\mathbf{x}), \mathbf{x}, -7, -5)$ $\textbf{Root}_{1c} = -6.00$

Unbracketed

$\mathbf{x} := -5$ New guess value for x

$\textbf{Root}_{1d} := \textbf{root}(\textbf{SampleFunction}(\mathbf{x}), \mathbf{x})$ $\textbf{Root}_{1d} = -6.00$

Check to see if results are zero at calculated roots.

$\textbf{SampleFunction}(\textbf{Root}_{1a}) = 0.00$ $\textbf{SampleFunction}(\textbf{Root}_{1c}) = 0.00$ **OK**

Here is a way to input your guess value as the argument of a function

$\textbf{Test}(\mathbf{x}) := \textbf{root}(\textbf{SampleFunction}(\mathbf{x}), \mathbf{x})$ — This function provides an initial guess to "SampleFunction."

$\textbf{Root}_{1e} := \textbf{Test}(0)$ $\textbf{Root}_{1e} = 1.00$ — Root of "SampleFunction" with x= 0 as the guess value.

$\textbf{Root}_{1f} := \textbf{Test}(-5)$ $\textbf{Root}_{1f} = -6.00$ — Root of "SampleFunction" with x= -5 as the guess value.

FIGURE 10.1 Using the function *root*

$F_2(x) := 2^x - 9x^2 + 120$

Define the function

From the plot, use the values 7 and 10 for a and b.

$Root_{2a} := root(F_2(x), x, 7, 10)$

$Root_{2a} = 9.40$

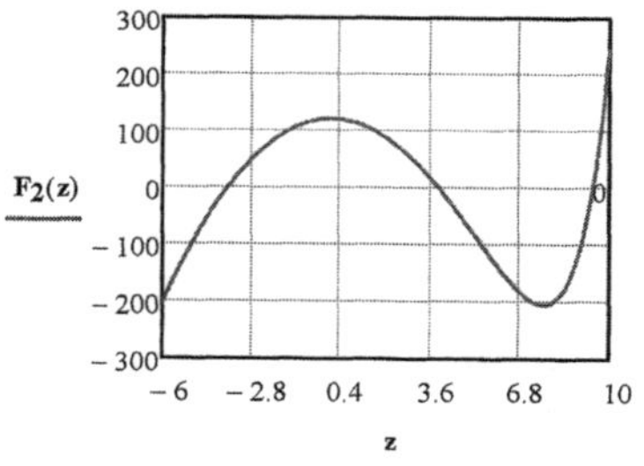

From the plot, use the values 3 and 4 for a and b.

$Root_{2b} := root(F_2(x), x, 3, 4)$

$Root_{2b} = 3.87$

From the plot, use the values -6 and -2 for a and b.

$Root_{2c} := root(F_2(x), x, -6, -2)$

$Root_{2c} = -3.65$

Check to see if results are zero at calculated roots.

$F_2(Root_{2a}) = 0.00$ $\quad F_2(Root_{2b}) = 0.00$ $\quad F_2(Root_{2c}) = -2.842 \times 10^{-14}$ **OK**

Input the guess value as part of a function

$TT(x) := root(F_2(x), x)$ This function provides an initial guess to "F_2".

$Root_{2d} := TT(-6)$ $\quad Root_{2d} = -3.65$ Root of "F_2" with x=-6 as the guess value.

$Root_{2e} := TT(3)$ $\quad Root_{2e} = 3.87$ Root of F_2" with x=3 as the guess value.

$Root_{2f} := TT(10)$ $\quad Root_{2f} = 9.40$ Root of "F_2" with x=10 as the guess value.

FIGURE 10.2 Another example of using the function *root*

The ***polyroots*** function takes the following form: ***polyroots***(v), where v is the vector of polynomial coefficients. The result is a vector containing the roots of the polynomial. Figure 10.3 shows the solution of the same simple quadratic equation as was used in Figure 10.1.

Figure 10.4 illustrates a larger polynomial. Notice that this polynomial has two imaginary roots. The solution to this polynomial is displayed to 12 decimal places to illustrate two different internal solution methods Mathcad uses to solve for

To use the *polyroots* function, create a vector of each of the coefficients of the polynomial. Begin with the constant term as the first element. Use zero for all zero coefficients.

Use *polyroots* to find the roots of the same equation as in Figure 10.1

$\mathbf{SampleFunction_{3a}(x)} := x^2 + 5x - 6$ $\quad \mathbf{V_{3a}} := \begin{pmatrix} -6 \\ 5 \\ 1 \end{pmatrix}$ Create vector of coefficients beginning with the constants term.

$\mathbf{Solution\_3a} := \mathbf{polyroots}(\mathbf{V_{3a}})$ $\quad \mathbf{Solution\_3a} = \begin{pmatrix} -6.00 \\ 1.00 \end{pmatrix}$

Check to see if results are zero at calculated roots.

$\mathbf{SampleFunction_{3a}}(\mathbf{Solution\_3a_1}) = 0.00$ $\quad \mathbf{SampleFunction_{3a}}(\mathbf{Solution\_3a_2}) = 0.00$

OK

FIGURE 10.3 Using the function *polyroots*

roots of a polynomial. These two internal solution methods are the LaGuerre Method and the Companion Matrix Method. This book will not discuss the differences between these solution methods. It just points out that if one method does not produce satisfactory results, there is another option available that might produce results that are slightly more accurate. The check results are displayed twice. The first result uses the General result format. The second result uses the Decimal result format with the number of decimal places set to 17. Notice that the results compare to at least the seventh decimal place.

Figure 10.5 uses live symbolics to have Mathcad create the polynomial vector automatically.

$\textbf{SampleFunction}_{4a}(x) := -x^5 - 2x^4 + x^3 - x^2 + 5$

Create a vector of coefficients.
Remember that the constant goes first.

$$\mathbf{V}_{4a} := \begin{pmatrix} 5 \\ 0 \\ -1 \\ 1 \\ -2 \\ -1 \end{pmatrix}$$

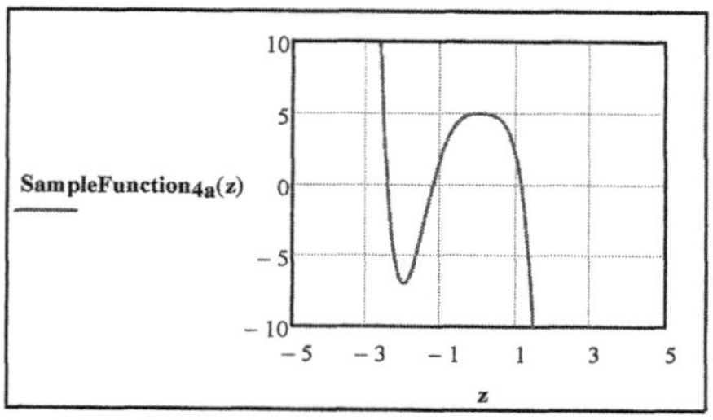

$\textbf{Solution\_4a} := \textbf{polyroots}(\mathbf{V}_{4a})$

$$\textbf{Solution\_4a} = \begin{pmatrix} -2.436964498961 \\ -1.216885522117 \\ 0.260206634327 + 1.191574629143i \\ 0.260206643463 - 1.191574625107i \\ 1.133436743287 \end{pmatrix}$$

Check $\textbf{Check}_{4a} := \textbf{SampleFunction}_{4a}(\textbf{Solution\_4a})$

$$\textbf{Check}_{4a} = \begin{pmatrix} 2.59 \times 10^{-8} \\ -7.68 \times 10^{-9} \\ -7.68 \times 10^{-9} - 1.22i \times 10^{-15} \\ 4.28 \times 10^{-8} - 1.90i \times 10^{-7} \\ 1.43 \times 10^{-7} \end{pmatrix}$$

Display results with decimals.

$$\textbf{Check}_{4a} = \begin{pmatrix} 0.00000002593219950 \\ -0.00000000767774910 \\ -0.00000000767775177 - 0.00000000000000122i \\ 0.00000004282237942 - 0.00000019017498676i \\ 0.00000014337904730 \end{pmatrix}$$

Note:
The above solutions are very close to zero. Because Mathcad uses an iterative process to solve for solutions, the results may not be exactly zero. Mathcad has two internal methods for finding polyroots. To select the second method, right click on the *polyroots* function and select Companion Matrix instead of LaGuerre. The following solution uses the Companion Matrix, which produces a slightly more accurate solution.

For most engineering applications, both solutions produce adequate results.

$\textbf{Solution\_4b} := \textbf{polyroots}(\mathbf{V}_{4a})$

$$\textbf{Solution\_4b} = \begin{pmatrix} -2.436964498276 \\ -1.216885521373 \\ 0.260206634267 - 1.191574628758i \\ 0.260206634267 + 1.191574628758i \\ 1.133436751116 \end{pmatrix}$$

Check

$\textbf{Check}_{4b} := \textbf{SampleFunction}_{4a}(\textbf{Solution\_4b})$

$$\textbf{Check}_{4b} = \begin{pmatrix} 5.86 \times 10^{-14} \\ 7.99 \times 10^{-15} \\ 1.78 \times 10^{-14} \\ 1.78 \times 10^{-14} \\ 4.44 \times 10^{-15} \end{pmatrix}$$

$$\textbf{Check}_{4b} = \begin{pmatrix} 0.00000000000005862 \\ 0.00000000000000799 \\ 0.00000000000001776 + 0.00000000000000067i \\ 0.00000000000001776 - 0.00000000000000067i \\ 0.00000000000000444 \end{pmatrix}$$

FIGURE 10.4 Using the function polyroots for a larger polynomial

Mathcad can automate the generation of the coefficient vector.
Use the following steps:
1. Type the name of the variable to be used for the vector, followed by a colon.
2. Open the symbolic toolbar.
3. Click the coeffs button.
4. Type the function name in the first placeholder using a previously undefined variable name as the argument.
5. In the second placeholder, type the name of the argument.
6. Click outside the region.

For this example, use the same equation as in Figure 10.4.

$$\mathbf{SampleFunction_{4a}(x)} := -x^5 - 2x^4 + x^3 - x^2 + 5$$

Rather than redefine $\text{SampleFunction}_{4a}$ we are just displaying it. (Remember - Properties, Calculation Tab, Disable Evaluation)

$$\mathbf{V_{5a}} := \mathbf{SampleFunction_{4a}(z)\ coeffs, z} \rightarrow \begin{pmatrix} 5 \\ 0 \\ -1 \\ 1 \\ -2 \\ -1 \end{pmatrix} \qquad \mathbf{V_{5a}} = \begin{pmatrix} 5.00 \\ 0.00 \\ -1.00 \\ 1.00 \\ -2.00 \\ -1.00 \end{pmatrix}$$

Mathcad generates the coefficient vector. If $\text{SampleFunction}_{4a}$ changes, the vector will update automatically.

$$\mathbf{Solution\_5a} := \mathbf{polyroots}(\mathbf{V_{5a}})$$

$$\mathbf{Solution\_5a} = \begin{pmatrix} -2.436964499 \\ -1.216885522 \\ 0.260206634 + 1.191574629i \\ 0.260206643 - 1.191574625i \\ 1.133436743 \end{pmatrix}$$

FIGURE 10.5 Using live symbolics to have Mathcad create the polynomial vector

SOLVE BLOCKS USING GIVEN AND FIND

Until now, we have been solving single equations and single variables. Mathcad is able to solve for multiple equations using multiple unknowns.

Several methods can be used to solve for multiple equations. The first method we will use is called a Solve Block using the keyword ***Given*** and the ***Find*** function. This is illustrated in Figure 10.6.

This method has the following steps:

1. Give an initial guess for each variable you are solving for.
2. Type the keyword `Given`.

Find the intersection points of a parabola and a line.

Use a Solve Block with *Given* and *Find*

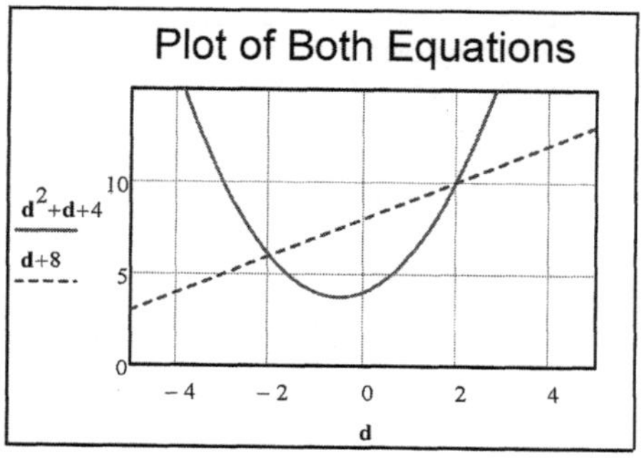

Give initial guesses for a and b.

a := 3   **b** := 12

Given Begin the Solve Block

List constraints.
Be sure to use the Boolean equal sign (CTRL+EQUAL).

$\mathbf{b} = \mathbf{a}^2 + \mathbf{a} + 4$ Parabola equation

$\mathbf{b} = \mathbf{a} + 8$ Line equation

Solution

$\begin{pmatrix} \mathbf{Solution\_6a} \\ \mathbf{Solution\_6b} \end{pmatrix} := \mathbf{Find(a, b)}$

The *Find* Function ends the solve block. Use both "a" and "b" in the *Find* function, because the constraint equations use both "a" and "b."

Solution_6a = 2.00   **Solution_6b** = 10.00

Check Result

$\mathbf{Solution\_6a}^2 + \mathbf{Solution\_6a} + 4 = 10.00$ **OK**

$\mathbf{Solution\_6a} + 8 = 10.00$ **OK**

Have Mathcad solve for the second solution. Change the initial guess and try again

a := −3   **b** := 6

Given   Begins Solve Block

Gives Constraints.

$\mathbf{b} = \mathbf{a}^2 + \mathbf{a} + 4$

$\mathbf{b} = \mathbf{a} + 8$

$\begin{pmatrix} \mathbf{Solution\_6c} \\ \mathbf{Solution\_6d} \end{pmatrix} := \mathbf{Find(a, b)}$ End Solve Block

Solution_6c = −2.00   **Solution_6d** = 6.00   Mathcad selects the other solution.

Check Result

$\mathbf{Solution\_6c}^2 + \mathbf{Solution\_6c} + 4 = 6.00$ **OK**

$\mathbf{Solution\_6c} + 8 = 6.00$ **OK**

FIGURE 10.6 Using a Solve Block to solve two equations

3. Below the word ***Given***, type a list of constraint equations using Boolean operators.
4. Type the function ***Find***() and list the desired solution variables as arguments.

If you are solving for n unknowns, you must have at least n equations between the ***Given*** and the ***Find***. Equality equations must be defined using the Boolean equals. This must be inserted from the Boolean toolbar, or use **CTRL+EQUAL**. If there is more than one solution, the solution Mathcad calculates is based on your initial guesses. If you modify your guesses, Mathcad might arrive at a different solution. If one of the solutions is negative and the other is positive, you can add an additional constraint telling Mathcad that you want the solution to be greater than zero. See Figure 10.7.

You can also assign the equations as functions and use the functions as the constraints. See Figure 10.8.

Solve Blocks should not contain the following:

- Range variables
- Constraints involving the "not equal to" operator

Find the intersection point of a parabola and a line.

Use a Solve Block with *Given* and *Find*

$a := -4$ Give initial guess for a.

Given Begins the Solve Block

List constraints.

$a + 8 = a^2 + a + 4$ In this case we set both equations equal to each other (b=b)

$a > 0$ We want to find the positive solution, so set "a" greater than zero.

Solution

$Result\_7 := Find(a)$ $Result\_7 = 2.00$

$a := Result\_7$ $a = 2.00$

Check Result

$a^2 + a + 4 = 10.00$ **OK**

$a + 8 = 10.00$ **OK**

The *Find* function ends the solve block. Because "a" is the only variable in the Solve Block you may use only "a" as the argument in the *Find* function. Mathcad finds the positive solution even though our initial guess was negative.

FIGURE 10.7 Forcing Mathcad to solve for a positive root

$\mathbf{f} := -100$ — Initial guess of solution

$\mathbf{SampleFunction_{8a}(d)} := \mathbf{d}^2 + \mathbf{d} + 4$ — Define two functions

$\mathbf{SampleFunction_{8b}(d)} := \mathbf{d} + 8$ — Remember that the argument "d" is only used to define the function. We can use "d" or any other argument when solving the functions. In this case we use the variable "f".

Given — Begin Solve Block

$\mathbf{SampleFunction_{8a}(f)} = \mathbf{SampleFunction_{8b}(f)}$ — Set constraints

$\mathbf{Result\_8a} := \mathbf{Find(f)}$ $\qquad \mathbf{Result\_8a} = -2.00$

Check

$\mathbf{SampleFunction_{8a}(Result\_8a)} - \mathbf{SampleFunction_{8b}(Result\_8a)} = 0.00$ **OK**

Now add an additional constraint so that "f" is greater than zero

Given

$\mathbf{SampleFunction_{8a}(f)} = \mathbf{SampleFunction_{8b}(f)}$

$\mathbf{f} > 0$

$\mathbf{Result\_8b} := \mathbf{Find(f)}$ $\qquad \mathbf{Result\_8b} = 2.00$ — Mathcad finds the positive solution

Check

$\mathbf{SampleFunction_{8a}(Result\_8b)} - \mathbf{SampleFunction_{8b}(Result\_8b)} = -0.00$ **OK**

FIGURE 10.8 Solve Block using functions instead of equations

- Other Solve blocks; each Solve Block can have only one ***Given*** and one ***Find***
- Assignment statements using the assignment operator (:=)

LSOLVE FUNCTION

The ***lsolve*** function is used similar to a Solve Block because you can solve for x equations and x unknowns. It is also similar to the ***polyroots*** function because the input arguments are vectors representing the coefficients of the equations. The use of this function is illustrated in Figure 10.9.

SOLVE BLOCKS USING *MAXIMIZE* AND *MINIMIZE*

The ***Maximize*** and ***Minimize*** functions allow you to find the maximum and minimum values that satisfy the constraints of the Solve Block. These function take the form of ***Maximize***(f, var1, var2, ...). The argument f is a previously

We can use a Solve Block to solve for four equations and four unknowns

Given $\quad w := 1 \quad x := 1 \quad y := 1 \quad z := 1$ Initial guess values

$$2.4\mathbf{w} + 6.1\mathbf{x} + 6.6\mathbf{y} + 2.1\mathbf{z} = 2.5$$

$$-7.5\mathbf{w} - 1.5\mathbf{x} - 2.3\mathbf{y} + 4.5\mathbf{z} = 14$$

$$4.3\mathbf{w} + 4.2\mathbf{x} + 1.8\mathbf{y} + 2.1\mathbf{z} = 10.2$$

$$5.8\mathbf{w} + 3.4\mathbf{x} + 7.8\mathbf{y} + 1.5\mathbf{z} = 0.1$$

$$\mathbf{Solution_A} := \mathbf{Find}(\mathbf{w}, \mathbf{x}, \mathbf{y}, \mathbf{z})$$

$$\mathbf{Solution_A} = \begin{pmatrix} 0.7686 \\ 0.3739 \\ -1.4480 \\ 3.7766 \end{pmatrix}$$

There is another way to solve for this condition. It is a function called *lsolve*.

This function is similar to the function polyroots, in that you use an array as input. To use this function, create a Matrix (M) of the coefficients of the four unknowns. Then create a vector of the constraints (V). The function takes the form *lsolve*(M,V)

$$\mathbf{Matrix} := \begin{pmatrix} 2.4 & 6.1 & 6.6 & 2.1 \\ -7.5 & -1.5 & -2.3 & 4.5 \\ 4.3 & 4.2 & 1.8 & 2.1 \\ 5.8 & 3.4 & 7.8 & 1.5 \end{pmatrix} \qquad \mathbf{Vector} := \begin{pmatrix} 2.5 \\ 14 \\ 10.2 \\ 0.1 \end{pmatrix}$$

With a Solve Block, you must give initial guess values. You do not need to give initial guesses with the *lsolve* function.

$$\mathbf{Solution_B} := \mathbf{lsolve}(\mathbf{Matrix}, \mathbf{Vector}) \qquad \mathbf{Solution_B} = \begin{pmatrix} 0.7686 \\ 0.3739 \\ -1.4480 \\ 3.7766 \end{pmatrix}$$

Check

$$\mathbf{Matrix} \cdot \mathbf{Solution_B} = \begin{pmatrix} 2.50 \\ 14.00 \\ 10.20 \\ 0.10 \end{pmatrix} \qquad \mathbf{Vector} = \begin{pmatrix} 2.50 \\ 14.00 \\ 10.20 \\ 0.10 \end{pmatrix}$$

OK

FIGURE 10.9 Using function *lsolve*

defined function that uses the arguments var1, var2, etc. The solution returns a vector with the values of var1, var2, etc, which satisfy the constraints in the Solve Block. Note that the solution does not return the actual maximum or minimum values of the function; it returns the values of the variables that when used in the function will return the maximum or minimum solution.

To use the ***Maximize*** function, follow these steps:

1. Define a function using the variables you are solving for.
2. Give initial guess values for the variables you are solving for.
3. Type the keyword ***Given***.
4. Below the word ***Given***, type a list of constraint equations using Boolean operators.
5. Type ***Maximize*** (then type the name of the function (do not include parentheses after the name of the function), then type the variables you are solving for separated by commas. Then type the closing parentheses). The ***Minimize*** function is used exactly the same way. See Figure 10.10. Note that these functions are very sensitive to boundary conditions and the initial guess value. In order to see this, try the equation shown in Figure 10.10 and vary the initial guess value and vary the boundary conditions. Additional examples of the Maximize and Minimize functions are found in Engineering Examples.

Tip! If ***Maximize*** or ***Minimize*** do not return a result, and the error box states, "This variable is undefined.", try adjusting the initial guess values. A plot may be helpful to select the initial guess values.

TOL, CTOL, AND MINERR

TOL and CTOL are built-in Mathcad variable names. They can be set in the Worksheet Options dialog box (**Worksheet Options** from the **Tools** menu). The values can also be redefined within the worksheet.

When using Solve Blocks and the ***root*** function, Mathcad iterates a solution. When the difference between the two most recent iterations is less than TOL (convergence tolerance), Mathcad arrives at a solution. The default value of TOL is 0.001. If you are not satisfied with the solutions arrived at, you can redefine TOL to be a smaller number. This will increase the precision of the result, but it will also increase the calculation time, or might make it impossible for Mathcad to arrive at a solution.

The CTOL (constraint tolerance) built-in variable tells Mathcad how closely a constraint in a Solve Block must be met for a solution to be acceptable. The default value of CTOL is 0.001. Thus, if a Solve Block Constraint is $x > 3$, this constraint is satisfied if $x > 2.9990$.

The ***Minerr*** function is used in place of the ***Find*** function in a Solve Block. You might want to use ***Minerr*** if Mathcad cannot iterate a solution. The ***Find*** function iterates a solution until the difference in the two most recent iterations is less than TOL. The ***Minerr*** function uses the last iteration even if it falls outside of TOL. This function is useful when ***Find*** fails to find a solution. If you use ***Minerr***, it is important to check the solution to see if the solution falls within your acceptable limits.

For more information on these topics, refer to the Mathcad Help. It has a good discussion of these topics in much greater depth.

USING UNITS

You can use units with ***root, polyroots***, and Solve Blocks. These will be illustrated in the following Engineering Examples.

ENGINEERING EXAMPLES

Engineering Example 10.1 summarizes six different ways of solving for the time it takes to travel a specific distance. Engineering Example 10.2 calculates the current flow in a multiloop circuit. Engineering Example 10.3 calculates the flow of water in a pipe network. Engineering Example 10.4 demonstrates how Mathcad can be used to solve chemistry equations. Engineering Example 10.5 uses a solve block to solve the same problem presented in Engineering Example 7.2. Engineering Example 10.6 uses the ***Maximize*** function to maximize the volume of a box with a fixed surface area. Engineering Example 10.7 uses given cost and revenue functions to calculate how many products to produce in order to obtain the maximum and minimum profits.

Engineering Example 10.1: Object in Motion

The general equation for the distance traveled by an object is given by the function $\mathbf{Dist} := \mathbf{v_0} \cdot \mathbf{t} + \frac{1}{2}\mathbf{a} \cdot \mathbf{t}^2$. Let's look at the different way of solving for this equation for the time needed to travel a specific distance.
Given: $v_0 = 10\text{m/s}$, $a = 6\text{m/sec}^2$, and Dist = 5000m

$$\mathbf{v_0} := 10\frac{\mathbf{m}}{\mathbf{sec}} \qquad \mathbf{a} := 6\frac{\mathbf{m}}{\mathbf{sec}^2} \qquad \mathbf{Dist} := 5000\mathbf{m}$$

First, Use Symbolics Menu

To do this, write the equation (collect terms on the same side of the equation and set it equal to zero) using the boolean equal sign (CTRL+PERIOD).
Select the variable t.
From the Symbolics toolbar, select Variable then Solve

$$\mathbf{v_0} \cdot \mathbf{t} + \frac{1}{2} \cdot \mathbf{a} \cdot \mathbf{t}^2 - \mathbf{Dist} = 0$$

$$\begin{bmatrix} \frac{1}{2 \cdot \mathbf{a}} \cdot \left[(-2) \cdot \mathbf{v_0} + 2 \cdot \left(\mathbf{v_0}^2 + 2 \cdot \mathbf{a} \cdot \mathbf{Dist} \right)^{\frac{1}{2}} \right] \\ \frac{1}{2 \cdot \mathbf{a}} \left[(-2) \cdot \mathbf{v_0} - 2 \cdot \left(\mathbf{v_0}^2 + 2 \cdot \mathbf{a} \cdot \mathbf{Dist} \right)^{\frac{1}{2}} \right] \end{bmatrix}$$

$$\mathbf{Time\_1_1} := \frac{1}{2\cdot \mathbf{a}}\cdot\left[(-2)\cdot \mathbf{v_0} + 2\cdot\left(\mathbf{v_0}^2 + 2\cdot\mathbf{a}\cdot\mathbf{Dist}\right)^{\frac{1}{2}}\right]$$

Hint: Copy and paste the equations rather than retype them.

$$\mathbf{Time\_1_2} := \frac{1}{2\cdot \mathbf{a}}\cdot\left[(-2)\cdot \mathbf{v_0} - 2\cdot\left(\mathbf{v_0}^2 + 2\cdot\mathbf{a}\cdot\mathbf{Dist}\right)^{\frac{1}{2}}\right]$$

$\mathbf{Time\_1_1} = 39.19\mathrm{s}$

$\mathbf{Time\_1_2} = -42.53\mathrm{s}$ $\quad \mathbf{Time\_1} = \begin{pmatrix} 39.19 \\ -42.53 \end{pmatrix}\mathrm{s}$

Second, Use Live Symbolics

$$\mathbf{Time\_2a} := \mathbf{v_0}\cdot\mathbf{t} + \frac{1}{2}\cdot\mathbf{a}\cdot\mathbf{t}^2 - \mathbf{Dist}\ \mathbf{solve,t} \rightarrow \begin{bmatrix} \dfrac{\mathbf{sec}^2\cdot\left(\dfrac{5\cdot\sqrt{601}}{\mathbf{sec}} + \dfrac{5}{\mathbf{sec}}\right)}{3} \\ \dfrac{\mathbf{sec}^2\cdot\left(\dfrac{5}{\mathbf{sec}} - \dfrac{5\cdot\sqrt{601}}{\mathbf{sec}}\right)}{3} \end{bmatrix} \mathbf{float} \rightarrow \begin{pmatrix} -42.525502240437542846\cdot\mathbf{sec} \\ 39.192168907104209513\cdot\mathbf{sec} \end{pmatrix}$$

Hide the first solution

$$\mathbf{Time\_2b} := \mathbf{v_0}\cdot\mathbf{t} + \frac{1}{2}\cdot\mathbf{a}\cdot\mathbf{t}^2 - \mathbf{Dist}\ \Bigg|\begin{matrix}\mathbf{solve,t}\\ \mathbf{float}\end{matrix} \rightarrow \begin{pmatrix} -42.525502240437542846\cdot\mathbf{sec} \\ 39.192168907104209513\cdot\mathbf{sec} \end{pmatrix} \quad \mathbf{Time\_2b} = \begin{pmatrix} -42.53 \\ 39.19 \end{pmatrix}\mathrm{s}$$

$\mathbf{Time\_2b_1} = -42.53\ \mathrm{s} \quad \mathbf{Time\_2b_2} = 39.19\ \mathrm{s}$

Third, Create a Function Using Live Symbolics

Define the function

$$\mathbf{TimeFunction\_3(v_0, a, Dist)} := \mathbf{v_0}\cdot\mathbf{t} + \frac{1}{2}\mathbf{a}\cdot\mathbf{t}^2 - \mathbf{Dist}\ \mathbf{solve,t} \rightarrow \begin{bmatrix} \dfrac{2\cdot\left(\dfrac{\mathbf{v_0}}{2} + \dfrac{\sqrt{\mathbf{v_0}^2 + 2\cdot\mathbf{Dist}\cdot\mathbf{a}}}{2}\right)}{\mathbf{a}} \\ \dfrac{2\cdot\left(\dfrac{\mathbf{v_0}}{2} - \dfrac{\sqrt{\mathbf{v_0}^2 + 2\cdot\mathbf{Dist}\cdot\mathbf{a}}}{2}\right)}{\mathbf{a}} \end{bmatrix}$$

$\mathbf{Time\_3a} := \mathbf{TimeFunction\_3(v_0, a, Dist)}$ Apply the arguments to the function

$$\mathbf{Time\_3a} = \begin{pmatrix} -42.53 \\ 39.19 \end{pmatrix}\mathrm{s}$$

This method allows you to use other input variables very easily.

$$\textbf{Time\_3b} := \textbf{TimeFunction\_3}\left(20\frac{\textbf{m}}{\textbf{s}}, 20\frac{\textbf{m}}{\textbf{s}^2}, 100000\textbf{m}\right)$$

$$\textbf{Time\_3b} = \begin{pmatrix} -101.00 \\ 99.00 \end{pmatrix}\textbf{s}$$

Fourth, Use the Function root

Solving for the positive root.

$$\textbf{Time\_4}_1 := \textbf{root}\left(\textbf{v}_0 \cdot \textbf{t} + \frac{1}{2}\textbf{a} \cdot \textbf{t}^2 - \textbf{Dist}, \textbf{t}, 0\textbf{sec}, 100\textbf{sec}\right) \quad \textbf{Time\_4}_1 = 39.19\,\textbf{s}$$

$$\textbf{Time\_4}_2 := \textbf{root}\left(\textbf{v}_0 \cdot \textbf{t} + \frac{1}{2}\textbf{a} \cdot \textbf{t}^2 - \textbf{Dist}, \textbf{t}, -100\textbf{sec}, 0\textbf{sec}\right) \quad \textbf{Time\_4}_2 = -42.53\textbf{s} \quad \textbf{Time\_4} = \begin{pmatrix} 39.19 \\ -42.53 \end{pmatrix}\textbf{s}$$

Fifth, Use the Function polyroots

$$\textbf{v}_0 = 10.00\frac{\textbf{m}}{\textbf{s}} \quad \textbf{Dist} = 5000.00\textbf{m} \quad \textbf{a} = 6.00\frac{\textbf{m}}{\textbf{s}^{2.00}}$$

$$\textbf{Vector} := \begin{pmatrix} \dfrac{-\textbf{Dist}}{\textbf{m}} \\ \dfrac{\textbf{v}_0}{\dfrac{\textbf{m}}{\textbf{sec}}} \\ \dfrac{\frac{1}{2} \cdot \textbf{a}}{\dfrac{\textbf{m}}{\textbf{sec}^2}} \end{pmatrix}$$

Remember that all units in a vector must be of the same unit dimension. Divide each variable by its units to create unitless numbers.

$$\textbf{Time\_5} := \textbf{polyroots}(\textbf{Vector}) \cdot \textbf{sec}$$

$$\textbf{Time\_5} = \begin{pmatrix} -42.53 \\ 39.19 \end{pmatrix}\textbf{s} \quad \textbf{Time\_5}_1 = -42.53\ \textbf{s}$$

$$\textbf{Time\_5}_2 = 39.19\ \textbf{s} \quad \textbf{Time\_5} = \begin{pmatrix} -42.53 \\ 39.19 \end{pmatrix}\textbf{s}$$

Sixth, Use a Solve Block

$\textbf{t} := 100\textbf{sec}$ Initial Guess for time to find positive solution

Given

$$\textbf{v}_0 \cdot \textbf{t} + \tfrac{1}{2} \cdot \textbf{a} \cdot \textbf{t}^2 - \textbf{Dist} = 0$$

$$\textbf{Time\_6}_1 := \textbf{Find(t)} \quad \textbf{Time\_6}_1 = 39.19\ \textbf{s}$$

$\mathbf{t} := -30\mathbf{sec}$ Initial Guess for time to find negative solution

Given

$$\mathbf{v_0} \cdot \mathbf{t} + \frac{1}{2} \cdot \mathbf{a} \cdot \mathbf{t}^2 - \mathbf{Dist} = 0$$

$\mathbf{Time\_6_2} := \mathbf{Find(t)}$ $\mathbf{Time\_6_2} = -42.53\ \mathbf{s}$

Given $\mathbf{t} := -15\mathbf{sec}$

$$\mathbf{v_0} \cdot \mathbf{t} + \frac{1}{2} \cdot \mathbf{a} \cdot \mathbf{t}^2 - \mathbf{Dist} = 0$$

An initial negative guess for time would normally find a negative solution, but adding an additional constraint finds the positive solution.

$\mathbf{t} > 0\mathbf{sec}$ Set constraint to find positive solution

$\mathbf{Time\_6_3} := \mathbf{Find(t)}$ $\mathbf{Time\_6_3} = 39.19\ \mathbf{s}$

Engineering Example 10.2: Electrical Network

Using Kirchoff's current law, calculate the current flowing in this network. Kirchoff's current law states that the algebraic sum of voltage drops around any closed path within a circuit equals the sum of the voltage sources.
$\sum V = \sum(I^*R)$.

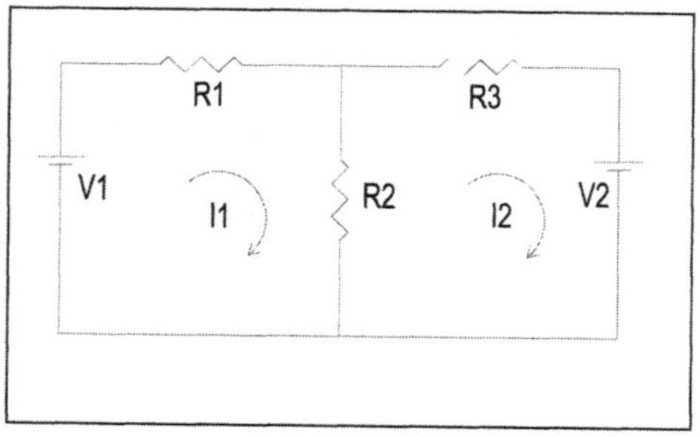

Resistance	Voltage	Initial Guess for Currents
$\mathbf{R_1} := 4 \cdot \Omega$	$\mathbf{V_1} := 75\mathbf{volt}$	$\mathbf{I_1} := 15 \cdot \mathbf{A}$
$\mathbf{R_2} := 15 \cdot \Omega$	$\mathbf{V_2} := 40\mathbf{volt}$	$\mathbf{I_2} := 15 \cdot \mathbf{A}$
$\mathbf{R_3} := 2 \cdot \Omega$		

Given Begins Solve Block

List constraints

$$\mathbf{V_1} = \mathbf{R_1} \cdot \mathbf{I_1} + \mathbf{R_2} \cdot (\mathbf{I_1} - \mathbf{I_2})$$
$$-\mathbf{V_2} = \mathbf{R_2}(\mathbf{I_2} - \mathbf{I_1}) + \mathbf{R_3} \cdot \mathbf{I_2}$$

$\mathbf{Current} := \mathbf{Find}(\mathbf{I_1}, \mathbf{I_2})$ $\quad \text{Current} = \begin{pmatrix} 6.89 \\ 3.72 \end{pmatrix} \mathbf{A}$ End solve block

$I_1 := \text{Current}_1$ $\quad I_2 := \text{Current}_2$ $\quad I_1 = 6.89\ \mathbf{A}$ $\quad I_2 = 3.72\ \mathbf{A}$

Check

$R_1 \cdot I_1 + R_2 \cdot (I_1 - I_2) - V_1 = 0.0000000\ V$ **OK**

$R_2(I_2 - I_1) + R_3 \cdot I_2 + V_2 = 7.1054274 \times 10^{-15} V$ **OK**

Engineering Example 10.3: Pipe Network

In a pipe network, two conditions must be satisfied: 1) The algebraic sum of the pressure drops around any closed loop must be zero (the pressure at any point is the same no matter how you get there), and 2) The flow entering a junction must equal the flow leaving it.

Use the Darcy-Weisbach equation for head loss $h_L := f \cdot \frac{L}{D} \cdot \frac{V^2}{2g}$. The factor f is a function of the relative roughness of the pipe. For this example assume f = 0.02.

$cfs := \frac{ft^3}{sec}$ $\quad f := 0.02$ $\quad TOL = 0.00100000000000000$. For this example, TOL = 0.001 is adequate.

Calculate the flow in each pipe given that:

Flow into A is 1.2 cfs
Flow out of C is 0.8 cfs
Flow out of E is 0.4 cfs

Pipe lengths and diameters Initial Guess

$L_{AB} := 3000ft$ $\quad D_{AB} := 4in$ $\quad Q_{AB} := .6cfs$

$L_{BC} := 4000ft$ $\quad D_{BC} := 5in$ $\quad Q_{BC} := .6cfs$

$L_{AE} := 2000ft$ $\quad D_{AE} := 3in$ $\quad Q_{AE} := .6cfs$

$L_{ED} := 3000ft$ $\quad D_{ED} := 8in$ $\quad Q_{ED} := .6cfs$

$L_{DC} := 5000ft$ $\quad D_{DC} := 6in$ $\quad Q_{DC} := .6cfs$

$L_{BD} := 3000ft$ $\quad D_{BD} := 7in$ $\quad Q_{BD} := .6cfs$

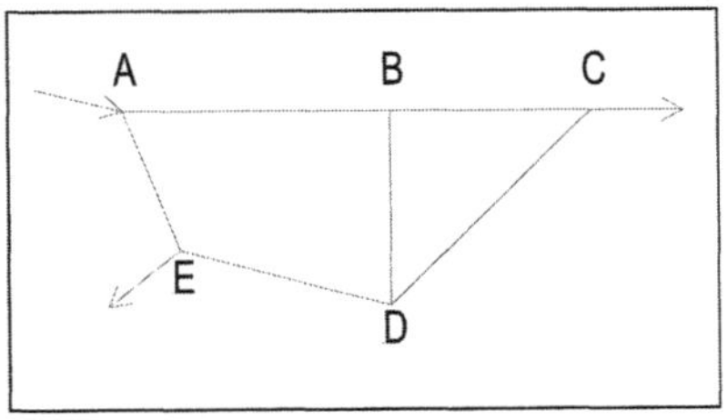

$$h_L(L, D, Q) := f \cdot \frac{L}{D} \cdot \frac{1}{2 \cdot g} \cdot \left(\frac{Q}{\frac{\pi \cdot D^2}{4}} \right)^2$$

Define a user-defined function for head loss based on the variables L, D, and Q. Velocity is a function of Flow and Area where V = Q/A

Given Begin Solve Block

Write equation for head loss at Point C.

$$(h_L(L_{AB}, D_{AB}, Q_{AB}) + h_L(L_{BC}, D_{BC}, Q_{BC})) - (h_L(L_{AE}, D_{AE}, Q_{AE}) + h_L(L_{ED}, D_{ED}, Q_{ED}) + h_L(L_{DC}, D_{DC}, Q_{DC})) = 0$$

Write equation for head loss at Point D.

$$(h_L(L_{AB}, D_{AB}, Q_{AB}) + h_L(L_{BD}, D_{BD}, Q_{BD})) - (h_L(L_{AE}, D_{AE}, Q_{AE}) + h_L(L_{ED}, D_{ED}, Q_{ED})) = 0$$

Write equation for head loss at Point B.

$$(h_L(L_{BC}, D_{BC}, Q_{BC})) - (h_L(L_{BD}, D_{BD}, Q_{BD}) + h_L(L_{DC}, D_{DC}, Q_{DC})) = 0$$

Write relationship between various flows into each of the points.

$$1.2\text{cfs} = Q_{AB} + Q_{AE} \qquad Q_{AB} = Q_{BD} + Q_{BC}$$
$$Q_{AE} = 0.4\text{cfs} + Q_{ED} \qquad Q_{ED} + Q_{BD} = Q_{DC}$$
$$Q_{BC} + Q_{DC} = 0.8\text{cfs}$$

$$\begin{pmatrix} Q_{AB} \\ Q_{BC} \\ Q_{AE} \\ Q_{ED} \\ Q_{DC} \\ Q_{BD} \end{pmatrix} := \text{Find}(Q_{AB}, Q_{BC}, Q_{AE}, Q_{ED}, Q_{DC}, Q_{BD})$$ End Solve Block

$Q_{AB} = 0.749210 \cdot \text{cfs}$

$Q_{BC} = 0.351147 \cdot \text{cfs}$

$Q_{AE} = 0.450790 \cdot \text{cfs}$

$Q_{ED} = 0.050790 \cdot \text{cfs}$

$Q_{DC} = 0.448853 \cdot \text{cfs}$

$Q_{BD} = 0.398063 \cdot \text{cfs}$

Check results

$$(h_L(L_{AB}, D_{AB}, Q_{AB}) + h_L(L_{BC}, D_{BC}, Q_{BC})) - (h_L(L_{AE}, D_{AE}, Q_{AE}) + h_L(L_{ED}, D_{ED}, Q_{ED}) + h_L(L_{DC}, D_{DC}, Q_{DC})) = 0.00 \cdot \text{ft}$$

$$(h_L(L_{AB}, D_{AB}, Q_{AB}) + h_L(L_{BD}, D_{BD}, Q_{BD})) - (h_L(L_{AE}, D_{AE}, Q_{AE}) + h_L(L_{ED}, D_{ED}, Q_{ED})) = 0.00 \cdot \text{ft}$$

$$(h_L(L_{BC}, D_{BC}, Q_{BC})) - (h_L(L_{BD}, D_{BD}, Q_{BD}) + h_L(L_{DC}, D_{DC}, Q_{DC})) = 0.00 \cdot \text{ft}$$

$Q_{AB} + Q_{AE} = 1.200000 \cdot \text{cfs}$ **Equals 1.2 cfs OK**

$Q_{AE} - Q_{ED} = 0.400000 \cdot \text{cfs}$ **Equals 0.4 cfs OK**

$Q_{BC} + Q_{DC} = 0.800000 \cdot \text{cfs}$ **Equals 0.8 cfs OK**

Engineering Example 10.4: Chemistry

You can get solutions to chemistry problems by writing out the needed equations and solving them using a solve block. See the example below.

The problem below involves a "closed system", that is, no exchange between the system and the outside, for example a closed beaker.

Find the concentration of carbonate $CO_3^{=}$ in equilibrium with a solution (*closed to the atmosphere*) containing 10^{-3} M soluble calcium [Ca^{+2}] and solid calcite, $CaCO_3$.

Problem interpretation: The calcite has dissolved up to the limit imposed by the solubility product. This is the point at which the [Ca^{+2}] is measured and found to be 10^{-3} M. The only other cations in the water are H^+ from the dissociation of water. The other anions in the water are hydroxide [OH^-] from water and the carbonate species created by the dissolution of $CaCO_3$. The dissolved carbonate molecules have reacted to form carbonate, bicarbonate and some carbonic acid according their respective equilibrium relationships.

First, define values for the constants to be used:

Soluble calcium in water $Ca := 10^{-3} \cdot \frac{\mathbf{mole}}{\mathbf{liter}}$

First dissociation constant for carbonic acid: $\mathbf{K_{a1}} := 10^{-6.3} \cdot \frac{\mathbf{mole}}{\mathbf{liter}}$

Second dissociation constant for carbonis acid: $\mathbf{K_{a2}} := 10^{-10.35} \cdot \frac{\mathbf{mole}}{\mathbf{liter}}$

Solubility product for calcite: $\mathbf{K_{so}} := 10^{-8.34} \cdot \left(\frac{\mathbf{mole}}{\mathbf{liter}}\right)^2$

Ion product of water: $\mathbf{K_w} := 10^{-14} \cdot \left(\frac{\mathbf{mole}}{\mathbf{liter}}\right)^2$

Unknowns: H^+, OH^-, H_2CO_3, HCO_3^-, $CO_3^{=}$. Therefore we need five equations.

Provide initial guesses for each unknown in the system of equations:

$\mathbf{H} := 10^{-7.0} \cdot \frac{\mathbf{mole}}{\mathbf{liter}}$ $\quad \mathbf{H2CO_3} := 10^{-9} \cdot \frac{\mathbf{mole}}{\mathbf{liter}}$ $\quad \mathbf{CO_3} := 10^{-2.9} \cdot \frac{\mathbf{mole}}{\mathbf{liter}}$

$\mathbf{OH} := \frac{\mathbf{K_w}}{\mathbf{H}}$ $\quad \mathbf{HCO_3} := 10^{-3.01} \cdot \frac{\mathbf{mole}}{\mathbf{liter}}$

Note: It sometimes helps to select initial guesses for H^+ and OH^- so that they satisfy the ion product of water:

Given

$\mathbf{H} \cdot \mathbf{OH} = \mathbf{K_w}$ Ionization of water

$\mathbf{H} \cdot \mathbf{HCO_3} = \mathbf{K_{a1}} \cdot \mathbf{H2CO_3}$ First dissociation for carbonic acid

$\mathbf{H} \cdot \mathbf{CO_3} = \mathbf{K_{a2}} \cdot \mathbf{HCO_3}$ Second dissociation for carbonic acid

$\mathbf{Ca + H = HCO_3 + 2 \cdot CO_3 + OH}$ Charge balance on the system

$\mathbf{Ca = H2CO_3 + HCO_3 + CO_3}$ Mass balance on carbon, all carbon species originated from calcium carbonate dissolution.

NOTE: Either the charge balance or mass balance on carbonate ions can be used as one of the equations to be solved

$\mathbf{Ca \cdot CO_3 = K_{so}}$ solubility product

$$\begin{pmatrix} H2CO3_{equil} \\ HCO3_{equil} \\ CO3_{equil} \\ H_{equil} \\ OH_{equil} \end{pmatrix} := \mathbf{Find}(H2CO_3, HCO_3, CO_3, H, OH)$$

$$H2CO3_{equil} = 0.00 \cdot \frac{mole}{liter}$$

$$HCO3_{equil} = 0.00 \cdot \frac{mole}{liter}$$

$$CO3_{equil} = 4.57 \times 10^{-6} \cdot \frac{mole}{liter}$$

$$H_{equil} = 9.67 \times 10^{-9} \cdot \frac{mole}{liter}$$

$$OH_{equil} = 1.03 \times 10^{-6} \cdot \frac{mole}{liter}$$

$$pH := -\log\left(\frac{H_{equil}}{\frac{mole}{liter}}\right) = 8.01$$

$$pOH := -\log\left(\frac{OH_{equil}}{\frac{mole}{liter}}\right) = 5.99$$

There are several ways to check the validity of our solution, for example the pH + pOH must equal 14

$$pH + pOH = 14.00$$

Now check to see if the solution is correct by substituting the values back into the original equations.

$$\frac{H_{equil} \cdot OH_{equil}}{K_w} = 1.00 \qquad \frac{Ca}{H2CO3_{equil} + HCO3_{equil} + CO3_{equil}} = 0.99$$

$$\frac{H_{equil} \cdot HCO3_{equil}}{K_{a1} \cdot H2CO3_{equil}} = 1 \qquad \frac{Ca \cdot CO3_{equil}}{K_{SO}} = 1.00 \qquad \frac{H_{equil} \cdot CO3_{equil}}{K_{a2} \cdot HCO3_{equil}} = 1.00$$

Note from Dr. Dixie Griffin: "The use of Mathcad to solve chemistry problems has been one of my most enlightening discoveries, as well as a time savings. An interesting aside here is that such problems often give answers varying over several to many orders of magnitude. Doing such problems by hand would be practically impossible. The graphical solutions give approximate solutions but require a knowledge of chemistry not posessed by the average civil engineering undergraduate."

Engineering Example 10.5: Determining the Flow Properties of a Circular Pipe Flowing Partially Full

In Engineering Example 7.2, we showed how to use plotting as a means to solve for the pipe diameter needed to carry 300 cfs flowing at a depth of 60% of the diameter.

In this example, we use a solve block to solve the same problem. Refer to Engineering Example 7.2 for a description of the problem and associated equations.

Known values

Pipe slope $S := 0.001 \cdot \frac{ft}{ft}$ Manning's coefficient $n := 0.013$ Flow $Q := 300 \frac{ft^3}{sec}$

Initial guesses for unknowns

$D := 15 \cdot ft$ $A_f := 42 \cdot ft^2$ $\alpha := 1 \cdot \pi$ $d := .6 \cdot D$ $R := \frac{D}{2}$

Begin solve block

Given Note: It sometimes works better to put the governing equations as a ratio equal to one.

$$\frac{Q}{\frac{1.49}{n} \cdot A_f \cdot \left(\frac{\frac{A_f}{R \cdot \alpha}}{ft}\right)^{\frac{2}{3}} \cdot S^{\frac{1}{2}} \cdot \frac{ft}{sec}} = 1 \qquad \frac{A_f}{\left[\frac{R^2}{2} \cdot (\alpha - \sin(\alpha))\right]} = 1 \qquad R = \frac{D}{2} \qquad \frac{d}{D} = 0.6 \qquad \frac{\cos\left(\frac{\alpha}{2}\right)}{\frac{R-d}{R}} = 1$$

$$\begin{pmatrix} A_{Solve} \\ d_{Solve} \\ \alpha_{Solve} \\ D_{Solve} \\ R_{Solve} \end{pmatrix} := \text{Find}(A_f, d, \alpha, D, R)$$

Pipe diameter $D_{Solve} = 9.42 \cdot ft$ Depth of flow $d_{Solve} = 5.65 \cdot ft$

Area of flow $A_{Solve} = 43.62 \cdot ft^2$ Relative depth $\frac{d_{Solve}}{D_{Solve}} = 0.60$

Angle $\alpha_{Solve} = 203.07 \cdot deg$ Radius $R_{Solve} = 4.71 \cdot ft$

Check the governing equations

$$\frac{Q}{\frac{1.49}{n} \cdot A_{Solve} \cdot \left(\frac{\frac{A_{Solve}}{\frac{D_{Solve}}{2} \cdot \alpha_{Solve}}}{ft}\right)^{\frac{2}{3}} \cdot S^{\frac{1}{2}} \cdot \frac{ft}{sec}} = 1.00 \qquad \frac{\cos\left(\frac{\alpha_{Solve}}{2}\right)}{\frac{R_{Solve} - d_{Solve}}{R_{Solve}}} = 1.00$$

$$\frac{A_{Solve}}{\left[\frac{R_{Solve}^2}{2} \cdot (\alpha_{Solve} - \sin(\alpha_{Solve}))\right]} = 1.00 \qquad \frac{R_{Solve}}{\frac{D_{Solve}}{2}} = 1.00 \qquad \frac{d_{Solve}}{D_{Solve}} = 0.60$$

The solution is the same as determined in Engineering Example 7.2, D=9.42 ft.

ENGINEERING EXAMPLE 10.6: BOX VOLUME

Maximize the volume of a box with an total surface area of 200 cm^2. The length of the top and bottom is to be twice the width.

Define Function

$$\mathbf{Volume(w, l, h)} := \mathbf{w \cdot l \cdot h} \qquad \mathbf{Area} := 200 \cdot \mathbf{cm^2}$$

Initial Guesses

$$\mathbf{w} := 4 \cdot \mathbf{cm} \qquad \mathbf{l} := 4\mathbf{cm} \qquad \mathbf{h} := 4\mathbf{cm}$$

Given

$$2 \cdot \mathbf{w} \cdot \mathbf{l} + 2 \cdot \mathbf{l} \cdot \mathbf{h} + 2 \cdot \mathbf{w} \cdot \mathbf{h} = \mathbf{Area} \qquad \mathbf{l} = 2 \cdot \mathbf{w}$$

$$\mathbf{Solution} := \mathbf{Maximize(Volume, w, l, h)} \qquad \mathbf{Solution} = \begin{pmatrix} 4.08 \\ 8.17 \\ 5.44 \end{pmatrix} \cdot \mathbf{cm}$$

$$\mathbf{w} := \mathbf{Solution_1} \qquad \mathbf{l} := \mathbf{Solution_2} \qquad \mathbf{h} := \mathbf{Solution_3}$$

$$\mathbf{Volume(w, l, h)} = 181.44 \cdot \mathbf{cm^3}$$

Verify Constraints $\quad 2 \cdot \mathbf{w} \cdot \mathbf{l} + 2 \cdot \mathbf{l} \cdot \mathbf{h} + 2 \cdot \mathbf{w} \cdot \mathbf{h} = 200.00 \cdot \mathbf{cm^2}$

$$2 \cdot \mathbf{w} = 8.17 \cdot \mathbf{cm} \qquad \mathbf{l} = 8.17 \cdot \mathbf{cm}$$

ENGINEERING EXAMPLE 10.7: MAXIMIZE PROFIT

The profit of certain manufactured product is determined by subtracting the cost of the product from the revenue of the product. The revenue function for a certain product was determined to be 10*n where n is the number of products produced in thousands.

The cost function for the same product was determined to be $n^3 - 7{*}n^2 + 18{*}n$.

Determine the number of products that should be produced to maximize the profit. Plot the revenue, cost, and profit to verify the solutions.

Define Functions

$$\mathbf{Rev(n)} := 10 \cdot \mathbf{n} \qquad \mathbf{Cost(n)} := \mathbf{n^3} - 7 \cdot \mathbf{n^2} + 18 \cdot \mathbf{n} \qquad \mathbf{Profit(n)} := \mathbf{Rev(n)} - \mathbf{Cost(n)}$$

Initial Guess **n**:= 5

In this example, there are no restraints, so a ***Given*** is not required.

$\mathbf{n_{max}} := \mathbf{Maximize(Profit, n)}$

$\mathbf{n_{max}} = 4.00$

$\mathbf{Profit(n_{max})} = 16.00$

$\mathbf{n_{min}} := \mathbf{Minimize(Profit, n)}$

$\mathbf{n_{min}} = 0.67$

$\mathbf{Profit(n_{min})} = -2.52$

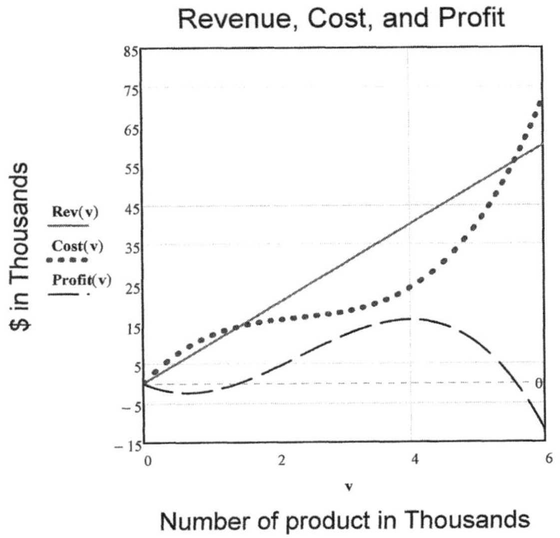

If we set restraints for n between 0 and 6, then the minimum will move to 6.

Given

$\mathbf{n} \geq 0$ $\mathbf{n} \leq 6$ $\mathbf{n_3} := \mathbf{Minimize(Profit, n)}$ $\mathbf{n_3} = 6.00$

$\mathbf{Profit(n_3)} = -12.00$

SUMMARY

The Mathcad Solve features are essential power tools to have in your toolbox. They should be kept sharp and used often.

In Chapter 10 we:

- Introduced the ***root*** function and showed how it can be used to solve for a single variable with a single function.
- Showed how the ***root*** function can be used with an initial guess prior to using ***root***, or how two guesses can be included as arguments in the ***root*** function.
- Encouraged the use of a plot to help define the initial guess values for the ***root*** function.
- Introduced the ***polyroots*** function and showed how it can be used to solve for all the roots of a polynomial equation.
- Briefly introduced the two solution methods used by the ***polyroots*** function (LaGuerre and Companion Matrix Methods).

- Introduced Solve Blocks and showed how they can be used to find solutions to multiple unknowns using multiple equations.
- Emphasized the rules of using ***Given***, ***Find***, ***Maximize***, and ***Minimize***.
- Discussed the built-in variables TOL and CTOL.
- Told how the ***Minerr*** function can be used in place of the ***Find*** function if Mathcad is not finding a solution.
- Provided several engineering examples to illustrate the use of the solve features.

PRACTICE

Additional problems and applications can be found on the companion site: www.elsevierdirect.com/9780123747839.

1. Create 10 polynomial equations. Use at least second- and third-order equations. Plot each equation. Use both the bracketed ***root*** function and unbracketed ***root*** function to solve for each equation.
2. Use the ***polyroots*** function to solve for each equation used in exercise 1.
3. Create three fifth-order equations, and use the ***polyroots*** function to solve for all roots. Use the live symbolics keyword "coeffs" to create the polynomial vector.
4. From your field of study, create four problems where you can use a Solve Block to solve for at least two unknowns with at least two constraints.

CHAPTER

Advanced Programming 11

Mathcad programs can be much more powerful than the logic programs discussed in Chapter 8. This chapter expands the topic of programming.

Chapter 11 will:

- Introduce the local definition symbol, and give examples of its use.
- Discuss the use of the other programming commands such as ***for***, ***while***, ***break***, ***continue***, ***return***, and ***on error***. Several examples will be given for the use of these features.
- Illustrate the use of programs within programs and the use of subroutines.
- Give examples that combine the features of programming, user-defined functions, and solving.

LOCAL DEFINITION

Mathcad allows you to define new variables within a program. These variables will be local to the program, meaning they will be undefined outside of the program. These are called local variables, and they are assigned with the local assignment operator. The local assignment operator is represented by an arrow pointing left ←. It can be found on the Programming toolbar, or inserted by pressing {. Let's look at a simple example of a program that uses local variables. See Figure 11.1.

As you can see from Figure 11.1, local variables can help simplify a program by breaking the program down into smaller pieces. In Figure 11.2, the function is very long, but each numerator is the same. By assigning a local variable, the function definition can be simplified.

You can also use local variables as counters to count how many times certain things are done in a program, or to count how many true statements are in a program. Figure 11.3 illustrates this concept.

Create a user-defined function that will find the two roots of a quadratic equation. Use local variables to simplify the program.

Use the left arrow button on the Programming toolbar to insert the local variable operator.

The local variables are only defined within the program. They are not defined outside the program.

$$\mathbf{QuadRoots}(\mathbf{a},\mathbf{b},\mathbf{c}) := \left| \begin{array}{l} \mathbf{Part} \leftarrow \sqrt{\mathbf{b}^2 - 4 \cdot \mathbf{a} \cdot \mathbf{c}} \\ x_1 \leftarrow \dfrac{-\mathbf{b} + \mathbf{Part}}{2\mathbf{a}} \\ x_2 \leftarrow \dfrac{-\mathbf{b} - \mathbf{Part}}{2\mathbf{a}} \\ \begin{pmatrix} x_1 \\ x_2 \end{pmatrix} \end{array} \right.$$

Define a local variable "Part."

Define a local variable "x_1."

Define a local variable "x_2."

Create an output vector comprising local variables x_1 and x_2. The values of x_1 and x_2 will be the output results of the function, but the local variables will not be recognized outside of the function.

$$\mathbf{QuadRoots}(1,5,6) = \begin{pmatrix} -2.00 \\ -3.00 \end{pmatrix}$$

Results of Function

$x_1 = ▮$ $\qquad$ $\mathrm{Part} = ▮$

The local variables are not recognized outside of the program.

$$\mathbf{Roots} := \mathbf{QuadRoots}(1,5,6)$$

$$\mathbf{Roots} = \begin{pmatrix} -2.00 \\ -3.00 \end{pmatrix} \qquad \mathbf{Roots}_1 = -2.00 \qquad \mathbf{Roots}_2 = -3.00$$

In order to make the results of the function available later on, the variable "Roots" was defined equal to the function QuadRoots.

FIGURE 11.1 Local variables

LOOPING

Looping allows program steps to be repeated a certain number of times, or until a certain criteria is met. This book introduces the concept of looping; an in-depth study of looping is beyond the scope of this book. The Mathcad Help and Tutorials provide excellent coverage on this subject.

There are two types of loop structures. The first allows a program to execute a specific number of times. This is called a ***for*** loop. The second type of loop structure will execute until a specific condition is met. This is called a ***while*** loop.

Local variables can help simplify complex equations

The numerators in this function are all the same.

$$G_1(x,y,z) := \frac{\left(\frac{x}{y}\right)+\left(\frac{y}{2x}\right)^2+\left(\frac{x}{3z}\right)^3+\left(\frac{z}{4x}\right)^4}{x}+\frac{\left(\frac{x}{y}\right)+\left(\frac{y}{2x}\right)^2+\left(\frac{x}{3z}\right)^3+\left(\frac{z}{4x}\right)^4}{y} \cdots$$
$$+\frac{\left(\frac{x}{y}\right)+\left(\frac{y}{2x}\right)^2+\left(\frac{x}{3z}\right)^3+\left(\frac{z}{4x}\right)^4}{z}$$

$G_1(2,6,3) = 2.61$

Note: Use CTRL+ENTER to break the expression at an addition operator.

If you define a local variable, the function is much simpler to follow.

$$G_2(x,y,z) := \left| \begin{array}{l} a \leftarrow \left(\frac{x}{y}\right)+\left(\frac{y}{2x}\right)^2+\left(\frac{x}{3z}\right)^3+\left(\frac{z}{4x}\right)^4 \\ \frac{a}{x}+\frac{a}{y}+\frac{a}{z} \end{array} \right.$$

$G_2(2,6,3) = 2.61$

FIGURE 11.2 Local variables

For Loops

You insert the ***for*** loop into a program by clicking the **for** button on the Programming toolbar. Do not type the word "for." A **for** loop takes the form

$$\begin{array}{l} \textbf{for } x \in y \\ \quad z \end{array}$$

Mathcad evaluates z for each value of x over the range y. The placeholders have these meanings:

- x is a local variable within the program defined by the ***for*** symbol. It initially takes the first value of y. The value of x changes for each step in the value of y.
- y is referred to as an iteration variable. It is usually a range variable, but it can be a vector, or a series of scalars or arrays separated by commas. Each time the program loops, the next value of y is used and this value is assigned to the variable x.
- z is any valid Mathcad expression or sequence of expressions.

The number of times a ***for*** loop executes is controlled by y.

Let's first look at two examples using range variables. In the examples, think of the ∈ symbol as a local variable definition. The local variable "i" is a range variable, and the value of variable "i" increments for each loop. Figure 11.4 shows two examples.

Local variables can be used as counters.

$$\mathbf{Counter}(\mathbf{x}) := \left| \begin{array}{l} \mathbf{m} \leftarrow 0 \\ \mathbf{m} \leftarrow \mathbf{m}+1 \ \ \textbf{if} \ \ \mathbf{x} > 4 \\ \mathbf{m} \leftarrow \mathbf{m}+1 \ \ \textbf{if} \ \ \mathbf{x} > 3 \\ \mathbf{m} \leftarrow \mathbf{m}+1 \ \ \textbf{if} \ \ \mathbf{x} > 2 \\ \mathbf{m} \leftarrow \mathbf{m}+1 \ \ \textbf{if} \ \ \mathbf{x} > 1 \\ \mathbf{m} \leftarrow \mathbf{m}+1 \ \ \textbf{if} \ \ \mathbf{x} > 0 \\ \mathbf{m} \end{array} \right.$$

Define a local variable "m" equal to zero.

This program adds 1 to the local variable "m" for every true statement.

$\mathbf{Counter}(-1) = 0.00$

$\mathbf{Counter}(1) = 1.00$

$\mathbf{Counter}(3) = 3.00$

$\mathbf{Counter}(5) = 5.00$

Check the program for different arguments.

In order to check a program for several different arguments, you can use a range variable or a vector. The example below use both a range variable and a vector to provide different arguments to the function "Counter."

Using a Range Variable

$\mathbf{i} := -1, 0..7$

$\mathbf{i} =$	$\mathbf{Counter}(\mathbf{i}) =$
-1.00	0.00
0.00	0.00
1.00	1.00
2.00	2.00
3.00	3.00
4.00	4.00
5.00	5.00
6.00	5.00
7.00	5.00

Using a Vector

The vectorize operator (CTRL+MINUS SIGN) must be used in order to use the vector with this program.

$$\mathbf{b} := \begin{pmatrix} -1 \\ 0 \\ 1 \\ 2 \\ 3 \\ 4 \\ 5 \\ 6 \\ 7 \end{pmatrix} \qquad \overrightarrow{\mathbf{Counter}(\mathbf{b})} = \begin{pmatrix} 0.00 \\ 0.00 \\ 1.00 \\ 2.00 \\ 3.00 \\ 4.00 \\ 5.00 \\ 5.00 \\ 5.00 \end{pmatrix}$$

FIGURE 11.3 Local variables as counters

Figure 11.5 shows a ***for*** loop using a list as the iteration variable. Notice that the list of numbers is not consecutive, and the numbers do not increment by the same value every time. When Mathcad loops, it moves to the next value in the list and assigns the value in the list to the local variable "i." Figure 11.6 is the same as Figure 11.5 except that the iteration variable is a previously defined vector containing the same numbers as the list in Figure 11.5.

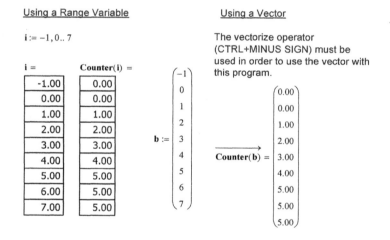

$j := 1, 2..5$

$$\text{ForLoop}_1(x) := \left| \begin{array}{l} a \leftarrow 0 \\ \text{for } i \in 1, 2..x \\ \quad a \leftarrow a + 1 \end{array} \right.$$

Define a local variable "a" equal to zero.

"i" is a local range variable that takes on the values 1,2,3... until x. For this example, i is not used anywhere. It only controls when the loop stops.

j =	$\text{ForLoop}_1(j)$ =
1.00	1.00
2.00	2.00
3.00	3.00
4.00	4.00
5.00	5.00

For the first loop, "a" takes the value 0+1=1.
For the second loop "i" increments to 2, and "a" becomes 1+1=2.
For the third loop "i" increments to 3, and "a" becomes 2+1=3.
The loop continues until "i" reaches the value of x.

For this next example, increment "a" by the value of i.

$$\text{ForLoop}_2(x) := \left| \begin{array}{l} a \leftarrow 0 \\ \text{for } i \in 1, 2..x \\ \quad a \leftarrow a + i \end{array} \right.$$

Define "a" as a local variable.

"i" is a local range variable that takes on the values 1,2,3... until x.

j =	$\text{ForLoop}_2(j)$ =
1.00	1.00
2.00	3.00
3.00	6.00
4.00	10.00
5.00	15.00

For the first loop, "i" is one, and "a" takes the value 0+1=1.
For the second loop, "i" is 2 and "a" becomes 1+2=3
For the third loop, "i" is 3, and "a" becomes 3+3=6.
For the fourth loop, "i" is 4, and "a" becomes 6+4=10.
For the fifth loop, "i" is 5, and "a" becomes 10+5=15.

FIGURE 11.4 ***For*** loop using range variables

Figure 11.7 uses two local variables. It uses the second variable "b" as a means to sum all the different values of "a."

Chapter 5 showed a method of converting a range variable to a vector. We did not tell how it was done; we only showed how it was possible. It is accomplished by using a ***for*** loop. Figure 11.8 shows how to convert a range variable to a vector. It also shows how to create a vector in a manner similar to creating a range variable.

$$\mathbf{ForLoop_3(x)} := \begin{array}{|l} \mathbf{a} \leftarrow 0 \\ \mathbf{for}\ \ \mathbf{i} \in 1,2,5,3 \\ \quad \mathbf{a} \leftarrow \mathbf{x} \cdot (\mathbf{a} + \mathbf{i}) \\ \mathbf{a} \end{array}$$

In this example the iteration variable is a list that controls the number of loops. The list also sets the assigned values of "i" .

The local variable "a" is incremented by the value of "i" and then multiplied by the argument "x".

$\mathbf{ForLoop_3}(2) = 58.00$

For x=2, the result is 2(0+1),2*(2+2),2*(8+5),2*(26+3), The final result is 2*(26+3)=58

j =	ForLoop3(j) =
1.00	11.00
2.00	58.00
3.00	189.00
4.00	476.00
5.00	1015.00

FIGURE 11.5 ***For*** loop using a list

$$\mathbf{Vector} := \begin{pmatrix} 1 \\ 2 \\ 5 \\ 3 \end{pmatrix}$$

$$\mathbf{ForLoop_4(x)} := \begin{array}{|l} \mathbf{a} \leftarrow 0 \\ \mathbf{for}\ \ \mathbf{i} \in \mathbf{Vector} \\ \quad \mathbf{a} \leftarrow \mathbf{x}(\mathbf{a} + \mathbf{i}) \end{array}$$

This is exactly the same as in the previous figure except that the iteration is controlled by a previously defined vector.

$\mathbf{ForLoop_4}(2) = 58.00$

j =	ForLoop4(j) =
1.00	11.00
2.00	58.00
3.00	189.00
4.00	476.00
5.00	1015.00

FIGURE 11.6 ***For*** loop using a vector

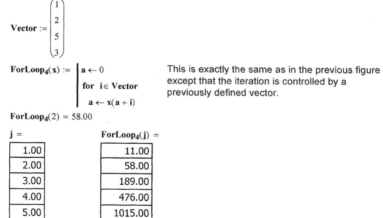

This example sums each value of "a"

$$\text{ForLoop}_5(x) := \left| \begin{array}{l} a \leftarrow 0 \\ b \leftarrow 0 \\ \text{for } i \in 1, 2..x \\ \quad \left| \begin{array}{l} a \leftarrow a + i \\ b \leftarrow b + a \end{array} \right. \\ b \end{array} \right.$$

j =	ForLoop5(j) =
1.00	1.00
2.00	4.00
3.00	10.00
4.00	20.00
5.00	35.00

Define "a" and "b" as a local variables.

"i" is a local variable that takes on the values 1,2,3 until "x."

For the first loop "i" is one, and "a" takes the value 0+1=1, Thus b=0+1=1.
For the next loop "i" is 2 and "a" becomes 1+2=3 and b=1+3=4.
For the third loop "i" is 3, and "a" becomes 3+3=6 and b=4+6=10.
For the fourth loop "i" is 4, and "a" becomes 6+4=10 and b=10+10=20.
For the fifth loop "i" is 5, and "a" becomes 10+5=15 and b=20+15=35.

FIGURE 11.7 ***For*** loop using two local variables

While Loops

The ***while*** loop is different from the ***for*** loop. A ***for*** loop executes a fixed number of times, based on the size of the iteration variable. A ***while*** loop continues to execute while a specified condition is met. Once the condition is not true, the execution stops. If the condition is always true, Mathcad will go into an infinite loop. For this reason, you must be careful when using a ***while*** loop. You insert the ***while*** loop into a program by clicking the **while** button on the Programming toolbar. Do not type the word "while."

Figure 11.9 shows a simple ***while*** loop. The program continues to loop until "a" is equal to or greater than the argument "x." When this happens, the loop stops and Mathcad returns the value of "a."

Figure 11.10 uses local variable "a" to control the number of loops. Local variable "b" is multiplied by 2 every time the program loops. Once "a" becomes equal to or greater than the argument "x," the loops stop and the program returns the value of "b."

Figure 11.11 shows three ways to create a user-defined function for factorial. Mathcad already has the ***factorial*** operator on the Math toolbar, but it is interesting to use factorial to demonstrate the ***while*** loop.

The ***while*** loop in Figure 11.12 will cause an infinite loop because the condition will always be true. If this condition happens, you can interrupt the loop by pressing the escape **ESC** key. Beginning with Mathcad version 13, you can also hover your mouse over **Debug** on the **Tools** menu and click **Interrupt**.

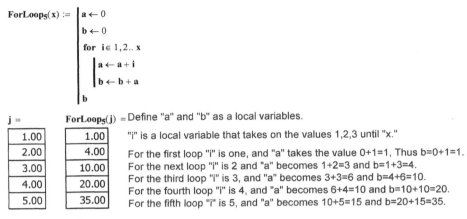

$$\text{Range2Vec(Range)} := \begin{vmatrix} \text{Count} \leftarrow \text{ORIGIN} \\ \text{for } i \in \text{Range} \\ \quad \begin{vmatrix} \text{Vec}_{\text{Count}} \leftarrow i \\ \text{Count} \leftarrow \text{Count} + 1 \end{vmatrix} \\ \text{Vec} \end{vmatrix}$$

This function converts the argument range variable to a vector by using the *for* loop. The counter is used as the vector index and begins at ORIGIN. This first value is assigned to the first value of the range variable. The process repeats until the end of the range variable. The vector "Vec" is a local variable.

$$\text{RV1} := 1, 1.2 .. 2.0 \qquad \text{NewVector\_1} := \text{Range2Vec(RV1)}$$

$$\text{NewVector\_1} = \begin{pmatrix} 1.00 \\ 1.20 \\ 1.40 \\ 1.60 \\ 1.80 \\ 2.00 \end{pmatrix} \qquad \text{NewVector\_1}_5 = 1.80$$

For loop to create a range vector

$$\text{RangeVec}(s_1, s_2, e) := \begin{vmatrix} \text{Count} \leftarrow \text{ORIGIN} \\ \text{for } i \in s_1, s_2 .. e \\ \quad \begin{vmatrix} \text{Vec}_{\text{Count}} \leftarrow i \\ \text{Count} \leftarrow \text{Count} + 1 \end{vmatrix} \\ \text{Vec} \end{vmatrix}$$

This function creates a vector with a range of values. This is based on a function in the February 2002 Mathcad Advisor Newsletter.
The *for* loop creates a range variable for iterating from the three input arguments. The vector "Vec" is a local variable.

Create a variable with range from 0 to 1 with 0.1 increment..

$\text{RangeVariable\_1} := 0, 0.1 .. 1.0$

$\text{NewVector\_2} := \text{RangeVec}(0, 0.1, 1.0)$

Refer to the discussion of range variables and vectors in Chapters 1 and 5.
These two variables appear to be identical, but they behave very differently.

$\text{RangeVariable\_1}_3 = \blacksquare$

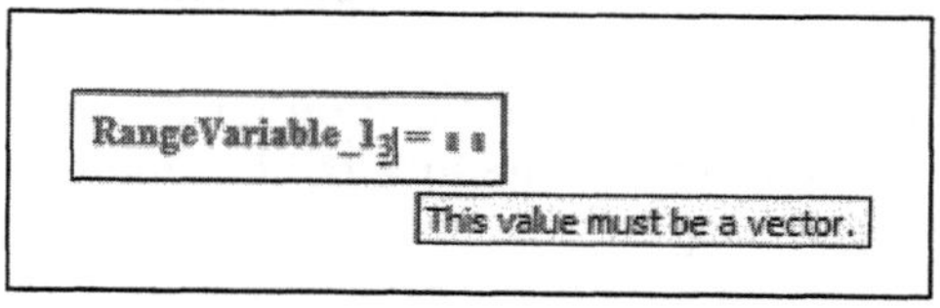

$$\boxed{\begin{array}{l} \boxed{\text{RangeVariable\_1}_3 = \blacksquare \blacksquare} \\ \qquad\qquad \boxed{\text{This value must be a vector.}} \end{array}}$$

$\text{NewVector\_1}_3 = 1.40$

RangeVariable_1 =	
	1
1	0.00
2	0.10
3	0.20
4	0.30
5	0.40
6	0.50
7	0.60
8	0.70
9	0.80
10	0.90
11	1.00

NewVector_2 =	
	1
1	0.00
2	0.10
3	0.20
4	0.30
5	0.40
6	0.50
7	0.60
8	0.70
9	0.80
10	0.90
11	1.00

FIGURE 11.8 *For* loop to convert a range variable to a vector

```
ForLoop1(x) := | a ← 0                 Defines local variable.
               | while a < x           Checks value of local variable.
               |   a ← a + 1           Increments local variable by one.
               | a                     Displays the value of "a" after the program stops.
```

j =	ForLoop$_1$(j) =
1.00	1.00
2.00	2.00
3.00	3.00
4.00	4.00
5.00	5.00

For the first loop "a" is zero, so the "while" statement is true and the program continues to the next line where "a" is incremented to the value 1.
On the second loop "a" is now 1 and the "while" statement is checked again. If it is true, then the program continues to the next line, and "a" is incremented to two. If the statement is false, then the program execution stops and the value of "a" is returned (1).
The loop continues until a >= x.

FIGURE 11.9 ***While*** loop

```
WhileLoop1(x) := | a ← 0               Defines local variables.
                 | b ← 1
                 | while a < x         Checks value of local variable "a".
                 |   | a ← a + 1       Increments "a" by one.
                 |   | b ← b · 2       Every time the program loops it multiplies "b" by 2
                 | b                   Displays the value of "b" after the program stops.
```

j =	WhileLoop$_1$(j) =
1.00	2.00
2.00	4.00
3.00	8.00
4.00	16.00
5.00	32.00

For the first loop "a" is zero, so the "while" statement is true and the program continues to the next lines where "a" is incremented to the value 1 and b is multiplied by 2.
On the second loop "a" is now 1 and the "while" statement is checked again. If a is less than x, then the program continues to the next line, and "a" is incremented to two and "b" becomes 2*2=4. If a is not less than x, then the program execution stops and the value of "b" is returned (4).

FIGURE 11.10 ***While*** loop

BREAK AND *CONTINUE* OPERATORS

The ***break*** and ***continue*** operators give more control to programs. They tell Mathcad what to do if specific conditions are met, and usually precede an ***if*** statement. These operators are used to check for specific conditions during the time that Mathcad is looping with either the ***for*** loop or the ***while*** loop.

The ***break*** operator stops the loop if the condition is true, and the ***continue*** operator allows Mathcad to skip the current loop iteration if a true statement

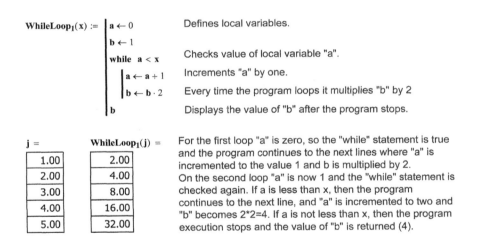

Mathcad already has a built in operator to calculate factorial, but let's look at a way to create a user-defined function for factorial.

Program	Comment
FactorialNew(x) := f ← x − 1	Assigns local variable "f" as one less than the argument "x"
while f > 0	Checks if "f" is greater than zero
x ← x · f	If "f" is greater than zero, reassigns "x" the value x*f
f ← f − 1	Reduces value of "f" by one
x	Returns x if "f" is not greater than zero

j =	FactorialNew(j) =
1.00	1.00
2.00	2.00
3.00	6.00
4.00	24.00
5.00	120.00

Assuming x=4:
On the first loop, "f" is assigned f=4-1=3.
Since "f" is greater than zero, x=4*3=12, and "f" is reduced by one to f=3-1=2, and the program returns to the while statement.
On the next loop, f=2 is greater than zero, so x=12*2=24, and f=2-1=1.
On the next loop, f=1 is greater than zero, so x=24*1=24, and f=1-1=0.
On the next loop, "f" is not greater than zero (the statement is false), so the execution stops and the value of x=24 is returned.

The above function is not perfect. If a negative number is entered, then the value of the negative number is returned. Some **if** statements are need to resolve this issue.

Note:
There are additional ways to create the factorial function.
The following are listed in Mathcad Help.

```
Fac1(n) := | f ← 1
           | while n ← n − 1
           |    f ← f · (n + 1)
           | f
```

$\mathbf{Fac_1}(4) = 24.00$

This while statement is not a conditional statement, but the loop stops when the value of n=0 (a false condition).

Assuming n=4:
On the first loop the argument "n" is reduced by 1 and becomes 3. The value of f becomes f=1*(3+1)=4.
On the second loop, n=3-1=2, and f=4*(2+1)=12.
On the third loop, n=2-1=1, and f=12*(2+1)=24.
On the fourth loop, n=1-1=0. This means that this is a false condition for the while loop and the loop stops and returns the value of f=24.

The following is a recursive function, which means that the function refers to itself. Refer to "Recursion" in Mathcad Help.

```
Fac2(n) := | n · Fac2(n − 1)  if n > 1
           | 1  otherwise
```

$\mathbf{Fac_2}(5) = 120.00 \qquad 5! = 120.00$

FIGURE 11.11 Factorial function

is encountered. The ***continue*** operator is useful if you want to continue looping even if a condition is true. For example, you might want to extract only even numbers. By using the ***continue*** operator, you can skip over odd numbers. Without the ***continue*** operator, Mathcad would stop if it encountered an odd number.

The following causes an infinite loop because "a" can never be anything but zero.

$$\text{WhileLoop}_2(x) := \left| \begin{array}{l} a \leftarrow 0 \\ \text{while } a \le x \\ \quad a \leftarrow a \cdot 2 \end{array} \right.$$

If this condition happens, press escape (esc) to stop the execution of the loop. Beginning with Mathcad 13, you can also click **Interrupt** by hovering your mouse over **Debug** on the **Tools** menu.

FIGURE 11.12 Avoid infinite loops

See Figure 11.13 for an example of the ***break*** operator. See Figure 11.14 for an example of the ***continue*** operator. These examples are relatively simple, and are intended to introduce the concept of using these two operators. More advanced examples are given in Mathcad Help.

This example is similar to Figures 11.9 and 11.10, but it allows you to give starting and ending numbers. In order to prevent an infinite loop, if the starting number is less than or equal to zero, the program is stopped by the break statement.

$$\text{WhileLoop}_3(w,x) := \left| \begin{array}{l} i \leftarrow 1 \\ \text{while } w \le x \\ \quad \left| \begin{array}{l} \text{break if } w \le 0 \\ a_i \leftarrow w \\ w \leftarrow w \cdot 2 \\ i \leftarrow i + 1 \end{array} \right. \\ a \end{array} \right.$$

Define local variable.

Continue looping as long as w <= x

Note: To add the *break* operator, add the **if** statement first, then click on the first placeholder and add the *break* operator. The break operator stops the loop if x<=0.

This creates a vector using the value of "i" as the vector subscript and assigns it the value of "w".

Changes the values of "w" and "i".

Returns the vector "a".

$$\text{WhileLoop}_3(1,32) = \begin{pmatrix} 1.00 \\ 2.00 \\ 4.00 \\ 8.00 \\ 16.00 \\ 32.00 \end{pmatrix}$$

$$\text{WhileLoop}_3(0,32) = 0.00$$

$$\text{WhileLoop}_3(-3,32) = 0.00$$

$$\text{WhileLoop}_3(5,100) = \begin{pmatrix} 5.00 \\ 10.00 \\ 20.00 \\ 40.00 \\ 80.00 \end{pmatrix}$$

FIGURE 11.13 Using the ***break*** operator to prevent infinite loops

This example illustrates how the continue operator can skip over certain conditions within a loop. In this example, the program skips over elements that are not integers. The function uses argument x to divide into integers from 1 to argument y, and selects only the integers.

```
WhileLoop4(x, y) := | i ← 0
                    | while i ≤ y
                    |   | j ← (i + 1)/x
                    |   | i ← i + 1
                    |   | continue  if floor(j) ≠ j
                    |   | a_i ← j
                    | a
```

Local variable "i" begins with 0, in order to increment before the continue statement.

Local variable "j" begins with j=(0+1)/x

Increment "i" by 1 to i=0+1=1

When you see the continue operator it means, "Continue with the loop if this statement is false. If the statement is true, then stop this loop and go back to the top of the loop." This statement is checking to see if "j" is an integer.
The local variable "a" is a vector with all the values of j.

Assume x=2, y=5:
On the first loop, i=0 and is less than 5 so the while loop is executed. The local variable j=(0+1)/2=0.5. Increment "i" by 1 to i=0+1=1. Check the continue statement. Floor(j)=0 is not equal to 0.5 so the statement is true. Since the statement is true, the continue operator tells Mathcad to stop the current loop and begin at the top of the loop.
On the second loop, i=1 and is less than 5 so the while loop is executed again. j=(1+1)/2=1.0. i=1+1=2. (Note: We had to increment i above the continue statement, because if the continue statement is true, the rest of the loop is skipped and "i" would not have been incremented. This would have created an infinite loop.) Floor(1)=1 is equal to 1.0 so the statement is false, and the loop continues. The vector "a" has element a_2 assigned the value 1.0. Because we did not assign element a_1, it is left at 0.0.
On the third loop, i=2 and is less than 5 so the while loop is executed again. j=(2+1)/2=1.5. i=2+1=3. Floor(1.5)=1 is not equal to 1.5. so the continue statement is true and the loop begins again at the top.
On the fourth loop, i=3 and is less than 5 so the while loop is executed again. j=(3+1)/2=2.0. i=3+1=4. Floor(2.0)=2 is equal to 2.0, so the continue statement is false, and the loop continues. The vector "a" has element a_4 assigned the value 2.0. Element a_3 is left at zero.
On the fifth loop, i=4 and is less than 5 so the while loop is executed again. j=(4+1)/2=2.5. i=4+1=5. Floor(2.5)=2 is not equal to 2.5. so the continue statement is true and the loop begins again at the top.
On the sixth loop, i=5 is less than or equal to 5 so the while loop is executed again. j=(5+1)/2=3.0. i=5+1=6. Floor (3.0)=3 is equal to 3.0, so the continue statement is false, and the loop continues. The vector "a" has element a_6 assigned the value 3.0. Element a_5 is left at zero.
On the next loop i=6 is not less than or equal to 5, so the while loop is stopped. The vector "a" is returned.

$WhileLoop_4(2,5) =$

	1
1	0.00
2	1.00
3	0.00
4	2.00
5	0.00
6	3.00

$WhileLoop_4(1,9) =$

	1
1	1.00
2	2.00
3	3.00
4	4.00
5	5.00
6	6.00
7	7.00
8	8.00
9	9.00
10	10.00

$WhileLoop_4(3,9) =$

	1
1	0.00
2	0.00
3	1.00
4	0.00
5	0.00
6	2.00
7	0.00
8	0.00
9	3.00

FIGURE 11.14 Using the ***continue*** operator

RETURN OPERATOR

We introduced the ***return*** operator in Chapter 8. This operator is used in conjunction with an ***if*** statement. It stops program execution when the ***if*** statement is true, and returns the value listed. See Figure 8.5 for an example.

ON ERROR OPERATOR

The on error operator is very useful to tell Mathcad what to do if it encounters an error. A very common error occurs when Mathcad tries to divide by zero. There will be many times when an input value to a function causes this to occur. The ***on error*** operator gives you the ability to tell Mathcad what to do when this occurs. See Figure 11.15 for several examples of using the ***on error*** operator.

The *on error* operator has the form x *on error* y. If y causes an error then x is evaluated.

$$\blacksquare \ \text{on error} \ \blacksquare$$

This figure give three methods of dealing with the function 24/x.

$\textbf{Result} := 0 \qquad \qquad j := -1, 0 .. 3$

Example 1   Example 2   Example 3

$\textbf{Example}_1(x) := \dfrac{24}{x}$ $\textbf{Example}_2(x) := \text{"Division by zero"} \ \ \textbf{on error} \dfrac{24}{x}$ $\textbf{Example}_3(x) := \infty \ \ \textbf{on error} \dfrac{24}{x}$

$\textbf{Example}_1(\textbf{Result}) = \blacksquare$ $\textbf{Example}_2(\textbf{Result}) = \text{"Division by zero"}$ $\textbf{Example}_3(\textbf{Result}) = 1.00 \times 10^{307}$

j =

-1.00
0.00
1.00
2.00
3.00

Note: To insert the built-in constant infinity type CRTL+SHIFT+z.

$\textbf{Example}_1(j) = \blacksquare$

$\boxed{\begin{array}{l} \textbf{Example}_1(j) = \blacksquare \ \blacksquare \\ \boxed{\text{Divide by zero.}} \end{array}}$

$\textbf{Example}_2(j) =$

	1
1	-24.00
2	"Division by zero"
3	24.00
4	12.00
5	8.00

$\textbf{Example}_3(j) =$

-24.00
$1.00 \cdot 10^{307}$
24.00
12.00
8.00

Example 1 produces errors,
Example 2 returns a string.
Example 3 returns Mathcad's numeric equivalent of infinity.

FIGURE 11.15 The ***on error*** operator

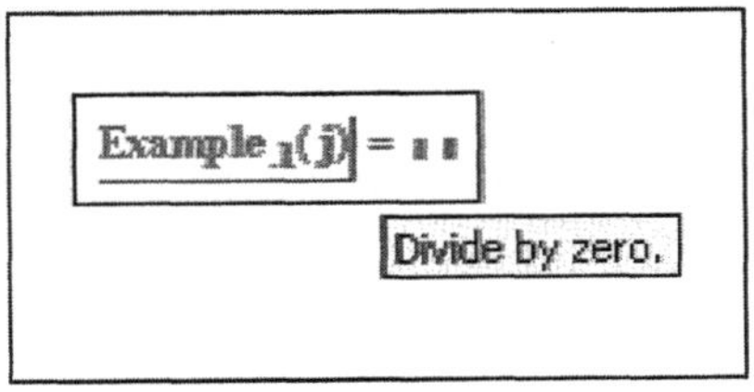

ENGINEERING EXAMPLE 11.1

Based on Problem 5.5 from CHIN, DAVID A., WATER RESOURCES ENGINEERING, 2nd, ©2007. Reproduced by permission of Pearson Education, Inc. Upper Saddle River, New Jersey.

DEVELOPMENT OF INTENSITY DURATION FREQUENCY (IDF) CURVES

Rainfall data (cm of rain) were compiled over several years by determining the maximum amount of rainfall that occurred in a given time period during each given year.

Use the rainfall data to develop IDF cures for 5-year, 10-year, and 25-year return periods.

The IDF curves should be fitted to a function of the form $\mathbf{i} = \frac{\mathbf{a}}{\mathbf{t} + \mathbf{b}}$.

This will occur in several steps:

Step 1. Calculate the rainfall intensity in cm/hr for each time period.

Step 2. Rank the rainfall intensity data by sorting heaviest to lightest for each time period. The lightest intensity will receive the highest rank and the highest intensity will receive the lowest rank. These numbers will be used to calculate the return period. The heaviest rains have a lower probability of occuring.

Step 3. Use the rankings to calculate return periods using the Wiebull formula.

Step 4. Plot the rainfall intensity vs return period for each rainfall time period.

Step 5. Create a data matrix giving rainfall intensities for 5-, 10-, and 25-year return periods and 1, 2, 4, 6, 10, 12, and 24 hour time periods.

Step 6. Plot a 5-year, 10-year, and 25-year IDF curve (using data points)

Step 7. Using the IDF curves, model the curves to fit a function of the form: $\mathbf{i} = \frac{\mathbf{a}}{\mathbf{t} + \mathbf{b}}$.

Step 8. Plot the IDF curves using the mathmatical formula and compare it to the data points.

The rainfall data for the years 1963 to 2007 is contained in an Excel spreadsheet file "RainfallData.xls." Use the Mathcad function READFILE to get the data and assign it to the variable "Rain."

Rain := **READFILE**("RainfallData.xls", "Excel")

Rain =

	1	2	3	4	5	6	7
1	NaN	"1-hr"	"2-hr"	"4-hr"	"6-hr"	"10-hr"	"12 hr"
2	1963.00	1.70	2.90	4.60	4.80	4.80	4.80
3	1964.00	1.90	2.60	2.60	3.00	3.70	3.70
4	1965.00	1.90	2.30	2.70	2.90	3.10	3.10
5	1966.00	1.90	2.40	2.80	4.00	4.70	4.90
6	1967.00	1.90	1.90	2.20	2.50	2.50	2.50
7	1968.00	2.10	2.40	3.00	3.50	4.00	4.20
8	1969.00	1.70	2.50	2.70	2.80	3.00	3.00
9	1970.00	1.70	3.60	3.60	3.60	4.30	4.30
10	1971.00	1.90	2.10	2.80	3.00	3.80	3.80
11	1972.00	2.40	3.70	4.50	4.50	4.60	4.60
12	1973.00	1.00	1.30	1.60	1.70	1.70	1.70
13	1974.00	2.70	2.80	3.20	3.70	3.70	3.80
14	1975.00	1.30	1.50	2.00	2.60	3.50	3.80
15	1976.00	2.80	3.20	3.40	4.10	5.20	5.70
16	1977.00	2.00	2.10	2.80	4.10	5.60	...

Use the ***submatrix*** function to extract information from the matrix.

rows(Rain) = 46.00

OneHr := **submatrix(Rain**, 2, **rows(Rain)**, 2, 2) · **cm**

TwoHr := **submatrix(Rain**, 2, **rows(Rain)**, 3, 3) · **cm**

FourHr := **submatrix(Rain**, 2, **rows(Rain)**, 4, 4) · **cm**

SixHr := **submatrix(Rain**, 2, **rows(Rain)**, 5, 5) · **cm**

TenHr := **submatrix(Rain**, 2, **rows(Rain)**, 6, 6) · **cm**

TwelveHr := **submatrix(Rain**, 2, **rows(Rain)**, 7, 7) · **cm**

TwentyFourHr := **submatrix(Rain**, 2, **rows(Rain)**, 8, 8) · **cm**

OneHr =

	1
1	1.70
2	1.90
3	1.90
4	1.90
5	1.90
6	2.10
7	1.70
8	1.70
9	1.90
10	2.40
11	...

· **cm**

Step 1: Compute the rainfall intensities (cm/hr)

Divide the rainfall amounts by the corresponding time.

$$\mathbf{OneHrRain} := \frac{\mathbf{OneHr}}{1 \cdot \mathbf{hr}}$$
$$\mathbf{TowHrRain} := \frac{\mathbf{TwoHr}}{2 \cdot \mathbf{hr}}$$
$$\mathbf{FourHrRain} := \frac{\mathbf{FoutHr}}{4 \cdot \mathbf{hr}}$$
$$\mathbf{SixHrRain} := \frac{\mathbf{SixHr}}{6 \cdot \mathbf{hr}}$$
$$\mathbf{TenHrRain} := \frac{\mathbf{TenHr}}{10 \cdot \mathbf{hr}}$$
$$\mathbf{TwelveHrRain} := \frac{\mathbf{TwelveHr}}{12 \cdot \mathbf{hr}}$$
$$\mathbf{TwentyFourHrRain} := \frac{\mathbf{TwentyFourHr}}{24 \cdot \mathbf{hr}}$$

TenHrRain =

	1
1	0.48
2	0.37
3	0.31
4	0.47
5	0.25
6	0.40
7	0.30
8	0.43
9	0.38
10	0.46
11	0.17
12	0.37
13	0.35
14	0.52
15	0.56
16	...

$\cdot \frac{\mathbf{cm}}{\mathbf{hr}}$

Step 2: Sort data and assign a rank

We need to compute the corresponding rank of each intensity with the largest one getting the lowest rank. First sort the data. The sort function sorts from smallest to largest. Then reverse the sort so the data is sorted from largest to smallest. Finally, compute the ranks of the data, sorted in this way. The rank is mean of the indeces corresponding to the given rainfall intensity. For example the intensity of 1.7 cm/hr is found in the 25th, 26th, 27th, 28th, and 29th element. The mean of these indeces is 27. Thus the intensity of 1.7 cm/hr is given the rank of 27. The intensity of 3.7 cm/hr is found in the 1st element only. Thus the intensity of 3.7 cm/hr is given the rank of 1.

Create a function to rank the data.

The ***match(z, A)*** function looks in the sorted vector, Sorted, for a given value, $\mathbf{x_i}$, from the original vector, x, and returns the index (indeces) of its position in sorted vector.

Note: Mathcad compares the values in its default values of m/s. In order for the match function to work, the Convergence Tolerance (TOL) for this worksheet needed to be changed to 1×10^{-10}. Go to **Worksheet Options** from the **Tools** menu.

```
RankData(x) := | Sorted ← reverse(sort(x))
               | for i ∈ 1 .. rows(x)
               |   | m ← match(x_i, Sorted)
               |   | Rank_i ← mean(m)
               | Rank
```

Test the function

Sorted := reverse(sort(OneHrRain))

Sorted =

	1
1	3.70
2	3.10
3	2.90
4	2.80
5	2.70
6	2.70
7	...

$\cdot \frac{cm}{hr}$

Check_1 := match(OneHrRain$_1$, Sorted)

$$\mathbf{Check\_1} = \begin{pmatrix} 25.00 \\ 26.00 \\ 27.00 \\ 28.00 \\ 29.00 \end{pmatrix} \quad \mathbf{mean(Check\_1)} = 27.00$$

$$\mathbf{Check\_2 := match}\left(3.7\frac{cm}{hr}, \mathbf{Sorted}\right)$$

Check_2 = (1.00)

RankOneHr := RankData(OneHrRain)

RankTwoHr := RankData(TwoHrRain)

RankFourHr := RankData(FourHrRain)

RankSixHr := RankData(SixHrRain)

RankTenHr := RankData(TenHrRain)

RankTwelveHr := RankData(TwelveHrRain)

RankTwentyFourHr := RankData(TwentyFourHrRain)

RankOneHr =

	1
1	27.00
2	17.50
3	17.50
4	17.50
5	17.50
6	10.50
7	27.00
8	...

RankTenHr =

	1
1	8.00
2	22.00
3	28.50
4	9.00
5	40.00
6	16.00
7	...

Step 3: Calculate return periods using the Wiebull formula

To calculate the return period using the Wiebull formula, use the number of data years plus 1.

$$\mathbf{Return(x)} := \frac{\mathbf{rows(x)+1}}{\mathbf{x}} \cdot \mathbf{yr} \qquad \mathbf{rows(RankOneHr)} = 45.00$$

ReturnOneHr := **Return(RankOneHr)**

ReturnTwoHr := **Return(RankTwoHr)**

ReturnFourHr := **Return(RankFourHr)**

ReturnSixHr := **Return(RankSixHr)**

ReturnTenHr := **Return(RankTenHr)**

ReturnTwelveHr := **Return(RankTwelveHr)**

ReturnTwentyFourHr := **Return(RankTwentyFourHr)**

ReturnOneHr = · **yr**

	1
1	1.70
2	2.63
3	2.63
4	2.63
5	2.63
6	4.38
7	1.70
8	1.70
9	...

ReturnTenHr = · **yr**

	1
1	5.75
2	2.09
3	1.61
4	...

Step 4: Plot the rainfall intensity vs return period

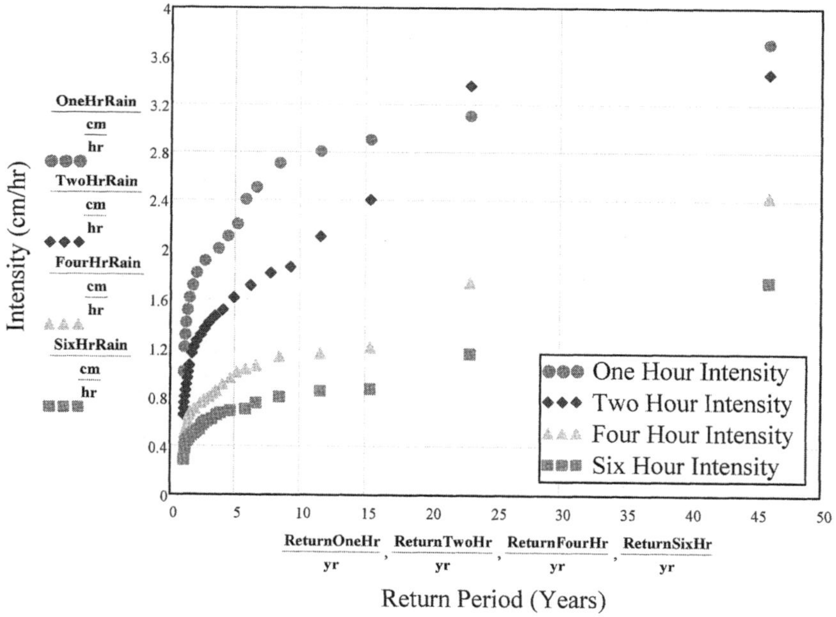

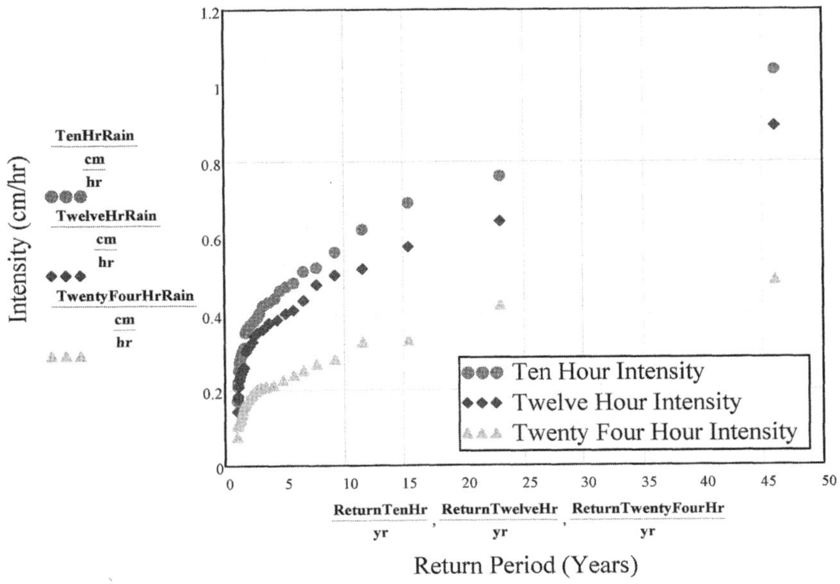

Step 5: Create a data matrix

For a return period of 5 years, read the intensity of 1hr, 2hr, 4hr, 6hr, 10hr, 12hr, and 24hr. Repeat for return period of 10 years, and repeat for return period of 25 years.
Put data into a matrix. This is done manually.

$$\mathbf{Data} := \begin{pmatrix} 1 & 2.2 & 2.73 & 3.2 \\ 2 & 1.6 & 1.98 & 3.35 \\ 4 & 1.01 & 1.12 & 1.85 \\ 6 & .696 & .83 & 1.24 \\ 10 & .478 & .590 & .81 \\ 12 & .41 & .499 & .689 \\ 24 & .23 & .297 & .431 \end{pmatrix}$$

Column 1 is duration in hours.
Column 2 is 5 year return period.
Column 3 is 10 year return period.
Column 4 is 25 year return period.

Step 6: Plot a 5, 10, and 25 year IDF curve (using data points)

$$\mathbf{Duration} := \mathbf{Data}^{\langle 1 \rangle} \cdot \mathbf{hr} \quad \mathbf{FiveYear} := \mathbf{Data}^{\langle 2 \rangle} \cdot \frac{\mathbf{cm}}{\mathbf{hr}} \quad \mathbf{TenYear} := \mathbf{Data}^{\langle 3 \rangle} \cdot \frac{\mathbf{cm}}{\mathbf{hr}}$$

$$\mathbf{TwentyFiveYear} := \mathbf{Data}^{\langle 4 \rangle} \cdot \frac{\mathbf{cm}}{\mathbf{hr}}$$

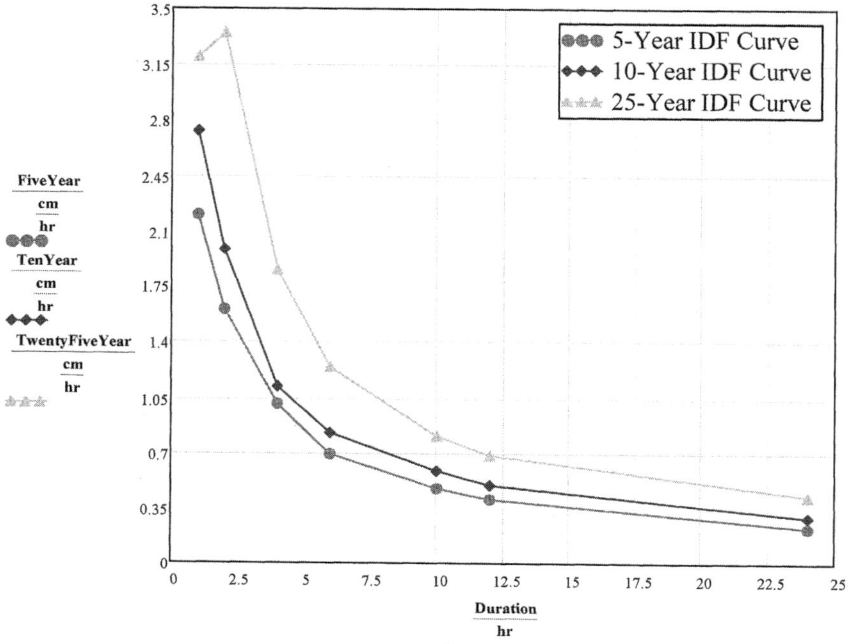

Step 7: Model IDF curves to fit a function of the form: $i = \frac{a}{t+b}$

Write the equation for the IDF curve in a linear form (y(x)=mx+b), where $\frac{1}{i}$ is the dependent variable, $\frac{1}{a}$ is the slope, and $\frac{b}{a}$ is the intercept.

$$\frac{1}{i} = \frac{t+b}{a} = \frac{t}{a} + \frac{b}{a}$$

$$\mathbf{SlopeIDF_5} := \mathbf{slope}\left(\mathbf{Duration}, \frac{1}{\mathbf{FiveYear}}\right) \qquad \mathbf{SlopeIDF_5} = 16.97\frac{1}{\mathbf{m}}$$

$$\mathbf{a_5} := \frac{1}{\mathbf{SlopeIDF_5}} \qquad \mathbf{a_5} = 5.89\cdot\mathbf{cm}$$

$$\mathbf{b/a_5} := \mathbf{intercept}\left(\mathbf{Duration}, \frac{1}{\mathbf{FiveYear}}\right) \qquad \mathbf{b/a_5} = 122088.31\frac{\mathbf{s}}{\mathbf{m}}$$

$$\mathbf{b_5} := \mathbf{b/a_5}\cdot\mathbf{a_5} \qquad \mathbf{b_5} = 2.00\cdot\mathbf{hr}$$

$$\mathbf{SlopeIDF_{10}} := \mathbf{slope}\left(\mathbf{Duration}, \frac{1}{\mathbf{TenYear}}\right) \qquad \mathbf{SlopeIDF_{10}} = 13.01\frac{1}{\mathbf{m}}$$

$$\mathbf{a_{10}} := \frac{1}{\mathbf{SlopeIDF_{10}}} \qquad \mathbf{a_{10}} = 7.69\cdot\mathbf{cm}$$

$$\mathbf{b/a_{10}} := \mathbf{intercept}\left(\mathbf{Duration}, \frac{1}{\mathbf{TenYear}}\right) \qquad \mathbf{b/a_{10}} = 121315.42\,\frac{\mathbf{s}}{\mathbf{m}}$$

$$\mathbf{b_{10}} := \mathbf{b/a_{10}} \cdot \mathbf{a_{10}} \qquad \mathbf{b_{10}} = 2.59 \cdot \mathbf{hr}$$

$$\mathbf{SlopeIDF_{25}} := \mathbf{slope}\left(\mathbf{Duration}, \frac{1}{\mathbf{TwentyFiveYear}}\right) \qquad \mathbf{SlopeIDF_{25}} = 9.13\,\frac{1}{\mathbf{m}}$$

$$\mathbf{a_{25}} := \frac{1}{\mathbf{SlopeIDF_{25}}} \qquad \mathbf{a_{25}} = 10.96 \cdot \mathbf{cm}$$

$$\mathbf{b/a_{25}} := \mathbf{intercept}\left(\mathbf{Duration}, \frac{1}{\mathbf{TwentyFiveYear}}\right) \qquad \mathbf{b/a_{25}} = 81193.25\,\frac{\mathbf{s}}{\mathbf{m}}$$

$$\mathbf{b_{25}} := \mathbf{b/a_{25}} \cdot \mathbf{a_{25}} \qquad \mathbf{b_{25}} = 2.47 \cdot \mathbf{hr}$$

Create Plot Range Variable $\mathbf{t} := 0\mathbf{hr}, 1\mathbf{hr}..\ 25\mathbf{hr}$

$$\mathbf{Prediction(SlopeIDF,x,a,b)} := \mathbf{SlopeIDF} \cdot \mathbf{x} + \frac{\mathbf{b}}{\mathbf{a}}$$

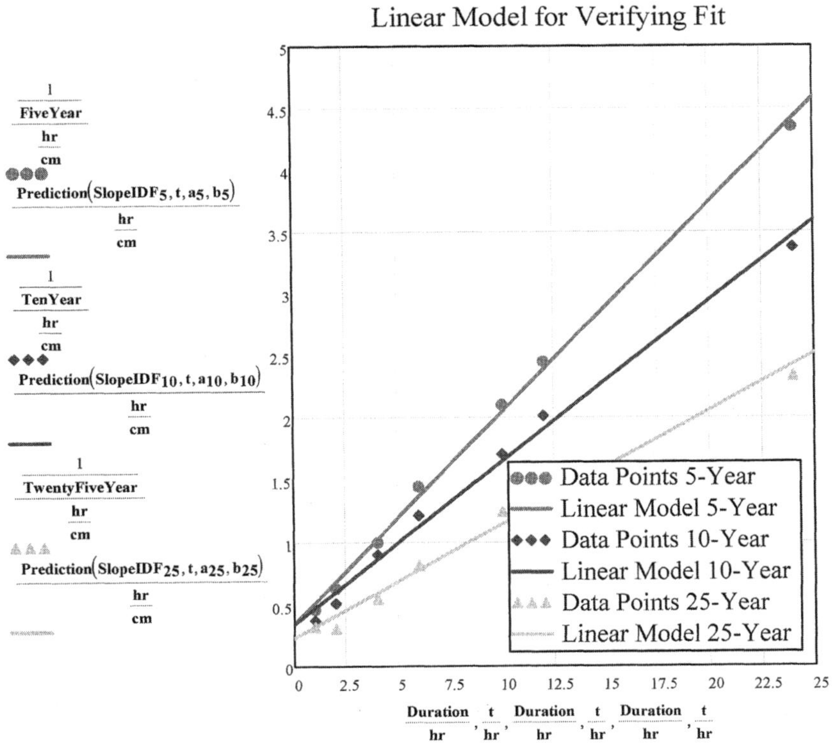

Step 8: Plot the IDF curves using the mathmatical formula

Using the values of a and b plot the 5-yr, 10-yr, and 25-yr IDF curves using $\mathbf{i} = \dfrac{\mathbf{a}}{\mathbf{t}+\mathbf{b}}$

$a_5 = 5.89 \cdot \mathbf{cm}$ $\quad b_5 = 2.00 \cdot \mathbf{hr}$ $\quad a_{10} = 7.69 \cdot \mathbf{cm}$ $\quad b_{10} = 2.59 \cdot \mathbf{hr}$ $\quad a_{25} = 10.96 \cdot \mathbf{cm}$ $\quad b_{25} = 2.47 \cdot \mathbf{hr}$

$\mathbf{Intensity}(t, a, b) := \dfrac{a}{t+b}$

$\mathbf{Intensity}(1\mathbf{hr}, a_5, b_5) = 1.97 \cdot \dfrac{\mathbf{cm}}{\mathbf{hr}}$ $\qquad \mathbf{Intensity}(10\mathbf{hr}, a_5, b_5) = 0.49 \cdot \dfrac{\mathbf{cm}}{\mathbf{hr}}$

$\mathbf{Intensity}(1\mathbf{hr}, a_{10}, b_{10}) = 2.14 \cdot \dfrac{\mathbf{cm}}{\mathbf{hr}}$ $\qquad \mathbf{Intensity}(10\mathbf{hr}, a_{10}, b_{10}) = 0.61 \cdot \dfrac{\mathbf{cm}}{\mathbf{hr}}$

$\mathbf{Intensity}(1\mathbf{hr}, a_{25}, b_{25}) = 3.16 \cdot \dfrac{\mathbf{cm}}{\mathbf{hr}}$ $\qquad \mathbf{Intensity}(10\mathbf{hr}, a_{25}, b_{25}) = 0.88 \cdot \dfrac{\mathbf{cm}}{\mathbf{hr}}$

Check Plot

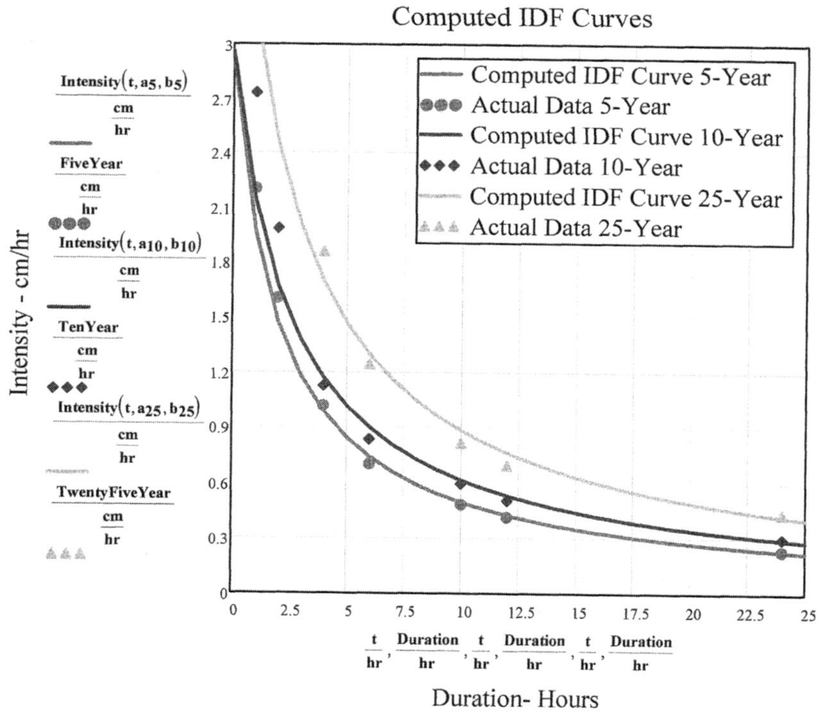

Summary

The two main concepts covered in this chapter are local variables and looping. These, added with the logical programming discussed in Chapter 8, can make some very powerful programs.

Local variables are useful to help simplify complex expressions. A local variable is a variable that is defined only within a program—it is not available outside the program. These local variables are essential when using loops, because they help control the loops.

There are two types of loops: ***for*** loops and ***while*** loops. The ***for*** loop loops a specific number of times and is controlled by an iteration variable. The ***while*** loop will loop until a specific condition is met. If the condition is never met, the loop will go indefinitely.

The ***break*** and ***continue*** operators are used to control when and how a program loops. The ***break*** operator stops the execution when a specific condition is met. The ***continue*** operator allows Mathcad to skip over a portion of a loop and move back to the beginning of the loop.

The ***return*** operator is used in conjunction with an ***if*** statement. It stops the execution of the program and returns the value following the ***return*** operator.

The ***on error*** operator is very useful when a particular operation could cause an error, because you can tell Mathcad what to do when the error occurs.

Practice

Additional problems and applications can be found on the companion site: www.elsevierdirect.com/9780123747839.

1. From your field of study, create 10 programs that use the features discussed in this chapter.
2. Create five functions or expressions that use the ***on error*** operator.

CHAPTER

Calculus and Differential Equations

12

Mathcad will easily perform differentiation and integration operations using either the numeric or symbolic processors. Mathcad can also perform a number of differential equation solutions for both ordinary differential equations (ODEs) and partial differential equations (PDEs). The following is a brief discussion of these topics.

DIFFERENTIATION

Let's start by using the symbolic processor. You can use the Symbolics menu if you are performing the operation only once. If you are performing the operation more than once, or if you want to use the result for further calculations, then use the live symbolics operator.

To differentiate an expression using the **Symbolics** menu, write the expression, select the variable, and then click **Variables>Differentiate** from the **Symbolics** menu. The result is added below the expression. Remember that the result is static; it does not change if you revise the expression. In order to keep the result dynamic, use the live symbolics operator. To do this, you need to use the derivative operator from the Calculus toolbar and the symbolics operator.

Figure 12.1 compares the **Symbolics** menu and live symbolics operator.

Now let's look at using the numeric processor. This processor gives the derivative at only a single point. It does not calculate the equation, as does the symbolic operator. Figure 12.2 shows how to use the numeric processor.

You can also take the nth derivative of an expression, and you can plot the results. See Figure 12.3.

Using the Symbolics Menu

To differentiate using the Symbolics menu, write the expression, select the variable, and select Symbolics>Variable>Differentiate. The result is static. It does not change if you change the expression.

$x^3 + 3x^2 + 4x - 2$ $\quad 2^x \quad$ $\sin(4x + 1)$ $\quad \dfrac{x^2+2}{x^3-1}$

$3 \cdot x^2 + 6 \cdot x + 4$ $\quad 2^x \cdot \ln(2) \quad$ $4 \cdot \cos(4 \cdot x + 1)$ $\quad 2 \cdot \dfrac{x}{x^3-1} - 3 \cdot \dfrac{x^2+2}{\left(x^3-1\right)^2} \cdot x^2$

Using Live Symbolics Operator

To differentiate using the live symbolics operator, use the Calculus tool bar to get the derivative operator, write the expression, type **CTRL+PERIOD**, then click outside the expression.

$\mathbf{Deriv_1(x)} := \dfrac{d}{dx}\left(x^3 + 3x^2 + 4x - 2\right) \rightarrow 3 \cdot x^2 + 6 \cdot x + 4$ $\qquad \mathbf{Deriv_1(2)} = 28.00$

$\mathbf{Deriv_2(x)} := \dfrac{d}{dx}2^x \rightarrow 2^x \cdot \ln(2)$ $\qquad \mathbf{Deriv_2(2)} = 2.77$

$\mathbf{Deriv_3(x)} := \dfrac{d}{dx}\sin(4x + 1) \rightarrow 4 \cdot \cos(4 \cdot x + 1)$ $\qquad \mathbf{Deriv_3(2)} = -3.64$

$\mathbf{Deriv_4(x)} := \dfrac{d}{dx}\dfrac{x^2+2}{x^3-1} \rightarrow \dfrac{2 \cdot x}{x^3-1} - \dfrac{3 \cdot x^2 \cdot \left(x^2+2\right)}{\left(x^3-1\right)^2}$ $\qquad \mathbf{Deriv_4(2)} = -0.90$

FIGURE 12.1 Symbolic differentiation

$x := 2$

Use the Calculus toolbar to get the derivative operator.

$\mathbf{Deriv_5} := \dfrac{d}{dx}\left(x^3 + 3x^2 + 4x - 2\right)$ $\qquad \mathbf{Deriv_5} = 28.00$

$\mathbf{Deriv_6} := \dfrac{d}{dx}2^x$ $\qquad \mathbf{Deriv_6} = 2.77$

$\mathbf{Deriv_7} := \dfrac{d}{dx}\sin(4x + 1)$ $\qquad \mathbf{Deriv_7} = -3.64$

$\mathbf{Deriv_8} := \dfrac{d}{dx}\dfrac{x^2+2}{x^3-1}$ $\qquad \mathbf{Deriv_8} = -0.90$

FIGURE 12.2 Numeric differentiation

$$F_1(x) := \frac{x^3+1}{x^3-8}$$

$$\text{FirstDeriv}(x) := \frac{d}{dx}F_1(x) \rightarrow \frac{3\cdot x^2}{x^3-8} - \frac{3\cdot x^2\cdot\left(x^3+1\right)}{\left(x^3-8\right)^2} \text{ simplify } \rightarrow -\frac{27\cdot x^2}{\left(x^3-8\right)^2}$$

$$\text{SecondDeriv}(x) := \frac{d^2}{dx^2}F_1(x) \rightarrow \frac{6\cdot x}{x^3-8} - \frac{18\cdot x^4}{\left(x^3-8\right)^2} + \frac{18\cdot x^4\cdot\left(x^3+1\right)}{\left(x^3-8\right)^3} - \frac{6\cdot x\cdot\left(x^3+1\right)}{\left(x^3-8\right)^2} \text{ simplify } \rightarrow \frac{108\cdot x\cdot\left(x^3+4\right)}{\left(x^3-8\right)^3}$$

$a := -4, -3.99 .. 4$ Set plot range

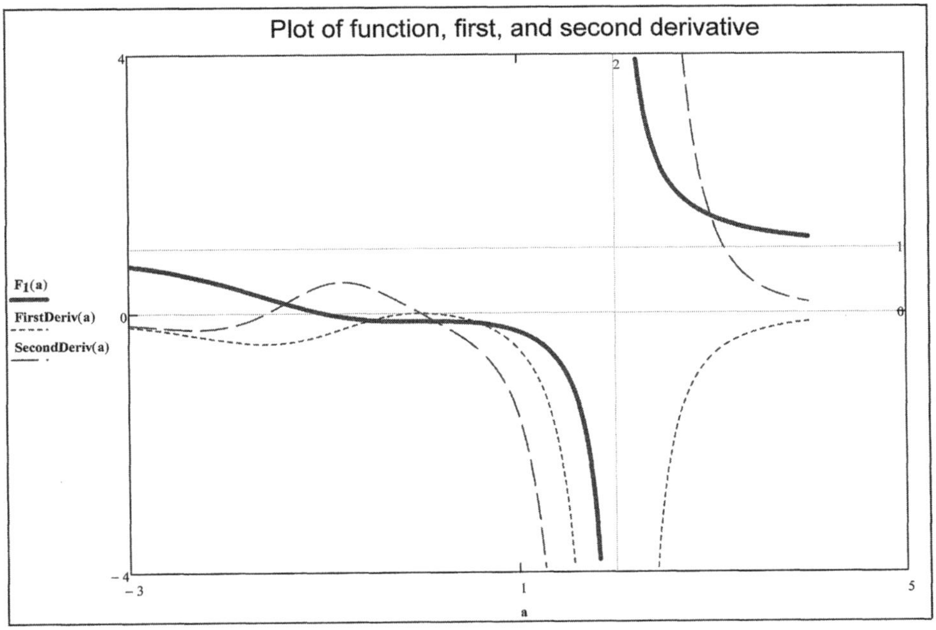

FIGURE 12.3 nth derivatives and plots

INTEGRATION

The numeric processor will perform only definite integrals. The symbolic processor will perform both definite integrals and indefinite integrals. The examples here are just an introduction. Mathcad can perform very sophisticated calculus operations. Refer to Mathcad Help for additional information.

Figure 12.4 gives examples of indefinite integrals using both the Symbolics menu and the live symbolics operator. Figure 12.5 gives examples of definite integrals using the live symbolics operator and using the numeric processor.

Using the Symbolics Menu

To integrate using the Symbolics menu, write the expression, select the variable, and select Symbolics>Variable>Integrate. The result is static. It does not change if you change the expression.

x^2 $\quad$ $x^3 + 2x^2 + x - 3$ $\quad$ $\sin(x)$

$\boxed{\frac{1}{3} \cdot x^3}$ $\quad$ $\boxed{\frac{1}{4} \cdot x^4 + \frac{2}{3} \cdot x^3 + \frac{1}{2} \cdot x^2 - 3 \cdot x}$ $\quad$ $\boxed{-\cos(x)}$

$(30 - x^2)^{\frac{3}{2}}$

$$\boxed{\frac{1}{4} \cdot x \cdot (30 - x^2)^{\frac{3}{2}} + \frac{45}{4} \cdot x \cdot (30 - x^2)^{\frac{1}{2}} + \frac{675}{2} \cdot \operatorname{asin}\left(\frac{1}{30} \cdot 30^{\frac{1}{2}} \cdot x\right)}$$

Using the Live Symbolics Operator

$$\text{Integrate}_1(x) := \int x^2\,dx \rightarrow \frac{x^3}{3}$$

$I_{1a} := \text{Integrate}_1(5)$ $\quad$ $I_{1a} = 41.67$ $\quad$ $I_{1b} := \text{Integrate}_1(1)$ $\quad$ $I_{1b} = 0.33$

$I_{1c} := I_{1a} - I_{1b}$ $\quad$ $I_{1c} = 41.33$

$$\text{Integrate}_2(x) := \int x^3 + 2x^2 + x - 3\,dx \rightarrow \frac{x^4}{4} + \frac{2 \cdot x^3}{3} + \frac{x^2}{2} - 3 \cdot x$$

$I_{2a} := \text{Integrate}_2(5)$ $\quad$ $I_{2a} = 237.08$ $\quad$ $I_{2b} := \text{Integrate}_2(1)$ $\quad$ $I_{2b} = -1.58$

$I_{2c} := I_{2a} - I_{2b}$ $\quad$ $I_{2c} = 238.67$

$$\text{Integrate}_3(x) := \int \sin(x)\,dx \rightarrow -\cos(x)$$

$I_{3a} := \text{Integrate}_3(5)$ $\quad$ $I_{3a} = -0.28$ $\quad$ $I_{3b} := \text{Integrate}_3(1)$ $\quad$ $I_{3b} = -0.54$

$I_{3c} := I_{3a} - I_{3b}$ $\quad$ $I_{3c} = 0.26$

$$\text{Integrate}_4(x) := \int (30 - x^2)^2\,dx \rightarrow \frac{x^5}{5} - 20 \cdot x^3 + 900 \cdot x$$

$I_{4a} := \text{Integrate}_4(5)$ $\quad$ $I_{4a} = 2625.00$ $\quad$ $I_{4b} := \text{Integrate}_4(1)$ $\quad$ $I_{4b} = 880.20$

$I_{4c} := I_{4a} - I_{4b}$ $\quad$ $I_{4c} = 1744.80$

FIGURE 12.4 Indefinite integrals

DIFFERENTIAL EQUATIONS

Mathcad has multiple functions to solve various differential equation systems. These solvers are numeric solutions and approximate the exact solutions. The solutions are a saved series of points and do not provide a smooth function. Units are not allowed in the differential equation solvers.

Using the Live Symbolics Operator

Compare from Figure 12.4

$\int_1^5 x^2\,dx \rightarrow \frac{124}{3}$ float $\rightarrow 41.333333333333333333$ $\qquad I_{1c} = 41.33$

$\int_1^5 x^3 + 2x^2 + x - 3\,dx \rightarrow \frac{716}{3}$ float $\rightarrow 238.66666666666666667$ $\qquad I_{2c} = 238.67$

$\int_1^5 \sin(x)\,dx \rightarrow \cos(1) - \cos(5)$ float $\rightarrow 0.25664012040491345293$ $\qquad I_{3c} = 0.26$

$\int_1^5 \left(30 - x^2\right)^2 dx \rightarrow \frac{8724}{5}$ float, 6 $\rightarrow 1744.8$

$I_{4c} = 1744.80$

Using Numeric integration

$a := 1 \quad b := 5$

Compare from Figure 12.4

$\text{Integrate}_5(a,b) := \int_a^b x^2\,dx \qquad I_5 := \text{Integrate}_5(a,b) \quad I_5 = 41.33 \quad I_{1c} = 41.33$

$\text{Integrate}_6(a,b) := \int_a^b x^3 + 2x^2 + x - 3\,dx \qquad I_6 := \text{Integrate}_6(a,b) \quad I_6 = 238.67 \quad I_{2c} = 238.67$

$\text{Integrate}_7(a,b) := \int_a^b \sin(x)\,dx \qquad I_7 := \text{Integrate}_7(a,b) \quad I_7 = 0.26 \quad I_{3c} = 0.26$

$\text{Integrate}_8(a,b) := \int_a^b \left(30 - x^2\right)^2 dx \qquad I_8 := \text{Integrate}_8(a,b) \quad I_8 = 1744.80 \quad I_{4c} = 1744.80$

FIGURE 12.5 Definite integrals

The two easiest-to-use functions, ***Odesolve*** and ***Pdesolve***, are used within Solve Blocks. These will be discussed shortly. There are many other functions that help solve systems of differential equations, including ***Adams***, ***AdamsBDF***, ***BDF***, ***Bulstoer***, ***bvalfit***, ***Jacob***, ***multigrid***, ***numol***, ***Radau***, ***relax***, ***Rkadapt***, ***rkfixed***, ***sbval***, ***statespace***, ***Stiffb***, and ***Stiffr***. The discussion of these functions is beyond the scope of this book, but the Mathcad Help and QuickSheets have excellent discussions of both.

ORDINARY DIFFERENTIAL EQUATIONS (ODEs)

The function ***Odesolve*** is used in a Solve Block to solve a single differential equation or a system of differential equations. It returns the solution as a function of the independent variable. It does this by saving solutions at a specific number of points (npoints) equally spaced in the solution interval, and then interpolating between these points using the function ***lspline***. The Solve Block sets the initial value or boundary constraints. The ODE must be linear in its highest derivative term, and the number of initial and boundary conditions must equal the order(s) of the ODE(s). The function has the form ***Odesolve***([vector], x, b, [npoints]). The arguments are as follows:

- Vector is used only for systems of ODEs and is a vector of function names (with no variable names included) as they appear within the Solve Block.
- x is the name of the variable of integration.
- b is the final point of the solution interval. (The initial point of the solution interval is specified by the initial conditions.)
- npoints is optional and is the integer number of equally spaced points used to interpolate the solution function. The default value of npoints is 1000. If npoints is increased, then the interpolated solution function is more accurate. The default value is usually adequate, but if you are solving over a large interval, set npoints to a value larger than 1000. Increasing npoints increases the calculation time.

The default solver for ***Odesolve*** is the ***Adams/BDF*** method. If you right-click the Odesolve function, you can select another solver for the ***Odesolve*** function. The choices are ***Fixed***, ***Adaptive***, and ***Radau***. Some functions use nonuniform step sizes internally when they solve the differential equations, adding more steps in regions of greater variation of the solution, but they return the solution at the number of equally spaced points specified in npoints. Figures 12.6, 12.7, and 12.8 provide examples of using the ***Odesolve*** function.

PARTIAL DIFFERENTIAL EQUATIONS (PDEs)

The most common function for solving partial differential equations is the ***Pdesolve*** function. This function returns a function or vector of functions that solve a one-dimensional nonlinear PDE or system of PDEs. Calculated values are interpolated from a matrix of solution points calculated using the numerical method of lines.

The ***Pdesovle*** function has the form ***Pdesolve***(u,x,xrange, t, trange, [xpts], [tpts]). The arguments are as follows:

- u is the scalar function name, or vector of function names (with no variable names included; that is, f instead of f(x,t)) as they appear within the Solve Block).
- x is the spatial variable name.

A tank is filled with 100 gal of salt solution containing 1 lbf of salt per gallon. Fresh brine containing 2 lbf of salt per gallon runs into the tank at a rate of 5 gal/min, and the mixture, assumed to be kept uniform by stirring, runs out at the same rate. Find the amount of salt in the tank at any time t, and determine the amount of salt after 15 minutes.

Let Q represent the total amount of salt (lbf) in solution in the tank at any point t; therefore, the amount of salt per gallon of solution is Q (lbf)/100 (gal). Salt enters the tank at a rate of (2lbf/gal)*(5gal/min)=10lbf/min. At any interval dt, the gain in salt is (10lbf/min)*(dt min) =10*dt*min.

Salt leaving the tank (at a rate of 5gal/min) in the interval dt is the same as the concentration of salt in the tank (Q (lbf)/100 (gal). Therefore (5gal/min)*(Qlbf/100gal)*dt (min)=(Qlbf/20)*dt.

dQ=10dt - (Q/20)dt=(10-Q/20)dt. It can be rewritten as dQ/dt+Q/20=10

Set up the differential equation solve block.

Given

$$\frac{d}{dt}Q(t) + \frac{Q(t)}{20} = 10 \qquad Q(0) = 100 \qquad \mathbf{Salt} := \mathbf{Odesolve}(t, 20)$$

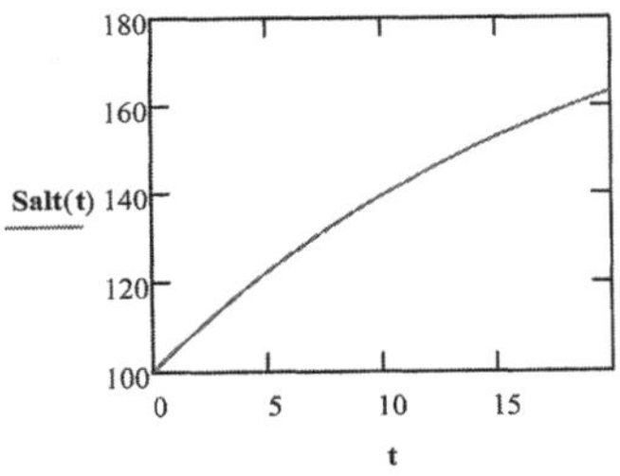

Amount of salt after 15 minutes.

$\mathbf{Salt}(15) = 152.76$

Graph of solution, showing amount of salt as a function of time.

FIGURE 12.6 Ordinary differential equation—Salt solution

- xrange is a two-element column vector containing the real boundary values for x.
- t is the time variable name.
- trange is a two-element column vector containing the real boundary values for t.
- xpts (optional) is the integer number of spatial discretization points.
- Tpts (optional) is the integer number of temporal discretization points.

Some points to remember are:

- PDE equations must be defined using Boolean equals.
- Second partial derivatives are not allowed on the left-hand side of equations; you must convert your equation to a system of equations in first derivates only. See Mathcad QuickSheets for examples.

The water flowing out of the bottom of a parabolic shaped tank is modeled by the function:
$\pi \cdot h^2 \cdot \frac{dh}{dt} = \frac{-2 \cdot \sqrt{h}}{5}$ where h is the height of water (meters) at time t (hours). The initial depth of water is h(0)=1. Find the depth of water after 2 hours.

Given

$$\pi \cdot h(t)^2 \cdot \left(\frac{d}{dt} h(t)\right) = \frac{-2 \cdot \sqrt{h(t)}}{5} \qquad h(0) = 1$$

Depth := **Odesolve**(**t**, 3.13)

Depth(2) = 0.67

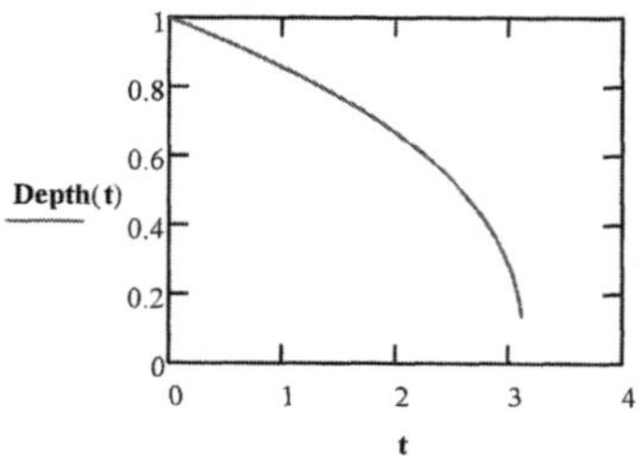

Graph of solution, showing depth of water as a function of time.

FIGURE 12.7 Ordinary differential equation—Water flow

- There must be an initial condition u(x,0), and n boundary conditions, where n is the order of the PDE, for each unknown function.
- Algebraic constraints are allowed.
- Inequality constraints are not allowed.
- Assign the output of the ***Pdesolve*** function to a function name, or vector of function names.
- The numerical method of lines is appropriate for solving only hyperbolic and parabolic PDEs. If you need to solve an equation in a program loop, use the function ***numol***. If you need to solve an elliptic PDE, such as Poisson's equation, use the functions ***relax*** or ***multigrid***.

Mathcad will display a partial derivative symbol. In order to do this, you need to change the display of the derivative operator by right-clicking the derivative operator and selecting **View Derivative As > Partial Derivative**. This will change the display of the derivative operator from $\frac{d}{d\blacksquare}\blacksquare$ here to $\frac{\partial}{\partial\blacksquare}\blacksquare$.

The Mathcad QuickSheets have examples of Partial Differential Equations.

The half-life of a substance in the process of transformation is the amount of time it take for one-half of the substance to change. The standard model for this process is: $\frac{dS}{dt} = k \cdot S$

where $S(0) = S_0$. $S(t)$ is the amount of the substance at time t, and $S(0)$ is the amount at time t=0. The model states that the rate of change is dependant upon the amount of of the substance. The smaller the substance, the slower the rate of change.

If the original amount is 10 grams and the half-life is 100 hours, find the amount left after 40 hours.

The challenge with this problem is that we do not know the value of 'k" because it is determined after the solving of the solution equation. In order to get around this, we will create a user defined function with "k" as the input variable. Once we have solved the solution equation, we can calculate the value of 'k," and then we can use the function with the correct value of 'k."

Given Note: Use CTRL+F7 to get the derivative symbol.

$S'(t) - k \cdot S(t) = 0$ $\quad S(0) = 10$ $\quad \mathbf{SubstanceEq}(k) := \mathbf{Odesolve}(t, 301)$

Create another function which will set the constraints for a solve block. The constraints are that (Substance at time 0) - 10 = 0, and (Substance at time 10) - 5 = 0

$$\mathbf{Constraints}(k) := \left| \begin{array}{l} y \leftarrow \mathbf{SubstanceEq}(k) \\ \overrightarrow{\left(y\left(\frac{Time}{hr}\right) - \frac{Sub}{gm} \right)} \end{array} \right. \qquad Time := \begin{pmatrix} 0 \\ 100 \end{pmatrix} hr \qquad Sub := \begin{pmatrix} 10 \\ 5 \end{pmatrix} gm$$

Set values for t=0,10 and Substance=10, 5.

Use a ***Minerr*** solve block to solve for the unknown "k." $\quad k := -0.1$ Initial guess

Given

$\mathbf{Constraints}(k) = 0$

$k := \mathbf{Minerr}(k)$ $\quad k = -0.006931$

Use the ***Minerr*** function, because we do not have an exact solution. The ***Find*** function would not arrive at an answer.

Now that we know "k," we can evaluate the function "Substance" with the value of "k."

$\mathbf{Substance} := \mathbf{SubstanceEq}(k)$

$\mathbf{Substance}(0) = 10.00$

$\mathbf{Substance}(40) = 7.58$ Substance after 40 hours.

$\mathbf{Substance}(100) = 5.00$

$\mathbf{Substance}(200) = 2.50$

$\mathbf{Substance}(300) = 1.25$

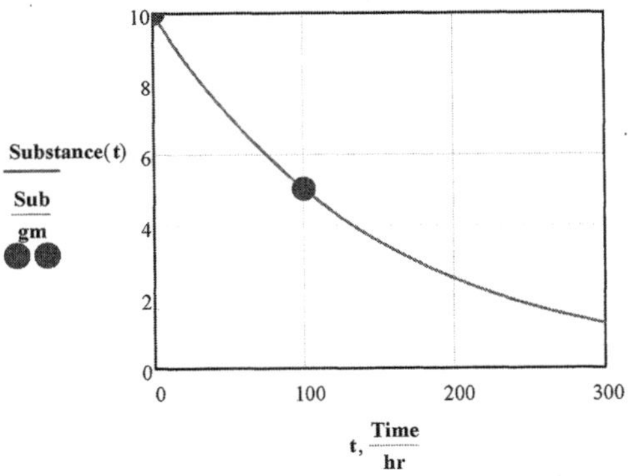

FIGURE 12.8 Ordinary differential equation—Half-life

ENGINEERING EXAMPLE 12.1

Integration

In Engineering Example 7.1, we plotted the shear, moment, and deflection of a beam with uniform load. In this example we derive the equations used to plot the shear, moment, and deflection.

Note: Mathcad does not include the integration constant C because Mathcad calulates a single solution. We are creating functions for multiple solutions, so we will maually add the integration constant C.

Shear

For this example the beam is loaded with a uniform load w. Calculate shear by integrating the area of the load under the loading diagram.

$\mathbf{Load(w)} := -\mathbf{w}$

Shear $\int \mathbf{Load(w)dx} \rightarrow -2 \cdot \mathbf{w}$

The integration constant C1 must now be considered and evaluated. The shear at x=0 is equal to w*L/2. Therefore, w*L/2=w*0 + C1, and C1 = w*L/2.

Now that we know the value of the integration constant C1, we can define the shear function.

$$\mathbf{Shear(w, L, x)} := \frac{\mathbf{w} \cdot \mathbf{L}}{2} - \mathbf{w} \cdot \mathbf{x}$$

Moment

Calculate moment by integrating the area under the shear curve.

Moment $\int \mathbf{Shear(w, L, x)dx} \rightarrow \dfrac{\left(2 \cdot \mathbf{x} - \dfrac{\mathbf{L} \cdot \mathbf{w}}{2}\right)^2}{2 \cdot \mathbf{w}}$

The integration constant must be considered. The moment at x=0 is equal to 0. Therefore, the integration constant C2 is equal to 0.

Now that we know that constant C2 is 0, we can define the moment function.

$$\mathbf{Moment(w, L, x)} := \frac{\mathbf{w} \cdot \mathbf{x}}{2} \cdot (\mathbf{L} - \mathbf{x})$$

Slope

From mechanics of materials, the relationship for the radius of curvature of a beam p and moment is defined as $\frac{1}{\mathbf{p}} = \frac{\mathbf{M}}{\mathbf{EI}}$, and the relationship of moment to slope θ is defined as $\mathbf{M} = \mathbf{EI} \cdot \frac{\mathbf{d\theta}}{\mathbf{dx}}$.

Calculate slope θ by integrating M/EI.

Slope $\int \frac{\mathbf{Moment(w, L, x)}}{\mathbf{E \cdot I}} \mathbf{dx\ simplify} \rightarrow \frac{\mathbf{w \cdot (3 \cdot L - 4)}}{\mathbf{12 \cdot E \cdot I}}$

$$\mathbf{Slope(w, L, x, E, I)} := \frac{\mathbf{-1 \cdot w \cdot x^2}}{\mathbf{12E \cdot I}} \cdot \mathbf{(2x - 3L) + C3}$$

The integration constant C3 must be considered. We will not calculate the constant here, but will calculate it later. It must be added to the Slope function.

Deflection

Calculate deflection by integrating the slope function.

Deflection $\mathbf{D(x)} := \int \mathbf{Slope(w, L, x, E, I)dx\ simplify} \rightarrow \frac{\mathbf{x \cdot (2 \cdot L \cdot w \cdot x^2 - w \cdot x^3 + 24 \cdot C3 \cdot E \cdot I)}}{\mathbf{24 \cdot E \cdot I}}$

The integration constants C3 and C4 must be considered. The deflection at $x = 0$ is zero; therefore, the integration constant C4 is equal to zero. Now let's solve for the integration constant C3. The deflection at $x = L$ is zero, so we can substitue L for x and solve for C3.

Set the function D(x) equal to 0 at $x = L$ and solve for C3.

$$0 = \mathbf{D(L)\ solve, C3} \rightarrow \frac{\mathbf{L^3 \cdot w}}{\mathbf{24 \cdot E \cdot I}}$$

Now that we know the value of C3 we can redefine the slope function and define the moment function.

$$\frac{\mathbf{x}}{\mathbf{24}} \cdot \left[\mathbf{2 \cdot w \cdot L \cdot x^2 - w \cdot x^3 + 24 \cdot \left(\frac{-w \cdot L^3}{24 \cdot E \cdot I}\right) \cdot E \cdot I}\right] \mathbf{simplify} \rightarrow \frac{\mathbf{w \cdot (L^3 - 8 \cdot L + 8)}}{\mathbf{12}}$$

$$\mathbf{Slope(w, L, x, E, I)} := \frac{\mathbf{-1 \cdot w \cdot x^2}}{\mathbf{12E \cdot I}} \cdot \mathbf{(2x - 3L)} - \frac{\mathbf{w \cdot L^3}}{\mathbf{24 \cdot E \cdot I}} \quad \mathbf{simplify} \rightarrow \frac{\mathbf{w \cdot (L^3 - 6 \cdot L \cdot x^2 + 4 \cdot x^3)}}{\mathbf{24 \cdot E \cdot I}}$$

$$\mathbf{Deflection(w, L, x, E, I)} := \frac{\mathbf{-w \cdot x}}{\mathbf{24 \cdot E \cdot I}} \cdot \mathbf{(L^3 - 2 \cdot L \cdot x^2 + x^3)}$$

Now plot similar values as in Engineering Example 7.1

$\mathbf{w_1} := 2500\frac{\mathbf{lbf}}{\mathbf{ft}}$ $\mathbf{L_1} := 20\mathbf{ft}$ $\mathbf{E} := 29000\mathbf{ksi}$ $\mathbf{I} := 428\cdot\mathbf{in}^4$ $\mathbf{x} := 0\mathbf{ft}, 1\mathbf{ft}..20\mathbf{ft}$

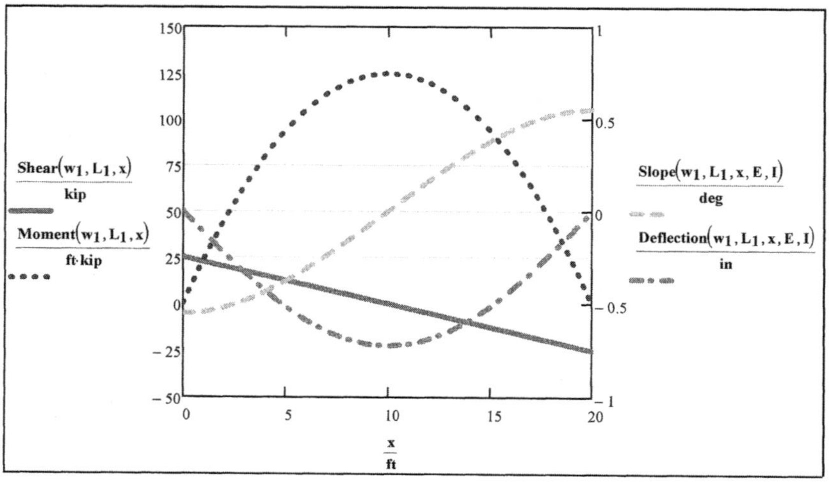

ENGINEERING EXAMPLE 12.2

2 Cell, Well Mixed, Lagoon System

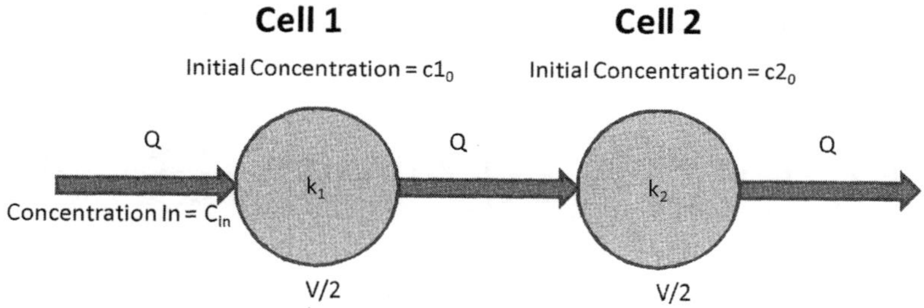

Problem:

A 2 cell, well mixed sewage lagoon that has a total surface area of 10 hectare and a depth of 1 m is receiving 8,640 m^3/day of waste containing 100 mg/liter of biodegradable contaminant. The contaminant is removed by bacteria via a first order reaction term with a reaction rate coefficient of $k_1 = 0.5$/day in Cell 1 and a reaction rate coefficient of $k_2 = 0.1$/day in Cell 2.

1. Write a mass balance equation on the contaminant for this system.

2. Solve the resulting ODE. The contaminant concentration in cell 1 at time = 0 is $C1_0 = 100$ mg/liter and the contaminant concentration in cell 2 at time = 0 is $c2_0 = 50$ mg/liter. Plot the contaminant concentration in the lagoon effluent vs time (hours) for 20 days.

Mass Balance Equation

A well mixed sewage lagoon implies that at any point in time the conditions everywhere in the lagoon are the same. The mass balance on biodegradable contaminant for a well mixed lagoon is:

Contaminant in - Contaminant out - Contaminant used by bacteria = Contaminant accumulation.

The concentration is equal to the Contaminant divided by the volume.
The volume in each lagoon is one-half the total volume.

Symbolically the change in Contaminant can be written as:

$$\frac{d}{dt}C(t) = C'(t) = Q \cdot c_{in} - Q \cdot \frac{C(t)}{Vol} - k \cdot C(t).$$

All terms in this form of the mass balance have units of mass/time

We are after the concentration in the lagoon effluent with concentration units of mg/liter. To get concentration, divide the Contaminant by the volume.

$$\frac{d}{dt}\left(\frac{C(t)}{Vol}\right) = \frac{Q \cdot c_{in}}{Vol} - \frac{Q \cdot \frac{C(t)}{Vol}}{Vol} - \frac{k \cdot C(t)}{Vol}.$$

Because c = C/Vol, the equation can be rewritten as:

$$\frac{d}{dt}c(t) = \frac{Q}{Vol} \cdot c_{in} - \frac{Q}{Vol} \cdot c(t) - k \cdot c(t).$$

Input Variables

Volume in each lagoon: $Vol := \frac{10 \cdot hectare \cdot 1 \cdot m}{2}$ $Vol = 5 \times 10^7 L$ $Vol = 5.00 \times 10^7 \cdot liter$

Flow rate: $Q := 8640 \cdot \frac{m^3}{day}$ $Q = 360000.00 \cdot \frac{liter}{hr}$

Initial concentration in cell 1: $c1_0 := 100 \frac{mg}{liter}$

Initial concentration in cell 2: $c2_0 := 50 \frac{mg}{liter}$

Influent concentration: $c_{in} := 100 \frac{mg}{liter}$

Bacteria reaction rate for cell 1: $k_1 := 0.50 \cdot \frac{1}{day}$ $k_1 = 0.02083 \cdot \frac{1}{hr}$

Bacteria reaction rate for cell 1: $k_2 := 0.10 \cdot \frac{1}{day}$ $k_2 = 0.00417 \cdot \frac{1}{hr}$

For this Engineering Example we will use two methods to solve the differential equations. The first method will use the ***Odesolve*** function. The second method will use the ***Radau*** function.

Method 1: *Odesolve*

Cell 1

Units cannot be used in ***Odesolve*** so divide values by the desired output units. Set the function and the constraints below the word Given.

Given

$$\frac{d}{dt}c(t) = \frac{\frac{Q}{\frac{liter}{hr}} \cdot \frac{c_{in}}{\frac{mg}{liter}}}{\frac{Vol}{liter}} - \frac{\frac{Q}{\frac{liter}{hr}}}{\frac{Vol}{liter}} \cdot c(t) - k_1 \cdot hr \cdot c(t) \qquad c(0) = \frac{c1_0}{\frac{mg}{liter}}$$

Concentration$_1$:= Odesolve(t, 480)

Display values of Concentration at various times.

Concentration$_1$(0) = 100.00 **Concentration$_1$(24)** = 63.61 **Concentration$_1$(96)** = 30.72
Concentration$_1$(200) = 25.96 **Concentration$_1$(480)** = 25.68

Plot the values of Concentration$_1$ over time.

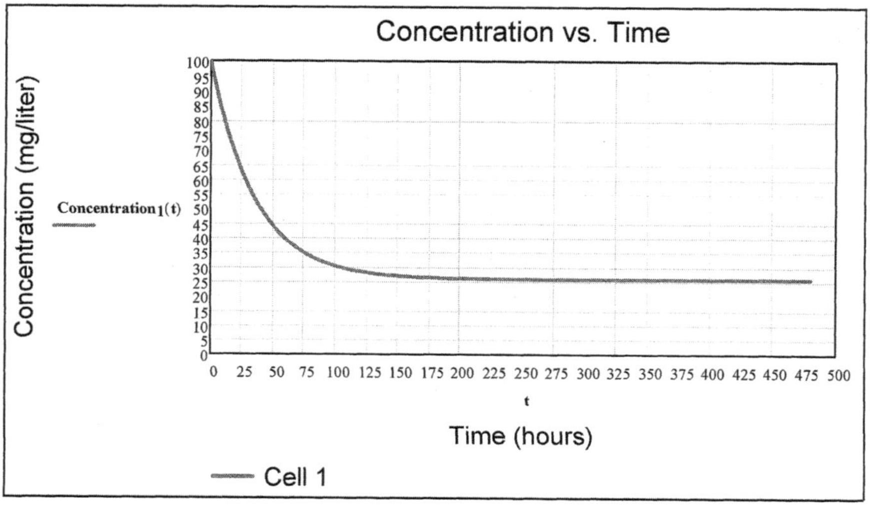

Cell 2

Bacteria reaction rate for cell 2: $\mathbf{k_2} = 0.00417 \cdot \dfrac{1}{\mathbf{hr}}$ $\quad$ $\mathbf{c2_0} = 50.00 \cdot \dfrac{\mathbf{mg}}{\mathbf{liter}}$

The input concentration for cell 2 is equal to the output concentration of cell 1. Therefore, use the solution to cell 1 as an input function for cell 2.

Given

$$\frac{d}{dt}\mathbf{c_2(t)} = \frac{\dfrac{\mathbf{Q}}{\frac{\mathbf{liter}}{\mathbf{hr}}} \cdot \mathbf{Concentration_1(t)}}{\dfrac{\mathbf{Vol}}{\mathbf{liter}}} - \frac{\dfrac{\mathbf{Q}}{\frac{\mathbf{liter}}{\mathbf{hr}}}}{\dfrac{\mathbf{Vol}}{\mathbf{liter}}} \cdot \mathbf{c_2(t)} - \mathbf{k_2} \cdot \mathbf{hr} \cdot \mathbf{c_2(t)} \qquad \mathbf{c_2(0)} = \frac{\mathbf{c2_0}}{\frac{\mathbf{mg}}{\mathbf{liter}}}$$

Concentration$_2$:= Odesolve (t, 480)[image]

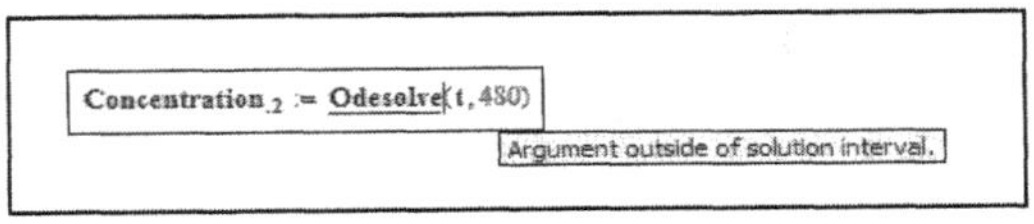

Solution cannot be found using the default solver (Adams/BDF). Try another solve method.

Given

$$\frac{d}{dt}\mathbf{c_2(t)} = \frac{\dfrac{\mathbf{Q}}{\frac{\mathbf{liter}}{\mathbf{hr}}} \cdot \mathbf{Concentration_1(t)}}{\dfrac{\mathbf{Vol}}{\mathbf{liter}}} - \frac{\dfrac{\mathbf{Q}}{\frac{\mathbf{liter}}{\mathbf{hr}}}}{\dfrac{\mathbf{Vol}}{\mathbf{liter}}} \cdot \mathbf{c_2(t)} - \mathbf{k_2} \cdot \mathbf{hr} \cdot \mathbf{c_2(t)} \qquad \mathbf{c_2(0)} = \frac{\mathbf{c2_0}}{\frac{\mathbf{mg}}{\mathbf{liter}}}$$

Right click the Odesolve function and select "Radau."

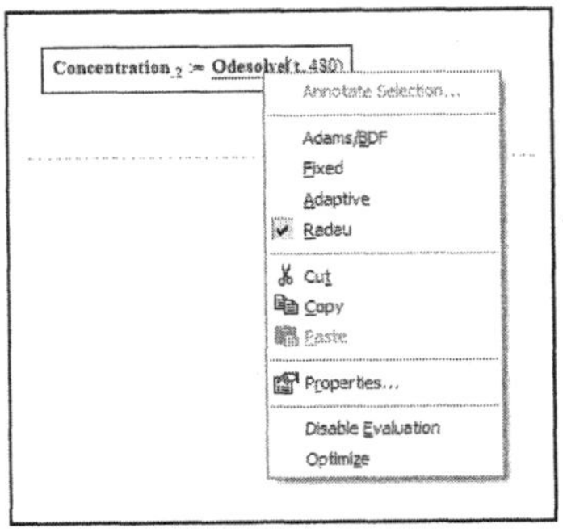

Concentration$_2$:= Odesolve (t, 480)

Concentration$_2$(0) = 50.00

Concentration$_2$(24) = 50.00

Concentration$_2$(96) = 36.20

Concentration$_2$(200) = 22.93

Concentration$_2$(480) = 16.55

Plot the values of $Concentration_1$ and $Concentration_2$ over time.

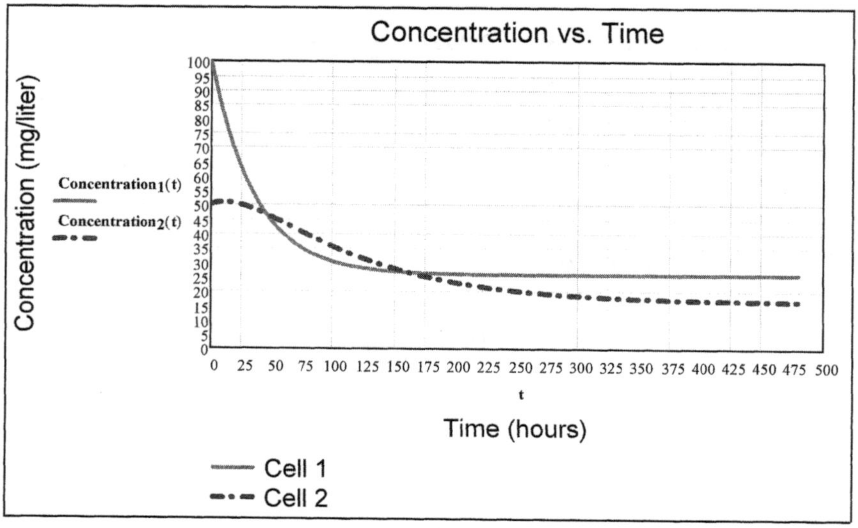

Method 2: *Radau*

Cell 1

We will now use the ***Radau*** function to solve the above problem. The other ODE functions are very similar. Mathcad defines the ***Radau*** function as follows:
Radau (▪,▪,▪,▪,▪,▪,▪,▪,)

Radau (y, x1, x2, npoints, D, [J], [M], [tol])

Returns a matrix of solution values for the stiff differential equation specified by the derivatives in D, and initial conditions y on the interval [x1, x2], using a RADAU5 method. Parameter npoints controls the number of rows in the matrix output. The arguments J, M, and tol are optional and are not needed for this solution.

For this solution we want to capture the concentration at time of 0, so we need to change ORIGIN from 1 to 0.

ORIGIN := 0

Define the function D for use in the ***Radau*** function. Note that in this function the arguments t and c are vector values, and c_0 is an array subscript, and not a literal subscript.

$$\mathbf{D1(t,c)} := \left[\frac{\frac{\mathbf{Q}}{\frac{\mathbf{liter}}{\mathbf{hr}}} \cdot \frac{\mathbf{c_{in}}}{\frac{\mathbf{mg}}{\mathbf{liter}}}}{\frac{\mathbf{Vol}}{\mathbf{liter}}} - \frac{\frac{\mathbf{Q}}{\frac{\mathbf{liter}}{\mathbf{hr}}}}{\frac{\mathbf{Vol}}{\mathbf{liter}}} \cdot \mathbf{c_0} - (\mathbf{k_1} \cdot \mathbf{hr}) \cdot \mathbf{c_0} \right] \qquad \mathbf{c1_0} = 100.00 \cdot \frac{\mathbf{mg}}{\mathbf{liter}}$$

Assign the result of the Radau function to the variable Concentration 1.

$$\textbf{Concentration1} := \textbf{Radau}\left(\frac{\textbf{C1}_0}{\frac{\textbf{mg}}{\textbf{liter}}}, 0, 480, 480, \textbf{D1}\right)$$

Input arguments for the ***Radau*** function

Radau(y, x1, x2, npoints, D, [J], [M], [tol])

y = Initial condition of 100 mg/liter

x1 is the first value of the interval (time = 0 hours.)

x2 is the last value of the interval (time = 480 hours.)

480**hr** = 20.00 · **day**

npoints is the number of points to solve for. In this case, we will solve for 480 points - one for each hour.

D is the previously defined function

The result is a matrix of two columns and 480 rows. The first column is the values of t and the second column is the values of c.

Concentration1 =

	0	1
0	0.00	100.00
1	1.00	97.95
2	2.00	95.95
3	3.00	94.01
4	4.00	92.12
5	5.00	

Extract the values of each column.

$\textbf{Time} := \textbf{Concentration1}^{\langle 0 \rangle}$

$\textbf{ConcentrationCell\_1} := \textbf{Concentration1}^{\langle 1 \rangle}$

Time1 =

	0
0	0.00
1	1.00
2	2.00
3	3.00
4	4.00
5	5.00
6	6.00
7	7.00
8	8.00
9	...

ConcentrationCell_1 =

	0
0	100.00
1	97.95
2	95.95
3	94.01
4	92.12
5	90.28
6	88.49
7	86.76
8	85.07
9	...

Plot the values of Cell 1

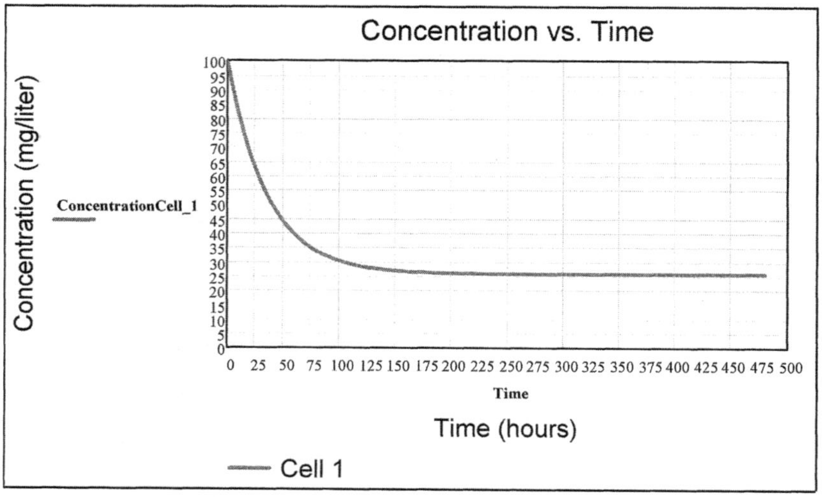

Cell 2

The output from Cell 1 is an array of data for time and concentration. The input for Cell 2 is based on the outflow from Cell 1. It is required to be in the form of function. An array will not work.

We need to create a function from the Cell 1 data. We will use the Mathcad curve fitting functions ***lspline*** and ***interp***.to create the function. Instructions on how to use these functions can be found in Mathcad Help.

Create the spline curve vector used by the ***interp*** function.
Here we use the linear spline function ***lspline***. The other spline functions available are cubic (***cspline***) and parabolic (***pspline***).:

$$\mathbf{S} := \mathbf{lspline(Time, ConcentrationCell\_1)}$$

Use the spline curve vector and the ***interp*** function to create a function to match the data:

$$\mathbf{ConcentrationFunction(t)} := \mathbf{interp(S, Time, ConcentrationCell\_1, t)}$$

Define the function D for use in the ***Radau*** function.
Note that in this function the value for c_{in} from Cell 1 is replaced by the ConcentrationFunction calculated above. The arguments t and c are vector values, and c_0 is an array subscript, and not a literal subscript.

$$\mathbf{D2(t, c)} := \left[\frac{\dfrac{\frac{Q}{liter}}{hr} \cdot \mathbf{Concentration\ Function(t)}}{\dfrac{Vol}{liter}} - \frac{\dfrac{\frac{Q}{liter}}{hr}}{\dfrac{Vol}{liter}} \cdot c_0 - (k_2 \cdot hr) \cdot c_0 \right] \qquad c2_0 = 50.00 \cdot \frac{mg}{liter}$$

$$\textbf{Concentration2} := \textbf{Radau}\left(\frac{\mathbf{c2_0}}{\frac{\mathbf{mg}}{\mathbf{liter}}}, 0, 480, 480, \mathbf{D2}\right)$$

$$\textbf{Time} := \textbf{Concentration2}^{\langle 0 \rangle}$$

$$\textbf{ConcentrationCell\_2} := \textbf{Concentration2}^{\langle 1 \rangle}$$

Time =

	0
0	0.00
1	1.00
2	2.00
3	3.00
4	4.00
5	5.00
6	6.00
7	7.00
8	8.00
9	...

ConcentrationCell_2 =

	0
0	50.00
1	50.14
2	50.27
3	50.38
4	50.48
5	50.56
6	50.63
7	50.69
8	50.73
9	...

<u>Input arguments for the ***Radau*** function</u>

Radau(y, x1, x2, npoints, D, [J], [M], [tol])

y = Initial condition of 50 mg/liter

x1 is the first value of the interval (time = 0 hours.)

x2 is the last value of the interval (time = 480 hours.)

480**hr** = 20.00 · **day**

npoints is the number of points to solve for. In this case, we will solve for 480 points - one for each hour.

D is the previously defined function

As you can see from the plot below, the solution using the ***Radau*** function matches the solution using the ***ODESolve*** function.

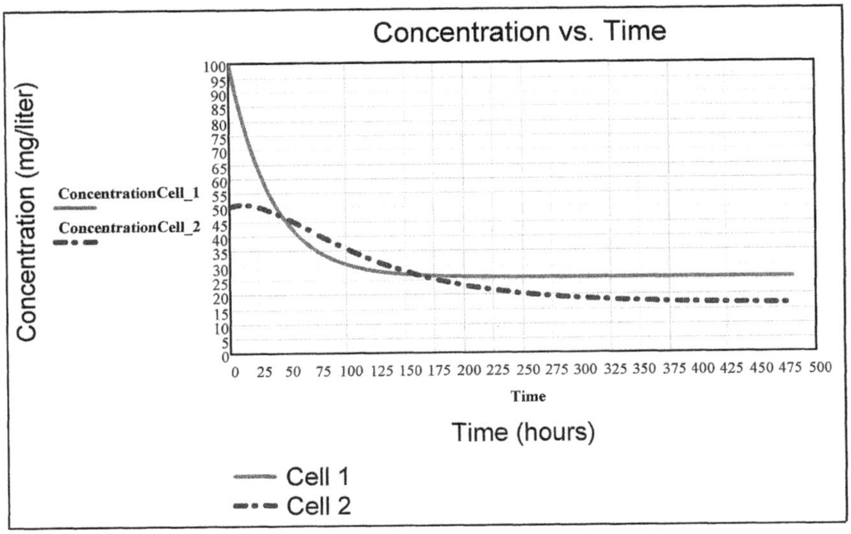

PRACTICE

Additional problems and applications can be found on the companion site: www.elsevierdirect.com/9780123747839.

1. From your calculus textbook, use the Mathcad Symbolics menu to solve 10 differentiation problems.
2. Use the live symbolics operator to solve the same problems as above.
3. From your calculus textbook, use the Mathcad Symbolics menu to solve 10 indefinite integral problems.
4. Use the live symbolics operator to solve the same problems as above.
5. From you calculus textbook, use the live symbolics operator to solve 10 definite integrals.
6. Use numeric integration to solve the same problems as above.
7. Use the Symbolics menu, the symbolics operator, and numeric differentiation (at x = 5) to differentiate the following (Note: Assign the numeric values at the end of your worksheet, so that the symbolic evaluations will not be affected by the numeric values.):

   a. acos(x)
   b. sec(x)
   c. x^n (assign n = 5 for the numeric differentiation)
   d. $3x^3 + 2x^2 - 4x + 3$
   e. e^x

8. Use the Symbolics menu and the symbolics operator to integrate the following:

   a. x^n (assign n = 5 for the definite integral in the next problem)
   b. e^x
   c. cos(x)
   d. x*sec(x^2)
   e. tan(a*x) (assign a = 0.8 for the definite integral in the next problem)

9. Evaluate the above integrals as definite integrals from the values of a = 1 and b = 5 (For x*sec(x^2 use a = 0.1 and b = 0.2).
10. From your differential equations text book, solve 5 first order ordinary differential equations using the ***Odesolve*** function.

PART

4

Creating and Organizing Your Engineering Calculations with Mathcad

Part IV is what sets *Essential Mathcad* apart from all other books written about Mathcad. This part is unique because it discusses the use of Mathcad to create a complete set of engineering calculations. This section uses the phrase "engineering calculations" extensively. The ideas and concepts covered are just as applicable to scientific or other technical calculations. I hope our friends from the scientific community will bear with the focus on engineering calculations.

You now have a toolbox full of simple tools and power tools. Many other tools still can be added, but it is time to discuss how to apply the tools you have. It is time to start building Mathcad calculations.

Part IV will begin with a discussion about customizing Mathcad settings and creating templates in order to get consistent worksheets. Part IV will also discuss embedding other programs into Mathcad (OLE). It will then discuss Mathcad components, especially the use of Microsoft Excel, including the use of your existing Microsoft Excel spreadsheets. We then discuss how to assemble calculations from multiple Mathcad files, and how to input and output data to your worksheets. This part will conclude with a chapter about adding hyperlinks and a table of contents to your worksheets.

CHAPTER

Putting It All Together

13

Calculations from traditional engineering projects use a combination of various computer programs, spreadsheets, and hand calculations. The printout from these computer programs are combined with the hand calculation sheets and placed into calculation binders. The paper binders then become the official calculations that are saved with the project.

Mathcad can be used to replace the book that contains the paper. All that information can reside within a Mathcad file: computer input and output files, printouts from computer programs, spreadsheets, PDF scans of hand calculations (why would you want to do hand calculations when you can use Mathcad?), and of course, your Mathcad calculations. The chapters in Part IV will discuss how to do all of this.

Creating an entire set of engineering project calculations with Mathcad is much more complex than writing a one- or two-page worksheet. Many issues need to be considered when creating a full set of engineering project calculations. The purpose of this chapter is to briefly review some of the topics covered in earlier chapters and to introduce the topics to be discussed in Part IV.

Chapter 13 will:

- Discuss the concept of the Mathcad toolbox and the tools in it.
- Introduce the concept of using the tools to build a project.
- Review the topics discussed in all the chapters up to this point.
- Paint a picture of what will be accomplished in Part IV, by briefly discussing the topics to be covered in Chapters 14 through 20.

INTRODUCTION

Let's go back to the analogy used at the beginning of this book—teaching you how to build a house. We have helped you build a toolbox—a place where you can store your tools. Your Mathcad toolbox is a thorough understanding of Mathcad basics.

We have also collected many tools and put them into your toolbox. We have taken classes in hammers, screwdrivers, pliers, power saws, power drills, and more. In Part IV, we teach you how to use these tools to build the house.

As you begin creating and assembling calculations, the variable names you choose to use will begin to be very important.

GUIDELINES FOR NAMING VARIABLES

When you create a simple one- or two-page worksheet, it usually does not matter what variable names you choose to use. Single-letter variables are more than adequate because the entire worksheet can be taken in context and you usually do not run the risk of redefining the variable.

When you create a complete set of project calculations that can be over 100 pages long, then variable names are critical. You might have many conditions where you want to use the same variable name, but you want each variable to be unique, and you do not want to redefine a previously used variable. The use of literal subscripts to make each variable name unique is very useful in this case.

Descriptive variable names are also very useful. They help you remember what the variable refers to, and they also help those reviewing the calculations know what the variable name means. For example, is it easier to understand d:=5 ft or DepthOfWater:=5 ft? In a single-page worksheet either would be adequate. When your calculations are 50 pages long, if you defined "d" on page 5 and do not use it again until page 50, it may be more useful to have a descriptive variable name.

Descriptive variable names also help prevent you from redefining a variable name. In the previous example, if you defined the depth of water as d=5 ft and then later defined the depth of soil as d=6 ft, you have redefined the variable d. This will cause an error in your calculations if later on you use the variable "d" thinking that it is the depth of water. It is much better to use DepthOfWater:=5 ft and DepthOfSoil:=6 ft.

The following naming guidelines are given to assist in choosing variable names to be used in your engineering calculations. There will be exceptions to each of these guidelines, but if they are consistently followed, they will help prevent problems that will be discussed in future chapters.

Naming Guideline 1: Use Descriptive Variable Names

Single letters are appropriate if you are just doing a quick calculation for something that won't be saved. They are also appropriate if the single letter represents a universal constant such as "e." They are also useful if you have an equation that uses single-letter variables. But, generally, it is better to use descriptive variable names with more than one letter.

Naming Guideline 2: Use a Combination of Uppercase and Lowercase Letters to Help Make Your Variable Names Easier to Read

Unless there is some specific reason not to, your variable names should begin with an uppercase letter. If your variable name is more than one word, use a combination of uppercase and lowercase letters to make the variable name easier to read. For example, instead of naming a variable "vesselpressure," name it "VesselPressure." See Figure 13.1 for some examples.

Naming Guideline 3: Use Underscores to Separate Different Names in Your Variable Names

In the previous guideline we discussed using a combination of uppercase and lowercase letters to make variable names easier to read. Another way to do this is to use an underscore to separate names. For example, instead of naming a variable "VesselPressure," you could name it "Vessel_Pressure." See Figure 13.2 for some examples.

Naming Guideline 4: Make Good Use of Subscripts in Your Variable Names

Literal subscripts are created by typing a period within your variable name. Subscripts are useful to distinguish variables that are very similar or related.

Buildingwidth	BuildingWidth
Maximumacceleration	MaximumAcceleration
Momentofinertia	MomentOfInertia
Transversewavevelocity	TransverseWaveVelocity
Areaofcircle	AreaOfCircle

FIGURE 13.1 Using uppercase letters to separate words in variable names

BuildingWidth	Building_Width
MaximumAcceleration	Maximum_Acceleration
MomentOfInertia	Moment_Of_Inertia
TransverseWaveVelocity	Transverse_Wave_Velocity
AreaOfCircle	Area_Of_Circle

FIGURE 13.2 Using an underscore to separate words in variable names

Area1	$Area_1$
Volume1	$Volume_1$
Velocity1	$Velocity_1$
Integral1	$Intergral_1$

FIGURE 13.3 Using subscripts in variable names

(f)	If you type f followed by typing the single quote key ' , you get this.
f\`	The \` mark is usually located next to the number 1 key.
f'	If you really want to use a single quote mark, then you can use the special text mode (CTRL+SHIFT+k) to include the single quote.

FIGURE 13.4 Options for using a prime symbol

Remember that once you use a subscript in a variable name, you cannot return to normal text size. Figure 13.3 shows a few examples of using subscripts.

Naming Guideline 5: Use the (') Key If You Need to Use a "Prime" (Single Apostrophe) in Your Variable Name

Sometimes your variable name has an apostrophe in it, such as the strength of concrete, f ′c. If you type the single quote key (′), then Mathcad will put parentheses around your variable name. In order to use the prime symbol in your calculations, use the (′) key in the upper left-hand corner of your keyboard. It is usually located on the same key as the (~) tilda key. See Figure 13.4.

MATHCAD TOOLBOX

Let's quickly review the skills and understanding you should have mastered from Part I.

Variables

By now, you are completely familiar with how to define and use variables, but do you remember how to use the special text mode (CTRL+SHIFT+K) to add symbols to your variable names? Do you remember the Chemistry Notation (CTRL+SHIFT+J)? If not, you might want to review these concepts from Chapter 2.

> Tip! It will be helpful to establish some personal or corporate naming guidelines. There have been many times when I have been very careful to use descriptive variable names, only to be caught later not remembering what I named a particular variable. Did I name it DepthOfWater or WaterDepth? Was it BuildingLength or LengthOfBuilding or Building_Length? It will be beneficial for you to create some guidelines for your specific needs so that the names used in your calculations have a consistent usage.

The figures for each chapter in this book are created in a single Mathcad file. This required the use of unique variable names for each figure. This book uses a variety of variable naming methods in the figures. This has been intentional; it is to expose you to several different ways to name variables.

Editing

You could not have gotten this far in the book without understanding how to edit an expression. We will assume that you have a good understanding of editing techniques.

User-Defined Functions

The discussion of user-defined functions in Chapter 3 was intentionally basic. The intent was to establish a solid understanding of user-defined functions prior to introducing the more powerful and complex features.

Each chapter has been built on this foundation of user-defined functions. In Chapter 4, we introduced units to user-defined functions. In Part II and Part III, we added arrays, programming, and live symbolics to user-defined functions.

Units!

Units are powerful, and essential to engineering calculations. Use them consistently.

You should be very comfortable with using units. If not, go back and review Chapter 4. Units are very easy to use with most engineering equations. For the few equations where units will not work (such as empirical equations), divide the variable by the desired unit to make it a unitless number of the expected value.

Remember to include units in the argument list for user-defined equations. When plotting, remember to divide the expressions by the units you want to plot. For example, if you want to plot in feet, divide the expression by feet. If you want to plot in meters, divide the expression by meters.

Mathcad Settings

Chapter 14 will discuss Mathcad settings and suggest some changes to Mathcad default values.

Customizing Mathcad with Templates

In Chapter 15, we will discuss styles for text and math variables. We will recommend changing some default Mathcad styles. We also recommended adding some new styles. We will discuss the use of headers and footers. In Chapter 16, we discuss saving these customizations into a template.

HAND TOOLS

Arrays are an essential tool for engineering calculations. We have used vectors in numerous examples in order to display results from multiple input variables. Range variables have been used in dozens of examples. If you have forgotten about the vectorize operator, review Chapter 5 for its use. These are essential tools as you begin to build engineering calculations.

Chapter 6 discussed a few selected functions. These are only the beginning of the functions you can add to your Mathcad toolbox. Review the functions and operators discussed in Chapter 6, and see which functions will be most useful to you.

Plotting is a very useful tool. If you are not comfortable with creating simple X-Y plots or simple Polar plots, review Chapter 7.

A thorough understanding of simple logic programming is essential. Engineering calculations require the use of logic programs to choose appropriate actions. If you are not yet comfortable with the use of programs in expressions or user-defined programs, review Chapter 8.

POWER TOOLS

Chapters 9 through 12 introduced some topics that might seem confusing at first. That is one reason why their discussion was held off until Part III.

The symbolic calculations, solving tools, programming tools, and calculus functions are extremely powerful. These tools will be very beneficial in your engineering calculations. When it comes to using Mathcad to solve for various equations, there are many ways to arrive at the same answer. We have tried to show you the various ways of solving for solutions. The engineering examples shown in Chapter 10 provide a good summary of how to use the solving functions.

LET'S START BUILDING

Now that you have your toolbox full of tools, it is time to start building. One of the primary purposes of this book is to teach you how to create and organize engineering calculations. The principals used can be used for any technical calculations, not just engineering calculations.

Mathcad is perfectly suited to create and organize your project calculations. The reasons for this include:

- It speaks our language—math.
- Calculations and equations are visible, and not hidden in spreadsheet cells.
- The calculations are electronic and can now be printed, archived, shared with coworkers around the world, searched, and reused.
- Mathcad calculations can be used over and over again.
- If input variables change, the calculations are updated automatically.
- Results from other software programs can be incorporated into your calculations.
- Mathcad can share information with other programs such as Microsoft Excel.

WHAT IS AHEAD

The chapters in Part IV will help you as you begin using Mathcad to create and organize your engineering calculations. In Chapter 17, we discuss how to save and reuse worksheets. It is most likely that Mathcad will not be the only software program you use. Therefore, it is important to discuss how to bring information from other software programs into your Mathcad worksheets. We discuss this in Chapter 18. A very powerful feature of Mathcad is its capbility to communicate with and transfer information between other programs. These programs are called components. In Chapters 19 and 20, we discuss the use of components—especially the use of Microsoft Excel. Finally, in Chapters 21 and 22 we discuss different ways of inputting and outputting information from your calculations, as well as ways to create hyperlinks to information in your current worksheet as well as other worksheets.

SUMMARY

This chapter is a review of the topics covered in previous chapters. We emphasized the importance of choosing suitable variable names in your calculations. We also encouraged you to go back and review any chapter with which you are not comfortable. We then looked to the future and highlighted the topics we are about to discuss.

PRACTICE

Additional problems and applications can be found on the companion site: www.elsevierdirect.com/9780123747839.

1. Create five variable names using the special text mode discussed in Chapter 2. Use different names than you used for the practice in Chapter 2.
2. Create five variable names using Chemistry Notation discussed in Chapter 2. Use different names than you used for the practice in Chapter 2.
3. Go back and briefly review the topics covered in each chapter. Look at the practice exercises to see how well you remember the topics.

CHAPTER

Mathcad Settings

14

This chapter will give you detailed information about many of the Mathcad settings. With this information you can make informed decisions as to what each different setting should be for your situation. The chapter will focus on the Preferences dialog box, the Worksheet Options dialog box, and the Result Format dialog box.

This chapter will discuss each setting in the order they appear in the dialog boxes. Skip over any item that is not clear or not useful. After you become more familiar with Mathcad you can refer back to this chapter to review the different settings. Much of the information in this chapter is taken directly from the Mathcad Help with additional comments added.

Chapter 14 will:

- Discuss the Preferences dialog box and show how different settings affect the way Mathcad starts up and how it operates.
- Discuss the Worksheet Options dialog box and show how different settings can affect a specific worksheet.
- Recommend specific settings for various features.
- Discuss the Result Format dialog box and show how to control the way results are displayed.

PREFERENCES DIALOG BOX

The Preferences dialog box sets global features. The changes made in this dialog box are effective for all Mathcad worksheets. They will be effective every time you open Mathcad.

To open the Preferences dialog box, click **Preferences** from the **Tools menu**. Figure 14.1 shows the Preferences dialog box.

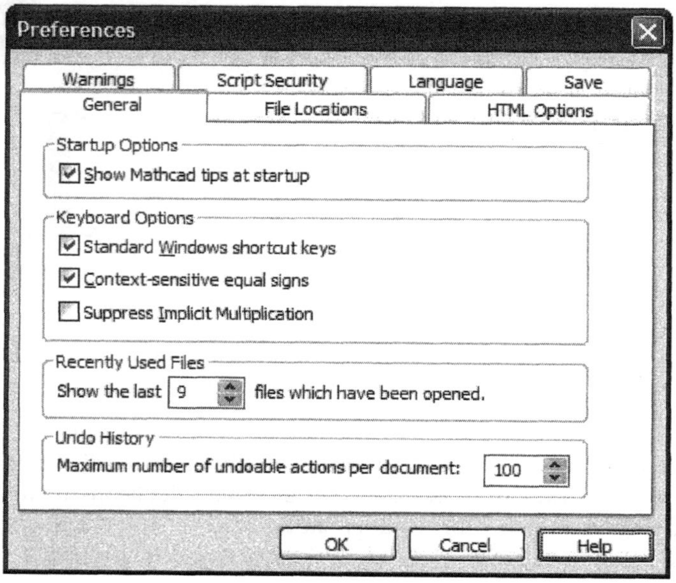

FIGURE 14.1 Preferences dialog box

General Tab

Startup Options

Checking the "Show Mathcad Tips" At Startup box will cause a Mathcad Tips dialog box to appear every time you start Mathcad. If you would like to see a Mathcad tip every time you open Mathcad, make sure that this box is checked. If you are tired of seeing the tips, uncheck this box. The tips are very useful if you are just starting to learn Mathcad. They give you many worthwhile suggestions concerning the many features of Mathcad.

> **Tip!** You can view and print the entire list of startup tips. The file is called mtips_EN.txt, and it is located in the main Mathcad program directory. This file can be opened in a word processor or in a text editor, and then printed.

Keyboard Options

Leave the "Standard Windows Shortcut Keys" box checked, unless you are a long-time Mathcad user and want to use some of the keystrokes from the early versions of Mathcad.

The "Context-Sensitive Equal Signs" box is checked by default. Leave this box checked. This feature allows the [=] key to insert the evaluation equal sign (=) when it is to the right of a defined variable and the definition symbol (:=) when it is to the right of an undefined variable.

Mathcad automatically inserts an invisible multiplication sign between numbers and letters. This box will turn off the feature. Leave this box unchecked.

Recently Used Files

This feature sets the number of recently opened worksheets displayed in the File menu. The default is four worksheets.

> **Tip!** I like to see as many files as possible, so I increase this number to 9.

Undo History

This feature controls how many actions can be undone when using the Undo command. The undo history can be set to any number between 20 and 200. The default value is 100. The higher the number, the more system memory is used. Unless you are doing some very critical calculations where you would need more than 100 undos, the 100 default should be adequate.

File Locations Tab

See Figure 14.2 for an example of the File Locations tab.

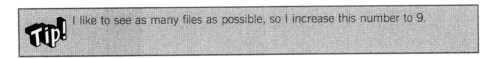

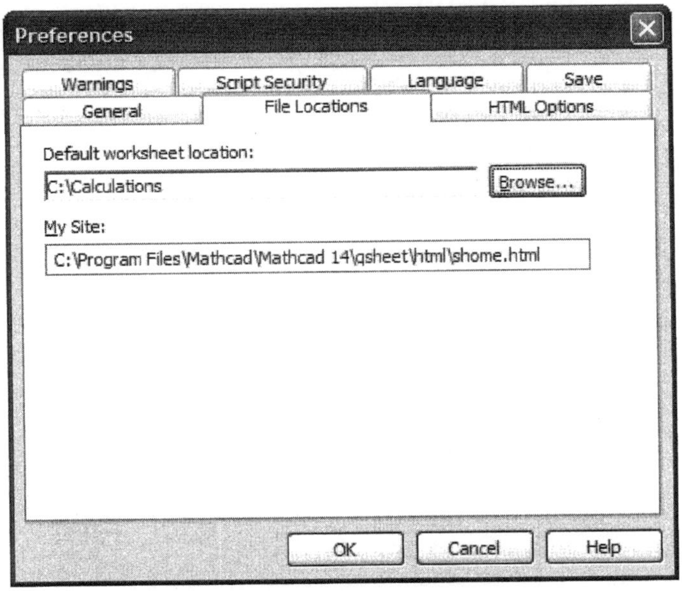

FIGURE 14.2 File Locations tab

Default Worksheet Location

This is where you set the directory location where Mathcad looks for files after starting. It is also the default location where files will be stored. Note that this location is the location just after Mathcad starts. Once Mathcad has started and you open a file from another location, or save a file to another location, Mathcad will go to the last used folder location, not this default location. The next time you start Mathcad this default location is used.

> **Tip!** I suggest changing the default worksheet location to a corporate calculation location or to a specific calculation folder.

My Site

This is the path that Mathcad uses to map to the My Site on the Resources toolbar (see Chapter 1). The default path opens a Mathcad window that has several useful links. Unless you have a reason to map to a specific site, stay with the Mathcad default.

HTML Options Tab

See Figure 14.3 for an example of the HTML Options tab.

With Mathcad you can export your files to HTML format to create web pages. This tab sets the way Mathcad saves the files as a web page. In a nutshell: if you want your web page to look like your Mathcad worksheet, select **PNG**. Your

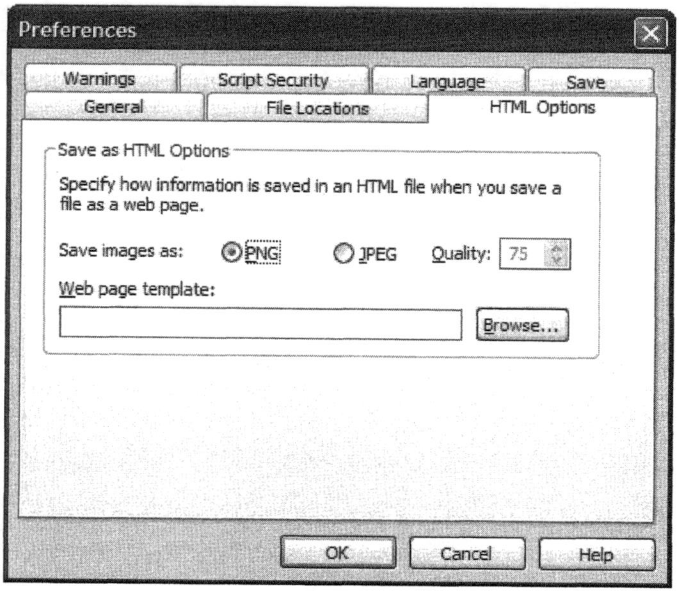

FIGURE 14.3 HTML Options tab

HTML worksheet will be an exact image of your Mathcad worksheet. The math cells, however, will not be interactive. They will be graphic images. After saving your Mathcad worksheet into HTML format, you can still use Mathcad to open the HTML file, and all the math cells become interactive again.

If you want the file size of your HTML document to be smaller than **PNG** provides, you can select **JPEG** and set **Quality** to about 50. This reduces the quality of the image in the HTML format, and reduces the file size. The lower the quality number, the lower the quality of the image in the HTML file. It becomes a trade-off between file size and image quality. If you can afford the larger file size, then always use **PNG**.

The HTML options on this tab set the defaults for when you select **Save As** from the **File** menu, and then choose **HTML File (*.htm)** from the "Save As Type" list. See Figure 14.4.

If you choose **Save As Web Page** from the File menu, you will be prompted for a file name. Once you click **Save**, a second dialog box will appear. From this dialog box, you can select the same features as from the HTML tab in the Preferences dialog box. See Figure 14.5.

Warnings Tab

The Warnings tab sets which items to highlight with the warning flag. See Figure 14.6. Make sure there is a check in the box next to "Show warnings on redefinitions of:."

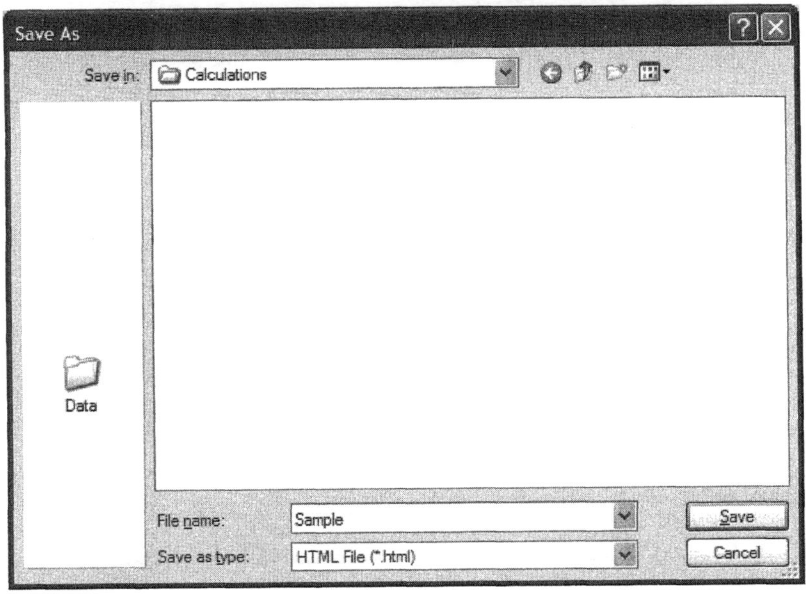

FIGURE 14.4 Using Save As HTML File.

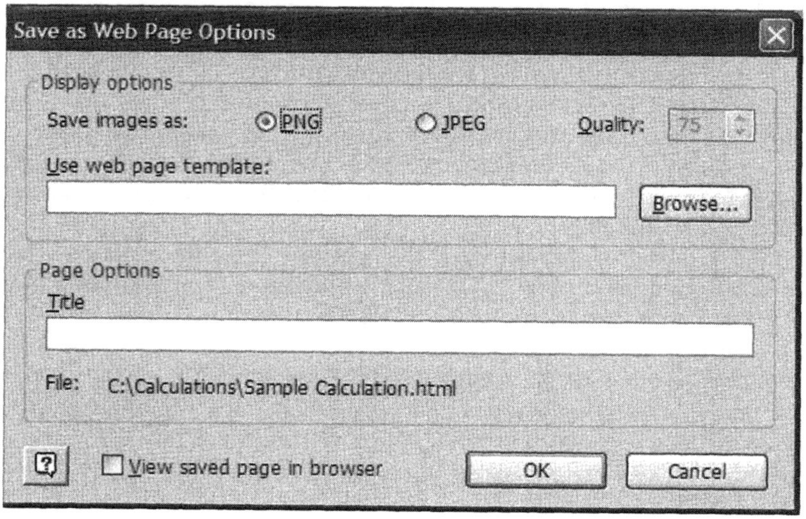

FIGURE 14.5 Save As Web Page Dialog Box

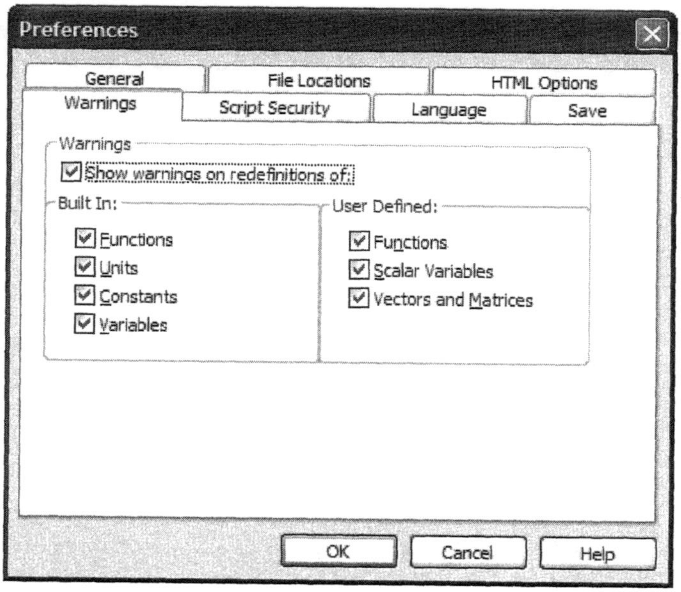

FIGURE 14.6 Warnings tab to set redefinition warnings

> **Tip!** When using Mathcad for scientific and engineering calculations, it is important to know when any variable or function is redefined. It is suggested that all the checkboxes be checked. This will better protect you as you prepare your calculations.

Built-in functions are listed in the Insert Function dialog box. Built-in constants include such things as e, pi, i, and j. For a complete list of all the Mathcad built-in constants, see Mathcad Help and search for "constants."

To see a list of the Mathcad built-in variables, click **Worksheet Options** from the **Tools** menu, and then select the Built-in Variables tab. This tab will be discussed later in this chapter.

The User-Defined Functions, Scalar Variables, and Vectors and Matrices are all definitions that you create in your worksheets. When these boxes are checked, Mathcad will give you a warning if you overwrite any of these definitions.

Script Security Tab

Mathcad worksheets may contain scriptable components. These components, such as VBScript, Jscript, or macros, may contain harmful code. This tab allows you to control how Mathcad deals with scriptable code when it opens a worksheet that contains scriptable code.

There are three security options from which to select:

High Security is the most secure. When this option is checked, Mathcad automatically disables all scripts when a worksheet is opened. In order to enable a scripted component when this option is selected, right-click the component and choose Enable Evaluation.

Medium Security has Mathcad prompt you whenever a worksheet containing scriptable components is opened. You are given an option of whether or not to disable the script.

Low Security allows Mathcad to enable all scripts whenever a Mathcad worksheet is open. This option is the least secure, because harmful scripts could be opened without your knowledge.

It is recommended you select Medium Security. This prevents unknown scripts from being opened, and it allows you to select which scripts to enable. If you have full confidence in the worksheet being opened, you can then choose to enable the scripts when opening the file.

If you have Microsoft Excel worksheets embedded into your Mathcad worksheets, and if your Excel worksheets have macros, you might also get an Excel warning. This warning is controlled by the Excel macro security level. The use of Microsoft Excel will be discussed in Chapter 20.

> **Tip!** If you have the Mathcad script security set to medium and you open a file containing several scripts, Mathcad stops at each instance of a script and asks whether to enable the script. This can be annoying as you try to scroll through the document after opening the file. If you are clicking the right scroll bar to move down through the document and Mathcad comes to a script, a dialog box is opened asking if you want to enable the script. The default is to disable the script. If you have a smart mouse enabled, your mouse might jump to the default button of the dialog box. If you are quickly clicking the scroll bar, you might accidentally disable a script you wanted to enable. In order to avoid this, type **CTRL+END**. This causes Mathcad to go to the end of the document. You may also type **CTRL+F9**. This forces Mathcad to calculate the entire worksheet. Now all instances of scripts will be encountered, and all the dialog boxes will appear one after another, so that you can answer all the questions at one time.

Language

Language Options

The options on the Language tab control how Mathcad displays languages. If additional languages are installed, you can select which language to display for menus and Mathcad's math language. English versions of Mathcad will have only the English option. Other language versions of Mathcad will have additional language options available.

Spell Check Options

These options set the language to use when spell-checking the Mathcad worksheet. Some languages have different spellings of words. Mathcad allows you to select which dialect to use for spell-checking. For example, in English you can choose from American English, Canadian English, or British English. In German, you can choose from old German and new German.

Save Tab

Default Format

Starting with version 12, Mathcad allows you to save in two new formats: **Mathcad XML Document** and **Mathcad Compressed XML Document**.

The default is **Mathcad XML Document**. This should be adequate for most applications. This option saves the worksheet as an XML document and allows the Mathcad worksheet to be read by any text editor or XML editor. There are many benefits to saving your documents in XML format that will be discussed later in Chapter 17. When files are saved in this format, the file extension is .XMCD.

If you select **Mathcad Compressed XML Document**, then the Mathcad file will be zipped. It will not be available as an XML file until it is unzipped. If you want to send a smaller size Mathcad file to someone else, you can select this option. Files saved in this format have the file extension .XMCDZ.

Autosave

Starting with version 13, Mathcad finally introduced an autosave function. Be sure to check this box and set a time interval for the autosave.

Summary of the Preference Tab

The items in the Preferences dialog box are Mathcad defaults. They are applicable to all new documents and all existing documents. Once the settings in the Preferences dialog box are set, they remain the same for all documents until the settings are changed. These settings affect the way Mathcad operates and how it saves documents.

WORKSHEET OPTIONS DIALOG BOX

We just discussed the Preferences dialog box, which sets global settings affecting all documents. We will now discuss the Worksheet Options dialog box. The settings in this dialog box affect only the current worksheet. Every worksheet can have different settings. This dialog box is opened by clicking **Worksheet Options** from the **Tools** menu.

Built-in Variables Tab

This is the location where built-in variables are defined. This tab is shown in Figure 14.7. The Mathcad default values for these built-in variables are noted in parentheses to the right of the boxes.

Array Origin

This option controls the value for the built-in variable "ORIGIN." This variable represents the starting index of all arrays in your worksheet. The Mathcad default for this variable is 0, but earlier in the book we changed this value to 1.

Convergence Tolerance (TOL)

This variable controls the length of the iteration in Solve Blocks and in the root function. This value was discussed in detail in Chapter 10. It also controls the precision to which integrals and derivatives are evaluated.

Use the Mathcad default for this variable.

Constraint Tolerance (CTOL)

This variable controls how closely a constraint in a Solve Block must be met for a solution to be acceptable. Solve Blocks were discussed in Chapter 10.

Use the Mathcad default for this variable.

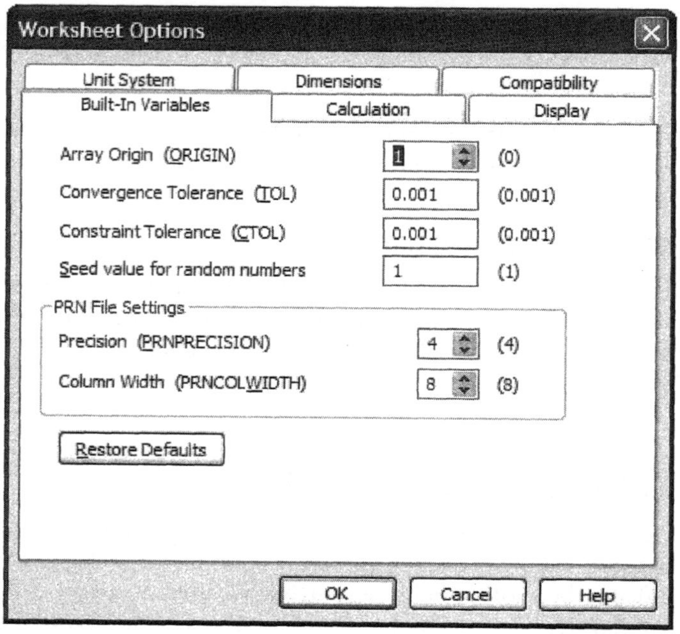

FIGURE 14.7 Built-in Variables tab

Seed Value for Random Numbers

This value tells a random number generator to use a certain sequence of random numbers. The default number is 1, but any number can be used to generate a different sequence of random numbers.

Precision (PRNPRECISION)

This variable controls the number of significant digits that are used when writing to an ASCII data file. Use the Mathcad default for this variable.

Column Width (PRNCOLWIDTH)

This variable controls the width of columns when writing to an ASCII data file. Use the Mathcad default for this variable. See Chapter 6 for a discussion about writing to files.

Calculation Tab

See Figure 14.8 for an example of the Calculation tab.

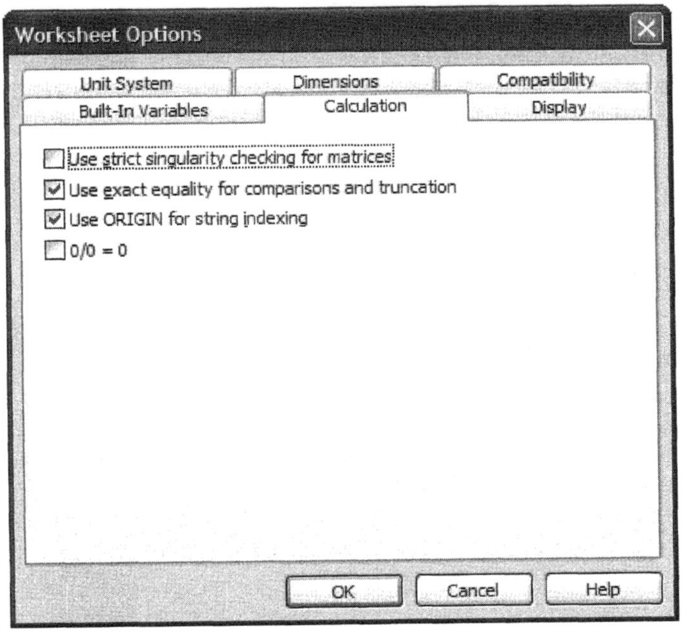

FIGURE 14.8 Calculation tab

Use Strict Singularity Checking for Matrices

This topic is beyond the scope of this book. The Mathcad default is unchecked. See Mathcad Help for more information.

Use Exact Equality for Boolean Comparisons

This check box controls the standard use of Boolean comparisons. The Mathcad default is checked. This is adequate for most applications.

Use ORIGIN for String Indexing

Mathcad defaults to considering the first character in a string as 0, similar to the default value for the built-in variable ORIGIN. Checking this box will consider the first character to be the value for ORIGIN. If you changed the default value of the built-in variable ORIGIN from 0 to 1, you can check this box and Mathcad will consider the first character in the string to be 1 instead of 0. If this box is unchecked, the integer associated with the first character in a string will be 0.

> **Tip!** I do not like the first character in a string to be considered 0; I like it to be 1. I like to check this box to match the change I make to ORIGIN.

0/0=0

When Mathcad divides by zero it returns an error. This box allows you to tell Mathcad that if you have a condition where you have 0/0, then return the value 0 rather than an error. The Mathcad default is unchecked.

Display Tab

The Display tab controls how various operators appear in your worksheet. Figure 14.9 shows the default Mathcad settings. Keep the default settings (except for the Currency symbol). However, if you are publishing or presenting your calculations, you might want to show different operators. For example, you might want to show ▤ instead of ⬚⬛ for the definition of an expression, or you might want to see × for multiplication instead of a dot. When you change the display, you have not affected the way Mathcad operates. The default operator will appear when you select the math regions. Refer to Mathcad Help for specific information on the different options.

The base currency symbol ¤ is inserted when you attach the unit of money to a number. This symbol will be displayed unless you select a preferred currency symbol. Change this symbol to your local currency.

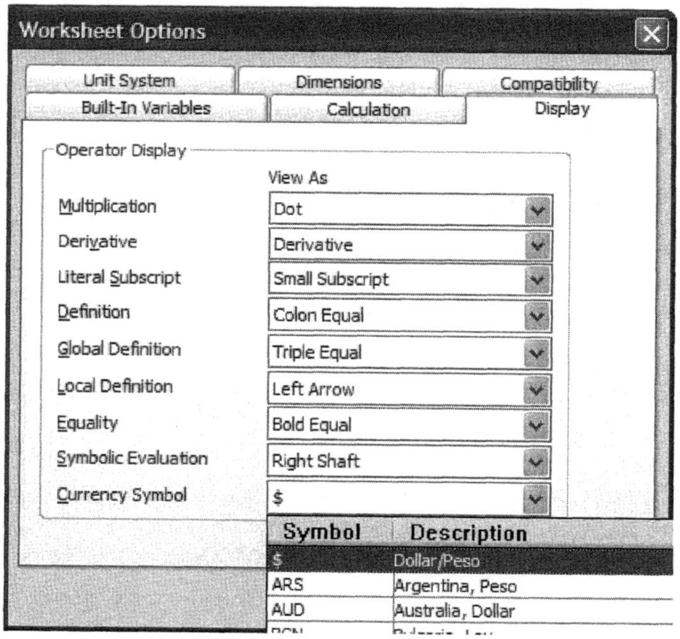

FIGURE 14.9 Display tab showing default Mathcad settings

Unit System Tab

This tab sets the default system of units used by Mathcad. The default system used by Mathcad is SI (International). Units were discussed at length in Chapter 4; refer to that chapter for additional information about this tab.

Dimensions Tab

This tab is not useful. Do not change the default options on this tab.

Compatibility Tab

These settings help in the transition between different versions of Mathcad. If you have worksheets from previous versions of Mathcad, the Mathcad Help associated with this tab will be very important.

RESULT FORMAT DIALOG BOX

The Result Format dialog box controls how results are displayed in Mathcad. For example, in this box you control how many decimals to display, whether to show trailing zeros, and what exponential form to use. To access the Result Format dialog box, select **Result** from the **Format** menu. Figure 14.10 shows a sample Result Format dialog box.

The changes you make in this dialog box change the display for all results in your current worksheet. Chapter 16 will discuss how to make the changes to all future worksheets. You can override the settings made in this dialog box for

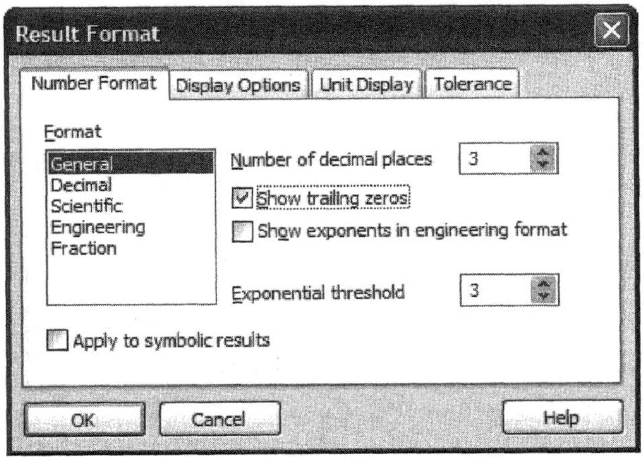

FIGURE 14.10 Result Format dialog box with General format selected

any specific result. To change the result format of a specific result, double-click the result and a similar dialog box will appear.

> **Tip!** If you change the result format of a specific region it supersedes the default settings. Be aware of this if you later change your default settings and some results are not updated to the default settings.

Number Format Tab

This tab controls the number of digits displayed to the right of the decimal point and controls how exponents are used in displaying your results. There are five different types of formats to select:

- When General is selected, the results are displayed in decimal format until the exponential threshold is reached. After the exponential threshold is reached, results are displayed in exponential notation.
- When Decimal is selected, the results are never in exponential notation. The number of decimal places displayed can be set.
- When Scientific is selected, the results are always in exponential notation.
- When Engineering is selected, the results are always in exponential notation and the exponents are in multiples of three.
- When Fraction is selected, the results are displayed as fractions. This also brings up a "Level Of Accuracy" box. This box controls how close the fraction approximation is to the decimal value. The higher the accuracy, the closer the approximation. The Mathcad default is 12. The "Use Mixed Numbers" check box allows the use of integers and fractions. It keeps the fraction part of the number less than one. See Figure 14.11.

The "Show Trailing Zeros" check box is unchecked by default. If you have a result of 1.6 and the "Number Of Decimal Places" is set to 3, Mathcad does not show trailing zeros and displays 1.6. If this box is checked, Mathcad will display trailing zeros to the number of decimal places. Thus the result would be displayed as 1.600.

> **Tip!** I like to check the "Show Trailing Zeros" check box, because if I see a number displayed as 1.6, I don't know how accurate the number is. The number could be 1.649 with the number of decimal places set to 1, or the number could be 1.600 with the "Show Trailing Zeros" box unchecked. If the box is checked, there is not a question.
>
> If the "Show Trailing Zeros" check box is checked, the exponents in units will also have trailing zeros. For example, area will be displayed as $m^{2.00}$. If you do not like this, double-click the result and uncheck the "Show Trailing Zeros" box. This will change the result for this specific region.

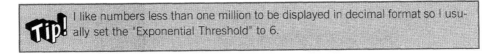

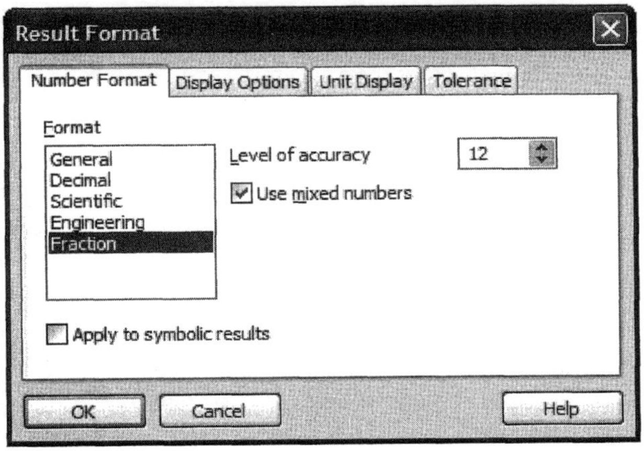

FIGURE 14.11 Result Format dialog box with Fraction format selected

The "Show Exponents In Engineering Format" is similar to the Engineering format just discussed.

The "Exponential Threshold" tells Mathcad when to start displaying results in exponential notation. This is used only when the General format is selected. The other formats either always display exponential notation or never display exponential notation.

> **Tip!** I like numbers less than one million to be displayed in decimal format so I usually set the "Exponential Threshold" to 6.

More information about the Number Format tab can be found in Mathcad Help.

Display Options Tab

See Figure 14.12 for a sample of the Display Options tab.

Matrix Display Style

This box selects whether an array is displayed in an output table or in matrix form. These options were discussed in Chapter 5.

Imaginary Value

This box tells Mathcad whether to use an i or j when displaying the imaginary part of an imaginary number.

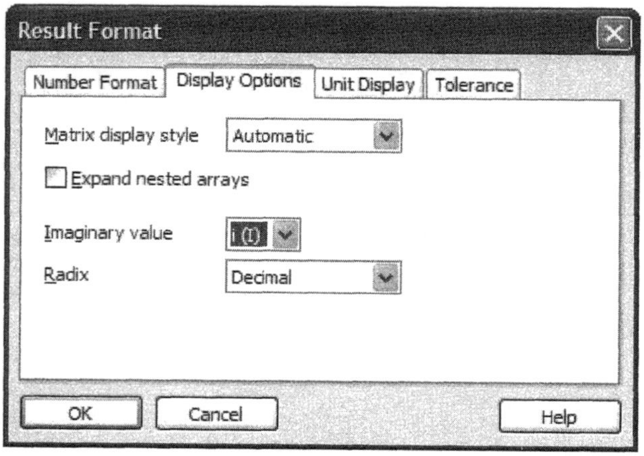

FIGURE 14.12 Display Options tab

Radix

This box allows you to display results as decimal, binary, octal, or hexadecimal numbers.

Unit Display Tab

See Figure 14.13 for a sample of the Unit Display tab.

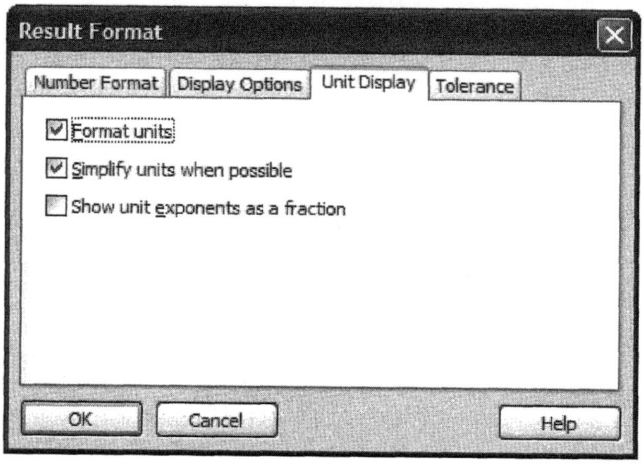

FIGURE 14.13 Unit Display tab

There are three check boxes on this tab. The first two are checked by default. The "Format Units" box reformats the units displayed to a more common notation. For example, sec^{-1} displays as 1/sec. The "Simplify Units When Possible" box displays the simplest unit possible. For example, kg*m/sec2 is simplified to N (Newton). The "Show Unit Exponents As A Fraction" box displays unit exponents as rational fractions when checked. Otherwise, they are displayed as decimals. This option can be set only for the whole worksheet, not on a region-by-region basis.

Tolerance Tab

See Figure 14.14 for a sample of the Tolerance tab.

The "Complex Threshold" box controls how much larger the real or imaginary part of a number must be before the display of the smaller part is suppressed. See Mathcad Help for examples.

The "Zero Threshold" box controls how close a result must be to zero before it is displayed as zero. This means that if the zero threshold is set to 10, numbers smaller than 10-10 will be displayed as zero, even if the decimal places is set to a number greater than 15. The Mathcad default is 15. See Mathcad Help for additional examples.

Individual Result Formatting

The preceding discussion relates to default formatting for the entire worksheet. You can overwrite the worksheet default settings and format the display of a

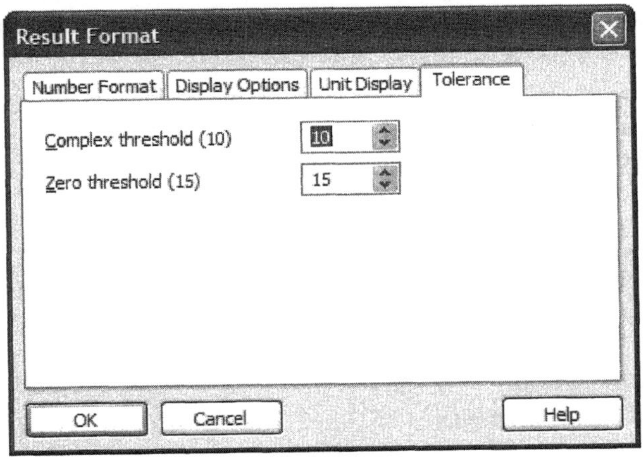

FIGURE 14.14 Tolerance tab

single result by double-clicking the result. This opens the Result Format dialog box. You may then set any of the formats discussed earlier. When you click **OK** the dialog box closes and only the selected result will change. All other results follow the worksheet default. If you later change the worksheet defaults, this result will not change because you have overwritten the worksheet defaults.

AUTOMATIC CALCULATION

By default, Mathcad automatically updates results any time an equation or plot is visible in the Mathcad window. Prior to printing, the entire worksheet is recalculated. After opening a document, the entire document is not calculated. Mathcad calculates only the visible portion of the worksheet. As you scroll down through the worksheet, it continually calculates as portions become visible.

It is recommended that you keep this setting. However, if you have some very mathematically intensive calculations that make it difficult to scroll through your worksheet, you can turn this feature off. To turn off Automatic Calculation select **Calculate>Automatic Calculation** from the **Tools** menu. If this feature has a check mark, selecting it will turn off Automatic Calculation. If this feature does not have a check mark, selecting it will turn on Automatic Calculation. If Automatic Calculation is turned on, you will see the word Auto on the information message line of the status bar located at the bottom of your Mathcad window. If Automatic Calculation is turned off, you will see the words Calc F9. This means that you need to press `F9` to update the displayed results. See Figure 14.15. You can also select **Calculate>Calculate Now** from the **Tools** menu. Remember that the results you see displayed might not be accurate until you recalculate.

If you are scrolling through a worksheet, it is easier to calculate the entire worksheet first. That way, you do not have to wait for each visible page to calculate as you scroll. To calculate the entire worksheet you can:

- Type `CTRL+F9`.
- Select **Calculate>Calculate Worksheet** from the **Tools** menu.
- Type `CTRL+END` to take you to the end of the worksheet. If Auto Calculate is turned on this will calculate the entire worksheet.

> **Tip!** A warning about manual calculation mode: If you print a worksheet prior to calculating the worksheet when in manual calculation mode, the results on the printout are not necessarily up to date. This can be a serious problem if someone is relying on your printed calculations. There is no warning on the printed page alerting you to the fact that the worksheet has not been calculated. For this reason, it is suggested that you stay in automatic calculation mode.

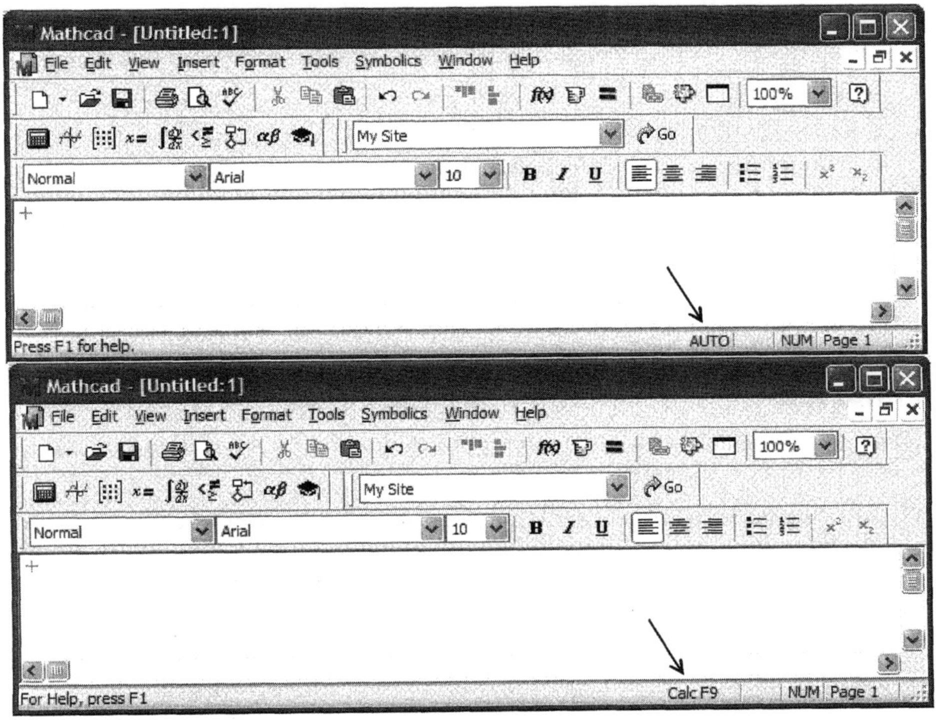

FIGURE 14.15 Automatic Calculation turned on (top) and turned off (bottom)

SUMMARY

Chapter 14, "Mathcad Settings" discussed the dialog boxes that affect how Mathcad functions and how Mathcad displays results. Some of the settings affect Mathcad globally and others affect only the specific worksheet. The Preferences dialog box settings affect all Mathcad worksheets. The Worksheet Settings dialog box and the Result Format dialog box affect only the current worksheet.

In Chapter 14 we:

- Showed how the Preferences dialog box sets global Mathcad settings.
- Discussed the general settings for how Mathcad operates and starts up.
- Set default locations for saving files.
- Explained the ways of saving files as web pages.
- Discussed how Mathcad can warn you when redefining variables.
- Discussed the settings for warning against scriptable components.

- Compared the settings for the default file format for saving Mathcad worksheets.
- Showed how the Worksheet Options dialog box can set worksheet-specific settings.
- Recommended the settings for values of built-in variables.
- Encouraged the use of the automatic calculation setting.
- Told how to set the default unit system.
- Showed how the Result Format dialog box controls how results are displayed.
- Showed how to set the number of displayed decimal places, and how to set the exponent threshold.

PRACTICE

Additional problems and applications can be found on the companion site: www.elsevierdirect.com/9780123747839.

1. Open the Preferences dialog box. Change the "Default Worksheet Location" to a location other than the Mathcad default. Next, set the Warnings tab so that all check boxes are checked.

2. Open the Worksheet Options dialog box. Set the Array Origin (ORIGIN) to 1. Next, on the Calculation tab place a check in the "Use ORIGIN For String Indexing" box.

3. Type the following on a blank Mathcad worksheet:

   a. 1.6=
   b. 1.06=
   c. 1.006=
   d. 1.0006=
   e. 1.00006=
   f. 1.000006=
   g. 1.0000006=
   h. 1.00000006=
   i. 1=
   j. 11=
   k. 111=
   l. 1111=
   m. 11111=
   n. 111111=
   o. 1111111=
   p. 11111111=

4. Now open the Result Format dialog box. Choose "General" format, and uncheck "Show Trailing Zeros." Change Number Of Decimal Places to zero and look at

the displayed results. Now change "Number Of Decimal Places" to 1 and look at the displayed results. Next change "Number Of Decimal Places" incrementally from 2 to 9 and see how each number affects the displayed results.

5. Open the Result Format dialog box. Place a check in the "Show Trailing Zeros" box, and look at the displayed results. Now change "Number Of Decimal Places" incrementally from 0 to 9 and see how each number affects the displayed results.

6. Open the Result Format dialog box. Change "Exponential Threshold" to zero and look at the displayed results. Next change "Exponential Threshold" incrementally from 1 to 7 and see how each number affects the displayed results.

7. Open the Result Format dialog box. Uncheck the "Show Trailing Zeros" check box and change the format to "Fraction." Set "Level Of Accuracy" to 12. Uncheck "Use Mixed Numbers" and look at the displayed results. Next, place a check in the "Use Mixed Numbers" box and look at the displayed results. Next, change "Level Of Accuracy" incrementally from 0 to 9 and see how each number affects the displayed results.

8. Open the Result Format dialog box. Experiment with the "Decimal," "Scientific," and "Engineering" formats. See how the different settings affect the displayed results.

CHAPTER

Customizing Mathcad

15

With some customizing, Mathcad calculations can look as nice as a published textbook. This chapter will teach you how to set up customizations to improve the appearance of your worksheets.

One way to achieve a consistent professional look to your calculation is by the use of styles. Styles allow numbers and constants to have different appearances. They also allow you to change the look of text for titles, headings, explanations, and conclusions. You can even change the look of specific variables, such as vectors, so that they stand out from other variables.

Headers and footers are critical to scientific and engineering calculations. They identify you, your company, the project information, and the date the calculations were performed.

Chapter 15 will:

- Discuss Mathcad styles for variables, constants, and text.
- Tell about the advantages of using styles.
- Show how to create and modify math and text styles.
- Explain how to create headers and footers.
- Describe how to create a standard header that includes a graphic logo.
- Give suggestions about information that should be included in headers and footers.
- Discuss margins, including how to use information located to the right of the right margin.
- Discuss how to customize the icons on the toolbar.

DEFAULT MATHCAD STYLES

Whether you know it or not, you always use styles when you use Mathcad. Every time you type a definition or enter text, Mathcad assigns a default style to the typed information. A style is a specific set of formatting characteristics associated with the items displayed on your Mathcad worksheet. The formatting characteristics of a style include such things as font type, size, and color; bold, underline,

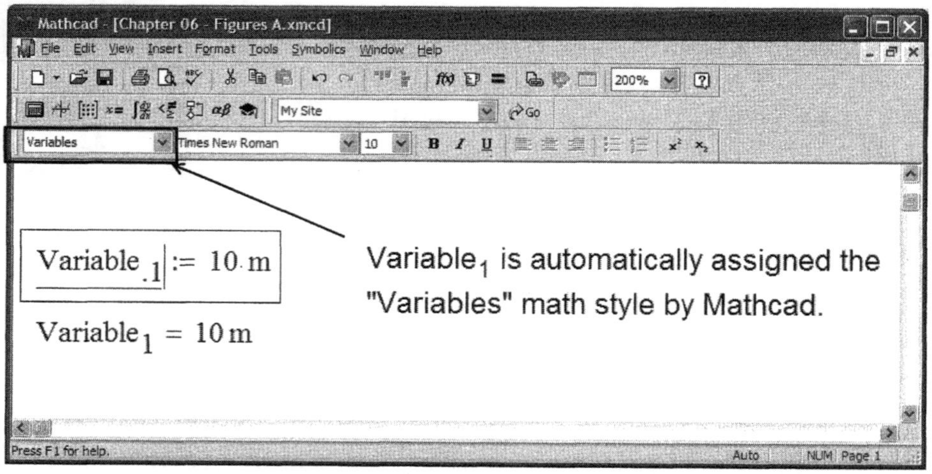

FIGURE 15.1 Variables math style for variable name

and italics; margins and indents; and more. There are two types of Mathcad styles: math styles and text styles. The style being used by Mathcad is shown on the left side of the Formatting toolbar. See Figure 15.1.

Default Math Styles

Math styles are associated with variable definitions and expressions. There are many possible math styles, but there are two default math styles: Variables and Constants. Variables are letters, Constants are numbers. Whenever you type letters for an expression (or a combination of letters and numbers), Mathcad assigns the Variables style. When you type numbers, or Mathcad displays numerical results, Mathcad assigns the Constants style. The Mathcad default for both these styles is Times New Roman, 12 point, black font, with no bold, italics, or underline. Let's look at some examples of the default Mathcad math styles. See Figures 15.1 through 15.4.

Additional math styles will be discussed later in this chapter.

Default Text Styles

Text styles control the appearance of text within Text Regions. Text styles control font and paragraph characteristics. When you create a text region, Mathcad assigns the Normal text style to the text in the region. The Mathcad default Normal text style is Arial, 10 point, black font, with no bold, italics, or underline. It has left justification, and there are no paragraph indents. See Figure 15.5.

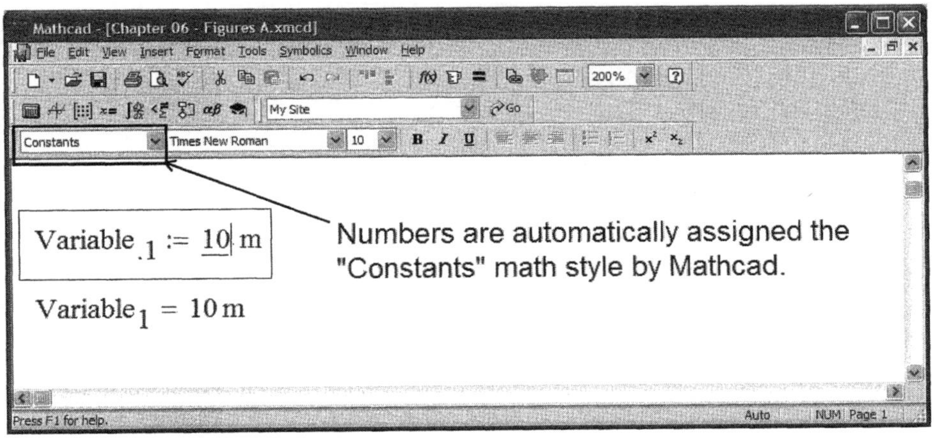

FIGURE 15.2 Constants math style for number

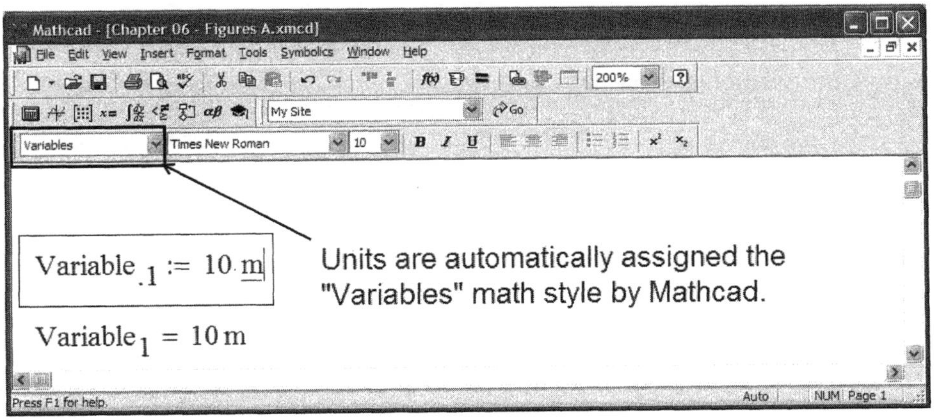

FIGURE 15.3 Variables math style for unit

Mathcad comes with several other styles that can be assigned to the text regions. These additional text styles will be discussed shortly.

ADDITIONAL MATHCAD STYLES

Mathcad comes with many styles in addition to the two default math styles and the one default text style just mentioned.

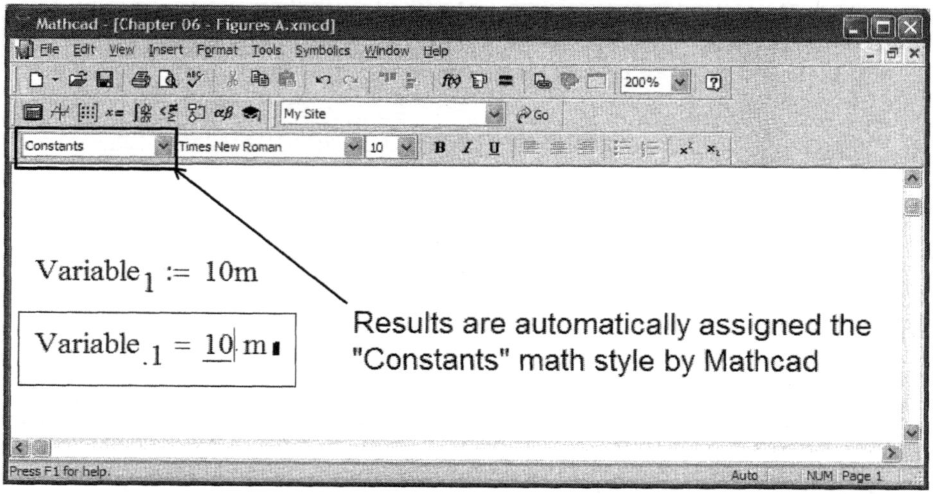

FIGURE 15.4 Constants math style for result

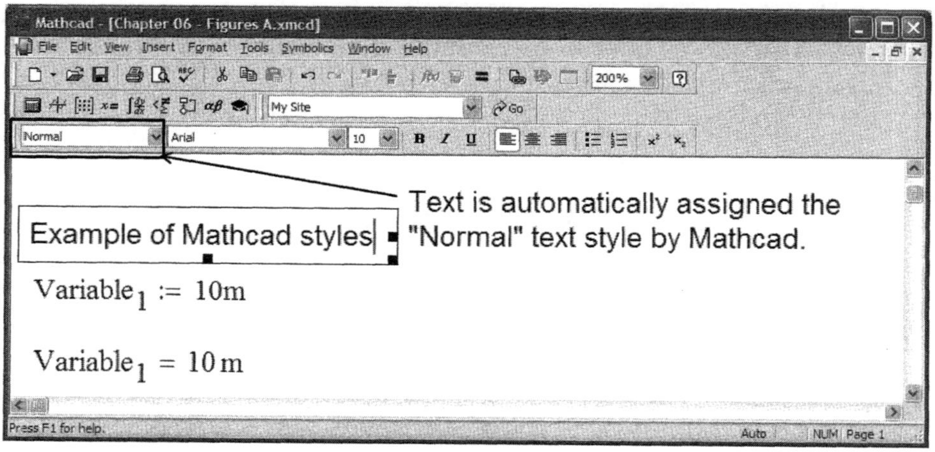

FIGURE 15.5 "Normal" text style

Additional Math Styles

If you click within any expression, the styles drop-down box on the formatting toolbar changes to show the available math styles. Clicking the drop-down arrow will reveal additional available math styles. Figure 15.6 shows that Mathcad comes

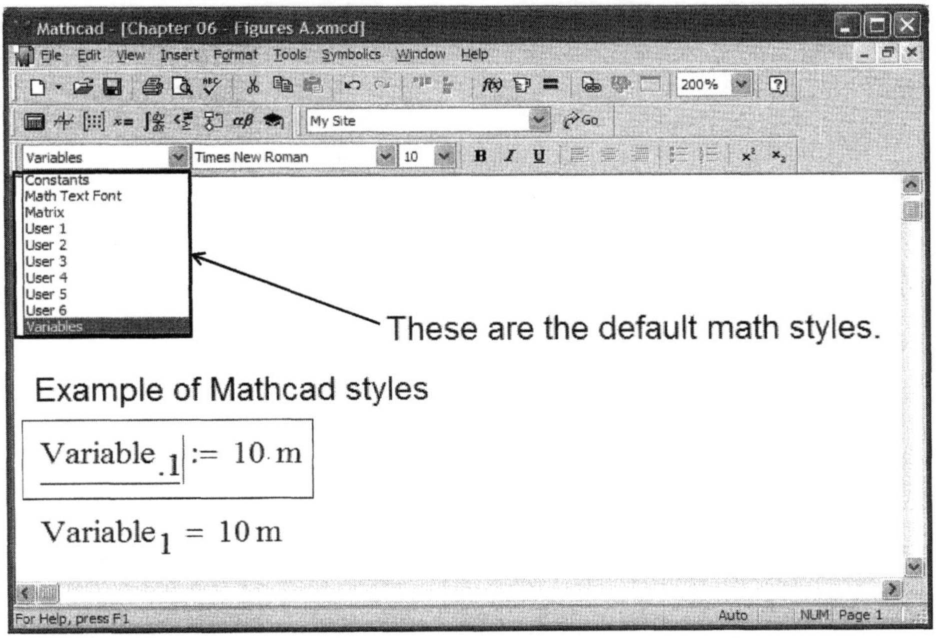

FIGURE 15.6 Default math styles

with 10 math styles: Constants, Math Text Font, User 1 through User 7, and Variables.

We have already discussed the Variables style and the Constants style. The Math Text Font style is the default style used in plot labels. The use of this style was discussed in Chapter 7.

The other math styles allow you to change the way specific variables look. For example, suppose you had a very important variable in your document that you wanted to stand out and look different from all the other variables. You can assign this variable to use User 3 style. This assigns the font characteristics associated with the style User 3 to the variable. To assign this style to a variable, click within the variable name, click the arrow in the drop-down styles box, and then click User 3. The variable name now takes on the font characteristics associated with the style User 3. See Figure 15.7 to see how the appearance of variables changes with the different styles.

It is important to remember that variable names are style sensitive. When you assign a style different from the default Variables style to a specific variable name, you must also assign the same style to all occurrences of that variable throughout your worksheet. Every time you type the variable you must stop and assign that

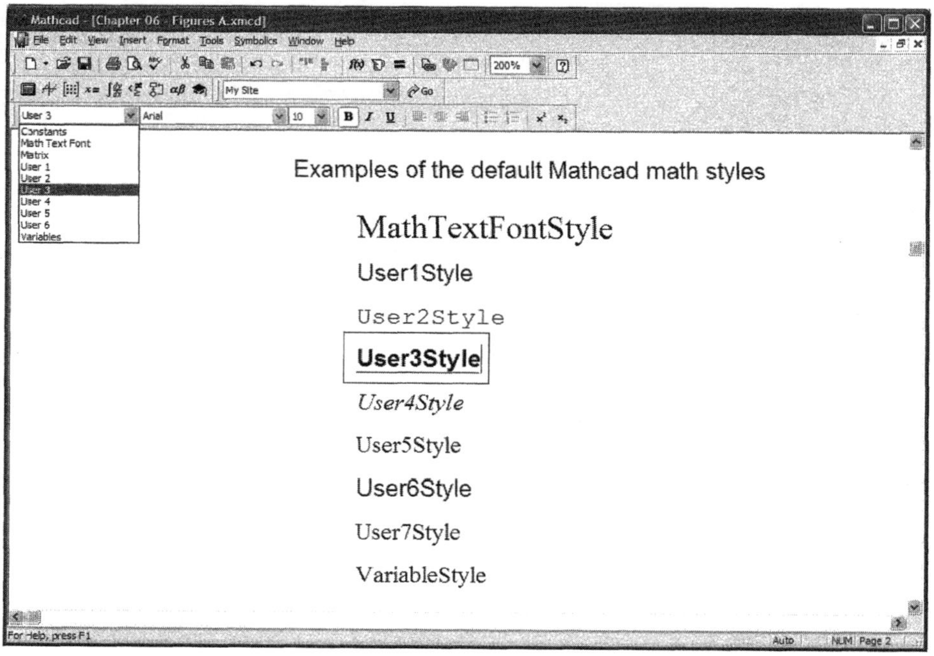

FIGURE 15.7 Default math styles

same style to the variable name. Mathcad will not recognize the variable until you assign the correct style to all occurrences of the variable. This can be very cumbersome if you have many variables with specific styles associated with them. You will constantly be stopping your typing and clicking the styles drop-down box. This can greatly reduce your efficiency in creating a worksheet. For this reason, use judgment in assigning styles to variable names. Use variable styles only when the added benefit of having a unique looking variable outweighs the extra effort of making it look unique. See Figure 15.8 to see how variables are style sensitive.

It is possible to have the same variable name use nine different math styles and have nine unique variables. See Figure 15.9 for an example of this.

If you have the same variable name with two or more styles, it can (and most likely will) introduce errors into your worksheet. You might forget to assign the correct style to the variable in your expression, and the result will be different from what you expect. This can be very dangerous in engineering calculations. For this reason, it is suggested to use only one style with each variable name.

$\mathbf{Variable_2}$:= 20N	$Variable_2$ is assigned to "User3" style.
$Variable_2$ = ▮	Mathcad doesn't recognize this variable because it is "Variables" style.
$\mathbf{Variable_2}$ = 20N	Mathcad recognizes the variable after changing to "User3" style.

FIGURE 15.8 Variable names are style sensitive

Because variable names are style sensitive, it is possible to use the same variable name yet have several different variables. The following example illustrates this. Notice how there are nine different variables with the same name.

WARNING: THIS IS FOR ILLUSTRATION PURPOSES ONLY. IT IS STRONGLY ADVISED NOT TO USE THE SAME VARIABLE NAME WITH DIFFERENT STYLES.

Variable Names		Style Used
$Variable_3 := 1$	$Variable_3 = 1$	Variables (Mathcad default)
$Variable_3 := 2$	$Variable_3 = 2$	User 1
$Variable_3 := 3$	$Variable_3 = 3$	User 2
$\mathbf{Variable_3} := 4$	$\mathbf{Variable_3} = 4$	User 3
$Variable_3 := 5$	$Variable_3 = 5$	User 4
$Variable_3 := 6$	$Variable_3 = 6$	User 5
$Variable_3 := 7$	$Variable_3 = 7$	User 6
$Variable_3 := 8$	$Variable_3 = 8$	User 7
$Variable_3 := 9$	$Variable_3 = 9$	Math Text Font

FIGURE 15.9 Variable names are style sensitive

Additional Text Styles

Clicking the drop-down arrow in the styles drop-down box on the formatting menu will list the text styles that come with Mathcad. The beauty of using different styles with your text is that you can create a different look for different parts of your calculations. You can create titles, subtitles, headings, and subheadings. You can also emphasize different parts of your calculations by creating special styles for input and output information.

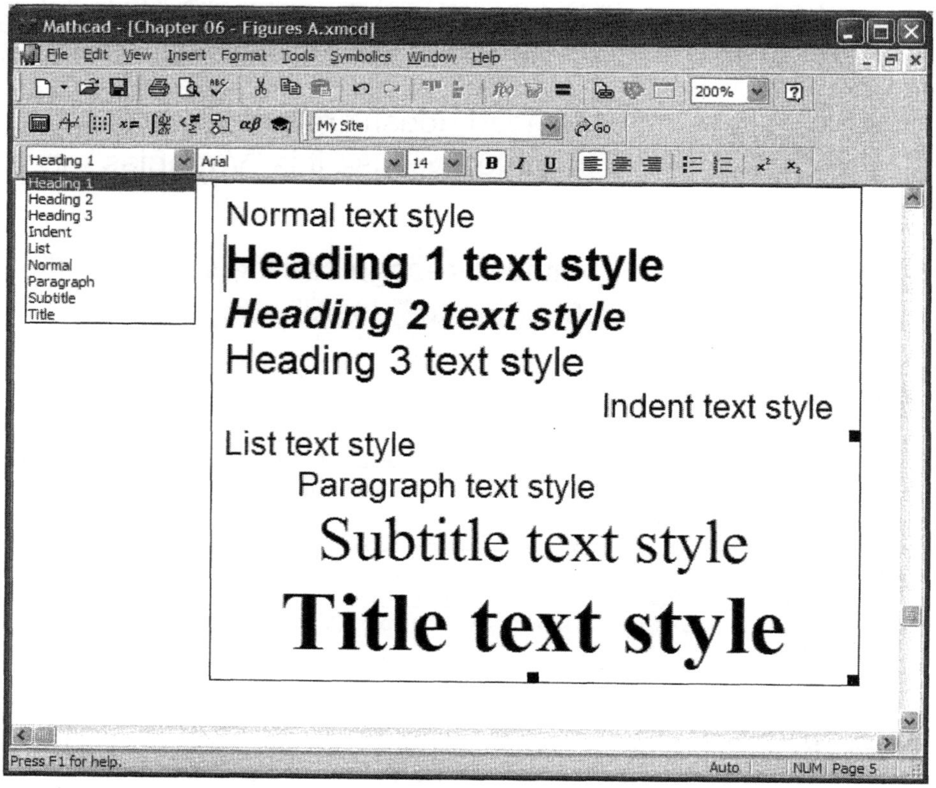

FIGURE 15.10 Default text styles

Figure 15.10 lists the text styles that come with Mathcad. It shows the different font and paragraph characteristics associated with each style.

The default font styles are only a beginning. The next section discusses how to change existing styles and how to create new styles.

CHANGING AND CREATING NEW MATH STYLES

We have just discussed the styles that come with the standard Mathcad installation. The real power in using styles is in creating styles that fit your own specific needs. For example, if you are part of a corporation, you can have customized styles so that all worksheets have a consistent look. If you are a student, you might want to create specific styles to give your worksheets a unique look. Professors might want homework to be submitted with consistent styles. This section focuses on teaching how to create and change styles. The changes you make to styles will be effective only in the current worksheet, not to all worksheets. Chapter 16 will discuss the ways to save your styles to be used over and over again as a template.

Changing the Variables Style

The Variables style controls how the text in your equation looks. A very quick way to change the Variables style is to select any portion of the text in any expression, and then from the formatting toolbar change the font style or add bold, italics, or underline. This will change the Variables style. Because most variables in your worksheet have been attached to the Variables style, they will all be changed to match the revised style. Another way to change the Variables style is to select **Equation** from the **Format** menu. You will see a dialog box similar to Figure 15.11. Select **Variables** from the Style Name, and click **Modify**.

From here you will be able to change font type, font style, font size, and font color. See Figure 15.12 for an example of the Variables font dialog box.

The changes you make here will occur for all variables in your worksheet.

Changing the Constants Style

The Constants style controls how numbers look in your calculations. You can change the Constants style in the same way that you change the Variables style. Simply select any number in an expression, and then from the formatting toolbar change the font style or add bold, italics, or underline. You can also select **Equation** from the **Format** menu, then select **Constants** from the Style Name. See Figure 15.13. Click **Modify** from this dialog box. Just as in the Variables style, you will be able to change font type, font style, font size, and font color. The changes you make will affect all constants in your worksheet.

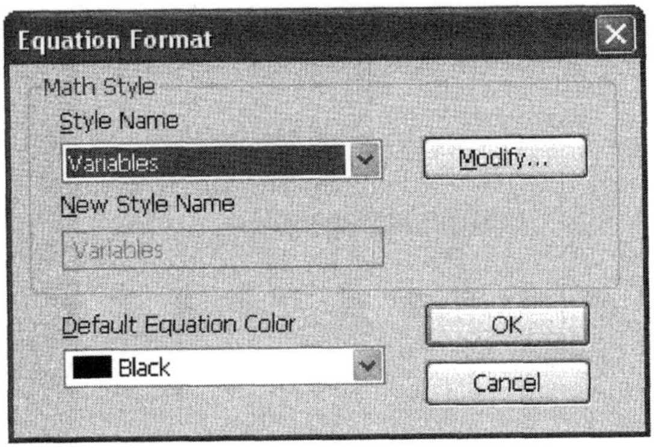

FIGURE 15.11 Changing the Variables style

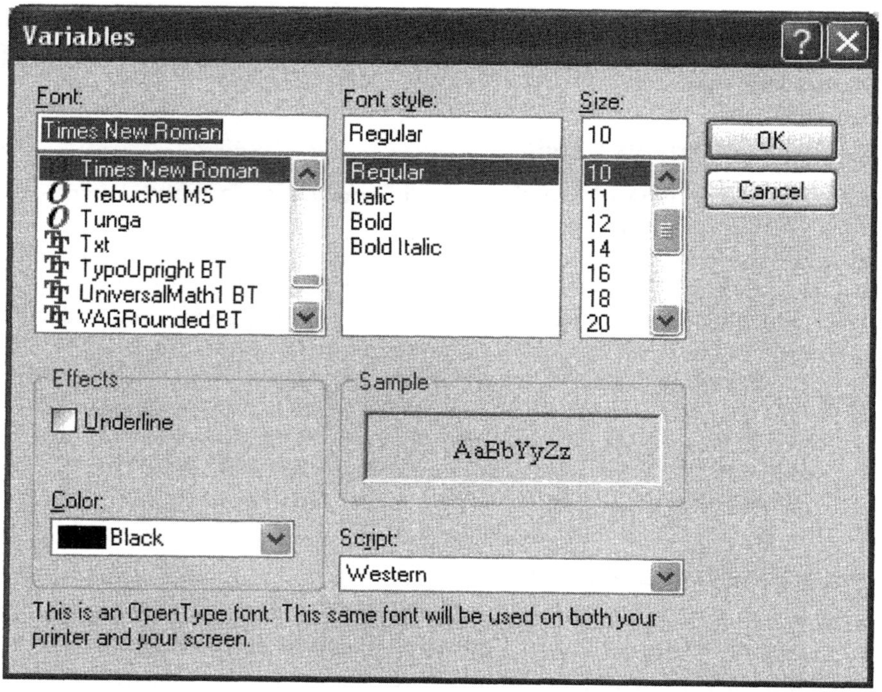

FIGURE 15.12 Variables dialog box

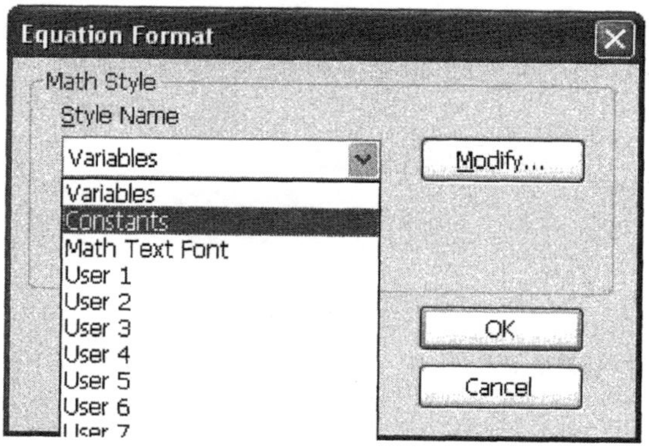

FIGURE 15.13 Changing the Constants style

Creating New Math Styles

You might want to have certain types of variables stand out from the other variables in your worksheet. You can either change the User 1 through User 7 styles mentioned earlier, or you can rename any one of the User styles to a new name and change the characteristics of the style. To change the name of a User style, select **Equation** from the **Format** menu. Click the drop-down arrow and select one of the User styles. The New Style Name box should now contain the name of the User style. You can now overwrite the name of the User style with your new style name. You can also now customize your new style by clicking the **Modify** button. In Figure 15.14, the User 1 style is renamed Sample Style 1.

Suppose you want all functions to be displayed differently from other variables. Let's create a new math style named Function. Select **Equation** from the **Format** menu. Click the drop-down arrow to the right of the box and select **User 7** from the list. Click the New Style Name box and type `Function`. Now click the **Modify** button. For this example we select Roman font, 12 point font size, with bold. Now when we attach this new math style to a variable it stands out from other variable names. See Figure 15.15.

Each time you have a function in your worksheet you can now attach this Function math style to the variable names.

CHANGING AND CREATING NEW TEXT STYLES

Having and using a good selection of text styles will add variety to your calculations, and make them easier to follow. This section discusses how to change the default Mathcad text styles, and how to create new text styles.

There are more characteristics that can be changed with text styles than with math styles. Math styles allow you to change the font, bold, italic, underline, font size, and color. With text styles, you can change all of these, but you also are

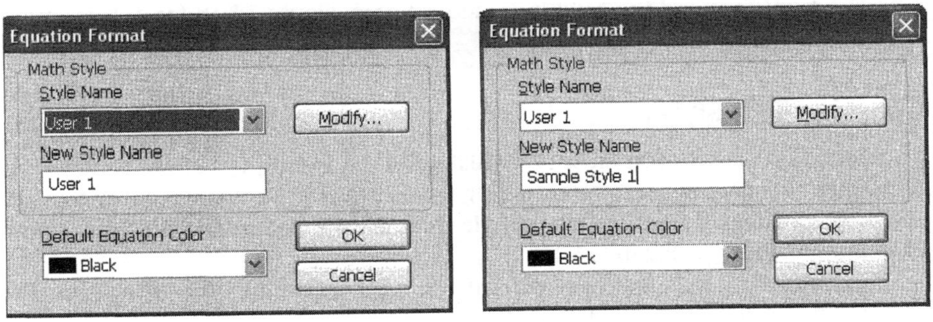

FIGURE 15.14 Renaming default style names

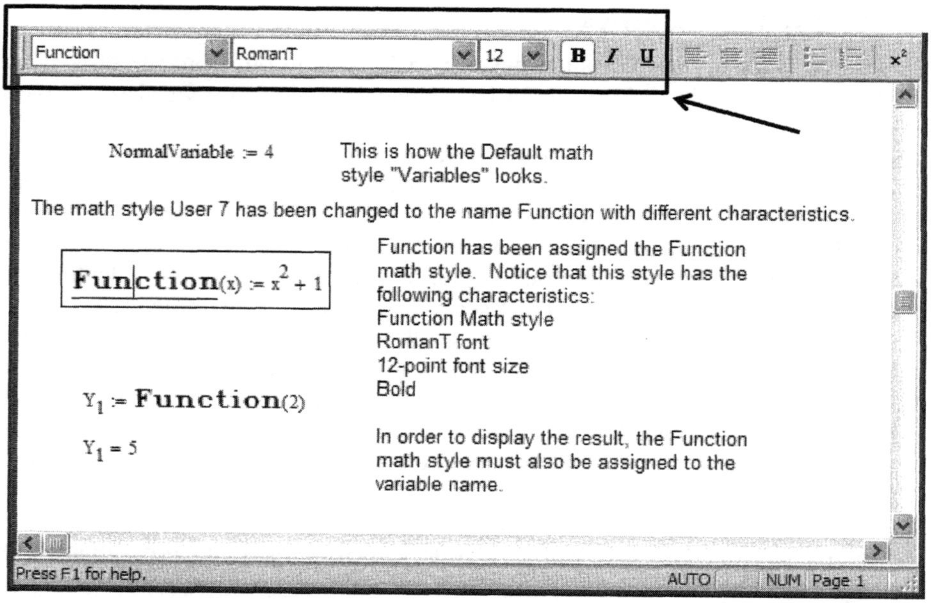

FIGURE 15.15 Variable with the new Function math style

allowed the following font characteristics: strikeout, subscript, and superscript. Text styles also allow you to set many paragraph features such as left indent, right indent, first line indent, hanging indent, bullets, numbering, paragraph alignment, and tab settings. Another feature about text styles is the ability to base one text style on another text style. See Figure 15.16. In this figure the style Subtitle is based on the style Title. This means that the Subtitle style will have all the font and paragraph characteristics of the Title style, except it will have 18 point font and the bold is turned off. (See Figure 15.18 for a description of the Title style.)

Changing Text Styles

To change a text style you must use a dialog box. It is not possible to change text styles just by selecting text and changing the font properties, as we did with math styles. To change a text style, select **Style** from the **Format** menu. This brings up a dialog box similar to Figure 15.17.

Click the style you want to edit, and then click **Modify**. This brings up the Define Style dialog box. See Figure 15.18.

In Figure 15.18, we have chosen to edit the Title text style. Notice the description of the font and paragraph characteristics of the style. The font characteristics are Times New Roman font, size 24, and bold. The paragraph characteristics are

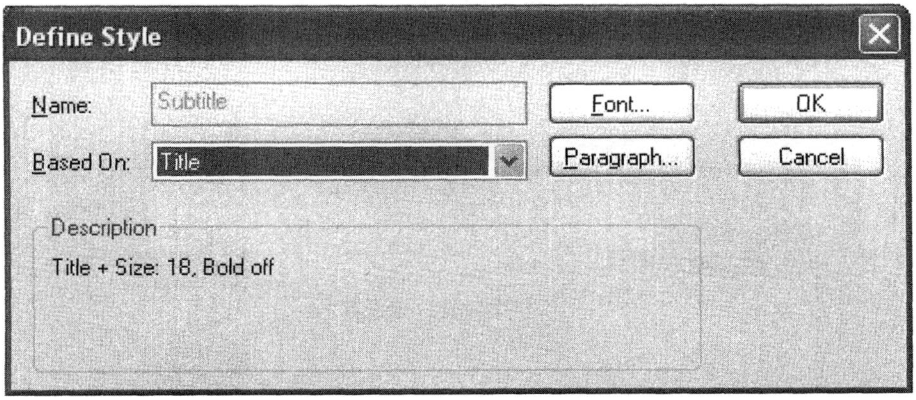

FIGURE 15.16 Basing one text style on another text style

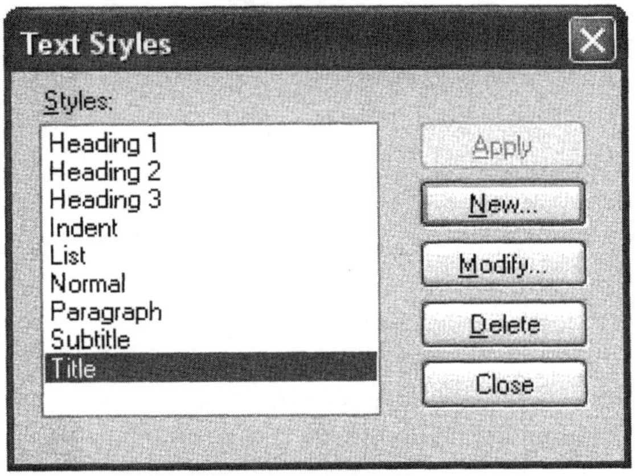

FIGURE 15.17 Text Styles dialog box

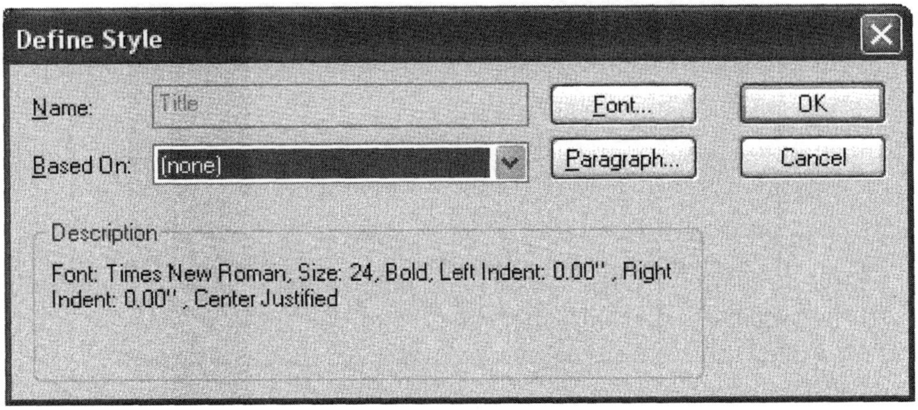

FIGURE 15.18 Define Style dialog box for text styles

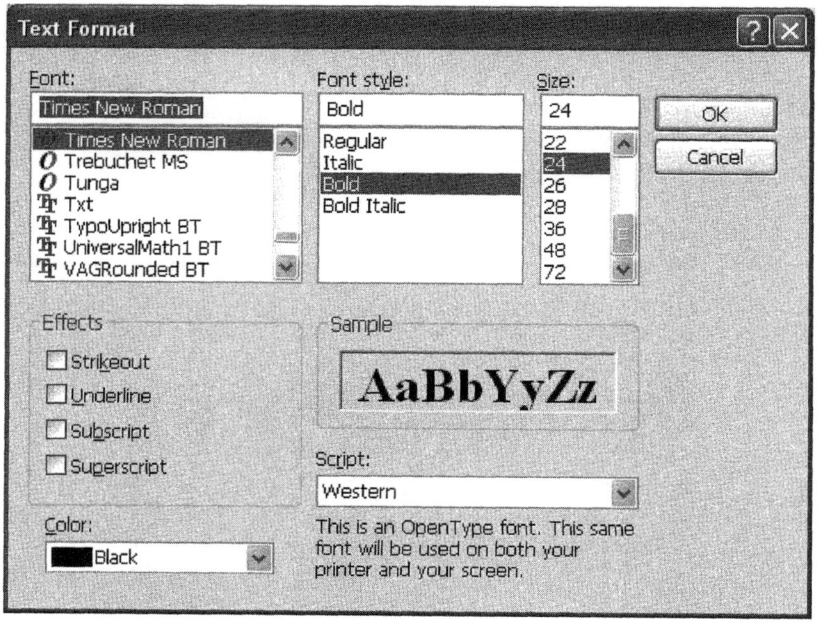

FIGURE 15.19 Text Format dialog box for text styles

no indent and center justified. If you click the **Font** button, Mathcad brings up the Text Format dialog box. See Figure 15.19.

From this box you can change many different font characteristics for the selected style. If you click the **Paragraph** button from the Define Style dialog box, Mathcad brings up the Paragraph Format dialog box. See Figure 15.20.

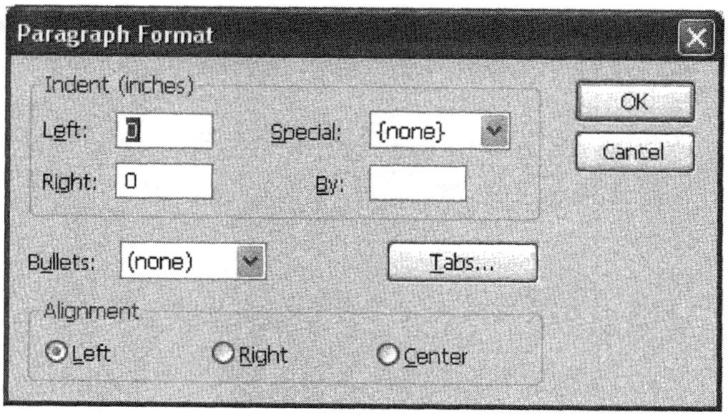

FIGURE 15.20 Paragraph Format dialog box for text styles

From this box you can change many different paragraph characteristics for the selected style. We will discuss more about the specifics of this dialog box later.

Creating New Text Styles

Mathcad comes with several default text styles. As you create your engineering calculations, you will most likely want to create some new text styles. Engineering calculations have many intermediate results. When you get to a result it is nice to be able to clearly identify this result from the intermediate results. We will create a style that can be used for that purpose. To create a new style select **Style** from the **Format** menu. Click **New**. This brings up the Define Style dialog box. Type a new style name in the Name box. For this example we will create a new style called Result Highlight. This style is based on None. Click the **Font** button. For this example, we have the following font characteristics: Arial font, bold, 20 point font, underline, and red color. See Figure 15.21.

When you are done changing the font characteristics, click **OK**. Next, click the **Paragraph** button from the Define Style dialog box. For this example **Bullets** is selected. All other paragraph characteristics are left at the default settings. See Figure 15.22.

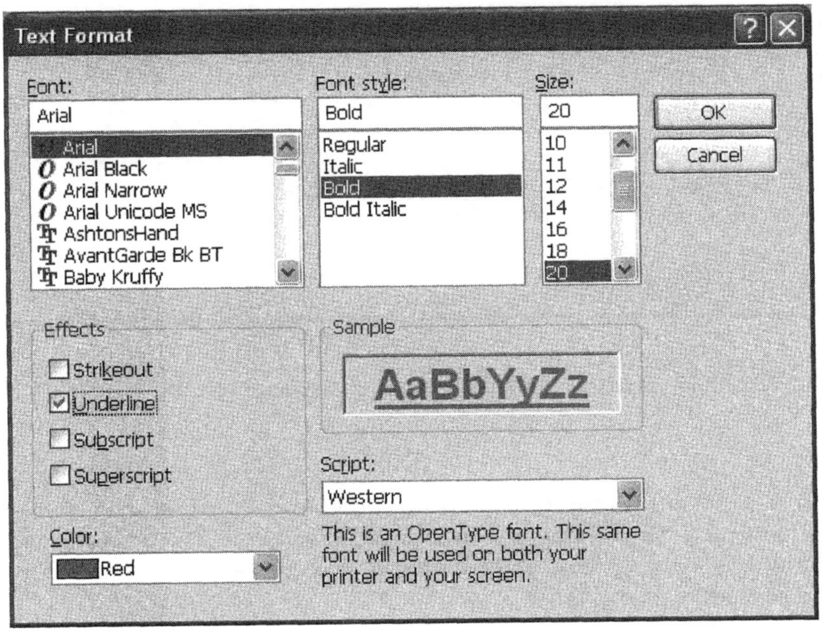

FIGURE 15.21 Font characteristics of the new text style Result Highlight

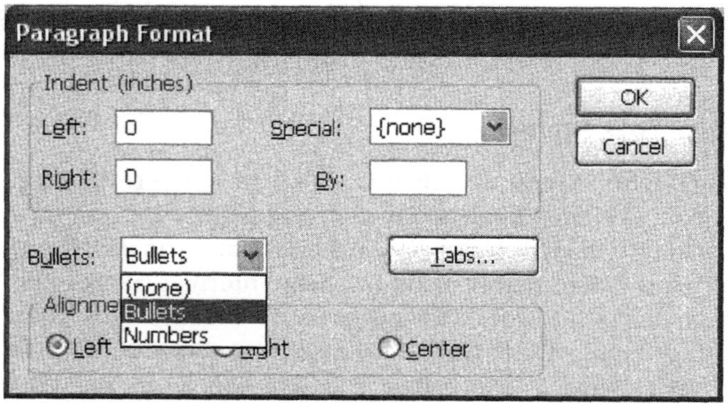

FIGURE 15.22 Paragraph characteristics for the new text style Result Highlight

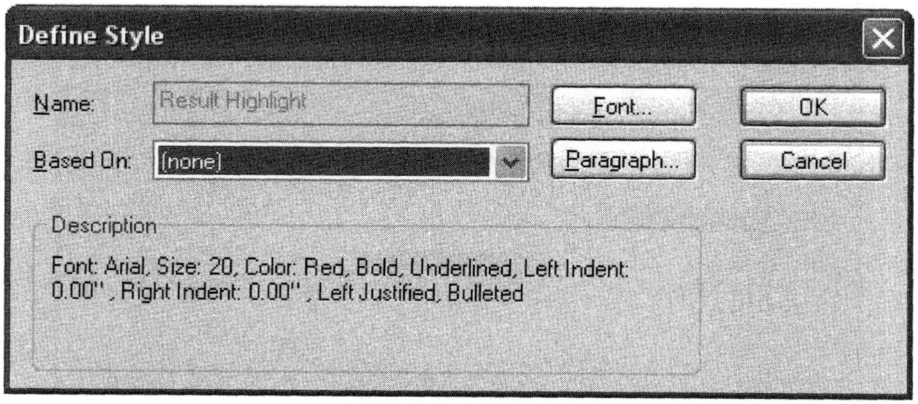

FIGURE 15.23 Style characteristics of the new text style Result Highlight

Figure 15.23 shows the characteristics now assigned to the new style. Click **OK** in the Define Style dialog box, and click **Close** in the Text Styles dialog box.

Figure 15.24 shows how this new text style can be used to emphasize the results of a series of equations.

$Sample_A := 5m + 3.25m + 0.50m$

$Sample_A = 8.75\,m$ Intermediate result

$Sample_B := 5N + 4.5N + 3.25N$

$Sample_B = 12.75\,N$ Intermediate result

$Sample_C := Sample_A \cdot Sample_B$

$Sample_C = 111.563\,J$

- **This is a final result**

(The above style is "Result Highlight")

FIGURE 15.24 Using the Result Highlight text style

HEADERS AND FOOTERS

We have just discussed creating, modifying, and using styles to add variety to your engineering calculations. Another essential feature in your engineering calculations is a way to clearly identify the project information at the top and bottom of each page. This can easily be done by using headers and footers. When using Mathcad to create and organize engineering calculations, headers and footers are critical. This section will discuss how to create headers and footers, and what information is important to include in the headers and footers.

Creating Headers and Footers

To create a header or footer, open the Header and Footer dialog box by clicking **Header and Footer** from the **View** menu. This will open a dialog box similar to Figure 15.25.

You will notice that there are left, center, and right sections. There are icons in the Tools section that can be used to insert codes that will automatically update as information changes. These icons allow you to insert the following: filename, file path, page number, total number of pages, save date, save time, current date, and current time. The **Format** button allows you to change font characteristics for your header or footer. The **Image** button allows you to insert a bit map (.bmp) image. The Start At Page Number allows you to tell Mathcad on which page to begin your header or footer. If there is a check in front of Different Header And Footer On First Page, then two additional tabs are added at the top for Page 1. See Figure 15.26. The check boxes in the Frame section put borders

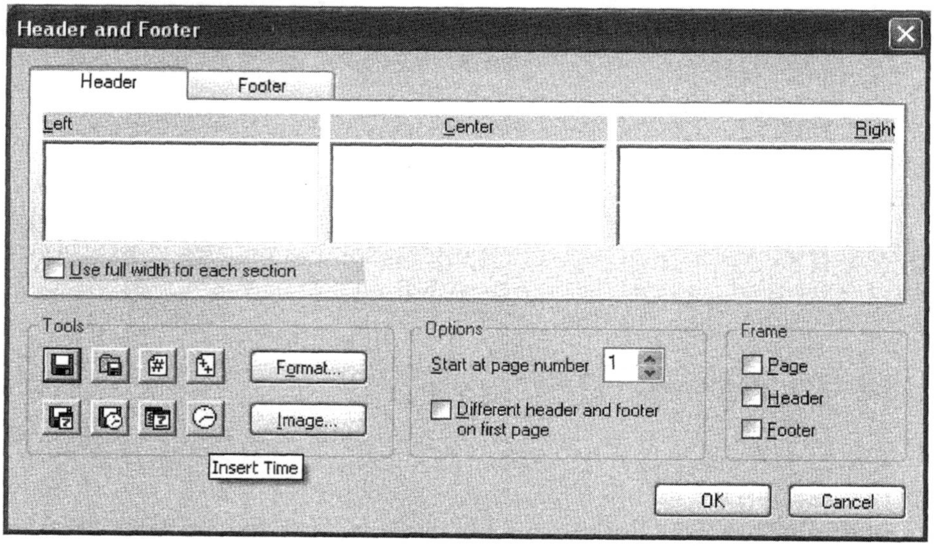

FIGURE 15.25 Header and Footer dialog box

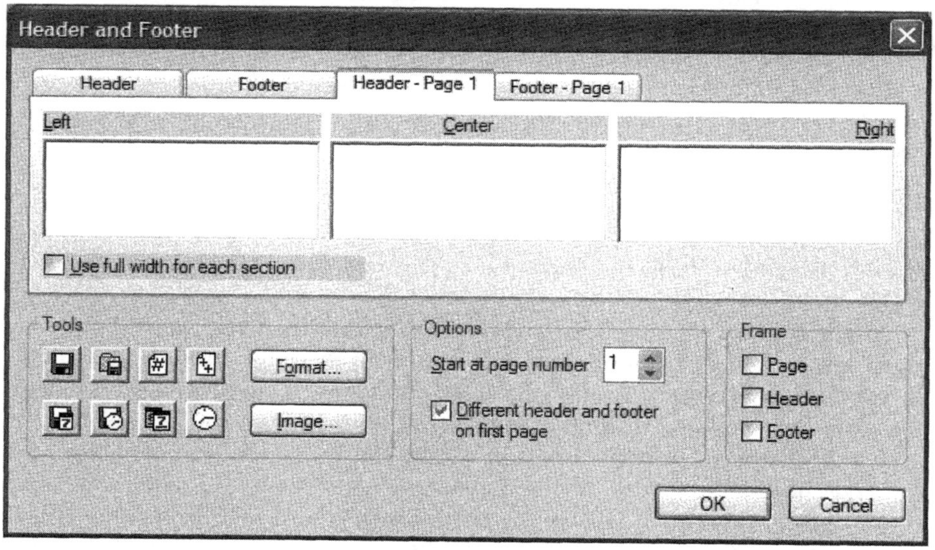

FIGURE 15.26 If there is a check by Different Header And Footer On First Page, two additional tabs appear at the top

around the page, the header, or the footer. The tab for Footer looks identical to the tab for Header.

Information to Include in Headers and Footers

Headers and footers are essential for printing engineering calculations created using Mathcad. Since Mathcad calculations can be changed very easily, it is important to know how current the printed calculations are. Thus, it is important to know when the file was saved, and when the file was printed. The save date is important because you will want to be able to make sure that the printed calculations are the same as the saved calculations. You can do this by comparing the save date on the printed calculations to the save date of the worksheet file.

The filename is also useful to have on printed calculations. This will help in locating the file for future reference. It will also help in making sure that the calculations are from the correct file. If you use several network drives, then the path is also a useful item to include on the printed calculations.

Page numbers are essential. It is suggested that you use this format: Page n of nn. This helps ensure that all printed calculations are kept together. If your calculations are contained in several files, it is also helpful to list a subject in front of the page number.

Company logos, photos, or scanned images can be included in a header or footer. The image must be a bit map (.bmp) file. The size of the bit map image must be adjusted to fit properly within the header or footer. You might need to use a separate imaging program to resize the image to best fit within your header or footer.

Other things you might want to consider adding to your header or footer include project title, project number, the part of the project the calculations are for, who created the calculations, who checked the calculations, among others.

Examples

Figures 15.27 and 15.28 show a sample header and footer.

Note that the center portion of the header in Figure 15.27 has some font format characteristics that were added. The image of the company logo was inserted from a bit map file. The image size needed to be adjusted to fit the header space. In order to fit the title in the center section on a single line, the check box Use Full Width For Each Section needed to be checked.

The footer in Figure 15.28 has both the save date and the print date. The save date is more important than the print date, because you will want to be able to compare how current the printed calculations are, compared to the saved worksheet file.

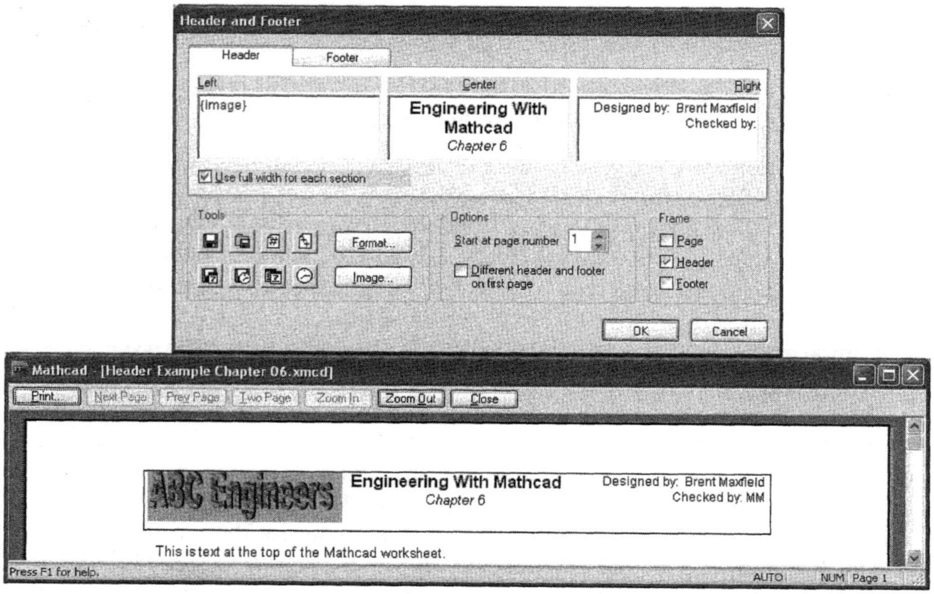

FIGURE 15.27 Sample header

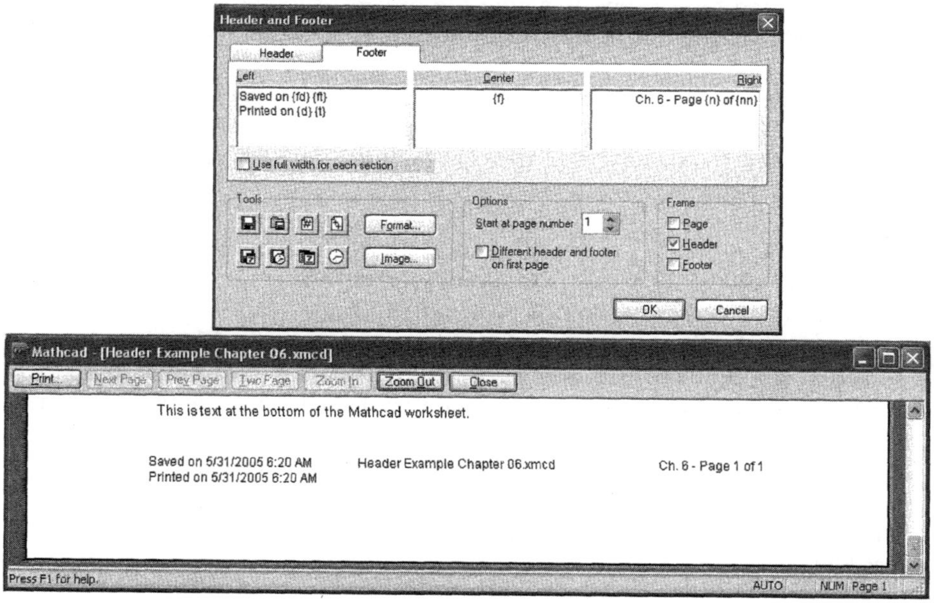

FIGURE 15.28 Sample footer

MARGINS AND PAGE SETUP

Headers fit above the top margin, and footers fit below the bottom margin. Setting paper size and margins is very similar to other Microsoft Windows-based programs. Select **Page Setup** from the **File** menu. This opens the Page Setup dialog box. From this dialog box, you can select your paper size and your page margins. See Figure 15.29.

The top and bottom margins you select will need to be large enough so that the amount of information contained in your header or footer will display. If you print a worksheet and you are missing information from the header or footer,

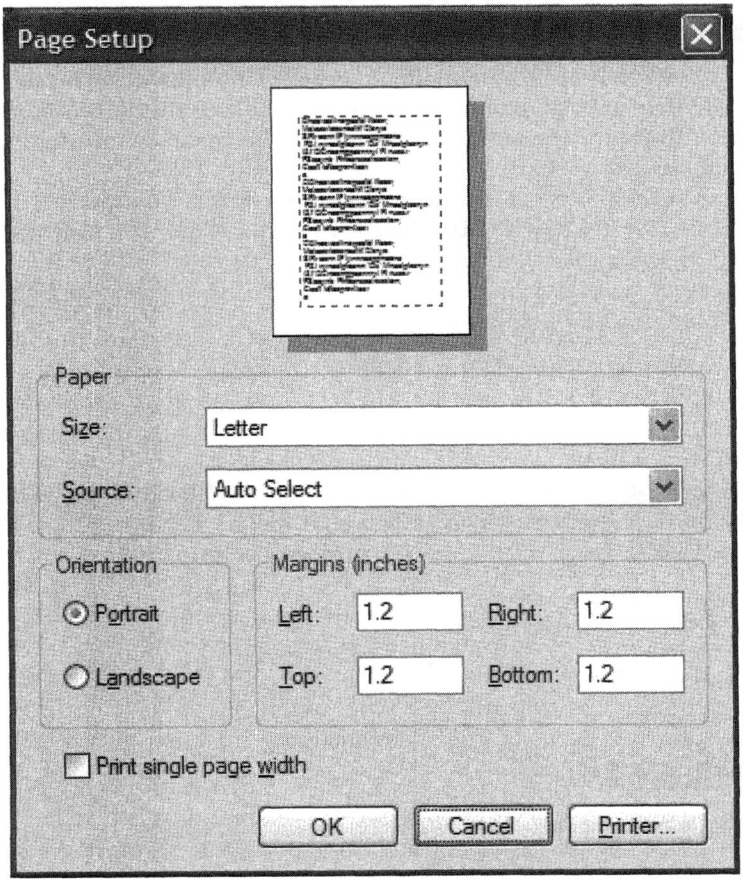

FIGURE 15.29 Page Setup dialog box

you will need to increase the top or bottom margin settings. You could also try to use a smaller font in your header or footer in order to fit the information within the specified margin.

If you put a check in the Print Single Page Width box, Mathcad will not print anything to the right of the right margin. This is a very useful feature for engineering calculations. It allows you to use the area to the right of the right margin for nonprinting information. You can include things such as notes to yourself, instructions for others using the worksheet, custom units that do not need to be printed, formulas that do not need to be printed, intermediate results, and more.

TOOLBAR CUSTOMIZATION

Mathcad allows you to customize the icon buttons that appear on the Standard toolbar and the Formatting toolbar. To customize a toolbar, right-click the toolbar and choose **Customize** from the pop-up menu. This opens the Customize Toolbar dialog box. See Figure 15.30.

From this dialog box you can select from the available buttons on the left and add them to the current toolbar buttons on the right. You can also remove buttons from the current toolbar.

A quick way to remove a button from a toolbar is to press the ALT key and drag the button off the toolbar.

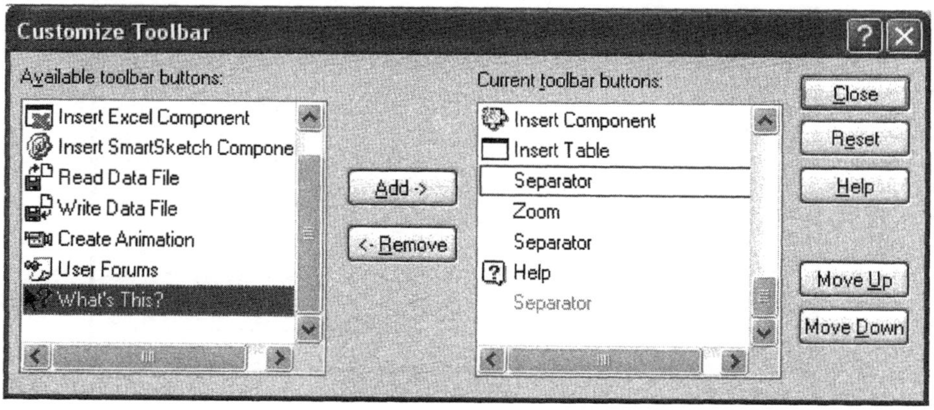

FIGURE 15.30 Customize Toolbar dialog box

SUMMARY

In Chapter 15 we:

- Learned that Mathcad uses styles for all variables, constants, and text.
- Showed how you can customize the default Mathcad styles.
- Discussed how to create your own custom styles.
- Explained when you might want to use a different math style.
- Explained how to use different text styles to highlight different parts of your calculations.
- Discussed headers and footers.
- Recommended what information is critical to have in your headers and footers.
- Showed how to customize the Mathcad toolbars.

PRACTICE

Additional problems and applications can be found on the companion site: www.elsevierdirect.com/9780123747839.

1. On a blank worksheet type `Sample. 1: 25kg`. Then type `Sample. 1=`. Next type `Sample. 2: 50kg` and `Sample. 2=`. Notice how all variable names have the Variables math style. Change the math style of Sample1 to User 1. Notice the difference between the variable names "Sample1" and "Sample2." Now change the style of Sample1 to User 2 through User 7. Compare the different look for each different style.
2. In a text box, type a heading and two paragraphs. Choose any topic to write about. Now experiment with the different text styles. Change the heading to each of the different heading and title styles. Change the two paragraphs to different text styles. Experiment with different combinations of styles.
3. Change all the variable names in exercise 1 back to the Variables style. Now experiment with changing the characteristics of the Variables style. Change the font style, the font size, the color, bold, and so on. If the regions overlap, use the **Separate Regions** command from the **Format** menu.
4. Create a new math style. Select a name for your style. Make it unique from the Variables style. List the characteristics of your new math style. Change the math style of Sample1 to your new style.
5. Change the characteristics of the text style to Normal. Change the font style, font color, and other characteristics. Type several text boxes to see how the new style looks.

6. Create five new text styles and assign these styles to several different text boxes.

7. Create a header and footer for use with your company or school. Include the items mentioned in this chapter. Also include a graphic with your header or footer. Place some text at the top and bottom of your worksheet. Print your worksheet page. Did the entire header and footer print? Did the text at the top and bottom of your page print?

8. Open the Page Setup dialog box. Adjust the top and bottom margins so that your full header and footer will print. Adjust the left and right margins to see how it affects your worksheet. Place a check in the Print Single Page Width checkbox. Now print your worksheet page. Did the entire header and footer print? Did the text at the top and bottom of your page print?

9. Add new icons to your Standard toolbar and to your Formatting toolbar.

CHAPTER

Templates

16

In the previous chapters we have discussed many different things that will make your worksheets unique. We have discussed how to change many of the default Mathcad features. We have discussed styles and how they give a consistent look to your calculations. We have also discussed how to set specific formats for your numerical results. In most cases, the changes made affected only the specific Mathcad worksheet you were working in. In Chapter 16, "Templates," we discuss how to save all these customizations so that they can be used over and over again. We do this through the use of templates.

Templates are essential in order to have a consistent look for all your engineering calculations, especially if you are working with other engineers. Templates allow Mathcad customizations to be applied consistently to all calculations. This chapter discusses the benefits of templates.

Chapter 16 will:

- Tell what a template is.
- Discuss the type of information that is stored in a template.
- Show the templates that are shipped with Mathcad.
- Explain when to make use of these templates.
- Review the items discussed in Chapters 4, 14, and 15, and show how to include these items in a customized template.
- Suggest items to include in a customized template.
- Create a sample template.
- Show how to modify an existing template.
- Discuss the normal.xmct file, and show how to store a customized template in this file, so that it opens whenever Mathcad is opened.

INFORMATION SAVED IN A TEMPLATE

A template is essentially a collection of information that Mathcad uses to set various settings when opening a new document. This information is stored in a template file. Every time Mathcad opens a file based on that template every thing is formatted the same way. A template can save the following information:

- Worksheet settings
- Headers and footers
- On-screen information
- Margins
- Unit settings
- Result display settings
- Styles
- Fonts
- Unit system

MATHCAD TEMPLATES

Mathcad comes with many templates. Every time you open Mathcad you are actually opening a new worksheet based on the normal.xmct template. You can also open worksheets based on the other templates that are shipped with Mathcad. To do this, select **New** from the **File** menu. This opens the New dialog box. See Figure 16.1.

From this dialog box, you can open a new file based on one of the many different templates. Take time to open worksheets based on each different template, and explore the differences between the worksheets opened with each template. Compare the math styles, text styles, headers, footers, margins, and

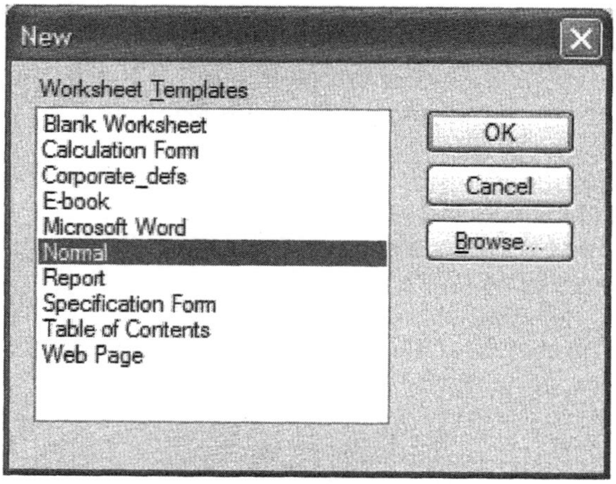

FIGURE 16.1 New dialog box for opening a new worksheet based on a template

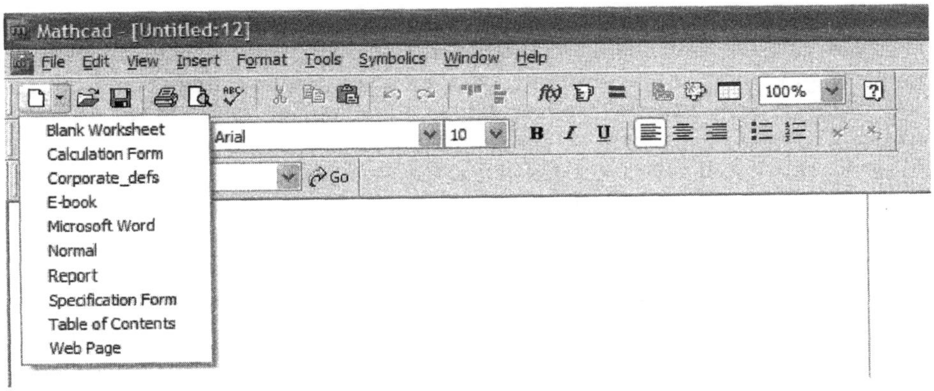

FIGURE 16.2 Using the new worksheet icon for opening a worksheet based on a template

tab settings. These templates are more useful as examples than as templates that you would use on a regular basis.

Another way to open a worksheet based on these templates is to click the down-arrow adjacent to the new worksheet icon. A list of available templates will be shown. Click one of these templates and a new worksheet based on that template will open. See Figure 16.2.

REVIEW OF CHAPTERS 4, 14, AND 15

Let's quickly review some of the things discussed in the previous three chapters. These are things that can be saved in a template.

- In Chapter 4 we discussed units. The default unit system can be saved in a template. Custom units you create can also be saved in a template.
- In Chapter 14 we discussed Mathcad settings. Any of the settings changed in the Worksheet Options dialog box can be saved in a template. Any of the result formats set from the Results Format dialog box can also be saved in a template.
- In Chapter 15 we discussed various math and text styles. We also discussed headers and footers. Each of these items can be saved in a template.

CREATING YOUR OWN CUSTOMIZED TEMPLATE

Saving your own custom template is easy. Simply open a new worksheet based on the normal.xmct or on another existing template. Next, make the changes to the worksheet as discussed earlier. Once you have the worksheet to a point where you want to use it for a template, simply select **Save As** from the **File** menu. From the "Save As Type" box choose **Mathcad XML Template (*.xmct)**. Mathcad versions earlier than version 12 will need to choose **Mathcad Template**. Type a filename for the template, and store the template file in the Template folder in the Mathcad program folder.

To open a new worksheet based on this new template, choose **New** from the **File** menu. The name of the template should appear in the list of templates. Click the template, and then click **OK**. A new worksheet should appear with the customizations included in the worksheet.

Now that we understand the concepts involved with customizing Mathcad, let's create some customized templates that will be used for the remainder of this book.

EM Metric

Let's call the first template EM Metric (for Essential Mathcad). Start by opening a new worksheet based on the Blank Worksheet template. This is easy to do—just click the down-arrow next to the new worksheet icon and select **Blank Worksheet**.

Worksheet Options

Let's start by setting worksheet options. Select **Worksheet Options** from the **Tools** menu. On the Built-In Variables tab, change Array Origin from 0 to 1. Next, click the Calculation tab. Add a check in the "Use ORIGIN For String Indexing" box. This tab should look like Figure 16.3.

Now let's create a Custom Unit System based on SI units. On the Unit System tab select the Custom button and choose Based On **SI**. Keep the same Base Dimensions. In Derived Units, highlight **Volume (Liter)** from the list and click **Remove**. Click **OK** to close the Worksheet Options dialog box.

Next, let's set the Result Format. Select **Result** from the **Format** menu. On the Number Format tab select **General**. Set "Number Of Decimal Places" to **2** and place a check in the "Show Trailing Zeros" box. Change "Exponential Threshold" to **6**. This tab should look like Figure 16.4.

Click the Unit Display tab and place check marks in both the Format Units and Simplify Units When Possible boxes. Click **OK** to close the dialog box.

Math Styles

We will now modify the math styles for our template.

Select **Equation** from the **Format** menu. Select **Variables** from Style Name and then click **Modify**. For Font, choose Times New Roman; for Font Style, choose bold; and for Size, choose 9. Leave Color black. The bold will make variables more prominent, and the 9-point font will allow more text to fit on the page. This tab should look like Figure 16.5. Click **OK**.

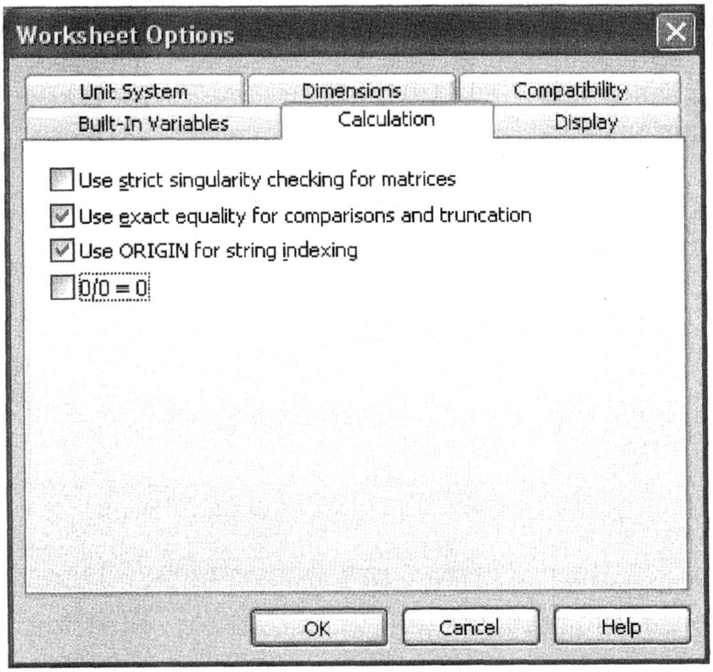

FIGURE 16.3 Settings for the Calculation tab

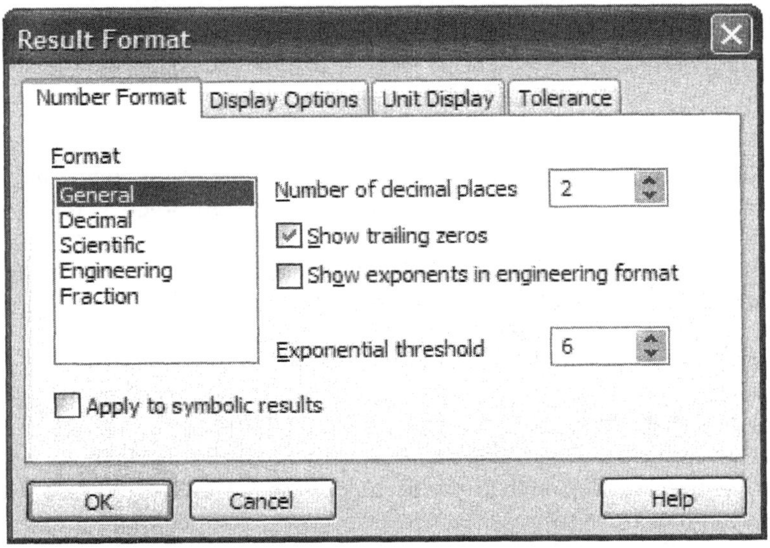

FIGURE 16.4 Settings for the Result Format tab

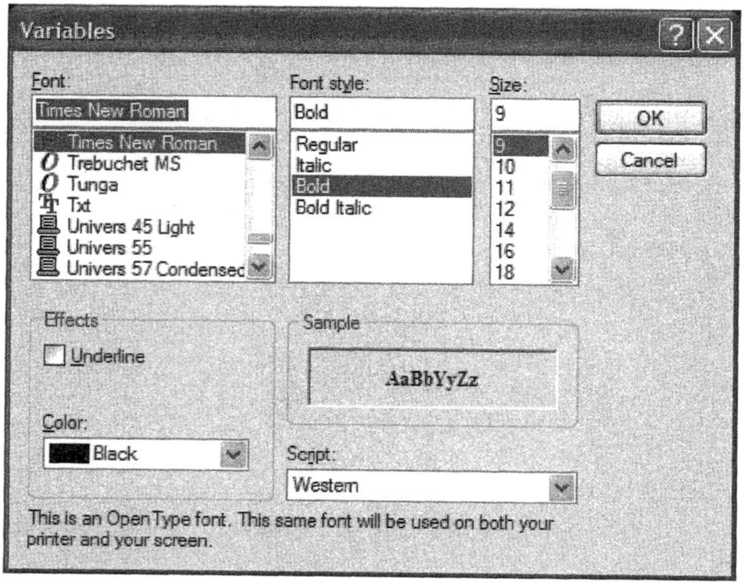

FIGURE 16.5 Variables math style characteristics

Now select **Constants** from the Style Name, and click **Modify**. Choose Times New Roman for Font, Regular for Font Style, and 9 for Size. Do not use bold. We want the variables to be more prominent than the constants. Click **OK** to close the dialog box.

Text Styles

Let's create some text styles for our template.

First, let's modify the normal text style. Select **Style** from the **Format** menu. Select **Normal** from the Styles list, and then click **Modify**. This style will be based on (**none**). Click the **Font** button. Choose Arial for Font, Regular for Font Style, and 10 for Size. Next click the down-arrow in the Color box and select Blue. This will make the text stand out from the variables and constants. Click **OK**. This style should now look like Figure 16.6. Click **OK** again to close the Define Style dialog box.

Now let's create some custom styles. Click the **New** button from the Text Styles dialog box. Type the name Heading 1 in the Name box. Select (**none**) from the Based On box. Now click the **Font** button. Choose Arial for Font, Bold for Font Style, and 18 for Size. Place a check mark next to Underline under the Effects. Next, click the down-arrow in Color, and select Green. This will make the heading stand out from other text. Click **OK**. Your Heading 1 style should now look like Figure 16.7. Click **OK** again to close the Define Style dialog box.

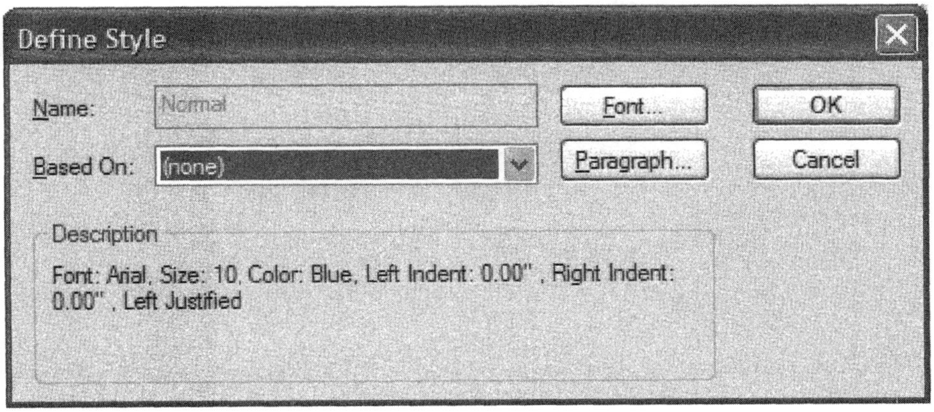

FIGURE 16.6 Normal text style characteristics

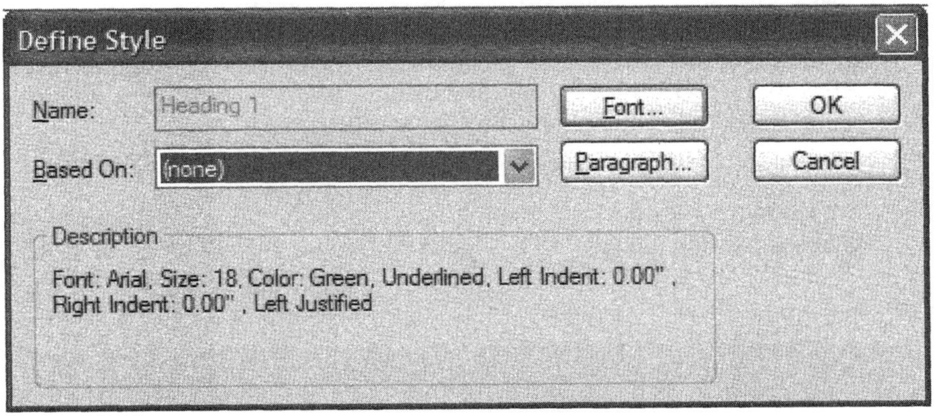

FIGURE 16.7 Heading 1 text style characteristics

We will now create a Heading 2 style. Select **New** from the Text Styles dialog box and type Heading 2 in the Name box. Select **Heading 1** from the Based On box. Click the **Font** button and change the font size from 18 to 16. Click **OK**. Now click the **Paragraph** button. Change the left indent from 0 to 0.5. Click **OK**. Your Heading 2 style should now look like Figure 16.8.

Because Heading 2 is based on Heading 1, if we change font characteristics of Heading 1, such as adding italics or changing the font color, the changes will also be reflected in Heading 2.

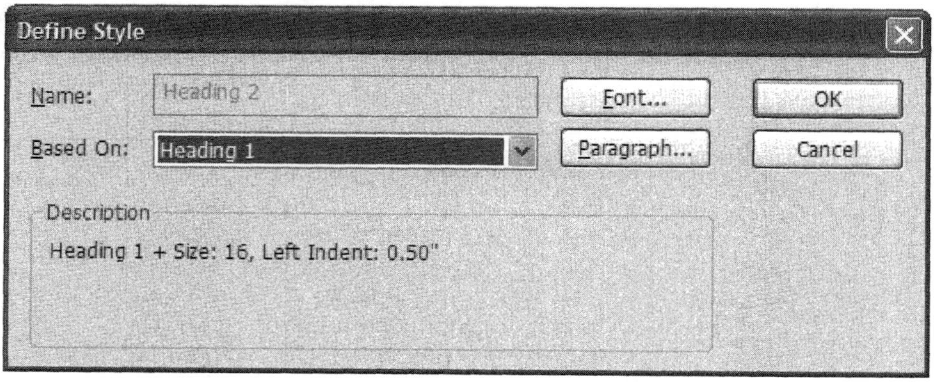

FIGURE 16.8 Heading 2 text style characteristics

For this example, let's add a couple more text styles. Add a new text style called Results. This should have the following characteristics: Based On (none), Arial Black font, Bold, 12 point size, and Red color. The Results style should look like Figure 16.9.

The next text style will be called Notes. This should have the following characteristics: Based On (none), Lucida Bright font, Italic, 12 point size, and Navy color. The Notes style should look like Figure 16.10.

Margins

Select **Page Setup** from the **File** menu. Change both the left and right margin to 0.5. Change the top margin to 1.0 and the bottom margin to 0.75. Place a

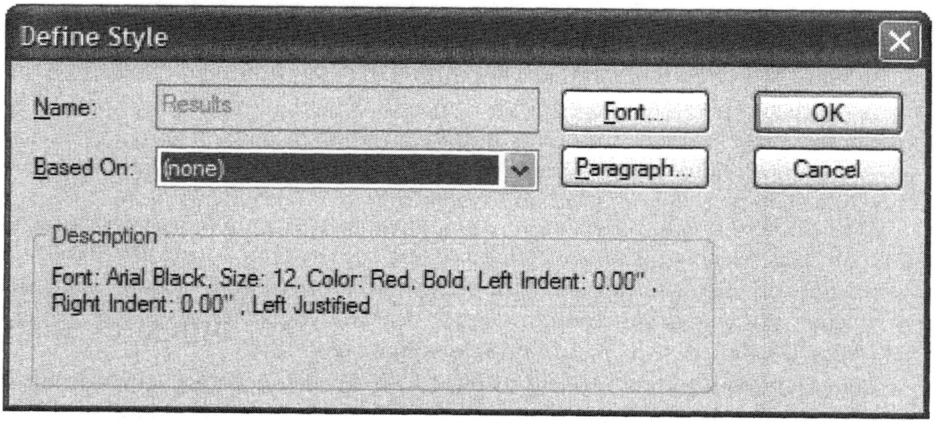

FIGURE 16.9 Results text style characteristics

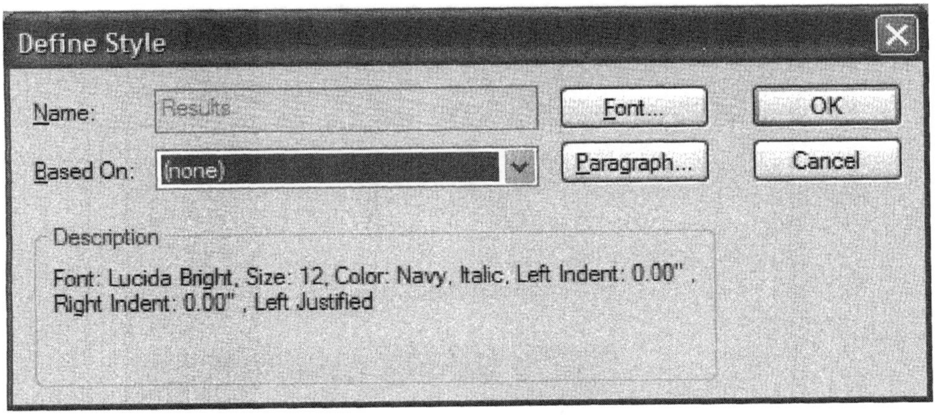

FIGURE 16.10 Notes text style characteristics

checkmark in the Print Single Page Width box. This will prevent Mathcad from printing information to the right of the right margin. You can now use the area on the right of the margin for notes and nonprinting information. Your Page Setup dialog box should look like Figure 16.11.

Headers and Footers

Finally, let's add a header and footer. Select **Header and Footer** from the **View** menu. On the Header tab, type `Essential Mathcad` in the Center section. Place a check in the "Header" check box under Frame. Your Header tab should look like Figure 16.12.

Now click the Footer tab. Click in the Left section, type `Printed`, and then in the Tools area click the icon showing a calendar, type a space, and then click the icon showing a clock. This inserts a {d} and {t} in the left section. This will print current date and time, thus printing the date and time the worksheet was printed. Press the `ENTER` key to get a new line in the left section. Type `Saved`, and then in the Tools area click the icon showing a floppy disk with a number, type a space, and then click the icon showing a floppy disk with a clock. This inserts a {fd} and {ft} in the left section. This will print the date and time the file was saved.

Click in the Center section. Type `Page`, and then click the icon showing a sheet with the # symbol on it. Next type `of` and then click the icon showing a sheet with two + symbols. This will print the current page number and the total page numbers.

Click in the Right section. Type `File:`, and then click the icon with a floppy disk. This inserts {f}. Mathcad will now print the name of the current file.

Now we will change the font size in the left and right sections. Select all the text in the Left section, and click **Format**. Place your cursor in the Size box and type `6`. (You cannot use the arrow keys to select 6.) Click **OK**. Now select the

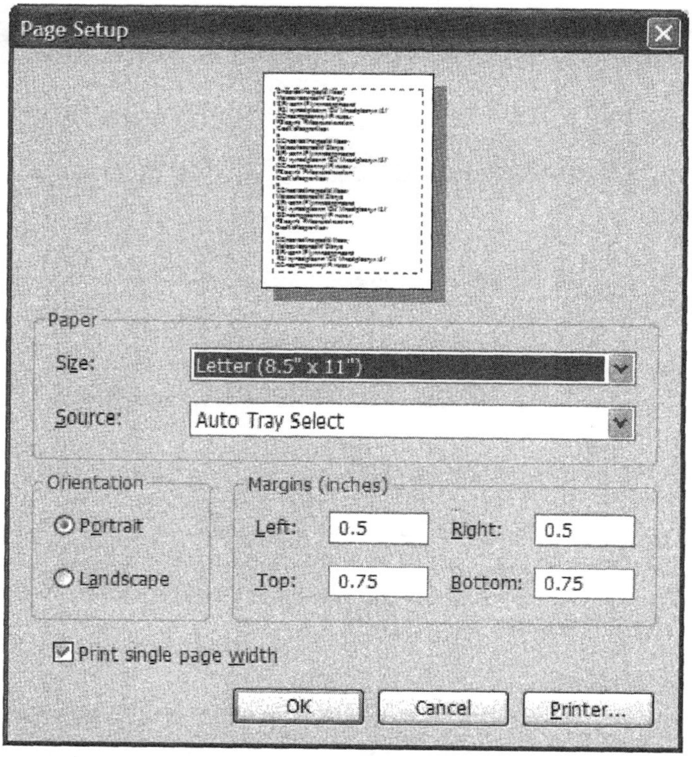

FIGURE 16.11 Page Setup settings

text in the Right section and click **Format**. Change the font size to 6 just as you did in the Left section.

See Figure 16.13.

Saving the Template

Now that we have made all these customizations, we are ready to save the template. Select **Save As** from the **File** menu. In the Save As Type box, select **Mathcad XML Template (*.xmct)**. For versions prior to version 12, select **Mathcad Template (*.mct)**. In the File name box type `EM Metric`. In order to have the template available with all other Mathcad templates, you need to save the template in the template folder of the Mathcad program folder. This folder might possibly have the following path: C:\Program Files\Mathcad\Mathcad 14\Template.

Once the template is saved, you can open a new document based on this template. Simply select EM Metric from the list of available templates as discussed earlier.

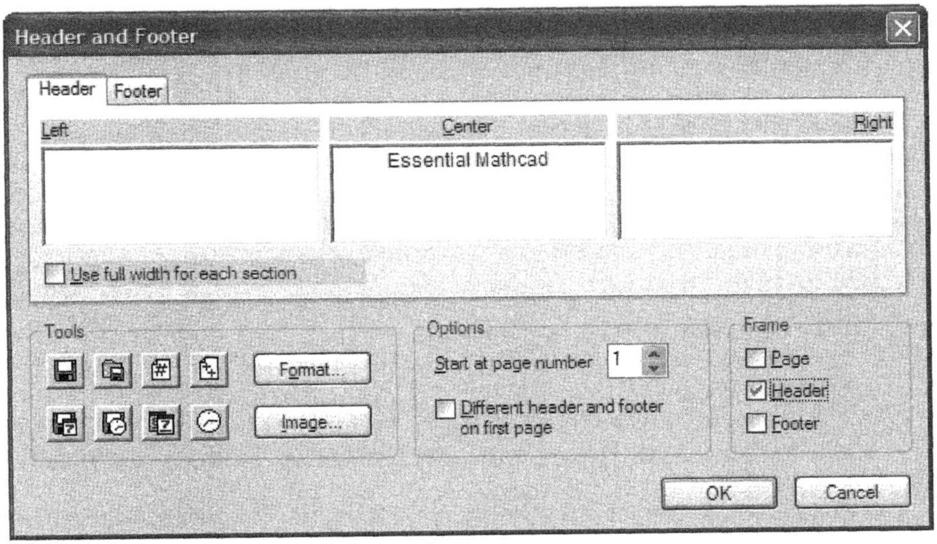

FIGURE 16.12 Header tab

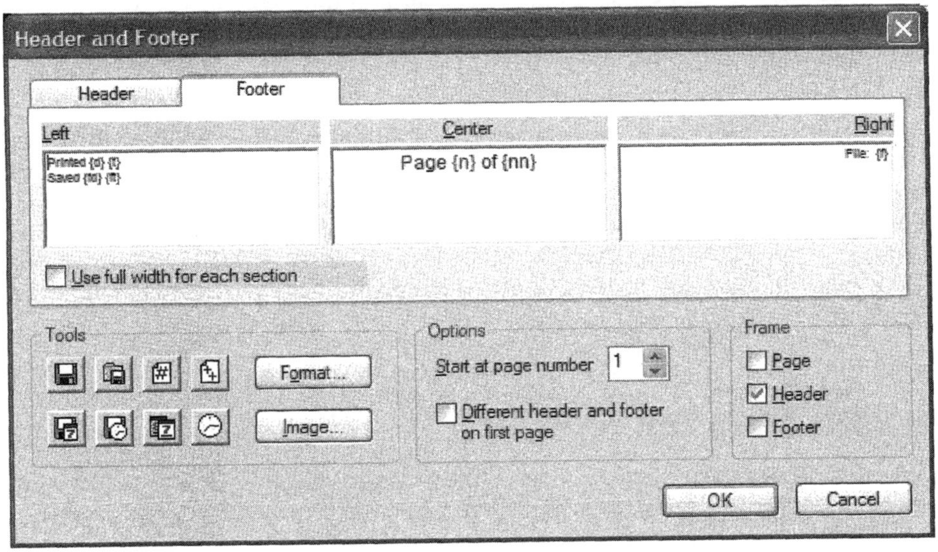

FIGURE 16.13 Footer tab

EM US

Now let's create a similar template for US units. This will be much simpler. We will open the EM Metric template, make a few changes, and then save it as EM US.

To open the EM Metric template, select **Open** from the **File** menu. Move to the Template folder of the Mathcad directory. The path might be as noted earlier. Select the EM Metric.xmct file and click **Open**. If the file does not appear, make sure that the Files Of Type box has either **All Mathcad Files** selected or **Mathcad Templates** selected. We will make the following changes:

1. Change the default unit system. Select **Worksheet Options** from the **Format** menu and click the Unit System tab.
2. The Custom Default Unit should be selected; however, select **U.S.** from the Based On list.
3. This will give you a warning stating, "Changing the 'Based on' unit system will discard changes to the custom unit system. Do you want to continue?" Click **OK**.
4. Remove the following from the Derived Units list: **Flow Rate (Gallons per minute)**, **Pressure**, and **Volume (Gallon)**.
5. Click **Insert**. This opens the Insert Unit dialog box.
6. In the upper Dimension list, scroll down to **Force Density**. Select **Pounds per cubic foot (pcf)** and click **OK**.
7. Click **Insert** again. In the upper Dimension list scroll down to the **Force per Length**. Select **Pounds Force per linear foot (plf)**, and click **OK**.
8. Click **Insert** again. Scroll down to **Pressure**, select **Pounds per square foot (psf)**, and click **OK**.
9. Click **OK** to close the Worksheet Options dialog box.

That is all we have to do. We want the rest of the template to remain the same. Now select **Save As** from the **File** menu. If you have not saved anything since opening EM Metric, the directory should already be in the Template folder. Make sure the Save As Type box reads **Mathcad XML Template (*.xmct)**. Change the file-name to EM US and click **Save**.

You should now have two customized templates that you can use anytime you want.

NORMAL.XMCT FILE

To create a new file based on one of the two customized templates, you must manually select the template as explained earlier. However, there is a way to have the EM Metric template be used every time you start Mathcad.

To have Mathcad always open with your customized template, open Windows File Manager or My Computer and then go to the Mathcad Templates directory.

Rename the Normal.xmct template to Original Normal.xmct. Then copy EM Metric.xmct and paste it in the same folder. This will create a copy of the file EM Metric.xmct. Rename the duplicate file Normal.xmct. When Mathcad starts, it uses Normal.xmct as the default template. Now the template EM Metric is your default template.

> **Tip!** We could have just renamed EM Metric.xmct to Normal.xmct, but I think it is best to keep the original filename in the directory. If you want to change the template, make a change to EM Metric.xmct and then make a copy again and overwrite the Normal.xmct file.

There is one drawback to having the Normal.xmct template stored on your local hard drive. If you want to have consistent templates in your organization, everyone must have the same Normal.xmct file stored on his or her local hard drive. If you change the default template very often, this makes it difficult to keep everyone using exactly the same default template. You can have a corporate template stored on a network drive, and everyone could browse to the corporate template when opening a new worksheet, but that is difficult to enforce. Hopefully in future releases, Mathcad will allow you to point to the location of Normal.xmct. This will allow you to have it stored on a network drive.

SUMMARY

Templates are essential to engineering with Mathcad. They will give your calculations a consistent look. They allow a consistent appearance for all corporate calculations. Instructors can create a template and have all students use the same template so that all assignments have a consistent appearance. You might want to create different templates for different clients. There are many uses for templates, but whatever the use, you must create your own template and begin using it.

In Chapter 16 we:

- Discussed what information is stored in a template.
- Told how a template works.
- Reviewed Chapters 4, 14, and 15.
- Showed how to save a template.
- Created a customized template with many unique styles and settings.
- Demonstrated how easy it is to create a new template when based on an existing template.
- Encouraged you to create your own customized template and store it as Normal.xmct.

PRACTICE

Additional problems and applications can be found on the companion site: www.elsevierdirect.com/9780123747839.

1. Create the EM Metric template discussed in this chapter.
2. Create the EM US template discussed in this chapter.
3. Create and save your own custom template. Start your template from a worksheet based on the Blank Worksheet template. Select a name for your template. Include the following in your template: header, footer, new math styles, new text styles, custom worksheet settings, and custom number format settings. List the characteristics of your new template.
4. Open the template file you created in exercise 3. Refer to the custom units you created in Chapter 4 practice exercises. Place these custom units on the right side of the right margin, and save your template. Open a new worksheet based on this template. Did the custom units come into the new worksheet?
5. Rename the Normal.xmct template file to Original Normal.xmct. Make a copy of your own custom template file. Rename this copied file Normal.xmct. Now open a new worksheet to see if the new worksheet is based on your custom template.

CHAPTER

Assembling Calculations from Standard Calculation Worksheets

17

Throughout this book, we have seen the benefits of using Mathcad to solve engineering problems. In this chapter, we will focus on the benefits of using Mathcad to organize your engineering calculations.

Once Mathcad worksheets are developed, they can be used over and over again. It is very useful to develop standard Mathcad calculation worksheets. These standard calculation worksheets can be used as independent files, or they can be copied and pasted into new Mathcad calculations. There are times when the Mathcad calculation file becomes so large that it might be necessary to divide the calculation into several different files. These files can be linked together, so they behave as if they were in one file.

Chapter 17 will:

- Explore ways of creating worksheets that can be used repeatedly.
- Demonstrate how to copy and paste these standard worksheets into new worksheets.
- Discuss the Mathcad features available to keep track of information such as Annotate, Comment, and Provenance.
- Show different ways of locking worksheets to protect some information from being changed while allowing other information to be edited.
- Explain the procedures for locking worksheets.
- Discuss ways of storing files and using files in web-based repositories.
- Show how to use the reference command to include information from worksheets located in different files.
- Discuss when it is appropriate to break calculations into different files.
- Discuss the potential problems that might occur when assembling calculations from different files.
- Show how variable definitions might change after pasting portions of another Mathcad worksheet.
- Provide examples showing some of the problems that occur after pasting other Mathcad worksheets into the calculations.

- Present different solutions to the problems discussed.
- Make recommendations as to how the calculations can be organized so that critical information is not redefined unexpectedly.
- Provide guidelines and examples to help avoid problems associated with variable redefinitions.
- Demonstrate the use of Find and Replace to help solve problems associated with variable names.

COPYING REGIONS FROM OTHER MATHCAD WORKSHEETS

The purpose of Part IV of *Essential Mathcad* is to teach you how to use Mathcad to create and organize your engineering calculations. In this chapter, we focus on Mathcad's capability to use previously created Mathcad calculations. You can easily copy and paste regions from previously created worksheets. You can copy a single region or an entire worksheet.

To reuse information from a previous Mathcad worksheet, open the worksheet, select the regions you want to copy, copy the regions, and then paste the regions into your current worksheet.

> **Tip!** If you are pasting between existing regions in your worksheet, you must add enough blank lines to make room for the pasted regions. Mathcad does not automatically add new lines. If you do not have enough space, Mathcad will paste regions on top of one another. If this occurs, undo the paste and add more blank lines.

It is possible to drag and drop regions from one worksheet to another, but this method is not recommended. The reasons will be discussed in the next section.

XML and Metadata

Beginning with Mathcad version 12, Mathcad worksheets are stored in XML file format. This file format is a text-based format that allows information to be tagged inside the Mathcad file. This data is called metadata, and it can be attached to the entire worksheet, a region, or portions of a region.

Worksheet metadata can include such things as:

- Worksheet title
- Worksheet comments
- Author
- Department
- Dates
- Any other custom data

Region metadata—called annotations—can be added to math regions, and can include such things as:

- Comments
- Notes about where the data in the region was derived
- Notes about the formulas being used

To attach metadata to a file, select **Properties** from the **File** menu. This opens the File Properties dialog box. The Summary tab is shown in Figure 17.1, and the Custom tab is shown in Figure 17.2. The information you add in this dialog box becomes permanent metadata stored with the file. Any text in the Description

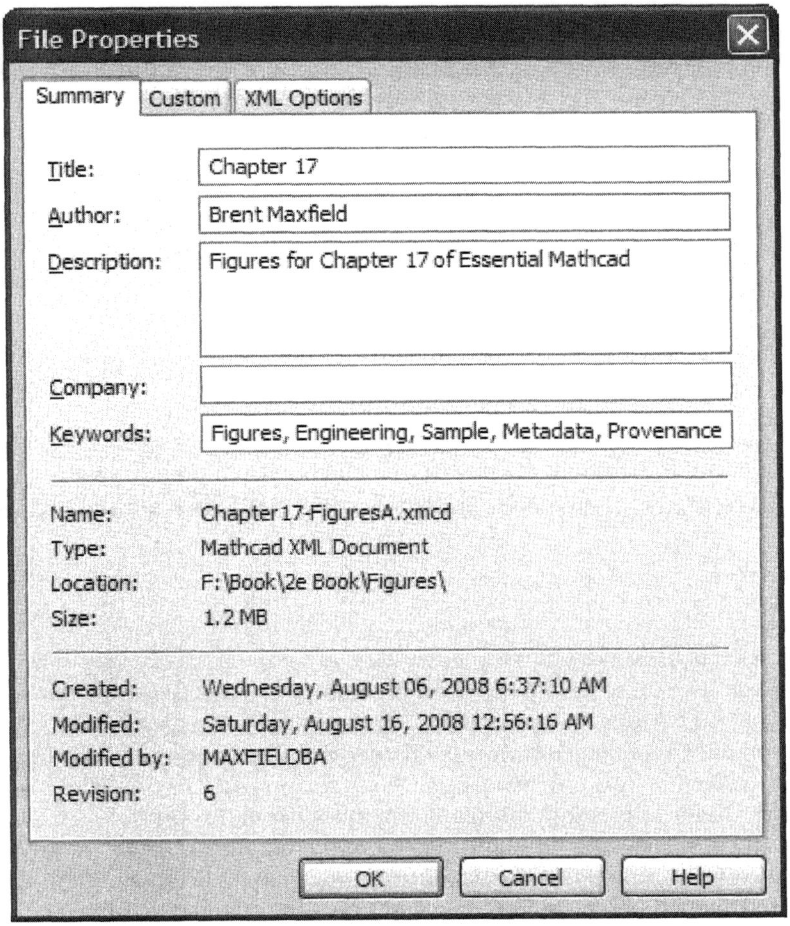

FIGURE 17.1 File Properties dialog box—Summary tab

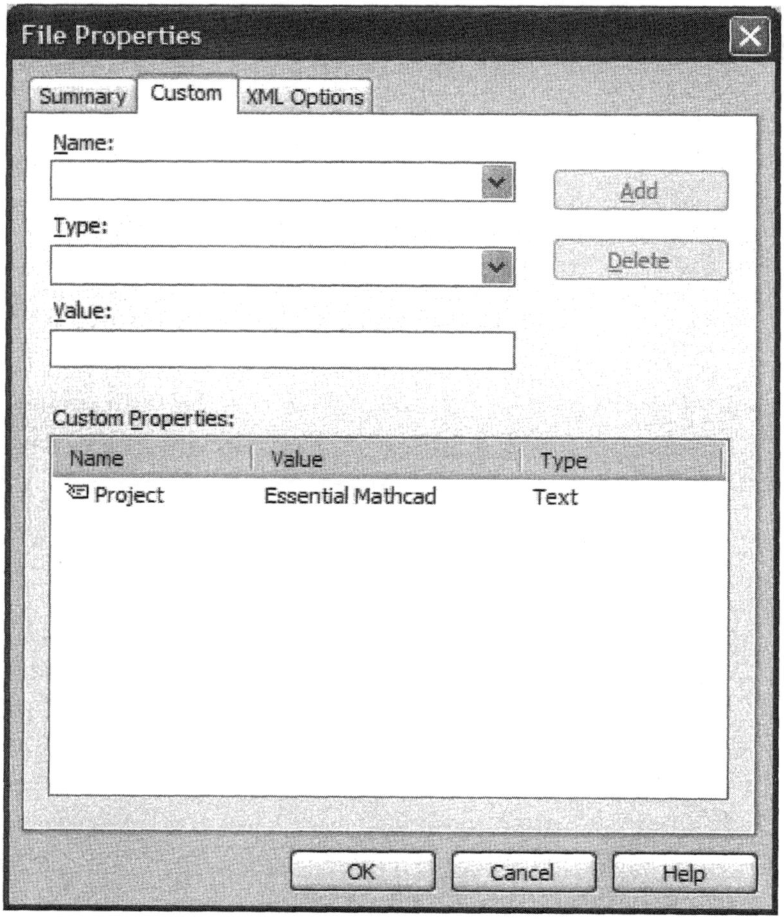

FIGURE 17.2 File Properties dialog box—Custom tab

box is automatically tagged to regions that are copied and pasted into another worksheet. To add information in the Custom tab, click the down-arrow under Name and select one of the options, or type your own description. The next thing you need to do is tell Mathcad what type of data to expect: Text, Date, Number, or Yes or No. You can then input your custom data in the Value box. Click the **Add** button to add the information to the Custom Properties box.

To attach metadata to a math region, select the entire region, right-click the region, and select **Annotate Selection** from the pop-up window. This opens the Selection Annotation dialog box. You can now add a comment about the selected region. It is also possible to annotate a portion of a region such as a variable name, a constant, or a constant with units. To do this, place the portion of the region you want to annotate between the blue editing lines and right-click the region. Select **Annotate Selection** from the pop-up window. Each region can have several annotations.

When you click a region that has annotations added, green parentheses appear around the region or portion of a region. These parentheses indicate the portion of the region that has the annotation. In order to view the annotation, enclose the parentheses between the blue editing lines, and right-click the region. If you have properly enclosed the parentheses, you should see a **View/Edit Annotation** menu option from the pop-up window. Clicking this will open the Selection Annotation dialog box. You can now view and/or edit the annotation. See Figure 17.3 for an example of creating an annotation, and see Figure 17.4 for an example of viewing an annotation.

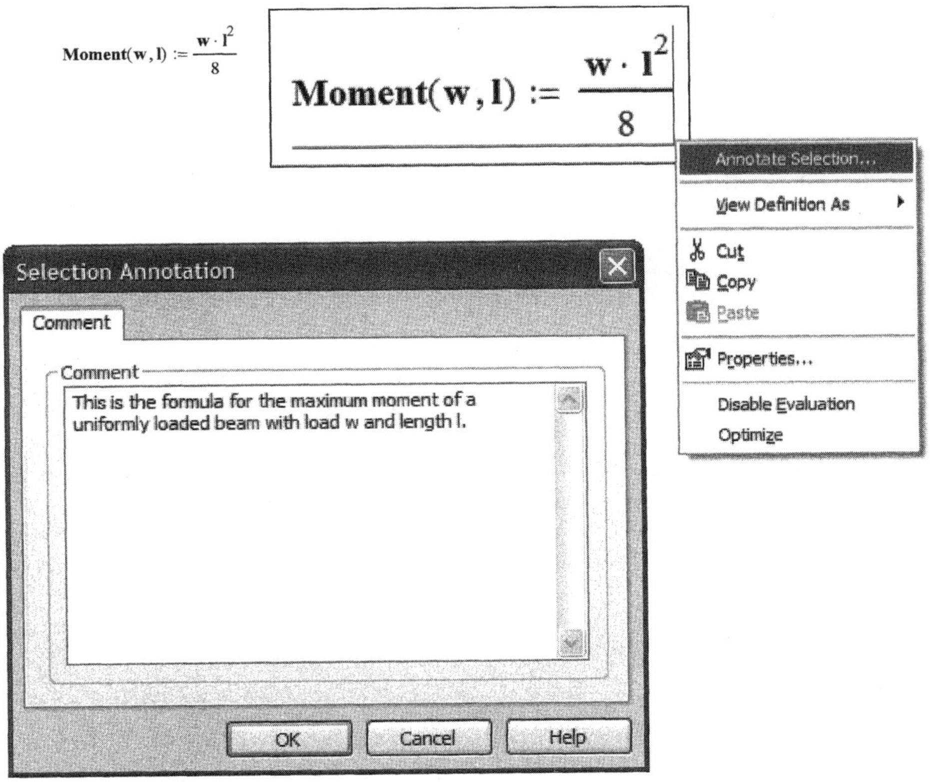

FIGURE 17.3 Adding comments to a region

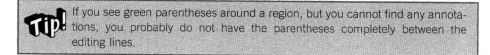

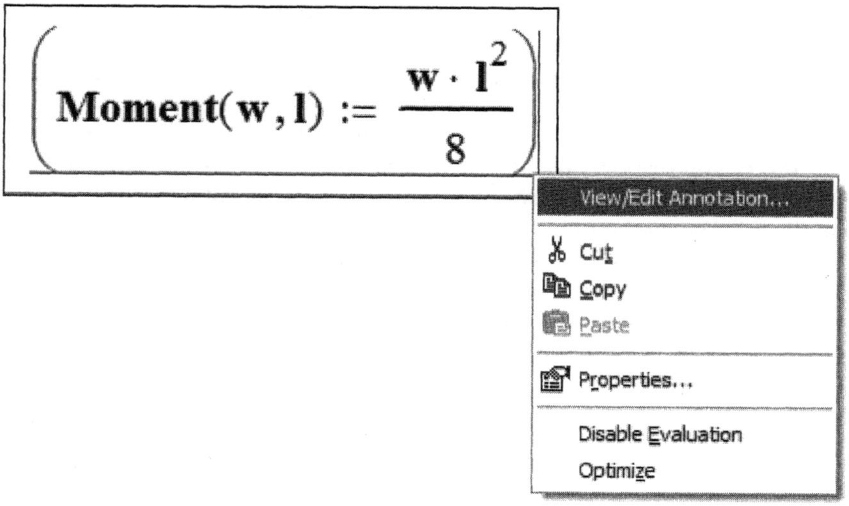

When you click on a region with annotations, green parenthesis appear. Enclose the parenthesis between the editing lines and right-click. Choose the View/Edit Annotation to see the comments.

FIGURE 17.4 Viewing annotations

> **Tip!** If you see green parentheses around a region, but you cannot find any annotations, you probably do not have the parentheses completely between the editing lines.

To view all annotation marks in a worksheet, select **Annotations** from the **View** menu. When this is turned on, all regions that have tagged annotations will have green parentheses showing. You can also change the color of the parentheses by selecting **Color** from the **Format** menu and then clicking **Annotation**.

It is also possible to attach custom metadata to a math region. To do this, right-click the region and click **Properties**. Select the Custom tab from the Properties dialog box. The information here is similar to the worksheet Custom tab, except the data is added only to the selected region. See Figure 17.5. This custom data seems to be redundant. The comment annotations usually should be adequate.

Provenance

When you copy and paste regions from one worksheet to another worksheet, Mathcad automatically attaches specific metadata information. The attached metadata includes the path and filename of the file from which the region was copied. It also includes annotations attached to the region. If there are comments

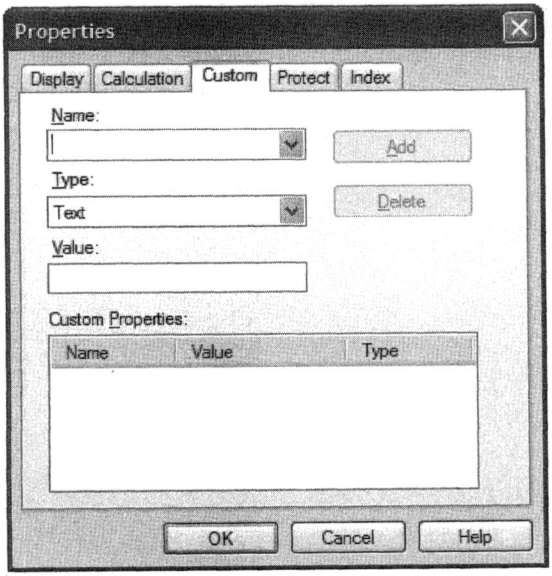

You can add custom metadata to a region, but this is redundant. The comments box, is usually adequate.

FIGURE 17.5 Adding comments to a specific region

attached to the worksheet, these worksheet comments are also attached. If the copied region was originally from a previous file, then the original worksheet path and filename are also included in the metadata. The custom metadata attached to specific regions is not copied. See Figure 17.6.

In versions 12 and 13 Mathcad called this transfer of metadata Provenance. It is a valuable way of tracing information in your calculations because it provides a history of the source data for copied regions. The Provenance metadata always includes the path and filename of the original source, as well as the path and filename of the most recent source. In some cases, these are the same.

> **Tip!** In previous versions of Mathcad, if you used the drag and drop method of moving regions, this Provenance data is not created. For this reason, it is much better to copy and paste regions when copying them from one worksheet to another.

If you modify a region after pasting it from another worksheet, the region is no longer exactly as it was in the previous worksheet, and Mathcad automatically removes the Provenance information, including any comments attached to the region.

CREATING STANDARD CALCULATION WORKSHEETS

You might already have created calculation worksheets that you use repeatedly. The results from these worksheets are undoubtedly printed and placed in your

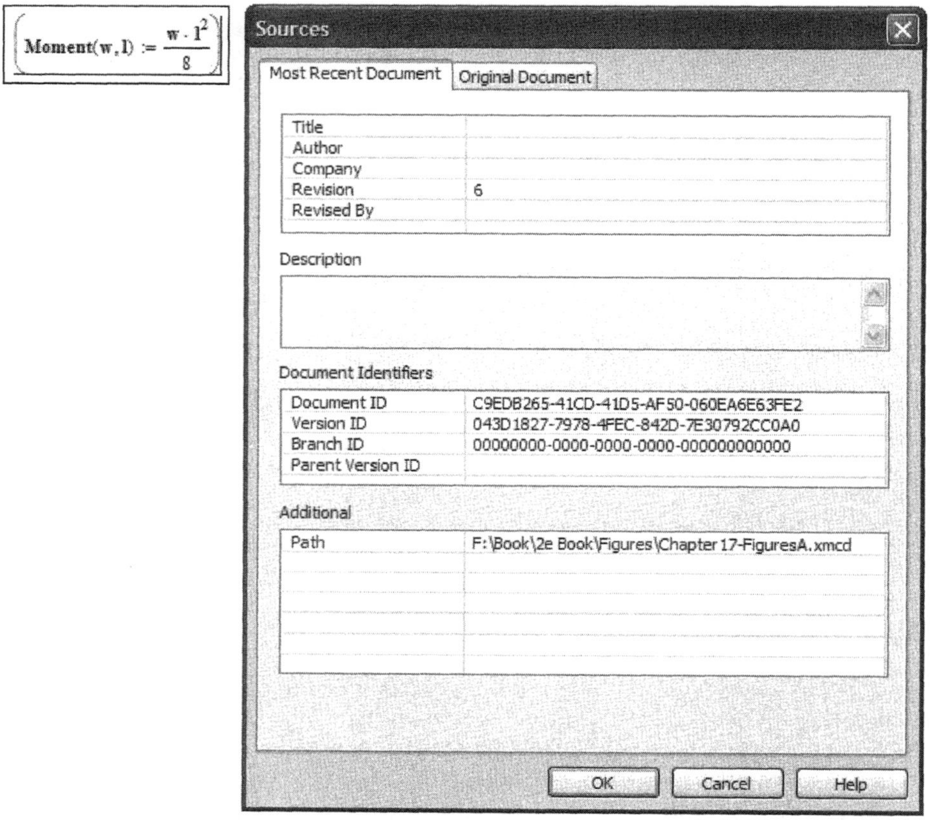

FIGURE 17.6 Provenance

calculation binder. The power of Mathcad is that you can assemble all these calculation worksheets into a single calculation file.

Mathcad makes it easy to create and organize the project calculations because you can reuse existing calculations and insert them anywhere into your current worksheet. Standard calculation worksheets can make it much easier to create a complete set of engineering project calculations. If you have commonly used worksheets stored as standard calculation worksheets, you can copy and paste from these standard worksheets. You can then assemble the calculations for an entire project into a single Mathcad worksheet, just as you used to assemble the paper calculations in a binder.

Standard calculation worksheet files can be as small as only a few regions, or they can contain multiple pages. You might only need to use a few regions from a standard calculation worksheet, or you might use the entire worksheet.

A few things should distinguish your standard calculation worksheets from other worksheets:

- They are designated as standard calculation worksheets.
- They have been thoroughly checked.
- They include a worksheet comment, listing who checked the calculation, and date the calculation was checked.
- Additional metadata can be added such as Author, Department, Dates, and so on.
- The file is locked to allow read access only, so that changes are not accidentally made to the checked worksheet file.
- They are stored in a common folder location, accessible to everyone in your organization.
- They are organized with thought about how they will be copied into other calculations.
- Portions of the worksheet can be protected or locked. (See discussion in next section.)

What type of worksheets should be saved as standard calculation worksheets?

- Repeatedly used equations and functions
- Repeatedly used calculation worksheets
- Cover sheets for project calculations
- Project design criteria
- Code equations
- Reference data
- Almost anything that will be used more than once

PROTECTING INFORMATION

After working hard to create and verify standard calculations, you need to be able to prevent unwanted changes from being made to your standard worksheets.

The first thing you need to do is write-protect the standard calculation files. This will prevent you from accidentally overwriting the standard calculation file. The easiest way to write-protect a file is to find the file in My Computer, then right-click the file and click **Properties**. On the General tab of the Properties dialog box, place a check mark in the Read-only box under Attributes. This will allow the file to be opened, but not saved.

Most standard calculation worksheets will have some variables that need to be changed and many equations and regions that you want to remain unchanged. There are two ways of protecting regions from being changed. First, you can lock complete areas of your calculations. Second, you can protect specific regions (similar to protecting cells in a spreadsheet). It is also possible to hide specific areas of the worksheet so they are not visible. Each of these protection means can be protected by a password.

Locking Areas

Let's first discuss Mathcad areas.

A Mathcad area is a portion of your worksheet defined between two horizontal lines. To insert an area into your worksheet, click **Area** from the **Insert** menu. This places two horizontal lines in your worksheet.

These lines behave as regions and can be moved up or down in your worksheet. If you move the top line up or the bottom line down, the area expands and includes additional regions. The regions between these lines constitute an area.

Once the area has been defined, you can then do several things with the area. You can lock the area so that no changes can be made to regions within the area. You can also collapse the area so that all the regions in the area are not visible. You can even hide the lines that define the collapsed area. (You can select the hidden lines by dragging your mouse over the region where the hidden lines are located.) There are also several formatting things that can be done to the area definition icons. These will be discussed shortly.

To lock an area, right-click one of the area boundaries and choose **Lock**. This opens the Lock Area dialog box. See Figure 17.7. From this dialog box you can enter a password, tell Mathcad to collapse the area when it is locked, tell Mathcad whether or not to allow the area to be collapsed or expanded while it is locked, and tell Mathcad to show a lock timestamp. The lock timestamp displays the date and time the worksheet was locked. To unlock an area, right-click an area boundary and choose unlock.

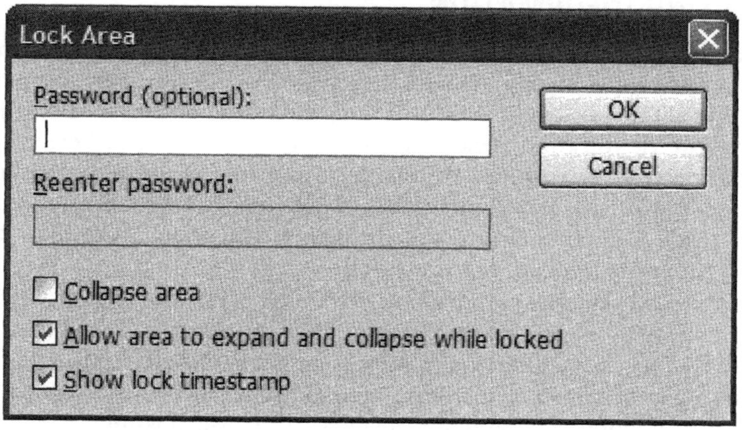

FIGURE 17.7 Lock Area dialog box

> **Tip!** If you enter a password and then forget the password, the area will be permanently locked. I usually do not enter a password. I only lock areas to prevent the accidental change of regions. If you do use a password, be sure to keep careful track of it.

If you right-click an area boundary and choose **Collapse**, the area between the boundaries disappears. A single horizontal line is shown, and all regions below the area are moved upward. All the regions within the collapsed area continue to function as before. You can use the collapse feature either to save space or to hide regions you do not want others to see.

Let's look at the formatting options associated with areas. If you right-click an area boundary and choose **Properties**, it opens a Properties dialog box. The Display tab is shown in Figure 17.8. If you place a check in the "Highlight Region" box, a colored band will appear on each of the area boundaries. The Mathcad default color is yellow, but you can change the color by selecting the Choose Color button. The "Show Border" check box will place a box around each of the area boundaries.

The Area tab is shown in Figure 17.9. This tab allows you to name the area. Using a name provides the opportunity to provide a description of what is being done within the area. This is especially helpful if the area is collapsed. You are not required to use a name. This tab also allows you to show four items. If you uncheck the "Line" box, the horizontal area boundary lines are not displayed. If you uncheck the "Icon" box, the small boxes with arrows are not displayed. If you uncheck the "Name" box, the name will not be used, even if you have used an area name. When the

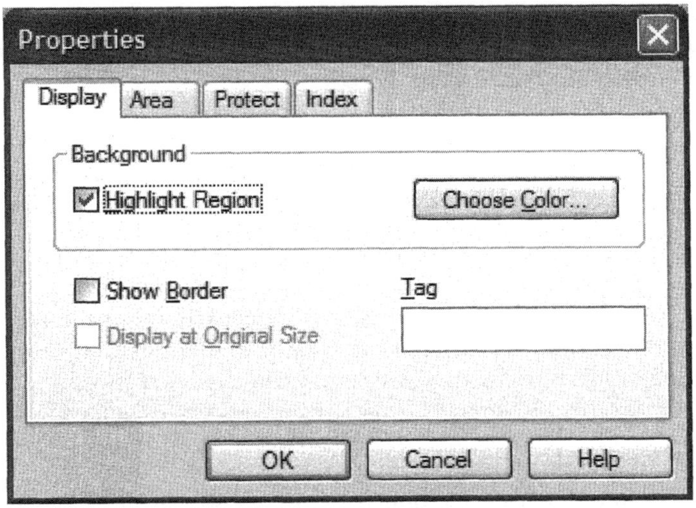

FIGURE 17.8 Display tab from area Properties dialog box

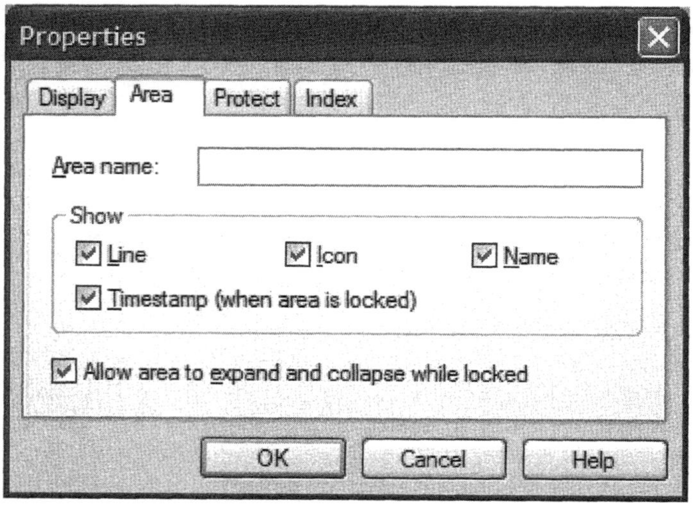

FIGURE 17.9 Area tab from the area Properties dialog box

"Timestamp" box is checked, the date and time that the area is locked will be displayed. If you have a collapsed region, and you uncheck all four boxes (assuming you have not used a highlight color and do not have a border), the collapsed region will not be visible. The only way you would see that there is a collapsed area is if you have View Region turned on (**Regions** from the **View** menu), or if you drag your mouse over the collapsed area. The last check box is similar to the check box in the Lock Area dialog box. It tells Mathcad whether to allow regions to be collapsed or expanded while they are locked.

Tip! In the context of this book, we are discussing areas as a way to protect standard calculations from being changed. I do not recommend using the collapse feature for standard calculations because it hides the regions that will need to be checked and reviewed before the project is finished.

You can delete unlocked areas by selecting an area boundary and deleting it. You can copy a locked region from one worksheet to another worksheet. The area will still be locked after it is pasted into the new worksheet.

Protecting Regions

Protecting regions in Mathcad is similar to protecting cells in a spreadsheet. Each region in Mathcad by default has an attribute that tells Mathcad to protect the region from editing when the worksheet protection is turned on. To prevent a region from being locked, you must tell Mathcad not to protect the region. You do this in the Properties dialog box. Select one or more regions, then right-click and choose **Properties**. Select the Protect tab and uncheck the Protect Region From Editing box.

In order to protect your worksheet it must be saved as a compressed XML file. To save a worksheet in this format choose **Mathcad Compressed XML (*.xmcdz)** from the Save As Type drop-down box of the Save As dialog box. Once you have saved your worksheet in the xmcdz format, you can protect your worksheet. There is a reason for this. Mathcad XML format can be opened and edited in programs other than Mathcad. The xmcdz format cannot be opened (and thus edited) in these other programs.

To protect your worksheet, choose **Protect Worksheet** from the **Tools** menu. This opens the Protect Worksheet dialog box. See Figure 17.10. You can choose a password, but the same rule applies as with Mathcad areas—if you forget the password, your worksheet will be permanently locked. You can choose from three levels of protection.

The most restrictive level of protection is Editing. When this option is selected from the Protect Worksheet dialog box, the worksheet cannot be edited in any way. You cannot move regions, add regions, or select any region, except those regions where the properties were changed to allow editing. This level of protection works well on worksheets that will be used by themselves. Because you cannot select regions, this type of protection does not allow you to copy the locked regions to another worksheet. It does not work well for a standard calculation worksheet.

The next level of protection is Content. This level of protection allows locked regions to be selected so that they can be copied, but you cannot change or

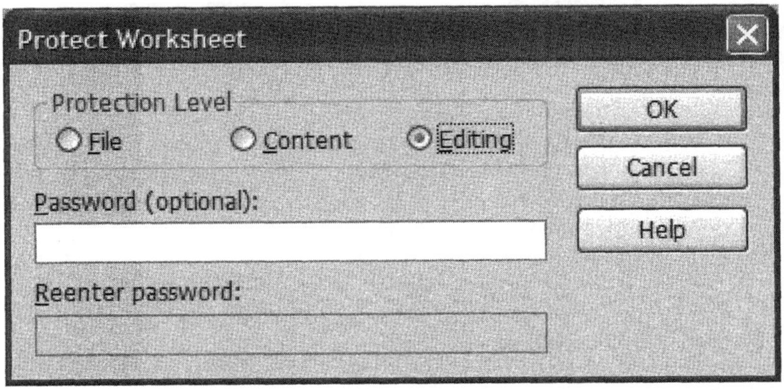

FIGURE 17.10 Protect Worksheet dialog box

move the selected regions. You can also add additional regions to the worksheet. This is the best level of protection to use for standard calculation worksheets.

Because you can select regions, you can also copy them and paste them into other worksheets. When you paste the regions into another worksheet, the regions retain their same properties. This means that if the worksheet where you paste the regions has protection turned on, the pasted regions will also be locked. If the worksheet does not have protection turned on, the regions will be unlocked until the protection is turned on.

The lowest level of protection is File. This means that the worksheet can be saved only in Mathcad Compressed XML (*.xmcdz)" format, or similar formats. It does not allow the worksheet to be saved in XML format.

Advantages and Disadvantages of Locking Areas versus Protecting Worksheets

Locked Area	**Advantages**	**Disadvantages**
	■ Easy to use ■ Easy to see which regions are locked ■ Can copy the entire locked region to a new region ■ All required calculations are kept within the locked area when pasting to a new worksheet	■ Cannot copy individual regions in a locked area ■ If a locked area is collapsed, you cannot see the regions, which makes it difficult to check the calculations
Protected Worksheet (File Level of Protection)	**Advantages**	**Disadvantages**
	■ More similar to a spreadsheet ■ When you copy to a new worksheet, the regions are not automatically locked ■ You do not need to copy all the regions, as you would need to do in a locked area	■ Cannot visually see which regions are unlocked ■ When you copy to a new worksheet, the regions are not automatically locked

POTENTIAL PROBLEMS WITH INSERTING STANDARD CALCULATION WORKSHEETS AND RECOMMENDED SOLUTIONS

There are many advantages to using standard calculation worksheets. However, there is also one very large disadvantage. When you paste a standard calculation worksheet that contains the same variable name(s) as your current worksheet, you redefine the variable definition(s). This can have disastrous consequences in your calculations.

When creating project calculations, it is wise to define the project criteria only once. This allows you to be able to change the criteria and have the entire project result updated automatically. For example, assume that you need to use the yield strength of steel in your calculations. You define this to be Fy:= 36 ksi on the first page of your calculations. Each time you use Fy in your calculations you do not need to redefine Fy:= 36 ksi. You simply use the variable Fy, and Mathcad knows that Fy = 36 ksi. Now, if you paste a standard calculation worksheet into your project calculations that has Fy:= 36 ksi you have a new variable definition in your worksheet. If the value is Fy:= 36 ksi, it appears that there is no problem—all your results are correct. But what happens if at some later time, you change the value of steel from Fy:= 36 ksi to Fy:= 50 ksi? You change this at the beginning of your calculation, but do not change the new definition added when you pasted the standard calculation worksheet. Now, all the results following the new definition are not updated to Fy = 50 ksi. They are still using Fy = 36 ksi.

Another scenario is if the standard calculation worksheet uses the value Fy:= 50 ksi. For our discussion, let's assume that Fy = 50 ksi is appropriate for the specific standard calculation worksheet so you do not change the value of Fy from 50 ksi to 36 ksi after pasting into your project calculation. If you happen to insert this standard calculation worksheet into the middle of your project calculation, what happens to all the expressions that use Fy for the remainder of the worksheet? They all use the value of Fy = 50 ksi. Now all the project calculation results for the remainder of the worksheet are incorrect.

We have been discussing only a single variable. It might be easy to solve this problem when you are dealing with only a single variable, but if the project calculations and the standard calculation worksheet have many identical variable names, it becomes very difficult to ensure that you do not have a problem.

This important issue needs to be considered whenever you create project calculations. It needs careful attention. The advantages of using Mathcad outweigh this disadvantage. Let's now look at some ways to avoid the problems discussed.

GUIDELINES

The following guidelines are suggested as a means to prevent unwanted changes to your project calculations.

1. Use Mathcad's redefinition warnings. To turn these on, click **Preferences** from the **Tools** menu, and then click the Warnings tab. See Figure 17.11. Place a checkmark in all the boxes.
2. After pasting a standard calculation worksheet into your project calculations, scan the pasted regions and look for redefinition warnings. If you find a redefinition warning, you need to decide if the variable that is being redefined has the exact same meaning as a previously defined variable.
   a. If the redefined variable has exactly the same definition and value as a previously defined variable, delete the definition operator : = and replace it with the equal sign. This way, the value of the earlier defined variable can be seen, but not overwritten.

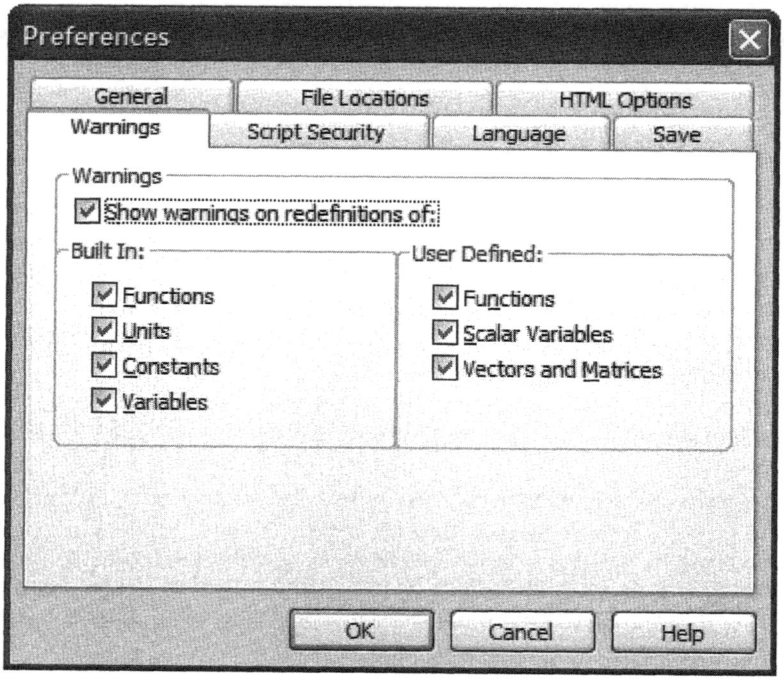

FIGURE 17.11 Turn on all redefinition warnings

  b. If the redefined variable has a different value or a different meaning than a previously defined variable, you need to revise either the project calculations or the standard calculation worksheet so that each variable has a unique name. See the following section on using the ***Find*** and ***Replace*** features.

  c. There might be a situation where redefined variables do not cause a problem. See the discussion in the following section.

3. If possible, place all variable definitions at the top of your standard calculation worksheets. This makes it easier to check for redefinitions after pasting into your project calculations. The most critical variables to place at the top are the ones that might overwrite your project design criteria.

4. Whenever you use previously defined variables in an expression, always display the value of the variables near the expression. For example, if you have an expression Area:= b*d, always type b= and d= near the expression. (This is assuming that b and d were defined on a previous page.) This makes it easier to check the calculations, and helps ensure that the proper values are being used in the expression. You can also use the "explicit" keyword with the symbolic equal sign. See Chapter 9 for a discussion of how to use this method of displaying previously defined variables.

5. Always display the resulting value of expressions or functions. For example, if you have an expression Area:= b*d, always type `Area=` following the definition. This makes it easier to check the calculations, and helps to ensure the results are what you expect.

6. Once you define a variable, do not redefine it (usually). Use subscripts or other means to make each variable unique. See the discussion in the next section on reusing variables.

7. Once a variable has been defined, use the variable name. In future expressions, do not use the variable value. This helps ensure the calculations are updated correctly if the variable changes.

HOW TO USE REDEFINED VARIABLES IN PROJECT CALCULATIONS

There are times when you need to use redefined variables in your project calculations. This section will give you some ideas on how to do this without causing problems in your project calculations.

Pasting the same standard calculation worksheet into your project calculations multiple times will cause redefinitions to occur. This could cause problems as discussed earlier. It is not practical to change the variables used in a standard

This standard calculation calculates the total head (Energy/Density of fluid) in a water system using the Bernoulli equation.

Input Variables:

Pressure	$P := 80\mathbf{psi}$
Density	$\rho := 62.4\mathbf{pcf}$
Velocity of fluid	$v := 5\frac{\mathbf{ft}}{\mathbf{sec}}$
Height above datum	$z := 25\mathbf{ft}$

Calculate total head in system

$$H := \frac{P}{\rho} + \frac{v^2}{2g} + z \qquad H = 210.00 \cdot \mathbf{ft}$$

FIGURE 17.12 Example of a standard calculation worksheet

calculation worksheet each time it is pasted into your project calculations. How do you prevent critical project calculation variables from being redefined?

Let's look at an example. Figure 17.12 shows a standard calculation worksheet. Figure 17.13 shows a project calculation with a standard calculation pasted into it two times. Notice that every variable in System 2 is redefining a previously used variable. None of the variables from System 1 can be reused later in the calculations. Figure 17.14 shows the same condition, but demonstrates how you can keep the variables unique in each system.

RESETTING VARIABLES

Another problem might occur in your calculations when you reuse standard calculation worksheets. If for some reason a redefined variable accidentally is deleted, Mathcad goes up and gets the previously defined value. This can occur without you knowing it. There is no error in Mathcad, but your result will be incorrect.

Here is another case that can cause errors. If your displayed result is just slightly higher in the worksheet than the definition for the result, Mathcad will go up and get the result from the previous variable instead of the adjacent variable. See Figure 17.15. There is a way to prevent this from happening. At the top of your standard calculation worksheet, you can define each variable to be a text string. Then, if a variable is deleted, Mathcad will use the text string instead of an unwanted number above. This will cause an error or display the text string.

Calculate the head for System 1

Input Variables:

Pressure	$P := 20\mathbf{psi}$
Density	$\rho := 62.4\mathbf{pcf}$
Velocity of fluid	$v := 10\frac{\mathbf{ft}}{\mathbf{sec}}$
Height above datum	$z := 40\mathbf{ft}$

Calculate total head in system

$$H := \frac{P}{\rho} + \frac{v^2}{2g} + z \qquad H = 87.71 \cdot \mathbf{ft}$$

Calculate head for System 2

Input Variables:

Pressure	$P := 30\mathbf{psi}$
Density	$\rho := 62.4\mathbf{pcf}$
Velocity of fluid	$v := 20\frac{\mathbf{ft}}{\mathbf{sec}}$
Height above datum	$z := 10\mathbf{ft}$

Calculate total head in system

$$H := \frac{P}{\rho} + \frac{v^2}{2g} + z \qquad H = 85.45 \cdot \mathbf{ft}$$

In this example the variables for System 2 redefined every variable for System 1. This may not be a problem. It only becomes a problem if one of the variables in System 1 needs to be used later on in the project calculations. Figure 17.14 shows one method of using standard calculation worksheets and still keeping unique variable names.

FIGURE 17.13 Inserting the same standard calculation worksheet multiple times

This way you will be notified that there is a problem, and it prevents Mathcad from using unwanted previously defined variables. See Figure 17.16.

USING USER-DEFINED FUNCTIONS IN STANDARD CALCULATION WORKSHEETS

When you define variables and equations in your standard worksheets, these variables and equations are then pasted into your project calculations. One way to avoid having all these additional variables added to your project calculations is to use user-defined functions.

This example is similar to Figure 17.13 except we attempt to prevent the redefinition of critical variables. Let's assume that the variables P, ρ, v, z, and H are not critical variables. They are only useful within the context of the standard calculation. If any of these variable names are critical to your project calculations, then you will need to change the variable names in either the project calculations or in the standard calculation worksheet.

For this example, let's assume that you either calculated or were given the input variables to be used in calculating the total head for System 1 and System 2. Let's also assume that you need to reuse these variables later on in the project so you do not want to redefine them.

Calculate the head for System 1

$P_{System1} := 20psi$ $\quad WaterDensity := 62.4\frac{lbf}{ft^3}$ $\quad V_{System1} := 10\frac{ft}{sec}$ $\quad Height_{System1} := 40ft$

We can now rewrite the standard calculation equation and use the above variables, but that would make the equation much more complex looking. A better way is to assign the above variables to standard calculation variables.

Input Variables:

Pressure	$P := P_{System1}$	$P = 20.00 \cdot psi$
Density	$\rho := WaterDensity$	$\rho = 62.40 \cdot pcf$
Velocity of fluid	$v := V_{System1}$	$v = 10.00 \cdot \frac{ft}{s}$
Height above datum	$z := Height_{System1}$	$z = 40.00 \cdot ft$

Calculate total head in system

$H := \frac{P}{\rho} + \frac{v^2}{2g} + z$ $\quad H = 87.71 \cdot ft$ $\quad Head_{System1} := H$ $\quad Head_{System1} = 87.71 \cdot ft$

Calculate the head for System 2

$P_{System2} := 30psi$ $\quad WaterDensity = 62.40 \cdot pcf$ $\quad V_{System2} := 20\frac{ft}{sec}$ $\quad Height_{System2} := 10ft$

Note: We do not want to redefine the variable WaterDensity so we only display it.

Input Variables:

Pressure	$P := P_{System2}$	$P = 30.00 \cdot psi$
Density	$\rho := WaterDensity$	$\rho = 62.40 \cdot pcf$
Velocity of fluid	$v := V_{System2}$	$v = 20.00 \cdot \frac{ft}{s}$
Height above datum	$z := Height_{System2}$	$z = 10.00 \cdot ft$

Calculate total head in system

$H := \frac{P}{\rho} + \frac{v^2}{2g} + z$ $\quad H = 85.45 \cdot ft$ $\quad Head_{System2} := H$ $\quad Head_{System2} = 85.45 \cdot ft$

Using this method, we are able use reuse the standard calculations, and still maintain the unique variable names from System 1 and System 2.

FIGURE 17.14 Inserting the same standard calculation worksheet multiple times and still keeping unique results

Look for the inaccurate displayed value of H for System 2

Calculate the head for System 1

Input Variables:

Pressure $P := 20\mathbf{psi}$

Density $\rho := 62.4\mathbf{pcf}$

Velocity of fluid $v := 10\frac{\mathbf{ft}}{\mathbf{sec}}$

Height above datum $z := 40\mathbf{ft}$

Calculate total head in system

$H := \frac{P}{\rho} + \frac{v^2}{2g} + z$ $\quad H = 87.71 \cdot \mathbf{ft}$

Calculate head for System 2

Input Variables:

Pressure $P := 30\mathbf{psi}$

Density $\rho := 62.4\mathbf{pcf}$

Velocity of fluid $v := 20\frac{\mathbf{ft}}{\mathbf{sec}}$

Height above datum $z := 10\mathbf{ft}$

Calculate total head in system

If H is just slightly above the redefined definition of H, then Mathcad selects the previous definition of H. This can cause errors to occur in your displayed calculations.

$H = 87.71 \cdot \mathbf{ft}$

$H := \frac{P}{\rho} + \frac{v^2}{2g} + z$

The correct result should be $H = 85.45 \cdot \mathbf{ft}$. Figure 17.16 shows one way to prevent this from occurring.

FIGURE 17.15 Potential problem to avoid when using standard calculations multiple times

By doing this, the standard worksheets become more compact, and the project calculations will not have so many additional variables. Let's revisit at Figures 17.12 and 17.14 and see how much cleaner it is to use a user-defined function. See Figures 17.17 and 17.18.

In Figure 17.15 we showed how the displayed value of H for system 2 was incorrect. In order to catch this error, you would need to do a very thorough check of the calculations. To prevent this error from occurring, reset the variables at the beginning of each System.

Calculate the head for System 1

Input Variables: $P := "*"$ $\rho := "*"$ $v := "*"$ $z := "*"$ $H := "*"$

Pressure $P := 20\mathbf{psi}$

Density $\rho := 62.4\mathbf{pcf}$

Velocity of fluid $v := 10\frac{\mathbf{ft}}{\mathbf{sec}}$

Height above datum $z := 40\mathbf{ft}$

Calculate total head in system

$$H := \frac{P}{\rho} + \frac{v^2}{2g} + z \qquad H = 87.71 \cdot \mathbf{ft}$$

Calculate head for System 2

Input Variables: $P := "*"$ $\rho := "*"$ $v := "*"$ $z := "*"$ $H := "*"$

Pressure $P := 30\mathbf{psi}$

Density $\rho := 62.4\mathbf{pcf}$

Velocity of fluid $v := 20\frac{\mathbf{ft}}{\mathbf{sec}}$

Height above datum $z := 10\mathbf{ft}$

Calculate total head in system

Now, if H is just slightly above the redefined definition of H, then Mathcad gives a text message instead of the previous definition of H. This way Mathcad will not use any previous definitions, and you can detect an error. If any of the variables in System 2 get deleted, you will get an error instead of using the value from System 1.

$$H := \frac{P}{\rho} + \frac{v^2}{2g} + z \qquad H = "*" \cdot \mathbf{ft}$$

The correct result should be $H = 85.45 \cdot \mathbf{ft}$.

FIGURE 17.16 Resolving a potential problem to avoid when using standard calculations multiple times

USING THE REFERENCE FUNCTION

As you assemble your project calculations, you do not need to have all the calculations included in a single Mathcad file. The ***reference*** function allows Mathcad to get variable definitions from another Mathcad file and use them as if they were

This standard calculation calculates the total head (H) (Energy/Density of fluid) in a water system using the Bernoulli equation.

Input Variables: Pressure (P), Density (ρ), Velocity of fluid (v), and Height above datum (z)

Calculate total head in system

$$H(P,\rho,v,z) := \frac{P}{\rho} + \frac{v^2}{2g} + z$$

$$H\left(80psi, 62.4pcf, 5\frac{ft}{sec}, 25ft\right) = 210.00 \cdot ft$$

This is an example and does not need to be copied to project calculations.

FIGURE 17.17 Example of a revised Figure 17.12

Calculate the head for System 1

(Values from Figure 17.14)

$P_{System1} = 20.00 \cdot psi$ $WaterDensity = 62.40 \cdot pcf$ $V_{System1} = 10.00 \cdot \frac{ft}{s}$ $Height_{System1} = 40.00 \cdot ft$

Input Variables: Pressure (P), Density (ρ), Velocity of fluid (v), and Height above datum (z)

Calculate total head in system

$$H(P,\rho,v,z) := \frac{P}{\rho} + \frac{v^2}{2g} + z$$

This formula is copied from the standard calculation worksheet in Figure 17.17.

$$Head_{system1} := H(P_{System1}, WaterDensity, V_{System1}, Height_{System1})$$

$$Head_{system1} = 87.71 \cdot ft$$

Calculate the head for System 2

(Values from Figure 17.13)

$P_{System2} = 30.00 \cdot psi$ $WaterDensity = 62.40 \cdot pcf$ $V_{System2} = 20.00 \cdot \frac{ft}{s}$ $Height_{System2} = 10.00 \cdot ft$

$$Head_{system2} := H(P_{System2}, WaterDensity, V_{System2}, Height_{System2})$$

$$Head_{system2} = 85.45 \cdot ft$$

Notice how much cleaner this worksheet is than Figure 17.14. The standard calculation only contained the user-defined function. There were not extra variables to copy.

Using this method, we are able to reuse the standard calculations, and still maintain the unique variable names from System 1 and System 2.

FIGURE 17.18 Inserting standard calculation containing user-defined equations

The values from this worksheet will be used in Figure 17.20.

Yield Strength of Steel $\mathbf{F_y} := 50\mathbf{ksi}$

Compressive Strength of Concrete $\mathbf{f'_c} := 4000\mathbf{psi}$

FIGURE 17.19 Using the *reference* function

included in the current worksheet. For example, you could have a project design criteria worksheet where all the key variables for a project are located. Then each project worksheet will reference this one worksheet for all project related criteria. You can also have a corporate worksheet that contains standard definitions and functions. This corporate worksheet can be referenced from your corporate template. Figures 17.19 and 17.20 give an example of using the reference function.

To insert the ***reference*** function, click **Reference** from the **Insert** menu. You can then browse to the file location, and select the file to be referenced. It is a good idea to place a check in the "Use Relative Path For Reference" check box. When this box is checked, it allows you to move both the worksheet and the

To access the Insert Reference dialog box, click **Reference** on the **Insert** menu.

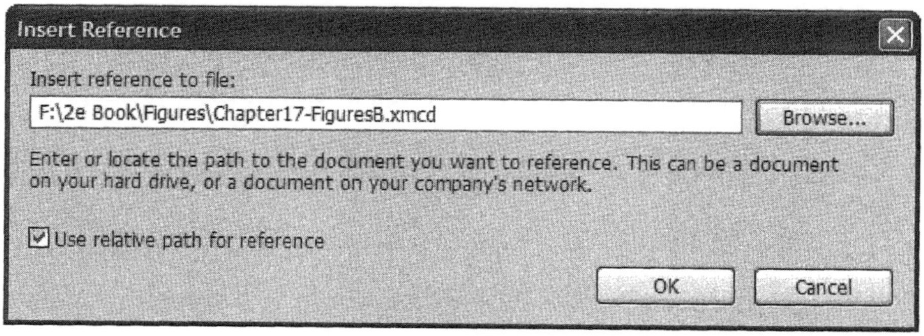

$F_y = ▮$ $f'_c = ▮$ These variables are not defined in this worksheet.

Reference:F:\2e Book\Figures\Chapter17-FiguresB.xmcd(R)

After adding the reference file, it is possible to access the variables defined in Figure 17.19.

$\mathbf{F_y} = 50.00 \cdot \mathbf{ksi}$ $\mathbf{f'_c} = 4000.00 \cdot \mathbf{psi}$

FIGURE 17.20 Using the *reference* function

referenced file together, and they will remain linked together. In order to use this feature, worksheets must reside on the same local or network drive, and both files must be stored prior to using the ***reference*** function.

WHEN TO SEPARATE PROJECT CALCULATION FILES

When is a project worksheet so large that it needs to be separated into two or more files? Try to stay with one file unless one or more of the following occurs:

- It takes too much time to recalculate the worksheet.
- Your worksheet has many graphics, which makes the files over 10 or 15 MB in size.
- Your worksheet gets to be over 100 pages in length.
- It makes sense to separate the project calculations into separate worksheets with related topics.

If you separate your project calculations, here are a few suggestions:

- Create a master worksheet that contains the critical project design criteria.
- Reference the master worksheet to get the critical project design criteria. Do not define the project design criteria in each worksheet because there is a risk of one worksheet using different criteria than another worksheet.
- If you need to use results from another file, use the ***reference*** function to add a reference to the file. Do not redefine the data in the new worksheet. If the data in the other file changes, you want to have up-to-date results in the current worksheet.
- After referencing a file, display key variables below the reference line. This will aid in checking calculations.
- Add page numbers to either the header or footer. Add prefixes to the page numbers for each file, so that paper copies of the calculations will have unique page numbers.

USING *FIND* AND *REPLACE*

If you find yourself in a situation where you have conflicting variable names that need to be changed, you can use the ***Find*** and ***Replace*** features to simplify the effort. The ***Replace*** feature, on the **Edit** menu, finds the requested variable and replaces it with another variable. See Figure 17.21.

There are a few things to keep in mind when using the ***Replace*** feature. The "Find In Math Regions" is turned off by default in older versions of Mathcad. Be sure to place a check in this box so Mathcad will search in Math regions.

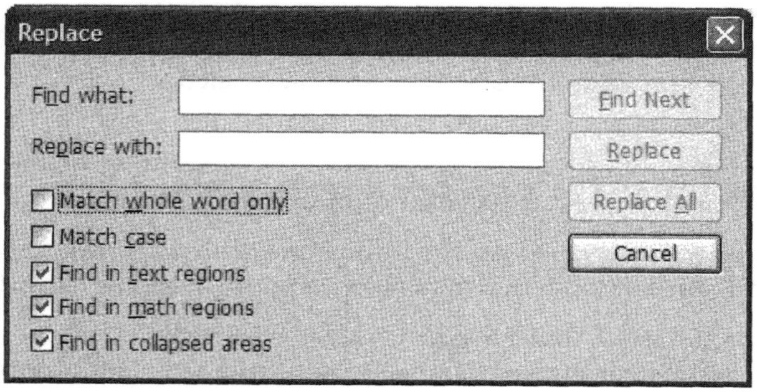

FIGURE 17.21 Replace dialog box

If you are searching for the variable "If," you might want to check the "Match Whole Word Only" box, so that you are not stopped in every word that contains the letters "if." If "Match Case" is checked, Mathcad will not stop at "if." If your variable name uses a literal subscript, be sure to include the period. For example F_y should be typed `F.y.`

SUMMARY

This chapter focused on using Mathcad to assemble project calculations. It showed how to use information from standard calculation worksheets. It also showed the advantages and disadvantages of pasting regions from standard calculation worksheets.

In Chapter 17 we:

- Demonstrated how to copy and paste regions from one worksheet to another.
- Discussed how to add comments to worksheets and regions.
- Introduced Provenance.
- Discussed creating standard calculation worksheets.
- Showed two ways of locking and protecting worksheets.
- Warned of potential problems that can occur when you paste regions from other worksheets.
- Gave suggestions on how to avoid the potential problems.
- Provided guidelines for using standard calculation worksheets.
- Showed how to use the ***reference*** function.
- Discussed when to split project calculations into separate files.

CHAPTER

Importing Files from Other Programs into Mathcad 18

The capability of Mathcad to import information from other programs is what makes Mathcad such an ideal program, not only to create engineering calculations, but also to organize the calculations. Most project calculations will be comprised of information generated from many different computer programs. It is possible to import and display the information from these programs within the Mathcad calculations.

Chapter 18 will:

- Discuss the linking and embedding of information from other software applications into your Mathcad worksheet.
- Identify the differences between linking and embedding.
- Discuss how the information from various software applications can be brought into Mathcad.
- Show how Mathcad can be used to store input files for software applications.
- Illustrate how to include output from other engineering programs into Mathcad.
- Show how to use Mathcad to calculate the input for engineering programs.

INTRODUCTION

In Chapter 13, we introduced the concept of using Mathcad to replace the binder that was used to hold all your paper calculations. In this chapter, we will show you how to accomplish that goal. Most project calculations are comprised of output from multiple software programs. Your Mathcad project calculations can be used to assemble the output from these programs.

The data we discuss in this chapter is not useable by Mathcad. In other words, Mathcad cannot use the data in its calculations. It can only display the data. If you want Mathcad to use any of the displayed data, you must manually input it into Mathcad. In the next chapter, we discuss the intelligent transfer of information between Mathcad and other software programs.

OBJECT LINKING AND EMBEDDING (OLE)

Object linking and embedding allows Mathcad to display information from other software programs within the Mathcad worksheet. When you double-click the displayed information, the application that created the result actually opens, and you can work on the result in the originating software program. If you are working on a worksheet that was created on another computer, you must have the application loaded on your computer in order to edit the result in the original application.

An embedded object means that the displayed data resides within the Mathcad file you are using. It is like having a photocopy of the original file. If you make changes to the copied data, the changes are not made to the original file. If you make a change to the original file, the changes are not reflected in the Mathcad file.

A linked object means that the source of the displayed data resides in another file. You have created a link to the original file, and the data does not reside in the Mathcad file. If the original file is changed, the displayed data in the Mathcad worksheet is also updated. If you double-click the data to open the application, you are actually changing the original data file. If you copy the Mathcad worksheet to a location where the linked object is not available, you will lose the displayed data. You can update, change, or delete links by selecting **Links** from the **Edit** menu.

The advantage of using a linked file is that the data in Mathcad is always up to date. The disadvantage is that if the original file gets deleted, Mathcad no longer has access to the data. You must also keep track of the original file if you are going to archive the Mathcad project calculations. The advantage to using an embedded file is that it always stays with the Mathcad worksheet. You do not need to worry about keeping track of it. The disadvantage is that if the original file is updated, the Mathcad file will be out of date until an updated file is embedded in the project calculations.

BRINGING OBJECTS INTO MATHCAD

There are two ways you can use embedded and linked objects. You can display the data from the object, or you can display an icon on the Mathcad worksheet. The icon can be either an embedded file stored within the Mathcad worksheet file, or it can be a link to an external file. When you double-click the icon, it opens the associated application and then opens the embedded file, or opens the linked file.

There are several ways of bringing objects into Mathcad. The easiest way is to copy the information from one application and then paste it into the Mathcad worksheet. You can also highlight the information from one application, and then drag and drop it onto the Mathcad worksheet. You can also use the Insert Object dialog box. To access this dialog box, click **Object** from the **Insert** menu. See Figure 18.1.

From this dialog box, you can insert an object from an existing file. Just click **Create from File**, then browse to the file and open it. Mathcad will insert the

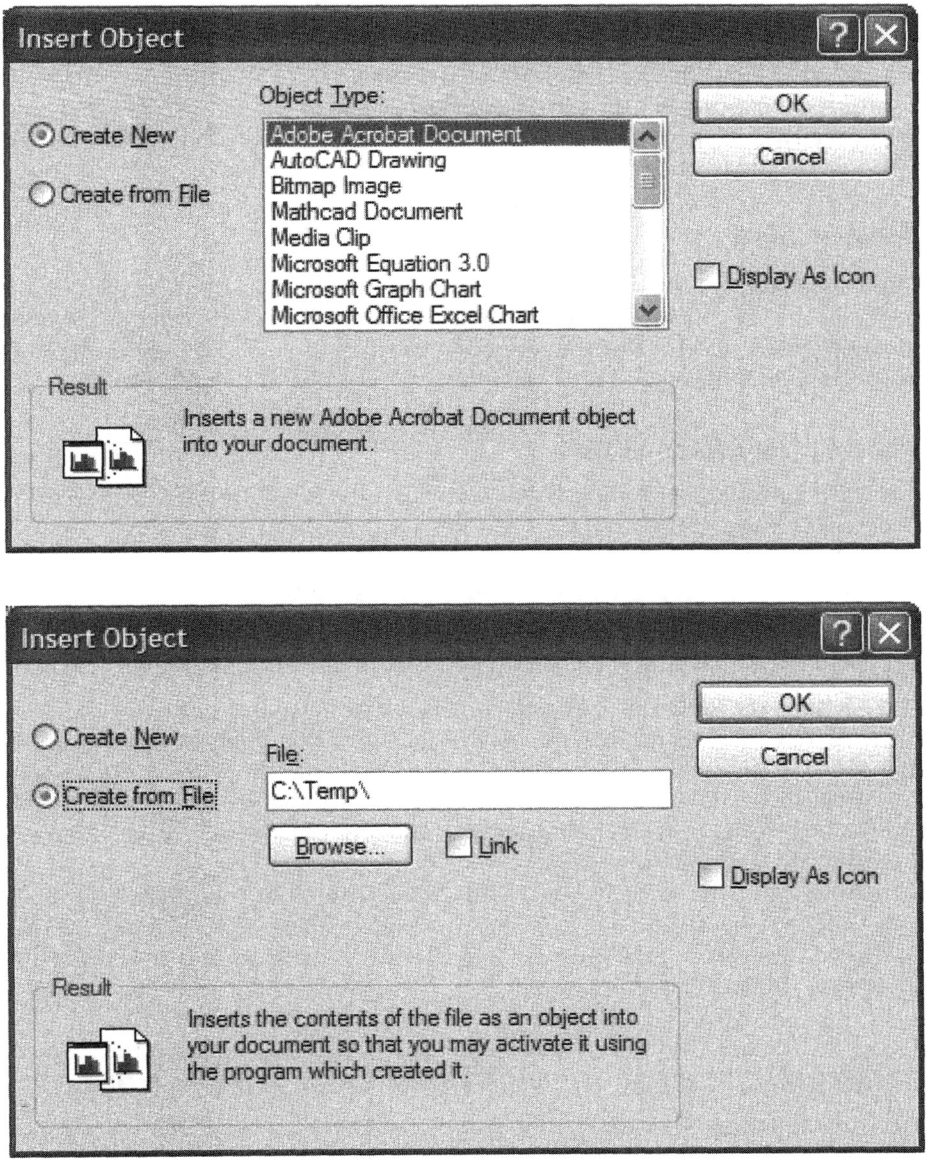

FIGURE 18.1 Insert Object dialog box

object into your Mathcad worksheet. If you want to link to the object, rather than embed the object, place a check in the "Link" box. If you only want to insert an icon, place a check in the "Display As Icon" box.

You can also use the Insert Object dialog box to insert a new object. Just select the type of application, and Mathcad will open an object box with the specified application, and you can create an object from scratch.

Drawing Tools

There are many times when you just want to create a simple quick bitmap sketch to include with your project calculations. You can do this by selecting **Bitmap Image** from the Insert Object dialog box. This will open a small Microsoft Paint object box in Mathcad and give you a few tools to create a simple sketch.

Use and Limitations of OLE

Object linking and embedding (OLE) has many advantages, but it is important to understand its limitations.

First, the data contained within the object is not recognized as intelligent data by Mathcad. It can only display the data, not use the data for other calculations. In the next chapter, we discuss the use of components, where Mathcad is able to communicate and share data with specific applications. However, the data from linked and embedded objects is only displayed.

Second, if the linked or embedded object is more than one page long, Mathcad displays only the first page of the document. When you double-click the object, you can view all pages of the object. You can even print the object from its original application. However, if you print the Mathcad document, only the visible information from the object will print. The entire object will not be printed when you print the Mathcad worksheet.

If you want to have each page of the object printed in your Mathcad calculations, the best way to do this is to create a bitmap image of each page and then paste each page as a bitmap object. This is time consuming and would need to be redone if the object changes.

COMMON SOFTWARE APPLICATIONS THAT SUPPORT OLE

Most Windows-based applications will support OLE. Let's discuss a few common software applications and how they can be used in your project calculations.

> **Tip!** When you copy data from these applications, it is always best to use the Paste Special command. This gives you control over what type of object Mathcad will create. The Paste Special command can be used by clicking **Paste Special** from the **Edit** menu, or you can right-click and choose **Paste Special**.

Microsoft Excel

This is a very popular application used for engineering calculations. Microsoft Excel files can be embedded or linked in your project calculations. However, the real power of Excel files is being able to have the Excel files communicate with your Mathcad worksheet. We devote an entire chapter to the discussion of Microsoft Excel spreadsheets in Chapter 20.

Microsoft Word or Corel WordPerfect

These word processing programs can be used if you have large text files that you want to include in your calculations. You can also bring text files from computer output into a word processor, and then paste the results into Mathcad. This way you can include the results from other programs in your Mathcad project calculations. The easiest way to bring data from a word processing program is to copy and paste. This will embed the information in Mathcad. If you want to link to a word processor file, you will need to insert the link using the Insert Link dialog box.

Adobe Acrobat

Many software applications do not provide output in a format that can be read into a word processor. It is difficult to include output from these programs into your Mathcad project calculations unless you have Adobe Acrobat. With Adobe Acrobat you can print your results to a PDF file. This PDF file can then be linked or embedded into Mathcad. If the PDF file is longer than one page, only the first page will print with the Mathcad worksheet, but the entire PDF file will be available to view when you double-click the object. If you want to show all pages of the PDF file in Mathcad, you can copy each page from the PDF file (use the Snapshot tool) and paste it into Mathcad as a bitmap object. The size of the bitmap image is a function of what zoom level Acrobat is set at when you copy the page.

If you do not have Adobe Acrobat and want to bring printed output into Mathcad, there are a couple of options. First, scan the printed output into a tif, jpg, or pdf format and embed these objects in the Mathcad worksheet. The second method is to do a print screen and paste the bitmap into Mathcad. If you first paste the object into another program, you can crop the object prior to pasting into Mathcad.

AutoCAD

AutoCAD files are very useful to bring into Mathcad calculations. There are unlimited uses to having AutoCAD drawings included in your calculations.

Here are a few tips to help bring AutoCAD images into your calculations:

1. In AutoCAD, select and copy the items you want to paste into Mathcad, and then paste the items into Mathcad.
2. The size of the AutoCAD window determines the size of the object brought into Mathcad. If the AutoCAD window is full-screen, a full-screen object will be created in Mathcad, even if the items you copy occupy only a small portion of the AutoCAD window.
3. It is best to size the AutoCAD window to the size of the object you want to paste into Mathcad. Once the AutoCAD window is the size you would like, zoom into the item you want to copy. You can select the items you want to copy before or after you zoom into the objects. Only the objects visible in the AutoCAD window will be copied into Mathcad, even though additional items are selected.
4. The AutoCAD drawing is pasted as an image by default. If you want to insert an AutoCAD object, which can be modified later, select Paste Special and choose AutoCAD Drawing.

Multimedia

You can include photos, video clips, sound clips, Microsoft PowerPoint presentations, and other multimedia in your project calculations. Many of these are identified by icons. When you double-click the icon the application runs. You can copy and paste photos from any photo editing program.

Data Files

You can store data files in your Mathcad worksheet. Simply drag and drop a data file from My Computer or Windows Explorer onto your Mathcad worksheet.

This creates an icon in your worksheet, and this file then becomes a part of the Mathcad project calculations. If you double-click the file, Mathcad will either open the data file in the associated application, or ask you which application you want to use to open the file. If you have included output results in your Mathcad project calculations (either as a displayed object or as an icon), it is an excellent idea to include the input file as well. Your Mathcad worksheet can contain numerous data files.

Tip! If the drag-and-drop method opens and displays the file in Mathcad instead of creating an icon, use the Insert Object dialog box and place a check in the "Display As Icon check" box.

SUMMARY

Mathcad makes it possible to include graphics, photos, and CAD drawings in your project calculations. You can also include MS Excel, Word, and PowerPoint files. This makes it possible to bring the results from other engineering software into your Mathcad project calculations. You can also include input and output data files within Mathcad.

In Chapter 18 we:

- Introduced object linking and embedding (OLE).
- Gave a brief description of different software applications that can be used in engineering calculations.
- Discussed the benefits and limitations of including output from other programs within Mathcad.

CHAPTER

19 Communicating with Other Programs Using Components

Components are a powerful way to place data from Mathcad into other programs and to place data from other programs into Mathcad. The previous chapter discussed object linking and embedding (OLE). The informatixon displayed by these objects is useful, but it is unintelligent. Mathcad can only display it, not use it. Mathcad does not know what the information is. However, by using components, Mathcad can give intelligent information to another program. That program can then use the input information, process it, and then return output values to Mathcad. Once Mathcad receives the output, the data becomes incorporated into the Mathcad worksheet. Components greatly expand the capability of Mathcad, by dynamically including other programs within the Mathcad calculations.

Mathcad also has an automation Application Programming Interface (API) that allows you to insert Mathcad into other software applications, and have Mathcad communicate with these programs.

Chapter 19 will:

- Introduce the concepts of components.
- Discuss the following components: Excel, ODBC, MathWorks MATLAB.
- Show how to incorporate these components into Mathcad.
- Discuss the different types of components: Application components, Data components, Controls.
- Discuss how to read from data files.
- Introduce the power of writing scripted objects.
- Show how scripting can add power to communications with other programs.
- Discuss Mathcad add-ins that allow Mathcad to be added as a component in other programs.

WHAT IS A COMPONENT?

Application components are similar to the simple OLE objects discussed in Chapter 18. The biggest difference between an application component and a simple OLE object is that components allow the intelligent transfer of data between

Mathcad and another application. This greatly expands the capability of using Mathcad for your project calculations. You can now have all the benefits of Mathcad in addition to the power and advantages of other software applications. In order to use an application component, the application must be installed on your computer.

Mathcad comes with several prebuilt easy-to-use components. There are three types of prebuilt components: Application components, Data components, and Scriptable Object components.

If you are familiar with C++ programming, you can build your own components using the Mathcad Software Development Kit (SDK). This can be downloaded from www.ptc.com. The SDK is .NET compliant and is intended for use with Microsoft's Visual Studio .NET.

We will first discuss the Application components and then the Data components. Since these are prebuilt components, they are easy to use and do not require any programming experience. We will then briefly introduce the Scriptable Object components. These components provide much added power to your project calculations, but they require some knowledge of scripting and Visual Basic programming.

APPLICATION COMPONENTS

The prebuilt Application components are Microsoft Excel, Microsoft Access, other ODBC database programs, and MathWorks MATLAB. Mathcad also integrates bi-diectionally with PTC's Pro/Engineer. In order to use these applications as a component, you must have the application installed on your computer. If you save a Mathcad worksheet with a component, and then give the file to someone else to use, that person must have the same application installed on his or her computer. Some Application components embed the data in Mathcad. Other components provide a link to the data.

Application components are inserted into Mathcad by clicking **Component** from the **Insert** menu. This opens the Component Wizard dialog box. See Figure 19.1.

Let's look at how to insert the Application components.

Microsoft Excel

Microsoft Excel is used widely in engineering project calculations. If you are like most engineers, you have multiple Excel spreadsheets that are used extensively. You can bring these Excel spreadsheets into Mathcad, have Mathcad provide the input to your spreadsheets, and then have Excel provide the results back to Mathcad.

Since Excel spreadsheets are so extensively used, Chapter 20 is dedicated to the use of Excel spreadsheets. We will delay our discussion of Excel until the next chapter.

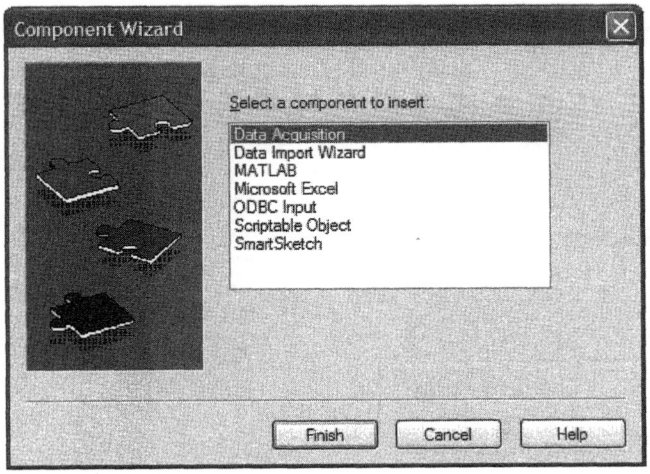

To open the Component Wizard dialog box, select Component from the Insert menu.

FIGURE 19.1 Component Wizard dialog box

Microsoft Access and Other Open DataBase Connectivity (ODBC) Components

The ODBC component allows you to read data from database programs such as dBASE, Microsoft Access, and Microsoft FoxPro. You can also read from Microsoft Excel. This component allows you to read from these programs, but you cannot write to them.

In order to use input from an ODBC component, Windows must establish a link to the database you want to use, prior to inserting the component. This is done through the Data Sources (ODBC) tool found in Windows Administrative Tools. You can access Administrative tools from the Control Panel.

When you open the Data Sources from Administrative tools, the ODBC Data Source Administrator dialog box opens. Once the box is open, select **Excel Files** and click the **Add** button. See Figure 19.2. For this example, we will use the sample database used in the Mathcad QuickSheet, "Using an ODBC Read Component." After clicking the **Add** button, you should see the Create New Data Source dialog box. From this box, select the driver for the database you will be using. In our example, select **Microsoft Access Driver (*.mdb)**, then click Finish. This opens the ODBC Microsoft Access Setup dialog box. See Figure 19.4. From this dialog box in the **Database** section, click the **Select** button. This opens a Select Database dialog box. From this box, find the database file you want to access. The file we want to use is in the same directory as your Mathcad installation in the path Mathcad 14\ qsheet\ samples\ ODBC. The filename is MCdb1. mdb. See Figure 19.3.

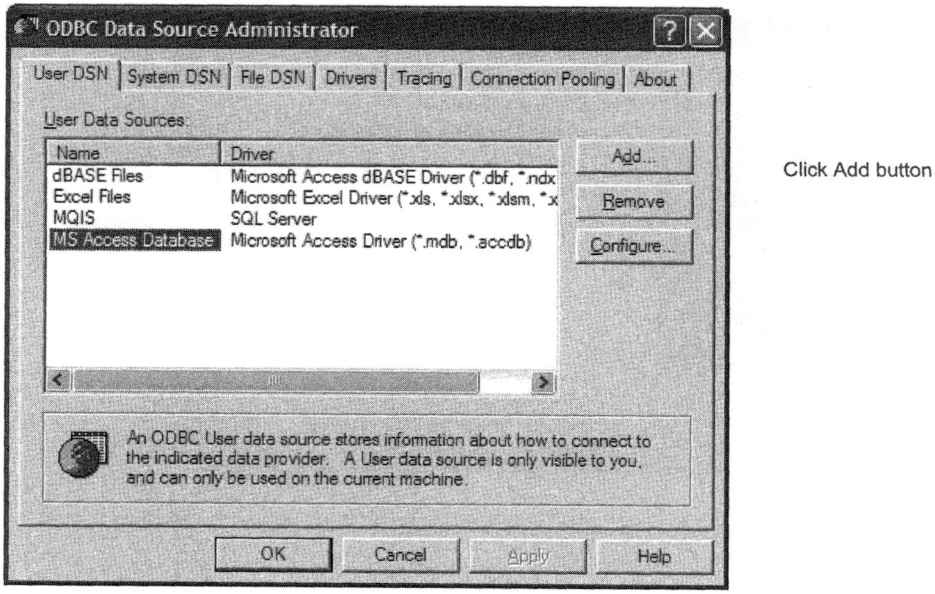

FIGURE 19.2 Configure data source in Windows

The final step is to provide a name for the data source. In our example, it is named Mathcad Sample Database. After inputting the name, click **OK**. The database should now appear in the ODBC Data Source Administrator dialog box. This name will be used to select the ODBC component. See Figure 19.4. Click **OK** to close the dialog box.

Once you have established a link in Windows to the data source, you can insert the ODBC component. You can only bring in data from one table at a time. If you want to import data from more than one table, you will need to insert multiple components. You are also able to choose which fields you want to import. If you know Structured Query Language (SQL), you can filter the data before bringing it into Mathcad.

Let's access data from the Mathcad Sample Database we just linked. Open the Component Wizard dialog box and select **ODBC Input**. This opens the ODBC Input Wizard dialog box. Find the Mathcad Sample Database from the drop-down list. See Figure 19.5.

The next dialog box will ask you to select a Table from the database file, and to select fields from the Table. Click the **Select All** button. See Figure 19.6.

After you select **Finish**, a region will appear with the name of the linked database. The placeholder is blank and needs to have a variable name added. When you display the value of the variable, a matrix appears, which is the data

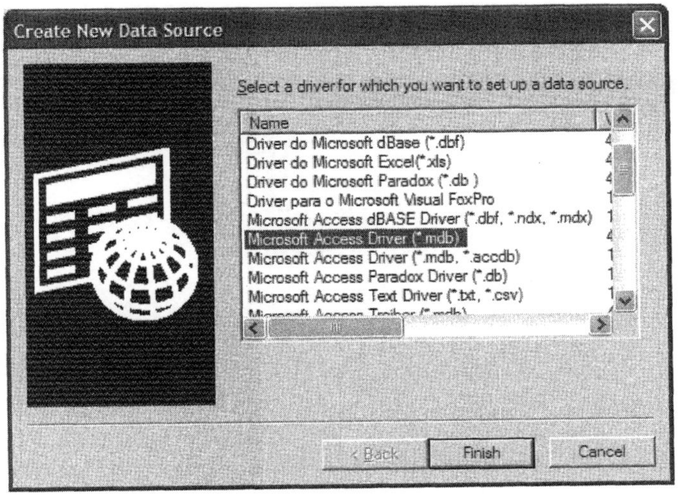

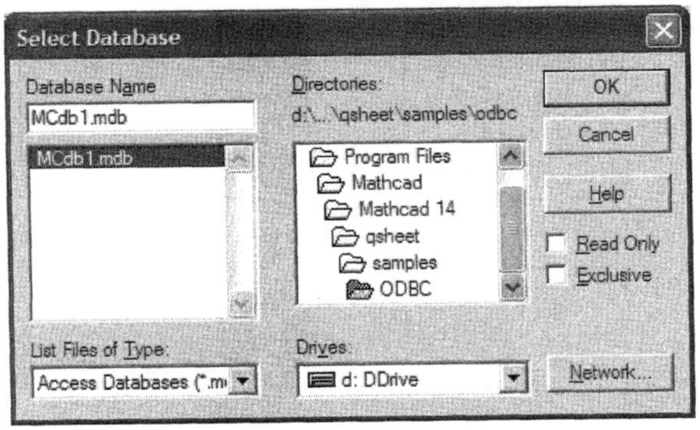

FIGURE 19.3 Choose Driver

from the Access database. See Figure 19.7. The data is now usable in your Mathcad worksheet. If the Access database is changed, the data in Mathcad is also updated.

To configure the component, right-click the component definition and select **Properties**. On the Advanced tab, check the "Include Field Labels in Retrieved Data" check box. If you know SQL, you can filter the data prior to importing. See Figure 19.8.

This is just a sample of what you can do with ODBC components. We are not able to give complete instructions on how to use all ODBC components. For more information, search for ODBC in Mathcad Help.

This database is located in the Mathcad program directory with the path ...Mathsoft\Mathcad 14\qsheet\samples\ODBC\MCdb1.mdb

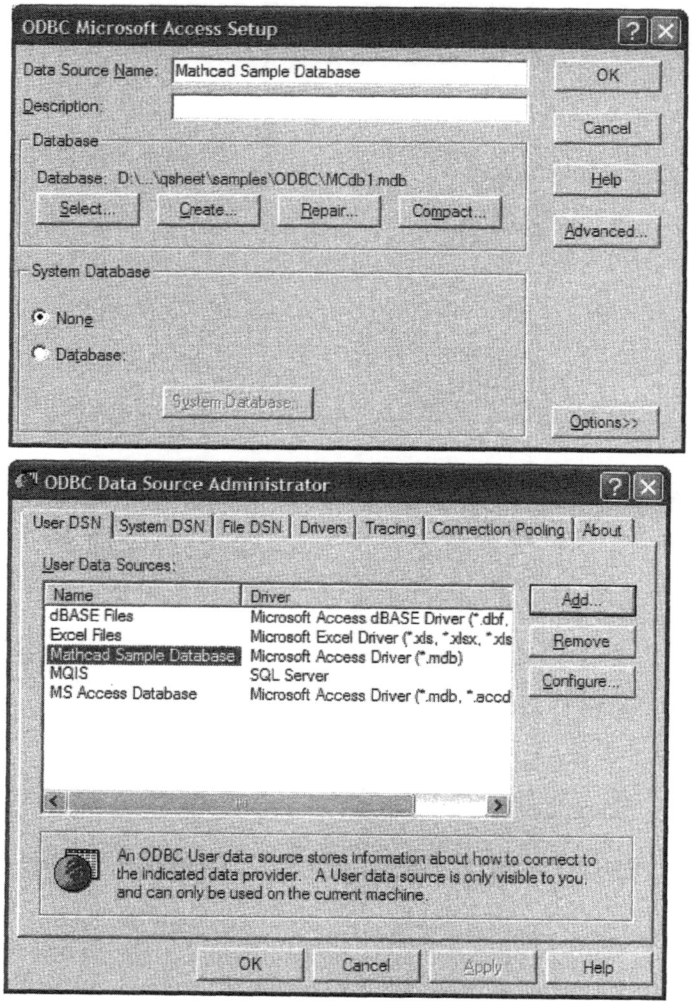

The database should now show on the ODBC Data Source Administrator dialog box.

FIGURE 19.4 Type a name for the data source

MathWorks MATLAB®

You can use the MATLAB component to execute MATLAB scripts using data from Mathcad, and then have MATLAB return the result back to Mathcad. The Mathcad component works with MATLAB.M files.

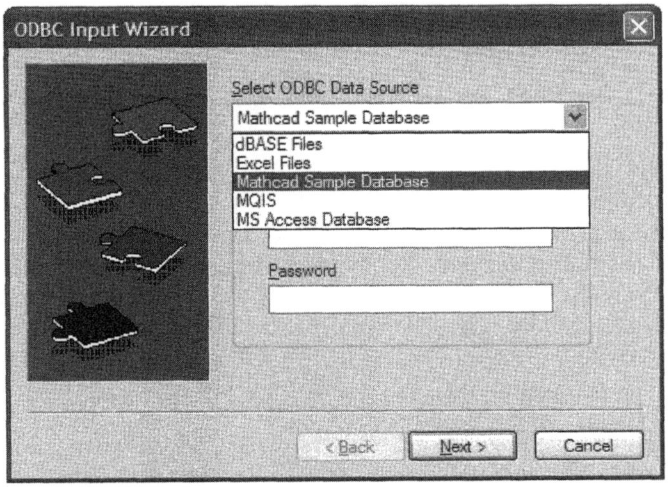

Find the Mathcad Sample Database from the drop down list.

FIGURE 19.5 Inserting Microsoft Access component

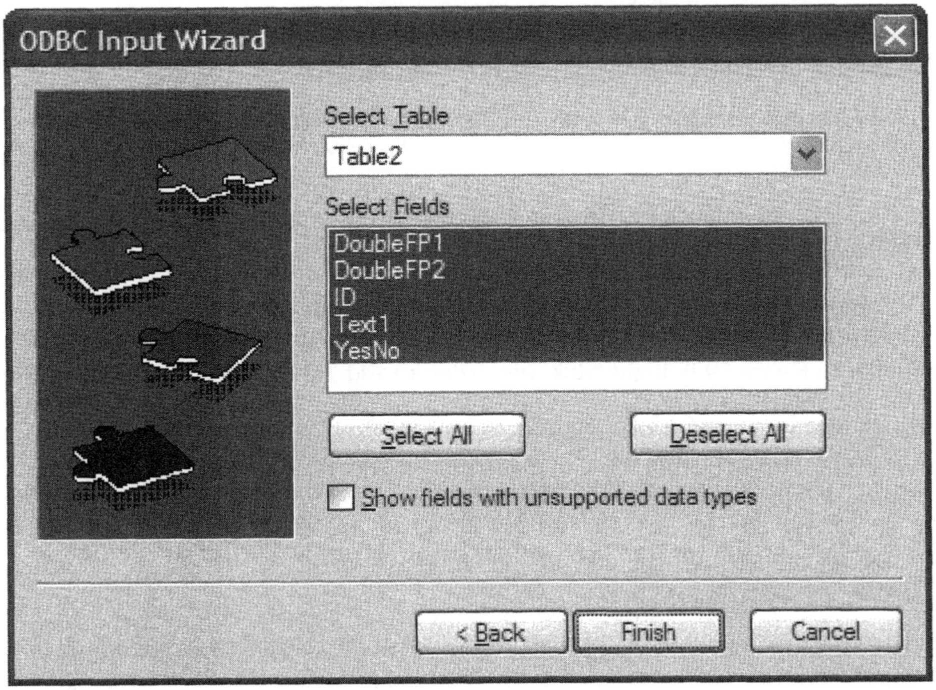

FIGURE 19.6 Select Table and Fields

■ :=
ODBC
athcad Sample Dat:

After you close the dialog box, this is what you will see. You will need to click in the placeholder and type in a variable name.

SampleData :=

ODBC
Mathcad Sample Database

Matrix output form

Note: See Figure 19.8 to turn on field labels.

SampleData =

"DoubleFP1"	"DoubleFP2"	"ID"	"Text1"	"YesNo"
1234.33	1234.57	1.00	"MyText1"	0.00
234.00	111.11	2.00	"23-45"	1.00
12.23	3424.00	3.00	"3.43.34"	0.00

Table output form

SampleData =

	1	2	3	4	5
1	"DoubleFP1"	"DoubleFP2"	"ID"	"Text1"	"YesNo"
2	1234.33	1234.57	1.00	"MyText1"	0.00
3	234.00	111.11	2.00	"23-45"	1.00
4	12.23	3424.00	3.00	"3.43.34"	0.00

You can now access individual elements of the matrix "SampleData"

$\mathbf{SampleData}_{1,1}$ = "DoubleFP1"

$\mathbf{SampleData}_{1,3}$ = "ID"

$\mathbf{SampleData}_{1,2}$ = "DoubleFP2"

$\mathbf{SampleData}_{3,3} = 2.00$

The above matrix "SampleData" is dynamically linked to the Microsoft Access database file "MCdb1.mdb. If the data in the database file is changed, the data in the matrix "SampleData" will also be changed.

FIGURE 19.7 Database example

Unlike other application components, the MATLAB component does not have a wizard. When you select MATLAB from the Component wizard, you get a region similar to Figure 19.9. Right-click this region to add input or output variables. You can have up to four input and four output variables. Right-click the region and select **Properties** to edit the input and output variable names for MATLAB. See Figure 19.10. Right-click the region and select **Edit Script** to open and edit the script for the MATLAB component. See Figure 19.11.

DATA COMPONENTS

Data components allow you to read from or write to data files. Once imported into Mathcad, you have full use of the data. You select a matrix variable name, and all data is accessible by using the matrix name and array subscripts.

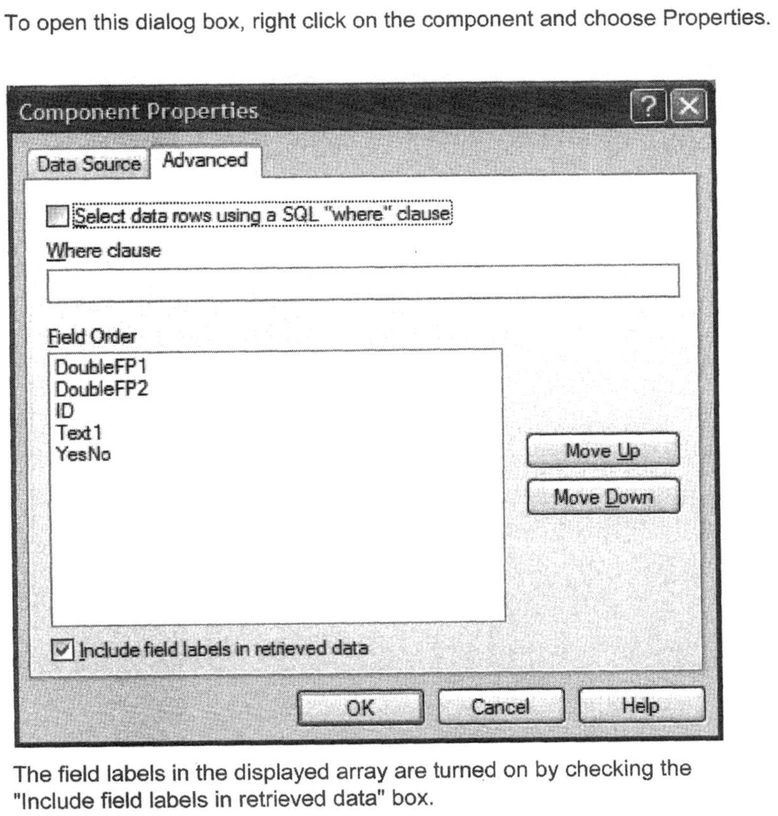

FIGURE 19.8 Configuring ODBC component

The ODBC component allows you to bring in data from database programs. The data components allow you to access data from a much larger number of format types.

There are several ways to import data from other files. The first methods establish a link with the data files, so that the information brought into Mathcad is always current. The second method brings in the data only once. Once brought into Mathcad, the data does not change.

Data Import Wizard Component

The Data Import Wizard is located in two places. You can access it by selecting **Components** from the **Insert** menu, or by selecting **Data** from the **Insert** menu. Both methods bring up the same dialog box. See Figure 19.12 for an example of the Data Import Wizard dialog box.

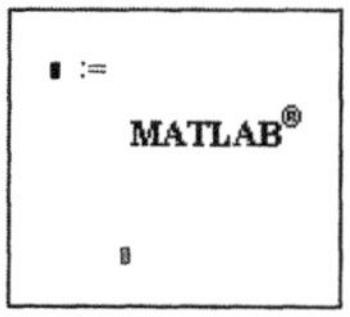

When you insert a MATLAB component, this is what is inserted into your worksheet.

$MATLAB_{Input1} := 10 \qquad MATLAB_{Input2} := 20$

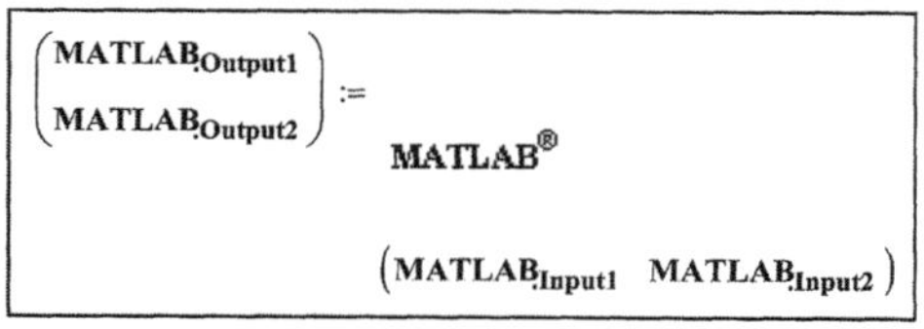

Right-click on MATLAB, and choose Add Input Variable.
Right-click on MATLAB again, and choose Add output variable.
In the placeholders, Type in the Mathcad variable names to be used as input to and output from MATLAB.

FIGURE 19.9 MATLAB component

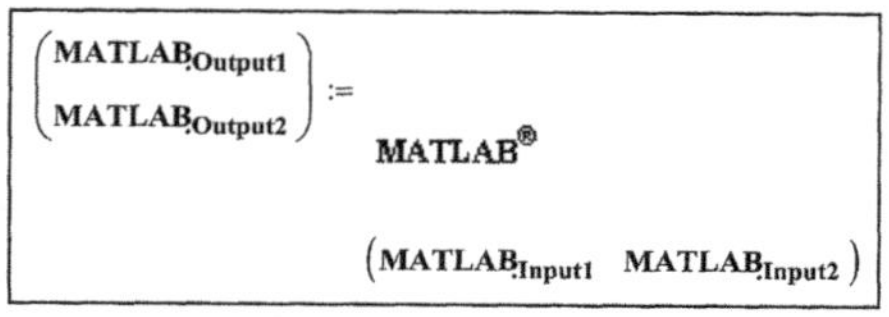

Right-click on MATLAB, and choose Properties. The default MATLAB variable names for input are in0, in1, in2, and in3. You are limited to four input variables. The default MATLAB variable names for output are out0, out1, out2, and out3. You are limited to four output variables. The value of in0 and in1 are taken from Mathcad variables $MATLAB_{Input1}$ and $MATLAB_{Input2}$. The MATLAB Variable out0 and out1 are transferred back to Mathcad variables $MATLAB_{Ouput1}$ and $MATLAB_{Output2}$.

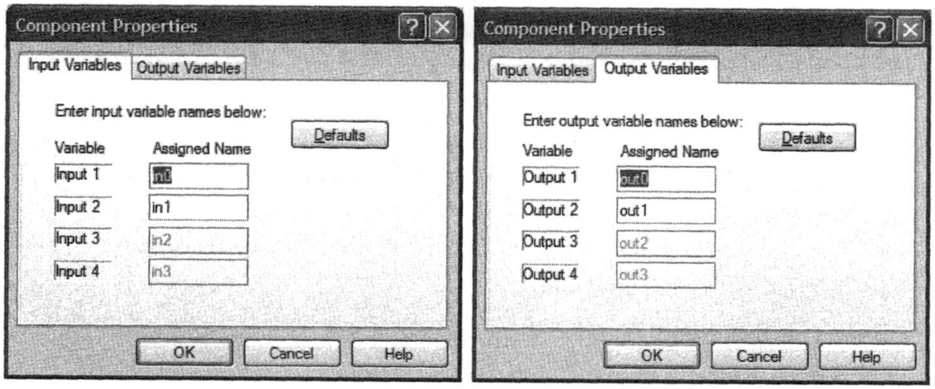

FIGURE 19.10 Naming MATLAB variables

$$\begin{pmatrix} MATLAB_{Output1} \\ MATLAB_{Output2} \end{pmatrix} := \text{MATLAB}^{®} \left(MATLAB_{Input1} \quad MATLAB_{Input2} \right)$$

Right-click on MATLAB, and choose Edit Script. A blank MATLAB Script Editor opens, where you can type your MATLAB script.

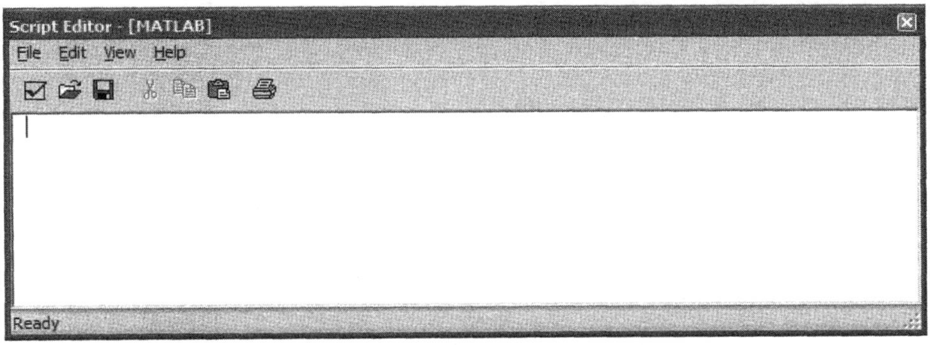

FIGURE 19.11 Writing MATLAB script

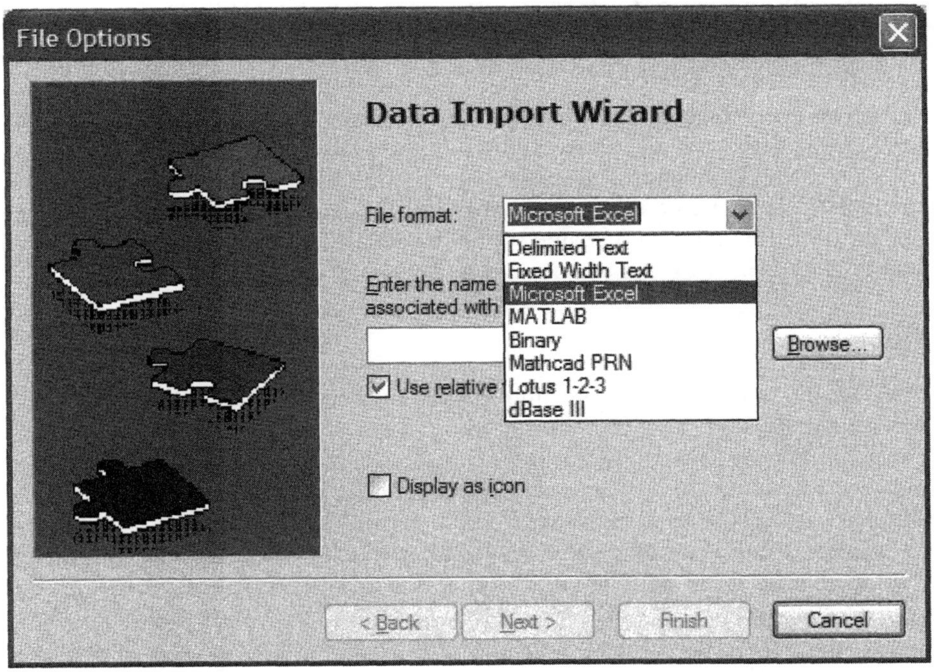

FIGURE 19.12 Data Import Wizard

The Data Import Wizard establishes a link between your Mathcad worksheet and the data file. Every time you recalculate your worksheet, Mathcad goes to the data file, gets the most up-to-date data, and imports it into the worksheet. This is similar to the ***READFILE*** function discussed in Chapter 6.

The Data Import Wizard allows you to access the following data types: Delimited text (you tell Mathcad the type of delimiter), Fixed-Width Text, Microsoft Excel, MATLAB, Binary, Mathcad PRN, and dBase. Once you select the file format and filename from the Data Import Wizard, Mathcad will begin asking you questions about the data file. You will be able to preview the data before you finish. You will be able to select things such as:

- Beginning row
- Ending row
- Rows to read
- Columns to read
- Delimiter type
- What to do with blank rows
- How to recognize text
- What to do with unrecognized or missing data
- Decimal symbol
- Thousands separator
- Named data range (Microsoft Excel)
- Worksheet name (Microsoft Excel)
- Binary data type including endian
- Column widths (for fixed-width data)

Once you are finished with the Import Data Wizard, Mathcad creates an output table showing all or only a portion of the imported data. You can view all the data by clicking the table and using the slider bars. If you do not want to display the imported data in a data table, you can check the Display as Icon box. This will display only the path and filename of the data file. All data is still accessible from within Mathcad.

File Input Component

The File Input is similar to the Import Data Wizard. The imported data is still linked and updated. The difference is in the control you have over the imported data. The Import Data Wizard asks you many different questions about the data file. The File Input option gives you limited control over the data. If you have a simple data file, File Input should be adequate for your needs.

The File Input is used by selecting **Data** from the **Insert** menu and then selecting **File Input**. This opens the File Options dialog box. See Figure 19.13.

Select a file type and a filename. When you click Next, Mathcad asks a much more abbreviated list of questions than the Import Data Wizard. The data is inserted into Mathcad as an icon. Once the icon is inserted, click the output

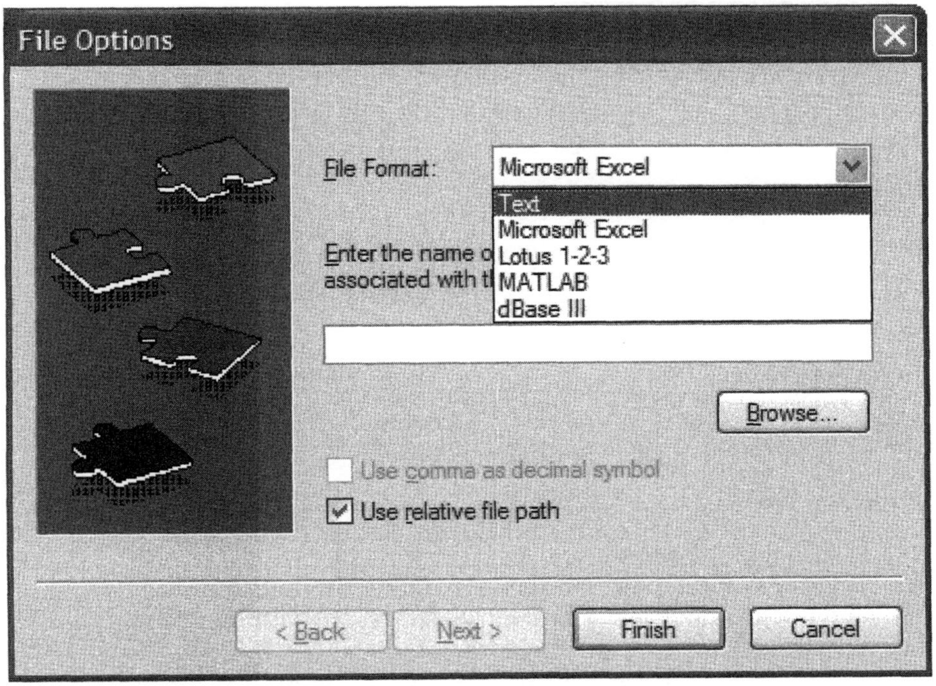

FIGURE 19.13 File Input component

placeholder and type the variable name of the matrix. You can view the data by typing the variable name followed by the colon. You have the same access to the data as you did with the Import Data Wizard.

Data Table

A data table is not a true component. It allows you to insert data from a file only once. The data is not updated, and it is not manipulated by another program. You can also use a data table to input your own data into a matrix rather than use the insert matrix dialog box discussed in Chapter 5.

To insert a data table into your worksheet, select **Data** from the **Insert** menu and then click **Table**. This inserts a blank data table. See Figure 19.14. Once the data table has been inserted into your worksheet, you can give the data table a variable name. You can now input data into the table, or you can import a data file into the data table. To import a data file, left-click the data table, and then right-click and select **Import**. This opens the File Options dialog box. This is the same dialog box as from the File Input component, except now you are embedding the file instead of linking to it.

■ :=

	0	1
0	0	
1		

After the data table is inserted, then you type the variable name in the placeholder.

Sample_Data_Table$_1$:=

	1	2
1	25	"text"
2	2	42

You can enter numbers or text in a data table.

$\mathbf{Sample\_Data\_Table_1}_{1,2}$ = "text"

You can import a data file into a data table. The following table is from the Mathcad program directory using the path C:\...Mathcadt\Mathcad14\qsheet\samples\excel\Thermocouple.xls

Sample_Data_Table$_2$:=

	1	2	3	4	5
1	nd calculation"	0	0	0	0
2	"Coefficients"	0	temperature."	0	0
3	0.039	0	sured Voltage"	ulated Temp."	"Date"
4	25.892	0	1.94	47.706	"05/19/2004"
5	11.87	0	1.84	45.358	"05/19/2004"
6	1.939	0	1.95	0	0
7	0.132	0	2.065	0	0
8	$3.152\cdot10^{-3}$	0	2.11	0	0
9	0.488	0	0	0	0
10	0.087	0	0	0	0
11	$6.795\cdot10^{-3}$	0	0	0	0
12	$2.1\cdot10^{-4}$	0	0	0	0
13	$8.635\cdot10^{-7}$	0	0	0	0
14	$-5.27\cdot10^{-8}$	0	0	0	0
15	$1.43\cdot10^{-9}$	0	0	0	...

$\mathbf{Sample\_Data\_Table_2}_{1,1}$ = "Type T Thermocouple calibration and calculation"

FIGURE 19.14 Inserting a data table

File Output Component

The File Output component allows you to save a data table to a file. You can save the data in a variety of file formats. You can choose from Formatted Text, Tab Delimited Text, Comma Separated Values, Microsoft Excel, Lotus 1-2-3, MATLAB, and dBase III. Using the File Output component is very similar to using the File Input component.

To insert a File Output component into your worksheet, hover your mouse over **Data** on the **Insert** menu and click **File Output**. This opens the File Options dialog box. See Figure 19.15. From this dialog box, choose the type of file format you want to use. Next, browse to the folder location where you want the file to be written, and then enter a filename.

Figure 19.16 provides an example of creating a data table, writing to a data file, reading from the data file and displaying the results. If the data in the table is updated, the file is updated, which is indicated by the variable InputFile1.

In order to update the worksheet results, type CTRL+F9 or hover your mouse over **Calculate** on the **Tools** menu and click **Calculate Worksheet**.

Read and Write Functions

Mathcad comes with numerous read and write functions that allow you to read from or write to various files in various file formats. These functions are discussed in Chapter 6.

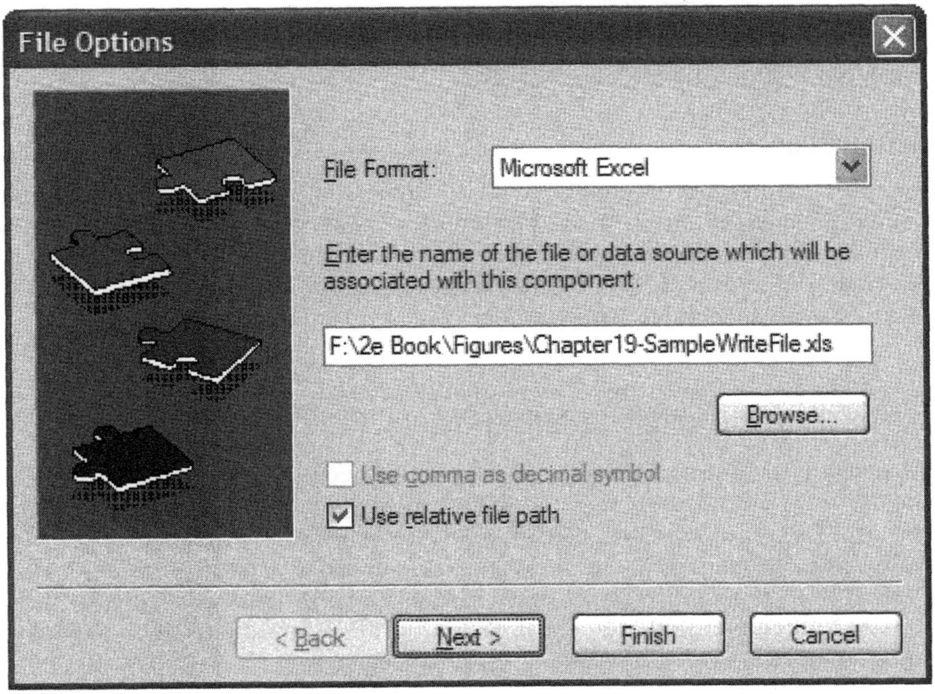

FIGURE 19.15 File Output component

$\mathbf{DataTable_1} :=$

	1	2
1	55	100
2	30	40

Create a simple data table. This is a static table. It only changes if you make a change.

Chapter19-SampleWriteFile.xls

$\mathbf{DataTable_1}$

Use the File Output component to write the data table to a file. In this case we use Microsoft Excel .xls format. It is stored in the file, "Chapter 19 Sample Write File.xls".

$\mathbf{InputFile_1} :=$

Chapter19-SampleWriteFile.xls

Use the File Input component to read from the file created above.

$\mathbf{InputFile_1} =$

	1	2
1	55.00	100.00
2	30.00	40.00

Display the result of the file just read. This is a dynamic table. The results change when the file, "Chapter19-SampleWriteFile" is updated.

If a number is changed in the DataTable_1, then the change is written to the file "Chapter19-SampleWriteFile". The region InputFile_1 reads the updated information from the file, which is displayed in InputFile_1. In order to refresh all displays, choose Calculate Worksheet from the Tools>Calculate menu, or type CTRL+F9.

FIGURE 19.16 Data input and output example

Data Acquisition

The Data Acquisition Component (DAC) allows you to read from and write to recognized measurement devices. With the DAC you can bring real-time data into Mathcad and perform real-time analysis of the data. Numerous settings can be set with the DAC. For more on the DAC, see Mathcad Help and the Mathcad Developer's Reference from the **Help** menu.

SCRIPTABLE OBJECT COMPONENT

We have just discussed the prebuilt Application components and Data components. These are relatively easy to use because they have been built and scripted for you. The Scriptable Object component can add increased functionality to your Mathcad worksheets, but you must have some knowledge of scripting and Visual Basic. The Scriptable Object component can exchange data between your Mathcad and any other application that supports OLE Automation. The component uses Microsoft's Active X scripting specification.

To get an idea of how many different applications support OLE Automation, open the Scripting Wizard. This is opened by clicking **Components** from the

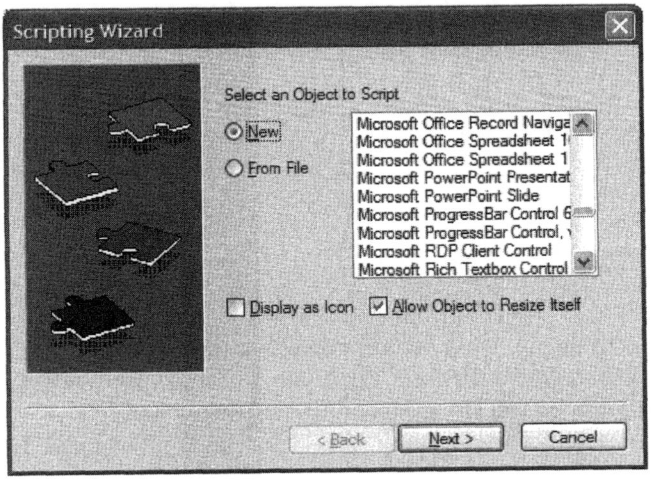

The Scripting Wizard lists the numerous software applications that can be scripted to interface with Mathcad.

FIGURE 19.17 Scripted Object component

Insert menu and then selecting **Scriptable Object**. The Scripting Wizard lists the many different applications that can be scripted to communicate with Mathcad. See Figure 19.17.

If you are familiar with scripting and with Visual Basic, the Mathcad Help provides complete instructions on how to include a Scriptable Object component into your Mathcad worksheet.

Once a component has been scripted, it can be exported as a customized component and reused in other Mathcad worksheets. Information on this feature can be found in Mathcad Help under Export as Component.

Controls

Mathcad controls are a Scriptable Object component. These will be discussed in Chapter 21.

INSERTING MATHCAD INTO OTHER APPLICATIONS

Mathcad has a built-in Application Programming Interface (API) that allows Mathcad to be used as an OLE Automation server from within another application. To use this interface, you need to have knowledge of scripting and Visual Basic. Information on this feature can be found in Mathcad Help or in the Mathcad Developer's Reference under the **Help** menu.

The Automation client interface already has been created for several applications including:

- Excel
- Pro/ENGINEER
- ANSYS
- AutoCAD
- LabVIEW
- Microstation
- SolidWorks
- ESRD StressCheck

These are called add-ins or plug-ins. To view and download these add-ins, go to www.ptc.com/community/free-downloads.htm. This can be found by going to PTC.com and selecting Resources and then selecting Free Downloads.

Pro/ENGINEER

Mathcad is fully integrated into the newest version of Pro/ENGINEER Wildfire. There is no need to use a plug-in. The benefits of using Mathcad within Pro/ENGINEER include:

- Create Pro/ENGINEER relations to identify and open Mathcad worksheets in a Mathcad session that runs in the background.
- Enable Mathcad calculations to drive Pro/ENGINEER models by creating relations to return Mathcad results variable values to Pro/ENGINEER for assignment to parameters or dimensions via relations.
- Enable Pro/ENGINEER model data to drive Mathcad calculations by creating relations to send Pro/ENGINEER parameter or dimension values to Mathcad as assignment variable values.
- Save time and reduce errors by direct bidirectional integration.

PTC has created several training videos and tutorials discussing the integration between Mathcad and Pro/ENGINEER. These can be found on the PTC Web site at www.PTC.com/products/tutorials/mathcad.htm.

SUMMARY

Components are a wonderful feature of Mathcad because they let you communicate and share information with other files and applications. This allows much greater power and flexibility when creating your project calculations.

In Chapter 19 we:

- Discussed the Application components: Microsoft Excel, Microsoft Access, and MathWorks MATLAB.

- Described the different ways to import data from data files.
- Explained the Data Import Wizard component.
- Explained the File Input component.
- Discussed data tables.
- Showed how to write data tables to a data file.
- Reviewed the read and write functions.
- Introduced the concept of Data Acquisition.
- Introduced the topic of Scriptable Object components.
- Introduced the integration of Mathcad into other software applications.

CHAPTER

Microsoft Excel Component

20

Because so many engineers have spent considerable time and money developing Microsoft Excel spreadsheets, this entire chapter will focus on integrating Microsoft Excel spreadsheets into Mathcad.

As discussed in earlier chapters, you can include your previously written spreadsheets within your Mathcad calculations. This chapter will show how to take a previously written Excel spreadsheet and have Mathcad provide the input to these spreadsheets, and how to have Excel pass the results back to Mathcad.

Chapter 20 will:

- Explore the use of Microsoft Excel as a Mathcad component.
- Discuss the advantages and disadvantages of Microsoft Excel.
- Provide several examples of incorporating input and output from existing spreadsheets into Mathcad.

INTRODUCTION

Mathcad communicates with the Excel component through input and output variables. You tell the component what the input variables are and in which cells to put the values. You also tell Mathcad in which cells to look for the output values, and what variable name to use for the output. Chapter 19 discussed the use of the Excel component to store and retrieve data. This chapter will focus on the use of Excel for its computational capability.

> **Tip!** Mathcad 14 version M020 is not fully compatible with MS Excel 2007. In order to function better, all Excel worksheets should be saved in the older 97-2003 format (.xls rather than .xlsx).

INSERTING NEW EXCEL SPREADSHEETS

To insert a new Excel spreadsheet component into your worksheet, click **Components** from the **Insert** menu, and then select **Microsoft Excel** and click the **Next** button. Select **Create an empty Excel worksheet**, and click the **Next** button. This opens the Excel Setup Wizard dialog box. See Figure 20.1. From this dialog box, you tell Mathcad how many input variables and output variables you will have. You also tell Mathcad where to put the input variables in the Excel worksheet, and where to get the output variables. Both the input and output variables can be arrays or matrices. If the input variable is a vector or

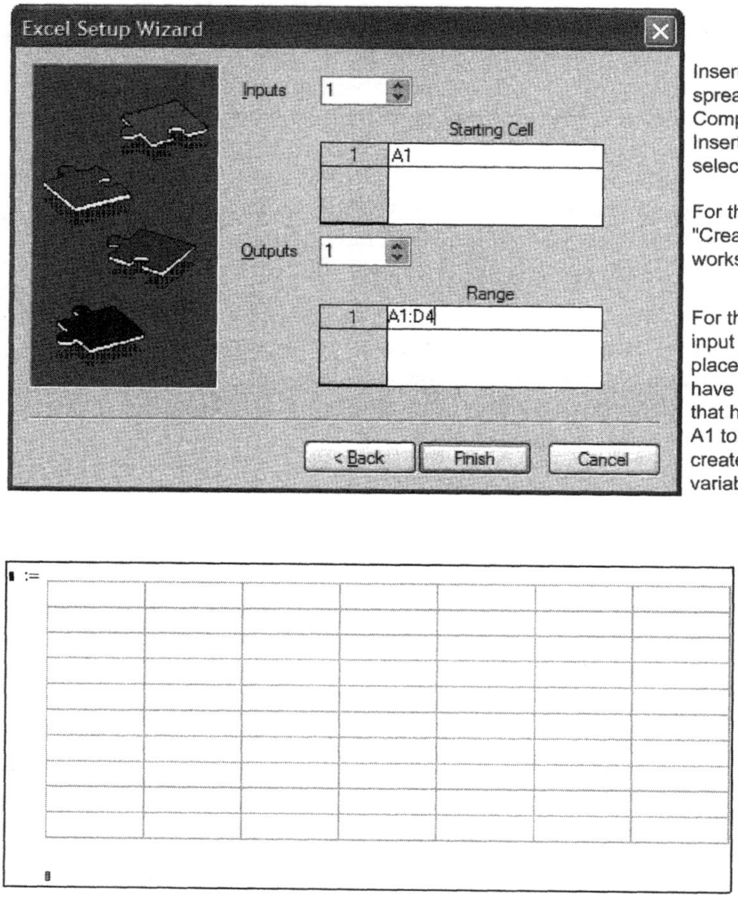
Excel Setup Wizard
Inputs
1
Starting Cell
1
A1
Outputs
1
Range
1
A1:D4
< Back
Finish
Cancel
Insert a Microsoft Excel spreadsheet by clicking Components from the Insert menu, and then selecting Microsoft Excel.
For this example, choose "Create an empty Excel worksheet."
For this example we have 1 input variable that will be placed in cell A1. We will have one output variable that has a range from cell A1 to cell D4. This will create a matrix output variable.
:=
The placeholder on the top left side is for the output variable.
The placeholder on the bottom is for the input variable.

FIGURE 20.1 Inserting a blank Microsoft Excel worksheet as a component

matrix, you enter the starting cell and Mathcad will fill adjacent cells to the right and below the starting cell. If you select a range for the output variable, the result will be a vector or a matrix. If you select more than one input or output, you will need to select a starting cell and range for each input and output. The output range can be a single cell. The example in Figure 20.1 has cell A1 as the input cell and the range A1:D4 for the output range.

If you do not know which cells you will use for your input or output, stay with the default. You will be able to change the inputs, outputs, and cell locations later. After you have entered the required cell locations, click the **Finish** button. This inserts a blank Excel spreadsheet into your Mathcad worksheet.

You can change how many cells are shown in the component by double-clicking the component to open Excel, and then dragging the handles on the side and bottom to show fewer or more cells. If you click the edge of the region, you can shrink or enlarge the size of the region by dragging the corner handles. This displays the same information, just smaller or larger. In order to keep the same proportions be sure to use the bottom right handle and drag diagonally.

> **Tip!** Mathcad 14 version M020 has a bug that prevents you from resizing the component region once you have opened Excel within the region. If you cannot resize the Excel region, close the Mathcad worksheet. When you reopen the worksheet you will be able to resize the region.

The placeholder on the bottom is for the input variable Mathcad provides Excel. The placeholder on the top left side is for the output variable name for the data Excel provides Mathcad.

Figure 20.2 gives an example of an Excel spreadsheet with input from Mathcad and output from Excel. If you change variable Excel_Input1 from 10 to 20, the data within Excel changes, and the output from Excel also changes.

In order to change the location of the input cells and output cells, right-click the component and select **Properties**. This opens the Component Properties dialog box. From this dialog box, you can change the number of input or output variables, and change the location of the input starting cell and the range of the output cells. See Figure 20.3.

Figure 20.4 warns of the problems that can occur if you change the output range after you have begun using the output data. A similar problem occurs if you add or remove rows or columns in the Excel spreadsheet. Mathcad does not adjust the input or output reference locations.

Multiple Input and Output

You can have an unlimited number of input and output variables in an Excel component. You are limited only by your system resources. You can add more input and output variables by right-clicking the component and selecting **Properties**.

$\mathbf{Excel\_Input_1} := 20$ This input will be placed in cell A1 of the Excel spreadsheet. If you change this value, the Excel table changes and the output to Mathcad changes.

$\mathbf{Excel\_Output_1} :=$

20	x*2	x^2	x^3
20	40	400	8000
21	42	441	9261
22	44	484	10648

$\mathbf{Excel\_Input_1}$

The formulas entered into this worksheet are listed at the top of the columns.

The input for this Excel component is the variable Excel_Input$_1$.

The output range A1:D4 is output to the variable Excel_Output$_1$ as a matrix.

$$\mathbf{Excel\_Output_1} = \begin{pmatrix} 20.00 & \text{"x*2"} & \text{"x^2"} & \text{"x^3"} \\ 20.00 & 40.00 & 400.00 & 8000.00 \\ 21.00 & 42.00 & 441.00 & 9261.00 \\ 22.00 & 44.00 & 484.00 & 10648.00 \end{pmatrix}$$

Output can be in matrix form or table form.

$\mathbf{Excel\_Output_1} =$

	1	2	3	4
1	20.00	"x*2"	"x^2"	"x^3"
2	20.00	40.00	400.00	8000.00
3	21.00	42.00	441.00	9261.00
4	22.00	44.00	484.00	10648.00

Individual elements of the output are accessible by using the array subscript operator.

You can assign variable names to specific output values.

$\mathbf{Excel\_Output_1}_{4,4} = 10648.00$ $\mathbf{Excel\_Output_1}_{2,3} = 400.00$

$\mathbf{Test} := \mathbf{Excel\_Output_1}_{3,3}$ $\mathbf{Test} = 441.00$

FIGURE 20.2 Using a blank Excel worksheet

Figure 20.5 gives an example of a component with four input variables and four output variables. Notice the Component Properties dialog boxes showing the cell assignments for the variables.

If you have multiple inputs, it might be easier to include all your input into a data table, a vector, or a matrix. Figure 20.6 shows the same component, except with a single input variable and a single output variable. Compare the Component Properties dialog boxes with Figure 20.5.

Hiding Arguments

It is possible to hide the display of the input and output variable names so that only the Excel spreadsheet is visible. To do this, right-click the component, and select **Hide Arguments**. See Figure 20.7.

In Figure 20.2, if you do not want to have the headers be a part of your output data, then you can change the output range from A1:D4 to A2:D4.

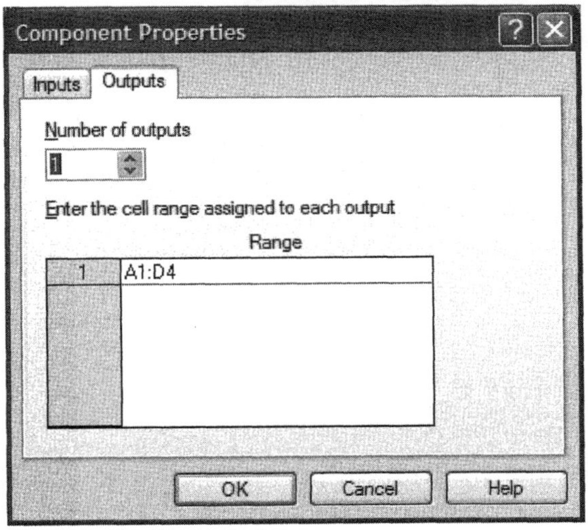

FIGURE 20.3 Modifying input and output properties

USING EXCEL WITHIN MATHCAD

When you double-click the Excel component, you are actually opening Excel inside of Mathcad. The scroll bar becomes active so that you can move up, down, left, and right within the Excel worksheet. The menu bar at the top changes to the Excel menu, and the toolbars change to the Excel toolbars. The component behaves as if you were working within Excel.

USING UNITS WITH EXCEL

Input

You are probably aware that Excel is unit ignorant—it does not know 12 ft from 12 m. If you have followed the recommendations in this book, all your Mathcad worksheets use units. The use of units in Mathcad presents a problem when passing values to Excel. In the case of passing values to Excel, what you see is not what you get. Mathcad does not necessarily pass the displayed value to Excel. It passes the internally stored default value. Figure 20.8 illustrates this problem.

It is important to understand what values you want to input into Excel, and it is equally important to understand how to get the proper values into Excel.

Be careful when changing the output range as this may have drastic impact on your worksheet.

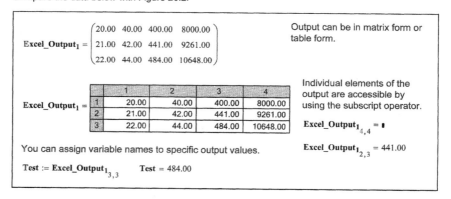

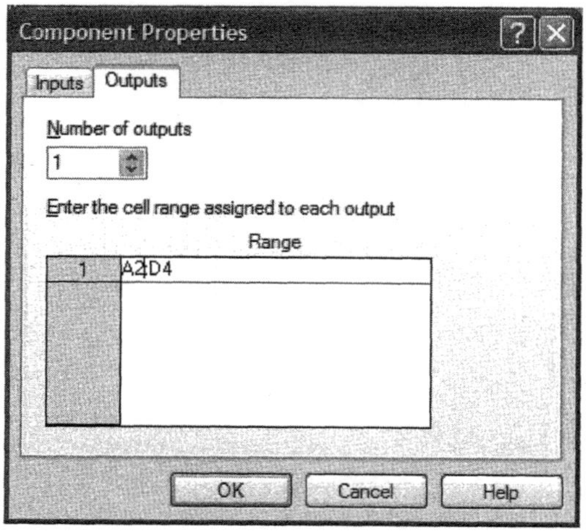

If the output range is changed from A1:D4 to A2:D4, look what happens to the output results. Compare these results to the results in Figure 20.2.

The numbers within Excel have not changed, but the reference to the numbers have changed. The first row in the output is now the first row of data, where before the first row in the output was a header row.

Be aware of this if you ever change the range assigned to the output. A similar problem occurs if you add or delete rows or columns in the Excel spreadsheet.

Compare the data below with Figure 20.2.

$$\mathbf{Excel\_Output_1} = \begin{pmatrix} 20.00 & 40.00 & 400.00 & 8000.00 \\ 21.00 & 42.00 & 441.00 & 9261.00 \\ 22.00 & 44.00 & 484.00 & 10648.00 \end{pmatrix}$$

Output can be in matrix form or table form.

Excel_Output₁ =

	1	2	3	4
1	20.00	40.00	400.00	8000.00
2	21.00	42.00	441.00	9261.00
3	22.00	44.00	484.00	10648.00

Individual elements of the output are accessible by using the subscript operator.

$\mathbf{Excel\_Output_1}_{4,4} = ▮$

$\mathbf{Excel\_Output_1}_{2,3} = 441.00$

You can assign variable names to specific output values.

$\mathbf{Test} := \mathbf{Excel\_Output_1}_{3,3}$ $\mathbf{Test} = 484.00$

FIGURE 20.4 Beware of changing the output range

The solution is similar to what you need to do for empirical equations and for plots. You divide the input variable by the units you want to be used in Excel. When the output comes back from Excel, you need to attach the proper units. See Figure 20.9.

Figure 20.10 illustrates the need to understand what values to input into Excel. There are times when it does not matter what units are input as long as consistent units are used. There are many other times when it does matter. The last two examples in Figure 20.10 illustrate this.

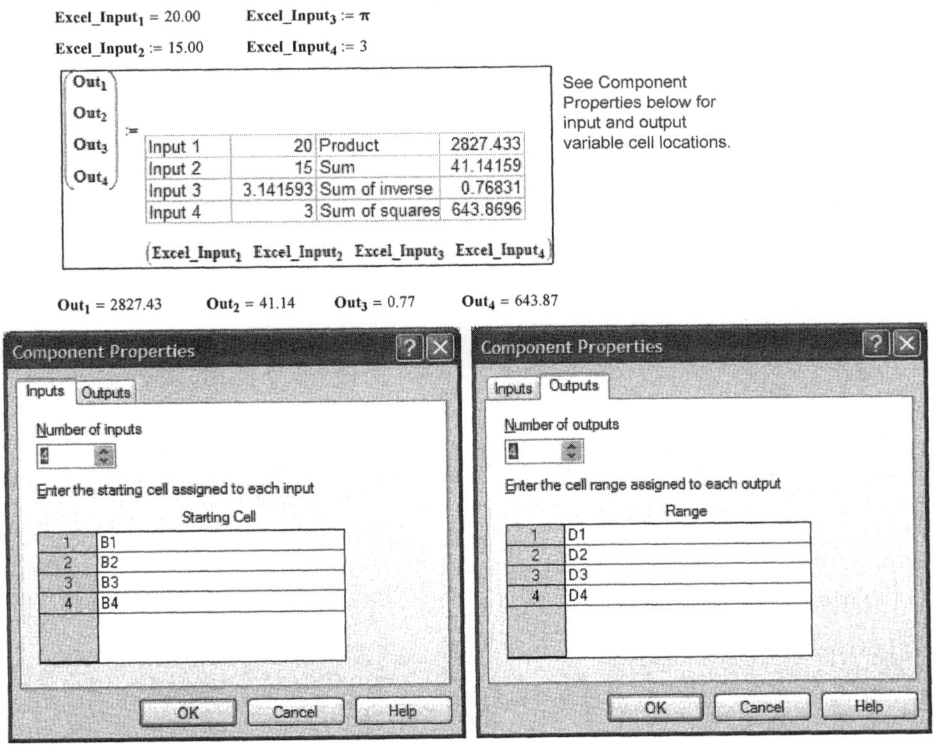

FIGURE 20.5 Multiple input and output

Output

When Excel provides output back into Mathcad there are no units attached. Just as it is important to understand what values Excel is expecting as input, it is equally important to understand what values Excel is providing as output.

Once you have assigned the Excel output to a Mathcad variable, you should attach units to the variable. The units you attach are dependent on the units the Excel value represents. You can assign a new variable with units attached, or you can redefine the same variable with units attached. See Figures 20.9 and 20.10 for examples.

INSERTING EXISTING EXCEL FILES

Mechanics

You have many existing Microsoft Excel spreadsheets. Let's learn how to insert these into Mathcad so that Mathcad can use the data. To insert an existing Excel

This example uses a data table for the input values. The Excel component is looking for a starting cell location. If you have a data table as an input variable, the data just fills in the adjacent cells to the right and below the starting cell. This example has the same data as Figure 20.5, but only a single input variable and single output variable.

$\mathbf{Out_5} :=$

Input 1	20	Product	2827.434
Input 2	15	Sum	41.14159
Input 3	3.141593	Sum of inverse	0.76831
Input 4	3	Sum of squares	643.8696

Excel_Input$_5$

Excel_Input$_5$:=

	1
1	20
2	15
3	3.142
4	3

See Component Properties below for input and output variable cell locations.

$$\mathbf{Out_5} = \begin{pmatrix} 2827.43 \\ 41.14 \\ 0.77 \\ 643.87 \end{pmatrix} \qquad \mathbf{Out_5}_1 = 2827.43 \qquad \mathbf{Out_5}_2 = 41.14 \qquad \mathbf{Out_5}_3 = 0.77 \qquad \mathbf{Out_5}_4 = 643.87$$

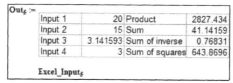

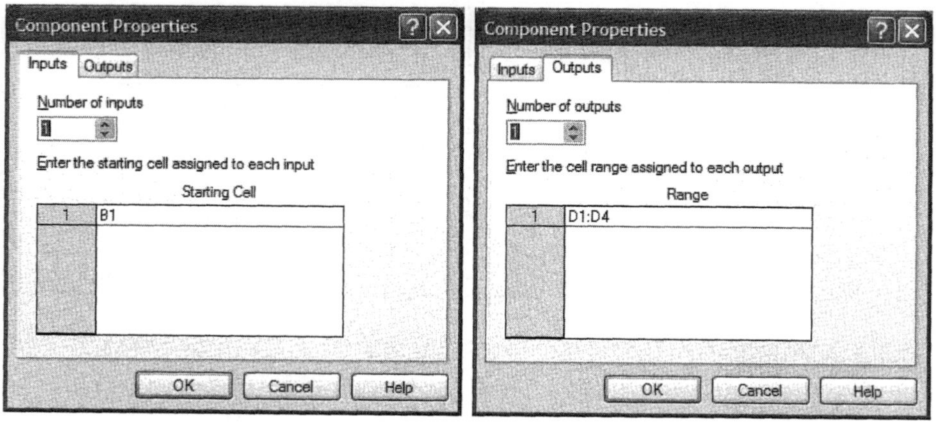

FIGURE 20.6 Multiple input and output

Input 1	20	Product	2827.433
Input 2	15	Sum	41.14159
Input 3	3.141593	Sum of inverse	0.76831
Input 4	3	Sum of squares	643.8696

This is what the component from Figure 20.5 looks like with hidden arguments.

You can hide the display of the input and output variables. To do this right click in the component and select "Hide Arguments."

To show the input and output variables again, right click and select "Show Arguments."

FIGURE 20.7 Hiding arguments

Let's look at some simple examples of using units.

Example 1

$Unit\_In_1 := 5m$

$Unit\_Out_1 :=$ 5 $Unit\_In_1$

$Unit\_Out_1 = 5.00$

This example inputs into Excel, a variable ($Unit\_In_1$) with units of length attached. The output uses the same cell as the input.

Microsoft Excel does not understand or recognize Mathcad units. The input into Excel is the number 5. The units are stripped off the number. The output is also unitless.

Example 2

$Unit\_In_2 := 5ft$

$Unit\_Out_2 :=$ 1.524 $Unit\_In_2$

$Unit\_Out_2 = 1.52$

Why does Excel not display the number 5?

$5ft = 1.524 \cdot m$

Excel does not understand units. Mathcad does not give the displayed unit to Excel; Mathcad gives the default unit to Excel. This worksheet has SI unit preferences. Thus the unit of length defaults to meters (See example above). Mathcad gives this default value of length to Excel.

Example 3

$Unit\_In_3 := 5ft$

$Unit\_Out_3 :=$ 5 $Unit\_In_3$

$Unit\_Out_3 = 5.00$

$5ft = 5.00\ ft$

If the preferences of this sheet were set to US, then the default unit of length is feet and the number 5 would be passed to excel.

Example 4

$Unit\_In_4 := 5ft$

$Unit\_Out_4 :=$ 60 $Unit\_In_4$

$Unit\_Out_4 = 60.00$

$5ft = 60.00\ in$

If the unit preferences of this sheet were set to Custom based on US and the default unit of length was set to inches, then this would be the result.

FIGURE 20.8 Using units with the Excel component

spreadsheet into your worksheet, click **Components** from the **Insert** menu, select **Microsoft Excel**, and click the **Next** button. Select **Create from file**. Next, browse to the location of the file you want to insert, and click the **Next** button. This opens the Excel Setup Wizard dialog box. See Figure 20.1. When you insert an Excel component from a file, you might not know where to put the input

Divide the input variable by the units expected by Excel. For example, if Excel is expecting kg, then divide the mass by kg. Multiply the output by the proper unit.

$\mathbf{Unit\_In_5} := 50\mathbf{kg}$

Example 5

$\mathbf{Unit\_Out_5} :=$ 50 $\dfrac{\mathbf{Unit\_In_5}}{\mathbf{kg}}$

$\mathbf{Unit\_Out_5} = 50.00$

$\mathbf{Out_{5a}} := \mathbf{Unit\_Out_5} \cdot \mathbf{kg}$ $\mathbf{Out_{5a}} = 50.00\,\mathbf{kg}$

Even though this spreadsheet has SI unit default and mass defaults to kg, it is still a good idea to divide the input by kg. This way, if the worksheet unit defaults change, you still have the correct input into Excel.

Example 6

$\mathbf{Unit\_Out_6} :=$ 110.2311 $\dfrac{\mathbf{Unit\_In_5}}{\mathbf{lbm}}$

$\mathbf{Unit\_Out_6} = 110.23$

$\mathbf{Out_6} := \mathbf{Unit\_Out_6} \cdot \mathbf{lbm}$ $\mathbf{Out_6} = 110.23 \cdot \mathbf{lbm}$ $\mathbf{Out_6} = 50.00\,\mathbf{kg}$

To have the value be input as pounds mass, divide the input by lbm.

$50\mathbf{kg} = 110.23 \cdot \mathbf{lbm}$

Example 7

$\mathbf{Unit\_In_5} = 50.00\,\mathbf{kg}$

$\mathbf{Unit\_Out_7} :=$ 1763.698 $\dfrac{\mathbf{Unit\_In_5}}{\mathbf{oz}}$

$\mathbf{Unit\_Out_7} = 1763.70$

$\mathbf{Out_7} := \mathbf{Unit\_Out_7} \cdot \mathbf{oz}$ $\mathbf{Out_7} = 1763.70 \cdot \mathbf{oz}$ $\mathbf{Out_7} = 50.00\,\mathbf{kg}$

To have the value be input as ounces, divide the input by oz.

$50\mathbf{kg} = 1763.70 \cdot \mathbf{oz}$

Example 8

$\mathbf{Unit\_In_5} = 50.00\,\mathbf{kg}$

$\mathbf{Unit\_Out_8} :=$ 50000 $\dfrac{\mathbf{Unit\_In_5}}{\mathbf{gm}}$

$\mathbf{Unit\_Out_8} = 50000.00$

$\mathbf{Out_8} := \mathbf{Unit\_Out_8} \cdot \mathbf{gm}$ $\mathbf{Out_8} = 50000.00 \cdot \mathbf{gm}$ $\mathbf{Out_8} = 50.00\,\mathbf{kg}$

To have the value be input as grams, divide the input by gm.

$50\mathbf{kg} = 50000.00 \cdot \mathbf{gm}$

FIGURE 20.9 Units continued

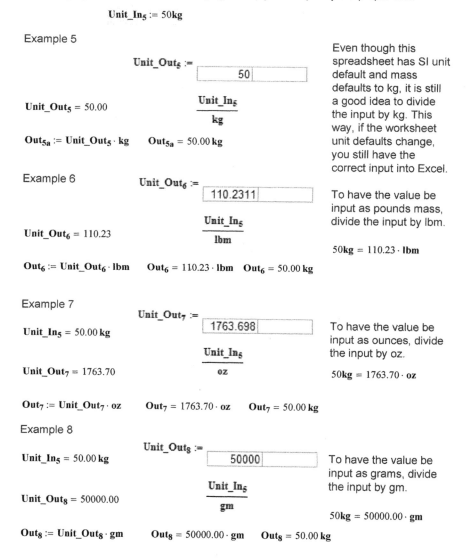

For this example, let's use Excel to calculate the velocity of an object given its acceleration (a), distance traveled (Dist) and initial velocity (V_0). (Yes, it is much easier to calculate this in Mathcad, but we are using this example to illustrate how important it is to understand units when using an existing Excel spreadsheet.)
Input values are in SI units. V_1 converts units to US units and Excel calculates the results in US units of ft/sec. V_2 uses SI units to calculate the velocity. Both results give the same final answer if the appropriate units are attached to the Excel output.

$\mathbf{a} := \mathbf{g}$ $\quad \mathbf{a} = 9.81 \dfrac{\mathbf{m}}{\mathbf{s}^{2.00}}$ $\quad \mathbf{Dist} := 100\mathbf{m}$ $\quad \mathbf{V_0} := 50 \dfrac{\mathbf{m}}{\mathbf{s}}$

$\mathbf{Velocity} := \sqrt{\mathbf{V_0}^2 + 2 \cdot \mathbf{a} \cdot \mathbf{Dist}}$ $\quad \mathbf{Velocity} = 66.79 \dfrac{\mathbf{m}}{\mathbf{s}}$ $\quad \mathbf{Velocity} = 219.14 \cdot \dfrac{\mathbf{ft}}{\mathbf{s}}$ $\quad \mathbf{Velocity} = 149.41 \cdot \mathbf{mph}$

$\mathbf{V_1} :=$

32.17405
328.084
164.042
219.1378

$\left(\dfrac{\mathbf{a}}{\frac{\mathbf{ft}}{\mathbf{s}^2}} \quad \dfrac{\mathbf{Dist}}{\mathbf{ft}} \quad \dfrac{\mathbf{V_0}}{\frac{\mathbf{ft}}{\mathbf{s}}} \right)$

$\mathbf{V_1} = 219.14$

$\mathbf{V_2} :=$

9.80665
100
50
66.79319

$\left(\dfrac{\mathbf{a}}{\frac{\mathbf{m}}{\mathbf{s}^2}} \quad \dfrac{\mathbf{Dist}}{\mathbf{m}} \quad \dfrac{\mathbf{V_0}}{\frac{\mathbf{m}}{\mathbf{s}}} \right)$

The values V1 and V2 are different, but when the proper units are attached the end results are the same.

$\mathbf{V_2} = 66.79$

$\mathbf{Velocity_1} := \mathbf{V_1} \cdot \dfrac{\mathbf{ft}}{\mathbf{s}}$ $\quad \mathbf{Velocity_1} = 219.14 \cdot \dfrac{\mathbf{ft}}{\mathbf{s}}$ $\quad \mathbf{Velocity_1} = 66.79 \dfrac{\mathbf{m}}{\mathbf{s}}$ $\quad$ Must attach ft/s units to V_1

$\mathbf{Velocity_2} := \mathbf{V_2} \cdot \dfrac{\mathbf{m}}{\mathbf{s}}$ $\quad \mathbf{Velocity_2} = 219.14 \cdot \dfrac{\mathbf{ft}}{\mathbf{s}}$ $\quad \mathbf{Velocity_2} = 66.79 \dfrac{\mathbf{m}}{\mathbf{s}}$ $\quad$ Must attach m/s units to V_2

These results work because the Excel formula is =SQRT(A3^2 +2*A1*A2) a formula that works for any consistent set of units.

If the Excel spreadsheet was written assuming units of feet and seconds in cell A3, and then converts the velocity to miles per hour (mph), then the results will be wrong if SI units are input into Excel.

$\mathbf{V_3} :=$

32.17405	
328.084	
164.042	
219.1378	ft/sec
149.4121	mph

$\left(\dfrac{\mathbf{a}}{\frac{\mathbf{ft}}{\mathbf{s}^2}} \quad \dfrac{\mathbf{Dist}}{\mathbf{ft}} \quad \dfrac{\mathbf{V_0}}{\frac{\mathbf{ft}}{\mathbf{s}}} \right)$

$\mathbf{V_3} = 149.41$

$\mathbf{V_4} :=$

9.80665	
100	
50	
66.79319	Excel expects this in ft/sec, but it is m/s
45.54081	mph is incorrect

$\left(\dfrac{\mathbf{a}}{\frac{\mathbf{m}}{\mathbf{s}^2}} \quad \dfrac{\mathbf{Dist}}{\mathbf{m}} \quad \dfrac{\mathbf{V_0}}{\frac{\mathbf{m}}{\mathbf{s}}} \right)$

$\mathbf{V_4} = 45.54$

$\mathbf{Velocity_3} := \mathbf{V_3} \cdot \mathbf{mph}$ $\quad \mathbf{Velocity_3} = 149.41 \cdot \mathbf{mph}$ $\quad \mathbf{Velocity_3} = 66.79 \dfrac{\mathbf{m}}{\mathbf{s}}$ $\quad$ Correct answer

$\mathbf{Velocity_4} := \mathbf{V_4} \cdot \dfrac{\mathbf{m}}{\mathbf{s}}$ $\quad \mathbf{Velocity_4} = 101.87 \cdot \mathbf{mph}$ $\quad \mathbf{Velocity_4} = 45.54 \dfrac{\mathbf{m}}{\mathbf{s}}$ $\quad$ Incorrect answer, because the spreadsheet was not set-up for SI units.

FIGURE 20.10 Example of using units in Excel

values, and where to get the output values. If you do not know, change both the input and output values to zero. You can modify the number and location of input and output values later using the Properties dialog box.

If you assign an input value to a specific cell, Mathcad will overwrite whatever was in that cell. If you accidentally assign the input value to a wrong cell, you could overwrite a very important cell in the spreadsheet. That is why it is a good idea to have no inputs until you know exactly where the input values need to go. Once you know where the input values need to go, open the Properties dialog box by right-clicking the component and selecting **Properties**.

> **Tip!** It is a good practice to turn on protection of your Excel worksheet prior to importing it into Mathcad. That way, you will not accidentally overwrite critical cells. You can always turn off the protection after you bring the component into Mathcad.

Embedding versus Linking

When you add an Excel component from an existing file, the original file becomes embedded in Mathcad. It is no longer linked to the original file. If you change the file later, the Mathcad component is not updated. If you change the Mathcad component, the file is not updated. It is not possible to link a file with the Excel component. If you want to keep a link to the existing file, use one of the data components discussed in Chapter 21. However, this will not allow you to use the computing power of Excel. You can also add an OLE link, but this will not give Mathcad access to the data in the Excel spreadsheet.

It is possible to save the Excel component spreadsheet to an Excel file. To do this, single-click the component, then right-click and select **Save As** from the pop-up menu. If you want to update the original file, select the name of the file and choose to overwrite the file.

Printing the Excel Component

Mathcad will print only what is visible within the component region. This has some drawbacks, especially if you want to print the entire spreadsheet. There is a way around this problem, however; simply do a Save As to an Excel file (as explained earlier), and then open the file in Excel and print it from Excel. If you print to a PDF file, you can cut and paste each page back into Mathcad as a bitmap image.

> **Tip!** Mathcad 14 Version M020 does not print some Excel regions. They appear as solid black boxes. In order to print, I have moved the Excel region to the right of the right margin, and made a graphic image of the Excel region and placed it to the left of the margin.

Getting Mathcad Data From and Into Existing Excel Spreadsheets

The quickest way to use your existing Excel spreadsheet is to include it as a component, then double-click the component to open it in Excel. This allows you to work in the spreadsheet just as you would in Excel. You can enter input and make changes just as you have always done. If you want to transfer data back to Excel, write down the cell addresses of the cells you want to transfer back to Mathcad. Be sure to write down the units so that you can attach units to the output. Now you can use the Properties dialog box to add outputs and cell locations. After you close the Properties dialog box, you should have a column of placeholders on the left side of the component. This is where you add the Mathcad variable names for your output. Make sure you have the variable names corresponding to the correct cell address. Unfortunately, you cannot attach units when you define the variable.

If you want Mathcad to provide the input information, follow the same procedure as discussed earlier. Activate the spreadsheet; write down the input locations and units expected; and use the Properties dialog box to add inputs and cell locations. This should add a row of placeholders below the component. You will now need to define the input variables above the Excel component (including units), and then include the variable names in the row of placeholders below the component. Be sure to divide each variable by the units expected by Excel. Also, be sure to put the variables in the same order you used in the Properties dialog box.

Warning! Once you have used the Mathcad input feature to add data to Excel, do not go into the spreadsheet and manually change the input cells. As soon as you click outside the component, Mathcad will update the component with input values, and overwrite any changes you just made.

> **Tip!** I recommend protecting the Excel worksheet, except for the cells that require input. Do this prior to importing the spreadsheet into Mathcad. This way, you do not accidentally overwrite a critical cell by providing a wrong cell address. You cannot protect the entire worksheet, because that would not allow Mathcad to change the input values.

Mathcad can only pass data to and retrieve data from the top Excel worksheet. If you have multiple sheets, make sure the sheet you need to access is on top (on the far left side of the sheets listed at the bottom of your workbook). If you need to get data from another worksheet, create a cell in the top worksheet that references the data on the other worksheet. If you need to input data into another worksheet, have Mathcad place the data somewhere in the top worksheet. You will then need to activate the component, and go to the other worksheet, and add a reference to look on the top worksheet for input value. Make sure you do not accidentally move the top sheet once you have set up your Mathcad component.

Another way to input and output data to your Excel spreadsheet is to create a block of cells somewhere in your existing top worksheet. You can have an input block and an output block. Mathcad can then place the input values in consecutive cells. You can also have an output block, where you have referenced all the output values. If you use this method, you will need to rewrite your spreadsheet so that the input values are placed in the proper location in the spreadsheet.

$\mathbf{DeadLoad} := 300\mathbf{plf}$

$\mathbf{LiveLoad} := 500\mathbf{plf}$

$\mathbf{BeamSpan} := 15\mathbf{ft}$

$\mathbf{BeamDepth} := 1.5\mathbf{ft}$

$\mathbf{BeamWidth} := .5\mathbf{ft}$

The moment in the beam is calculated from the formula M=w*L^2/8. The beam stress is calculated from the formula Stress=M/S, where S= (1/6)*b*d^2.

The following component was input from the file Chapter 22 - Excel Example - Beam Stress.xls. See Figure 20.12 for a list of the input and output properties.

$\mathbf{BeamStress_1} :=$

Calculate bending stress in a solid beam		
Dead Load lbf/ft	4378.171	lbf/ft
Live Load lbf/ft	7296.951	lbf/ft
Beam Span (ft)	4.572	ft
Beam depth (inches)	0.4572	inches
Beam width (inches)	0.1524	inches
DL Moment (W*L^2/8)	11439.71	ft*kip
LL Moment (W*L^2/8)	19066.19	ft*kip
Total Moment (DL+LL)	30505.9	ft*kip
Section Modulus b*d^2/6	0.005309	in^3
Bending Stress (M*12/6)	68947573	psi

(**DeadLoad LiveLoad BeamSpan BeamDepth BeamWidth**)

Notice how easy this is in Mathcad!

$$\mathbf{M} := \frac{(\mathbf{DeadLoad} + \mathbf{LiveLoad}) \cdot \mathbf{BeamSpan}^2}{8}$$

$\mathbf{M} = 22500.00 \cdot \mathbf{ft} \cdot \mathbf{lbf}$

$$\mathbf{Section} := \frac{1}{6} \cdot \mathbf{BeamWidth} \cdot \mathbf{BeamDepth}^2$$

$\mathbf{Section} = 324.00 \cdot \mathbf{in}^3$

$$\mathbf{F_b} := \frac{\mathbf{M}}{\mathbf{Section}} \qquad \mathbf{F_b} = 833.33 \cdot \mathbf{psi}$$

$\mathbf{BeamStress_1} := \mathbf{BeamStress_1} \cdot \mathbf{psi}$ $\qquad \mathbf{BeamStress_1} = 6.89 \times 10^7 \cdot \mathbf{psi}$

Results are incorrect because input values were not divided by the units that Excel was expecting.

$\mathbf{BeamStress_2} :=$

Calculate bending stress in a solid beam		
Dead Load lbf/ft	300	lbf/ft
Live Load lbf/ft	500	lbf/ft
Beam Span (ft)	15	ft
Beam depth (inches)	18	inches
Beam width (inches)	6	inches
DL Moment (W*L^2/8)	8437.5	ft*kip
LL Moment (W*L^2/8)	14062.5	ft*kip
Total Moment (DL+LL)	22500	ft*kip
Section Modulus b*d^2/6	324	in^3
Bending Stress (M*12/6)	833.3333	psi

$$\left(\frac{\mathbf{DeadLoad}}{\mathbf{plf}} \quad \frac{\mathbf{LiveLoad}}{\mathbf{plf}} \quad \frac{\mathbf{BeamSpan}}{\mathbf{ft}} \quad \frac{\mathbf{BeamDepth}}{\mathbf{in}} \quad \frac{\mathbf{BeamWidth}}{\mathbf{in}}\right)$$

Remember: If you change any of the input cells within Excel, Mathcad will overwrite them once you click outside of the Excel component.

$\mathbf{BeamStress_2} := \mathbf{BeamStress_2} \cdot \mathbf{psi}$ $\qquad \mathbf{BeamStress_2} = 833.33 \cdot \mathbf{psi}$

Note that this result is correct even though beam depth and width were input in feet.

FIGURE 20.11 Inserting an Excel component from a file

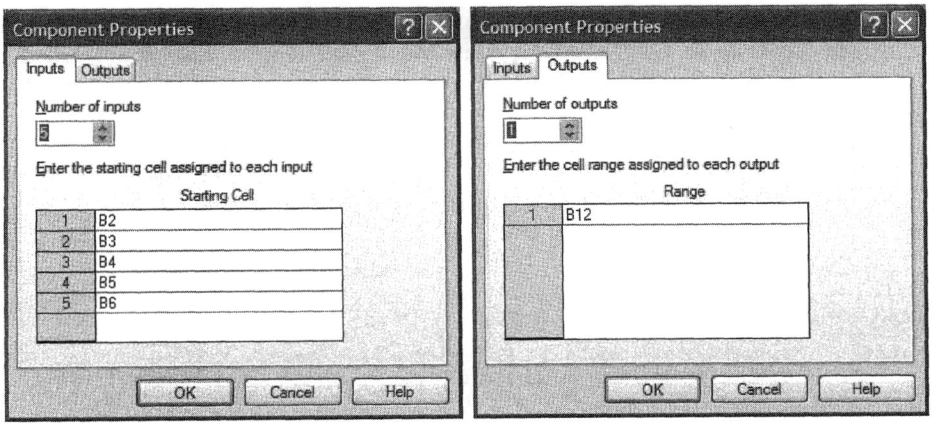

FIGURE 20.12 Component properties from Figure 20.11

Example

Figure 20.11 is an example of an existing Excel spreadsheet being brought in Mathcad. The spreadsheet calculates the uniform moment on a beam, and then calculates the bending stress in a solid beam. The first example returns incorrect results because the input values were not divided by the expected units. The second example returns the correct results. The input and output properties are shown in Figure 20.12.

SUMMARY

The Excel component makes the transition to Mathcad much easier because you do not have to start over. You can easily bring your existing Excel spreadsheets into your project calculations. You can also create new Excel spreadsheets in your project calculations. The two programs can work together by exchanging data back and forth.

In Chapter 20 we:

- Learned how to insert a blank Excel worksheet into Mathcad.
- Learned how to reference the Excel cells to transfer information into Excel and extract information from Excel.
- Showed how to get the proper values into Excel when using Mathcad units.
- Discussed ways of attaching units to Excel output.
- Explained how to bring in your existing Excel files into Mathcad including:
  - How to size the region and worksheet
  - How to reference the correct cell addresses

   - How to save and print the component
- Encouraged you to protect your spreadsheet prior to inserting it into Mathcad.
- Warned about manually changing variables once you have added Mathcad inputs.
- Warned about mistakes that can occur if you do not divide your input by the proper units.
- Provided an example to illustrate the concepts.

PRACTICE

Additional problems and applications can be found on the companion site: www.elsevierdirect.com/9780123747839.

1. From your field of study, write five Excel components. Add input from Mathcad, and provide output to Mathcad. Use units.
2. Bring two existing Excel worksheets into Mathcad. Be sure that Mathcad inserts information in the correct cells. List items that need to be considered when using units.

CHAPTER

Inputs and Outputs

21

It is important to have a simple way of organizing and inputting design criteria and assumptions into Mathcad. Calculations for engineering projects can become rather large. Having a clear way to identify input and output information is very useful. This chapter will explore different ways to include input information in Mathcad worksheets. It will also give suggestions on ways of making results stand out.

Chapter 21 will:

- Discuss the need to highlight the calculation input variables to make them stand out from other variables.
- Discuss various ways of inputting information.
- Discuss the various ways of making output information stand out.
- Introduce the concept of Mathcad controls.
- Explain how to use Web Controls for calculation input.

EMPHASIZING INPUT AND OUTPUT VALUES

Input

In project calculations, some regions are more important than other regions. Regions that require user input can often get lost amid the surrounding regions. It is very helpful to have a consistent means of identifying the variable definitions that need to be input or changed by the user of the worksheet. Some variables are input at the beginning of a project and should not be changed. You should have a means of protecting these variables.

There are various means available to make variables stand out. The easiest method is to highlight the region. This places colored shading inside the region. The default color is yellow, but you can change the color. If each region that requires user input is shaded, it is very easy to quickly scan a worksheet and review the input values. To highlight a region, right-click the region, and select

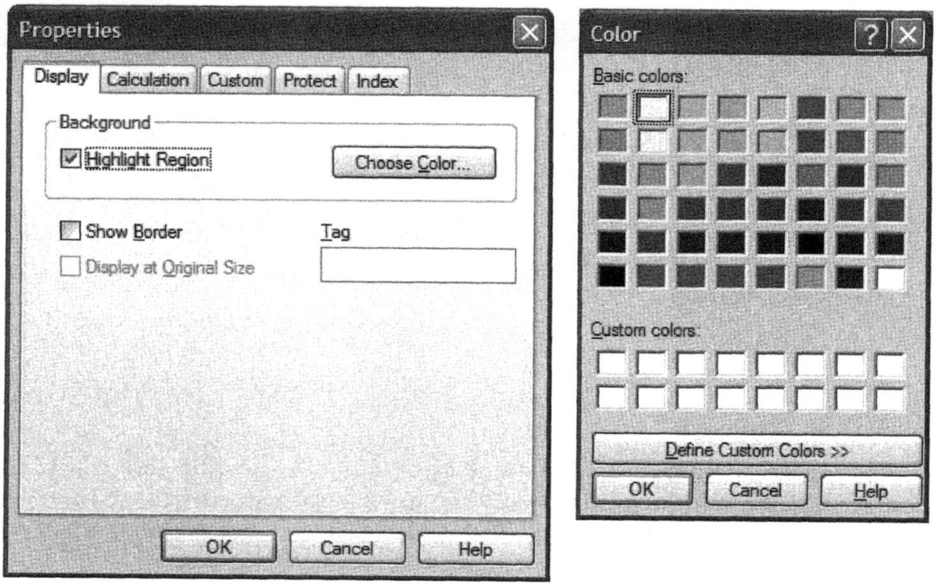

FIGURE 21.1 Highlight region

Properties. This opens the Properties dialog box. See Figure 21.1. Place a check mark in the "Highlight Region" check box. If you click the **OK** button, the default yellow color is used. If you click the **Choose Color** button, then the Color dialog box opens. You can now select one of the basic colors, or you can define your own custom color. You can change the default color by selecting **color > highlight** from the **Format** menu.

Another method of making an input region stand out is to put a box around the region. To do this, right-click in the region, open the Properties dialog box and click the "Show Border" box. This places a black border around the region.

You can also place a graphics symbol adjacent to regions requiring user input. Figure 21.2 provides some examples of ways to highlight input.

Output

You can highlight output in the same way as your input. In some ways, it is even more important to emphasize your output more than your input. In your project calculations, you will have numerous equations, formulas, results, and conclusions. It helps when reviewing the calculations to quickly identify the important conclusions or results.

Figure 21.3 provides some examples of ways to highlight results.

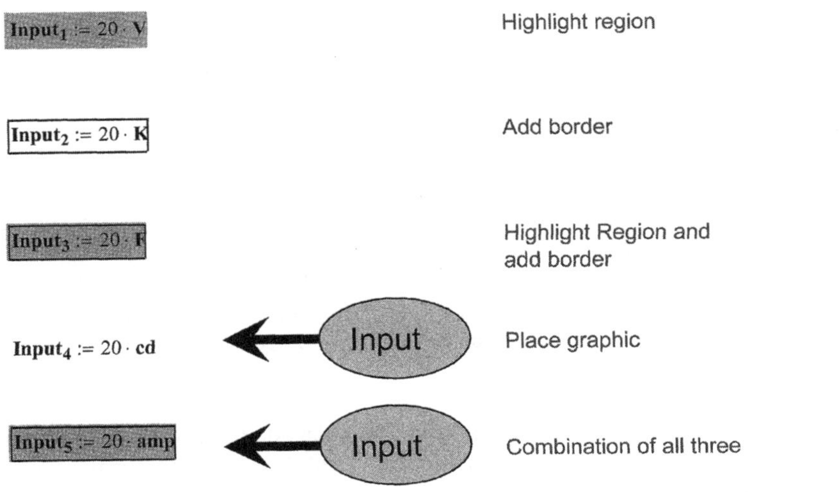
Input1 := 20 · V
Highlight region
Input2 := 20 · K
Add border
Input3 := 20 · F
Highlight Region and add border
Input4 := 20 · cd
Input
Place graphic
Input5 := 20 · amp
Input
Combination of all three

FIGURE 21.2 Ways to emphasize input

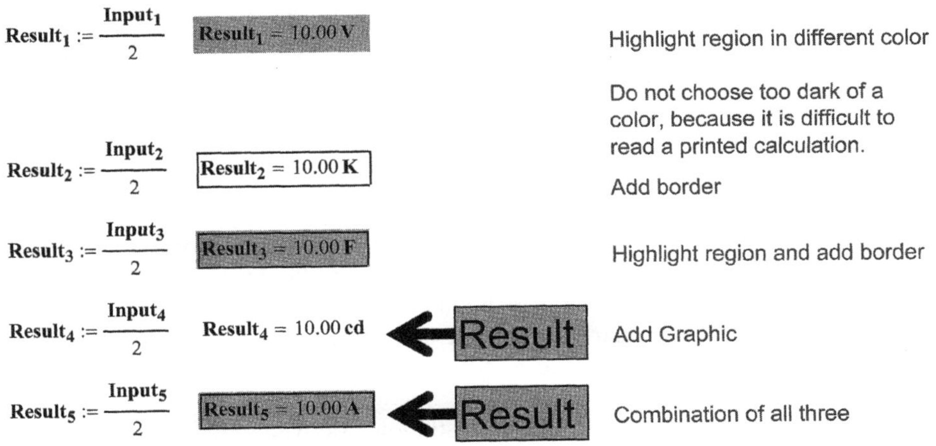
Result1 := Input1 / 2
Result1 = 10.00 V
Highlight region in different color
Do not choose too dark of a color, because it is difficult to read a printed calculation.
Result2 := Input2 / 2
Result2 = 10.00 K
Add border
Result3 := Input3 / 2
Result3 = 10.00 F
Highlight region and add border
Result4 := Input4 / 2
Result4 = 10.00 cd
Result
Add Graphic
Result5 := Input5 / 2
Result5 = 10.00 A
Result
Combination of all three

FIGURE 21.3 Ways to emphasize results

PROJECT CALCULATION INPUT

There are many different ways to organize the input variables needed for your project calculations. For our discussion, let's establish two different types of variables: Project Variables and Specific Variables. Project Variables will be variables that will be used repeatedly throughout the project calculations. These variables include things such as design criteria, material strengths, loading criteria, design parameters, and more. Specific Variables will be variables necessary to calculate a specific element in the calculations. These variables can be used in several calculations, but are not applicable to the entire project.

The easiest method of inputting variables is to create a variable whenever and wherever it is needed in the project calculations. This does not require much planning or forethought. Just start working and whenever you need a variable input, just create it. This method works, but if your project calculations are very long, it makes it difficult to find the location where the variable was defined. It also makes it more likely that a variable will be redefined or defined again with a slightly different name.

If all your Project Variables are placed at the top of your worksheet, these will be easy to find and will be prominently displayed. They will not become hidden in your calculations. It also causes you to think about the organization of your project calculations.

If your project calculations will include several different Mathcad files, then create a specific project input file. All other project worksheets should reference this file. (See Chapter 17 for a discussion of the ***reference*** function.) The beginning of each of these worksheets should display, but not redefine, the Project Variables. This way, all project worksheets will be using the same project criteria. It might be a good idea to protect the project input file from being changed accidentally.

VARIABLE NAMES

We have had several discussions about variable names. It is important to be consistent in how you name your variables. They need to be descriptive, but not too long. They also need to be easy to remember. The names need to be able to distinguish between very similar values. For example, if you have different strengths of steel, you will need to create a different variable name for each different strength of steel.

Literal subscripts are very useful to distinguish between closely related input variables. There are some cases where array subscripts will be even more useful. (Remember that array subscripts will create an array and are defined using the [key.) Figure 21.4 gives an example where array subscripts will be more useful than literal subscripts. By using array subscripts in this example, it is possible to sum all the variables and get a total.

CREATING INPUT FOR STANDARD CALCULATION WORKSHEETS

A standard calculation will be used repeatedly, so it is important to consider which variables will be changed when the worksheet is reused. This section

Let's suppose you are inputting and calculating the loads on the roof of a structure.

Using Literal Subscripts		**Using Array Subscripts**
Roofing material	$\mathbf{DeadLoad_1 := 10psf}$	$\mathbf{DL_1 := 10psf}$
Ceiling weight	$\mathbf{DeadLoad_2 := 5psf}$	$\mathbf{DL_2 := 5psf}$
Piping	$\mathbf{DeadLoad_3 := 3psf}$	$\mathbf{DL_3 := 3psf}$
Roof equipment	$\mathbf{DeadLoad_4 := 7psf}$	$\mathbf{DL_4 := 7psf}$

$\mathbf{TotalWeight := DeadLoad_1 + DeadLoad_2 + DeadLoad_3 + DeadLoad_4}$

$\mathbf{TotalDL := \sum DL}$

The summation operator is on the Matrix menu.

$\mathbf{TotalWeight} = 25.00 \cdot \mathbf{psf}$

$\mathbf{TotalDL} = 25.00 \cdot \mathbf{psf}$

When using array subscripts, you are actually making the array variable DL larger each time you add a new variable. Let's illustrate using a new variable name.:

$\mathbf{Dead_1 := 10psf}$ $\quad \mathbf{Dead} = (10.00) \cdot \mathbf{psf}$

$\mathbf{Dead_2 := 5psf}$ $\quad \mathbf{Dead} = \begin{pmatrix} 10.00 \\ 5.00 \end{pmatrix} \cdot \mathbf{psf}$

$\mathbf{Dead_3 := 3psf}$ $\quad \mathbf{Dead} = \begin{pmatrix} 10.00 \\ 5.00 \\ 3.00 \end{pmatrix} \cdot \mathbf{psf}$

$\mathbf{Dead_4 := 7psf}$ $\quad \mathbf{Dead} = \begin{pmatrix} 10.00 \\ 5.00 \\ 3.00 \\ 7.00 \end{pmatrix} \cdot \mathbf{psf}$

FIGURE 21.4 Using array subscripts

discusses the different methods available for inputting information into your standard calculation worksheets.

We have discussed locating these types of variables at the top of the worksheet, so that they are all accessible in one place. This way, you do not need to scan the worksheet looking for variables to change. It also allows you to protect or lock the remainder of your worksheet to prevent unwanted changes.

Let's look at different methods you can use to create input for your standard calculation worksheets.

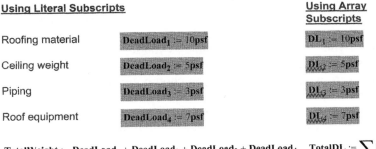

Inputting Information from Mathcad Variables

The most common way of inputting information into your Mathcad worksheet is by creating variable definitions. You can then click the variable definition value and change the value. This has a big advantage over other methods, because you can input your units at the same time as you input your data. If your input has the units of length, you can attach units of meters, mm, feet, or inches. Mathcad does not care what unit of length you choose. If you use another method of input, as will be shown in the next section, it is critical to input the data with specific units of lengths. For example, if your data is input in a data table or in Microsoft Excel, it is critical to know if you need to input meters, mm, feet, inches, or some other length unit. You will need to do the conversion before inputting the value.

Another advantage of inputting information from Mathcad is that it usually takes less space. Your definitions are already made. When you use a data table or Microsoft Excel, you might need to reassign or redefine the variable in order to attach units to it. This adds more variable definitions to your worksheet.

A disadvantage of inputting information by individual variable definition is that it is sometimes much more time consuming. You will need to click every definition and change the value. If you use a data table or Microsoft Excel, you can quickly input large amounts of data without using your mouse to click each input variable. If you have large amounts of data to input or change, you might want to consider using a data table or Excel.

Data Tables

If you have considerable input that does not have units attached, or that has consistent units, a data table is a useful way to input and change many variables very easily. A data table creates an array, thus every cell in the table has a unique variable name using array subscripts.

The data table allows for easy input because you can use the **ENTER** key to move to the next input cell. You can also use the up and down arrows to scroll through the input. You do not need to click every variable and change it, as you would have to do in Mathcad. You can also include a column of text to describe what values to input.

The biggest disadvantage to using a data table is that you cannot attach units to the input. All units are attached after the data is exported to Mathcad. This can lead to errors if the proper values are not entered. Another disadvantage to using a data table is that all the output is contained in a single variable matrix. There are ways of taking values from this matrix and assigning them to specific variables. Figure 21.5 gives two examples of how you can take the input data from the data table and assign it to variable names and attach units to the input.

$\mathbf{Table_1} :=$

	1	2
1	20	30
2	10	40

$$\mathbf{Table_1} = \begin{pmatrix} 20.00 & 30.00 \\ 10.00 & 40.00 \end{pmatrix}$$

$\mathbf{Resistance_1} := \mathbf{Table_1}_{1,2} \cdot \Omega$ $\qquad \mathbf{Resistance_1} = 30.00\,\Omega$

$\mathbf{Voltage_1} := \mathbf{Table_1}_{2,2} \cdot \mathbf{V}$ $\qquad \mathbf{Voltage_1} = 40.00\,\mathbf{V}$

After data is input, you can use Mathcad to reassign names and to add units.

Note: The data table below has font and number format changes made to the default data table.

$\mathbf{Table_2} :=$

	1	2
1	"Roofing Material"	30.0000
2	"Ceiling Weight"	5.0000
3	"Piping"	5.0000
4	"Roof Equipment"	10.0000
5		

$$\mathbf{Table_2} = \begin{pmatrix} \text{"Roofing Material"} & 30.00 \\ \text{"Ceiling Weight"} & 5.00 \\ \text{"Piping"} & 5.00 \\ \text{"Roof Equipment"} & 10.00 \end{pmatrix}$$

$\mathbf{DeadLoad} := \mathbf{Table_2}^{\langle 2 \rangle} \cdot \mathbf{psf}$

Use the matrix column operator [CTRL+6] to extract the column data from the matrix "$Table_2$." Also attach units to the data.

$$\mathbf{DeadLoad} = \begin{pmatrix} 30.00 \\ 5.00 \\ 5.00 \\ 10.00 \end{pmatrix} \cdot \mathbf{psf}$$

$$\mathbf{TotalDeadLoad} := \sum \mathbf{DeadLoad}$$

$$\mathbf{TotalDeadLoad} = 50.00 \cdot \mathbf{psf}$$

FIGURE 21.5 Using data tables for input

You can format the font, number format, trailing zeros, and tolerances of the values in a data table. To do this, single-click the table, then right-click and select **Properties**.

Another thing to be aware of with data tables is that if you enter a value in a cell, that cell will always retain a value. You can change the value in the cell, but you cannot delete the cell. Mathcad reads all cells with values (including zero) in a data table. Thus, if you add unnecessary cells in your data table, your input array will also contain unnecessary values. For this reason, it is usually best to use a Microsoft Excel component (if you have Excel loaded on your computer) rather than using a data table.

MICROSOFT EXCEL COMPONENT

A Microsoft Excel component can also be used to input values into Mathcad. This is very similar to a data table, except that you now have the full functionality of Microsoft Excel. You can:

- Format and highlight individual cells.
- Protect specific cells.
- Use lookup tables.
- Use controls such as drop-down boxes.
- Adjust the width of each column.
- Place borders around specific cells.

Another advantage of Excel over a data table is that you can assign variable names to each cell in the Excel component rather than having a single variable name, as required for a data table.

Figure 21.6 gives an example of using Excel to input values. Notice the following in this example:

- Some cells are shaded.
- The input cells are highlighted.
- Different fonts are used.
- Bold is used in some columns.
- The column widths are different.
- A border is used.
- There are four outputs associated with the four input values.
- Units had to be attached outside of the component definition.

In this example, the spreadsheet also is protected. You can change only the input values; the other cells are locked. It is also critical to input the proper units. For example, the pipe diameter must be input as mm, not meters.

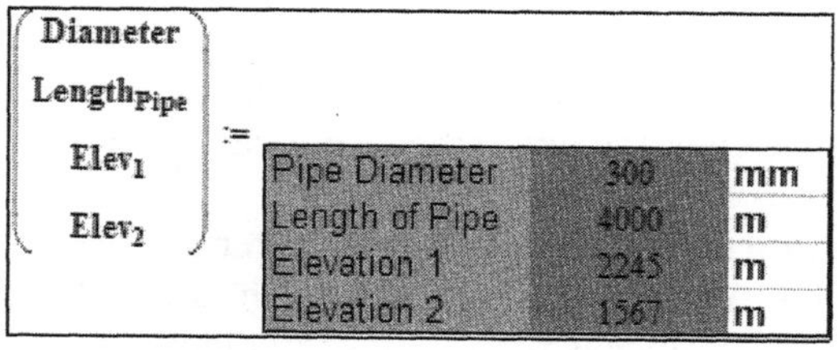

With Excel, you can highlight cells that require input, center the data, add borders and shading, adjust column widths, etc.

$\text{Diameter} := \text{Diameter}\cdot \text{mm}$ $\quad \text{Length}_{\text{Pipe}} := \text{Length}_{\text{Pipe}}\cdot \text{m}$ $\quad \text{Elev}_1 := \text{Elev}_1\cdot \text{m}$ $\quad \text{Elev}_2 := \text{Elev}_2\cdot \text{m}$

$\text{Diameter} = 0.30\,\text{m}$ $\quad \text{Length}_{\text{Pipe}} = 4000.00\,\text{m}$ $\quad \text{Elev}_1 = 2245.00\,\text{m}$ $\quad \text{Elev}_2 = 1567.00\,\text{m}$

FIGURE 21.6 Using Microsoft Excel for input

Which Method Is Best to Use?

Which input method is the best method to use for standard calculation worksheets? It depends. Data tables are not a good option if you have Microsoft Excel available. There are advantages to each method. The answer to the question depends on how much input is necessary. Does the extra work of creating an Excel input component outweigh the extra time of clicking and changing individual variable definitions? The answer also depends on whether you want the flexibility of inputting data using different units, rather than being forced to use a specific unit as you would be required to do in Excel.

The examples used in Figures 21.5 and 21.6 are very short. Your input would need to be much more extensive before an Excel input component would be worthwhile. The examples are only for demonstration.

SUMMARIZING OUTPUT

If you have a long calculation worksheet, you might want to summarize the results at the end of the worksheet.

You can display each result again by typing the variable name followed by the equal sign. By listing all the variables, you will be able to see a summary of the results. You might also want to use one of the method discussed earlier to highlight the results.

You can also put the results into an Excel spreadsheet. This, however, is much more work. In order to do this, you will need to create an Excel component, list each result variable as an input variable, and then tell Mathcad into which Excel cell to put it. You will also have to divide the result by the proper unit so that Excel displays the proper value. This is quite a bit of work, but there might be a time when having the results in Excel outweighs the effort of creating the output component.

CONTROLS

Controls are programmed boxes and buttons that you can add to your worksheets to help automate or limit input. They are best used in worksheets that will be developed by one person and used by another, or in worksheets that will be used repeatedly. Controls are useful in some standard calculation worksheets because you can limit input to just a few selected items. You can also use controls to have the user select if a specific condition is met.

The standard Mathcad controls use a scripted language. If you are familiar with VBScript or Jscript, these controls will be easy for you to learn and to use. They are very powerful and useful. If you are not familiar with scripting, the Mathcad Developer's Reference provides a good description of how to use the standard controls.

For those who do not want to use or learn scripting, Version 12 of Mathcad added what are called Web Controls. The Web Controls have been prescripted, and are easy to use. They were developed for web-based document servers such as the Mathsoft Application Server, but they work fine in Mathcad. They are limited in their application, but can still be useful to those who do not want to use scripted language.

Web Controls are added by selecting **Controls > Web Controls** from the **Insert** Menu. This opens the Web Control Setup Wizard. See Figure 21.7.

Web Controls do not require input values, and each control provides one output value that is assigned to a Mathcad variable. Thus, Web Controls give a means of providing input to your calculations. Let's look at each of the different Web Controls provided in Mathcad.

Text Box

The Text box control provides a box where the user can input text. You can set the limit on how many characters are visible, and you can also limit how many characters can be entered. The Mathcad limit is 255 characters. The output from a text box is a text string. This text string is assigned to a variable.

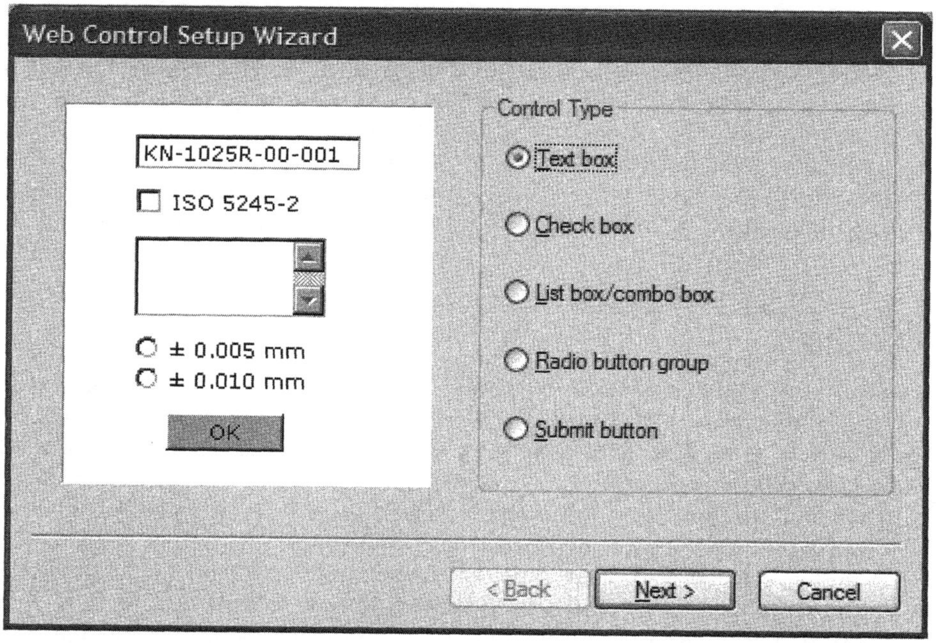

FIGURE 21.7 Web Controls

When would a Text box be used?

- You could have the user input his or her name. This name could then be displayed as a text string anywhere in the calculations.
- You could have the user input today's date. This date could then be added to the calculations so that you know when the calculations were modified.
- If a teacher were creating a test, the text string could be used to ask the student to respond to a question.

Figure 21.8 gives an example of how to create a text string and how to apply it.

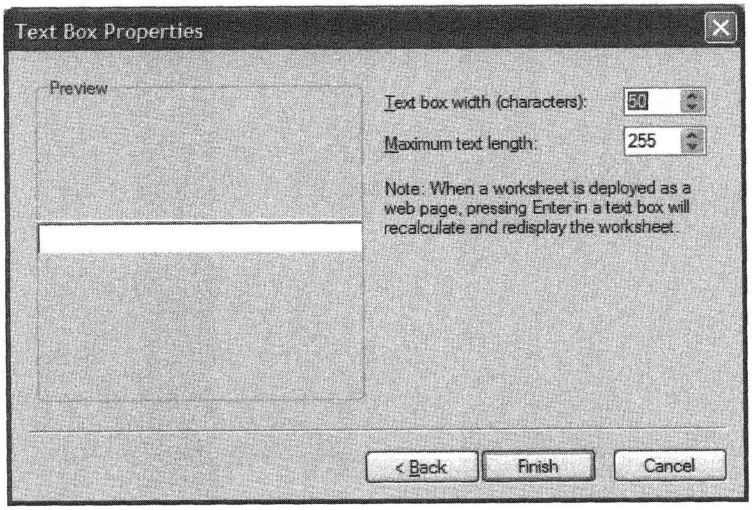

Let's suppose that you have a project calculation input file that all project files reference. The following could be added to the top of the project input file.

Enter your name in the text box below. Enter today's date in the text box below.

Name := Brent Maxfield   Date := October 13, 2008

Name = "Brent Maxfield"   **Date** = "October 13, 2008"

The following statement could be added to the top of each worksheet that references the input file.

The input file was last modified by **Name** = "Brent Maxfield" on **Date** = "October 13, 2008" .

Note: Use the Math Region from the Insert menu to add math regions to a text box.

If you are creating a test you could have similar boxes where students can input their names and the date that the test was taken.

FIGURE 21.8 Text box

Check Box

The Web Controls check box only allows for a single check box in the control. The output is a 1 if the box is checked, and a 0 if the box is unchecked. This makes the box useful to ask if a specific condition is met. You can then use a logical program to determine the result.

See Figure 21.9 for an example.

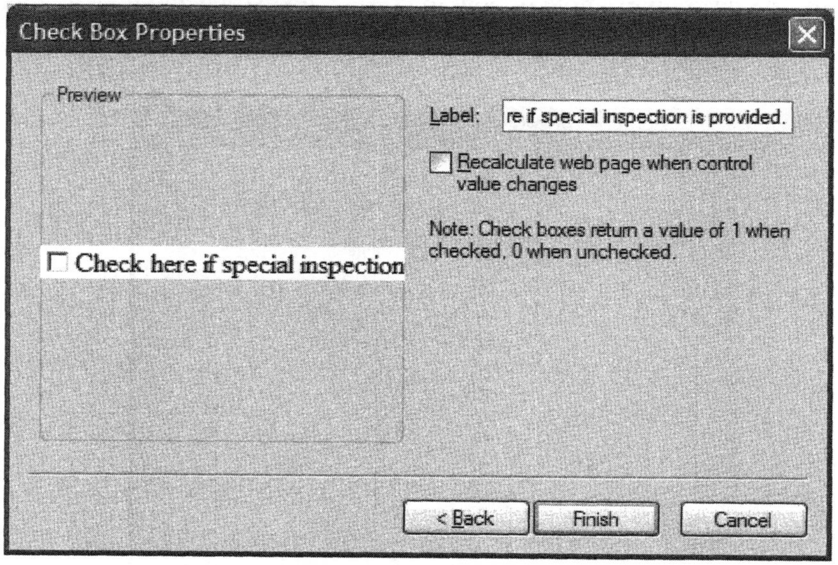

Some older building codes required that only one-half of the allowable stresses could be used in masonry design if special inspection was not used during construction You can create similar check boxes with associated logic programs.

SpecialInspection := ☑ Check here if special inspection is provided.

SpecialInspection = 1.00

$\mathbf{F_v} := 50\mathbf{psi}$

$$\mathbf{AllowableShearStress} := \begin{cases} \mathbf{F_v} & \textbf{if SpecialInspection} = 1 \\ \dfrac{\mathbf{F_v}}{2} & \textbf{otherwise} \end{cases}$$

AllowableShearStress = 50.00 · **psi** The value of this variable changes depending on whether or not the check box has a check in it.

FIGURE 21.9 Check box

List Box

A List box allows the user to select from a list of items. The list can include up to 256 items. The selected item is returned to Mathcad as a variable.

The way a list box appears in the worksheet depends on how many rows are displayed. You can tell Mathcad how many rows to display. If the display is set to show one row, the box displays as a standard combo box where all the items appear when you select the down arrow. This condition is illustrated in Figure 21.10.

If the display is set to two rows or higher, the list appears as a scrolling list box if there are more items than can be displayed. If the display is set to show more rows than are in the list, the box shows empty unselectable lines below the list. The scrolling list box is illustrated in Figure 21.11.

Radio Box

A radio box is similar to a list/combo box. Both types of boxes select an input from a list. The list/combo box item is selected by selecting the item from a list. A radio box has small circles to the left of the listed items. You select an item by clicking the circle adjacent to the item.

Figure 21.12 gives an example of a radio box.

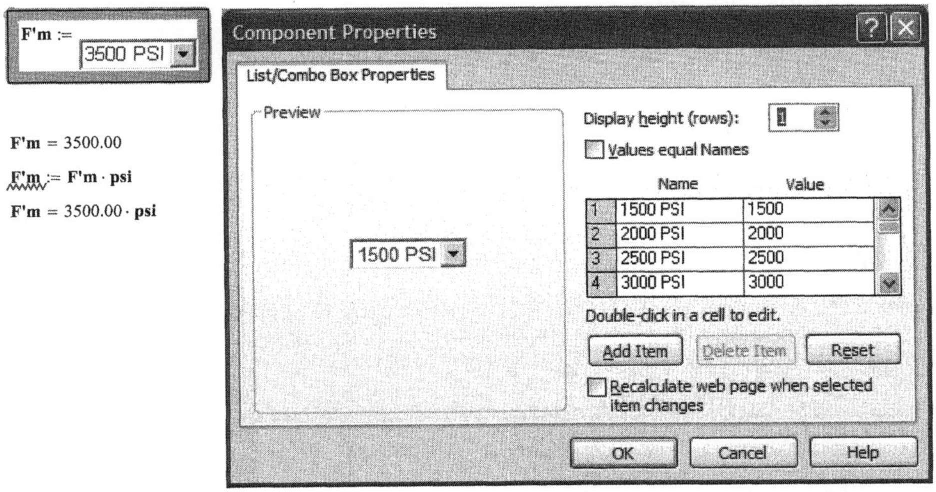

FIGURE 21.10 List box

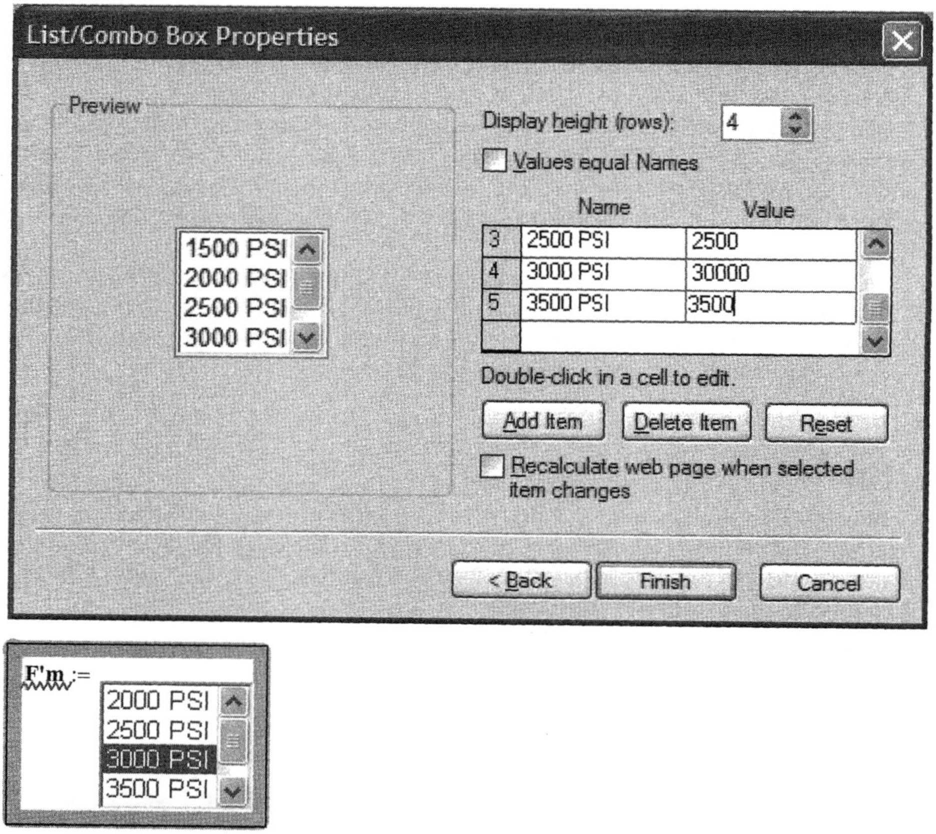

$\mathbf{F'm} = 3000.00$ $\mathbf{F'm} := \mathbf{F'm} \cdot \mathbf{psi}$ $\mathbf{F'm} = 3000.00 \cdot \mathbf{psi}$

FIGURE 21.11 List/combo box

Submit Box

The submit button is used only when a worksheet is published on the Web. It does not have any affect for normal worksheets. See Figure 21.13.

SUMMARY

This chapter discussed the various way of getting information into your Mathcad worksheets and how to make your critical values stand out from the rest of your worksheet.

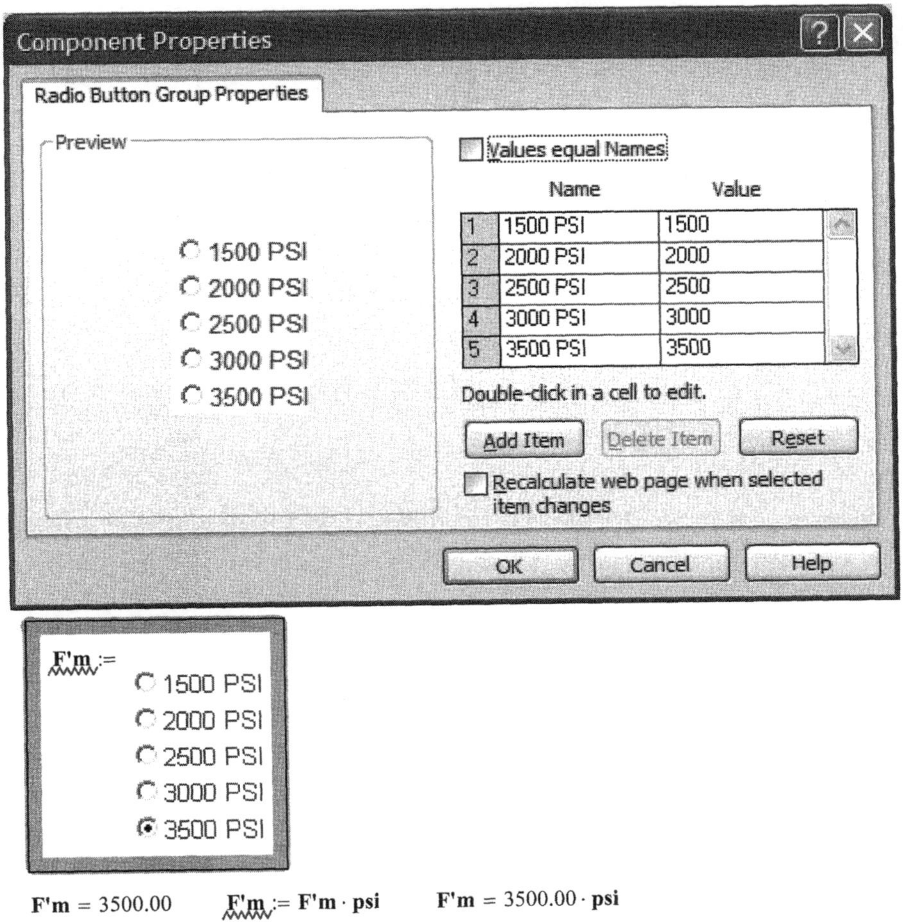

FIGURE 21.12 Radio button

In Chapter 21 we:

- Showed how to highlight input variables so they will be easy to see.
- Gave other ways to identify input variables, such as using borders or placing a graphic adjacent to the variables.
- Suggested using different means to highlight results so that the critical results are clearly identified.
- Discussed having a project calculation input file, and emphasized the need to protect the file, and reference it from other project files.

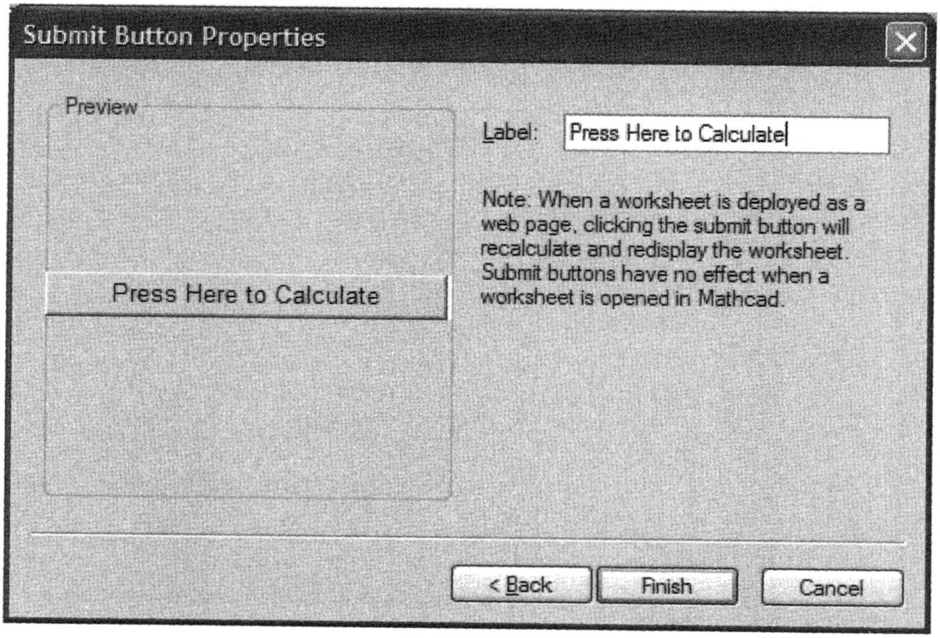

Calculate = 3.00

FIGURE 21.13 Submit button

- Showed how to use array subscripts for input values that will be summed.
- Explained the various methods for inputting values into a worksheet.
- Discussed the advantages and disadvantages of using Microsoft Excel to input variables.
- Introduced Mathcad Controls and encouraged you to research more in the Mathcad Developer's Reference.
- Showed how to use the prescripted Web Control in standard calculation worksheets.

CHAPTER

Hyperlinks and Tables of Contents

22

A hyperlink is a special region (text or graphic) that when double-clicked, opens the document or location referenced by the hyperlink. You can create hyperlinks that will take you to regions within the current worksheet, regions within other worksheets, a Web page on the Internet, or even a file created by another program.

By using hyperlinks, you can refer the user of your worksheet to reference material, other worksheets, or locations within your current worksheet. By using hyperlinks, you are able to create a Table of Contents and link all files in your project calculations. This will be illustrated shortly.

Chapter 22 will:

- Define hyperlinks.
- Show how hyperlinks are created.
- Discuss the use of hyperlinks in text regions and nontext regions.
- Encourage the use of hyperlinks only in text regions.
- Explain the use of relative path.
- Tell how to create a Table of Contents.
- Explain Author's Resources.
- Introduce how to create a calculation E-book.

HYPERLINKS

A hyperlink is created from selected text in a text box or from a graphic image. You can select the entire content of a text box or only a single character. Once you have selected the text or graphic you want to become your hyperlink, select **Hyperlink** from the **Insert** menu. You can also right-click and select **Hyperlink** from the pop-up menu. This opens the Insert Hyperlink dialog box. In Figure 22.1, the link is to the Mathcad web site. Be sure to include the http: before the www. You can add a text message that will appear at the bottom of your screen in the status line when your mouse is placed over the hypertext.

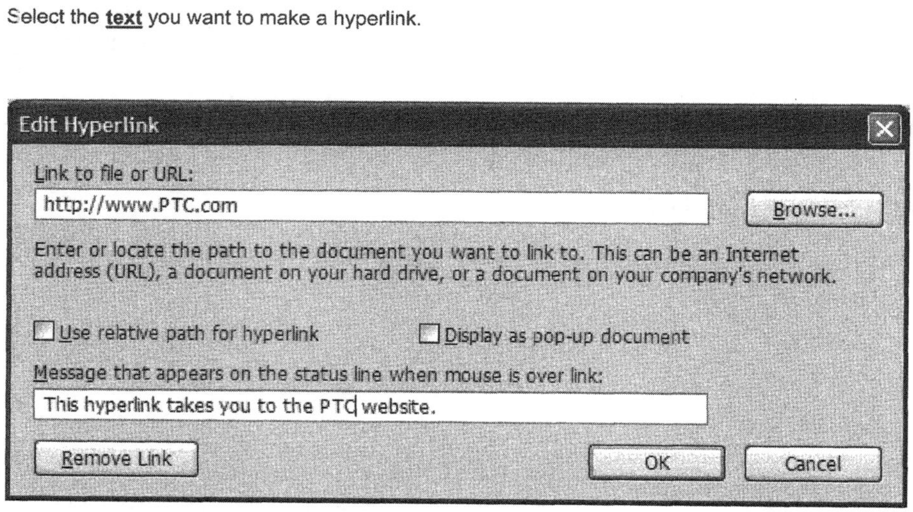

FIGURE 22.1 Inserting a hyperlink

When you click **OK** to close the box, the selected text becomes bold and underlined. When you place your cursor over the hyperlink text, the cursor changes to a hand symbol. You need to double-click to activate the hyperlink. If you used a graphic as your hyperlink, there is no change in the appearance of the graphic, but if you place your cursor over the graphic, the cursor changes to a hand icon.

> **Tip!** You might want to highlight your graphic regions with a color, so you can easily identify the graphics with hyperlinks attached.

The hyperlink can be to any location where you can navigate on your computer. If you link to a file that has a software program assigned to the file type, the file will open in the associated program. If you link to a Mathcad worksheet, the worksheet will open and your cursor will be placed at the top of the worksheet.

Linking to Regions in Your Current Worksheet

You can also link to a region in your current worksheet or to a region in another worksheet. In order to link to a specific region, you need to assign a tag to the region to which you want to link. This is similar to attaching a bookmark in a word processing program. To do this, right-click the region and select **Properties**. On the Display tab, add a name or phrase in the Tag box. A hyperlink will refer to this tag name. The tag name cannot have a period in it such as "Voltage. 1."

Tip! There is no way to browse to a region, so you must remember the exact name of a tag. For this reason, keep your tag names simple and easy to remember.

Once you have a tag name assigned to a region, you can create a hyperlink to that region. To do this, highlight the chosen text or select a graphic, and open the Insert Hyperlink dialog box. In the Link to file or URL: box, type the # symbol followed by the name of the tag. See Figure 22.2 for an example.

Linking to Region in Another Worksheet

If you want to link to a region in another worksheet, open the Insert Hyperlink dialog box. The "Use Relative Path for Hyperlink" box is checked by default. Keep this box checked unless the file you are referring to is a permanent location. If you intend to copy or move the current worksheet and the hyperlinked worksheet, you will want to use the relative path. Click the **Browse** button to browse to the worksheet, and then click **Open**. If the "Use Relative Path for Hyperlink" box is checked, and the file is in the same folder, the path name will not be added. If the file is in a different folder, or the relative path box is not checked, a path will be added. Once the filename is added, type a # and the name of the tag. See Figure 22.3.

When you double-click the hyperlink, the worksheet will be opened and the linked region will appear at the top of your screen.

Notes about Hyperlinks

It is important to note that when you link to a region, the screen moves to the region, but the cursor does not. If you link to a region in your current worksheet, the cursor remains in the hyperlink. If you link to a region in another worksheet, the cursor is moved to the top of the worksheet, even though the region is visible.

It is possible to create a hyperlink from a math region or from a plot region. Do not do this. It is much better to use a text box or a graphic. For example, if you create a hyperlink from a plot, when you double-click the plot, it opens the link associated with the hyperlink, rather than opening the Plot Format dialog box. A hyperlink in a math or plot region does not appear any different, so it is also difficult to distinguish the hyperlink. When you move your cursor over a math or plot region that has a hyperlink assigned, the cursor changes to a hand symbol.

To remove a hyperlink, right-click the hyperlink, select **Hyperlink**, and click the **Remove Link** button. This will remove the link and associated bold and underlined text. See Figure 22.4.

Create a tag in the following region. Right-click the region and add a name in the Tag box. Do not use a period "." in the tag name.

$\mathbf{Voltage_1 := 900amp}$

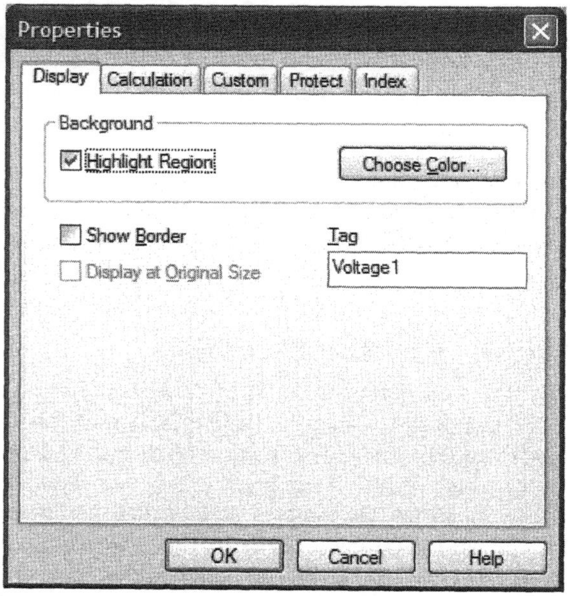

The above cell region now has a tag called, "Voltage1." You can now hyperlink to this region.

Hyperlink to **Voltage1**

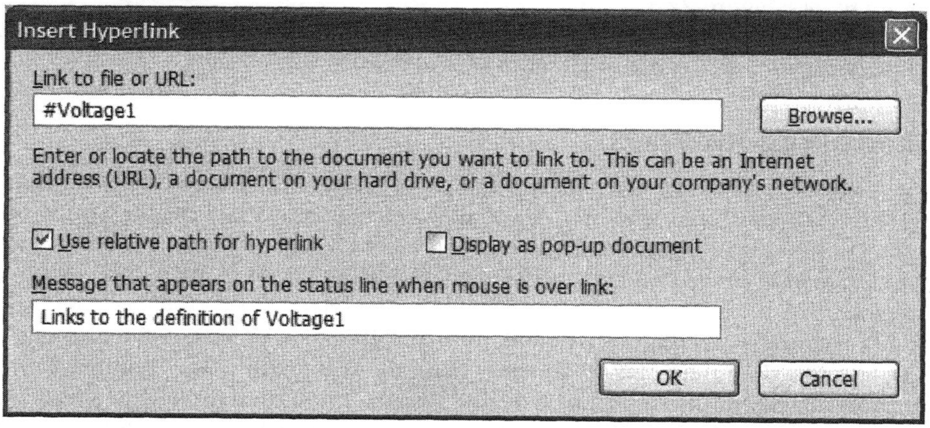

FIGURE 22.2 Adding tags to regions

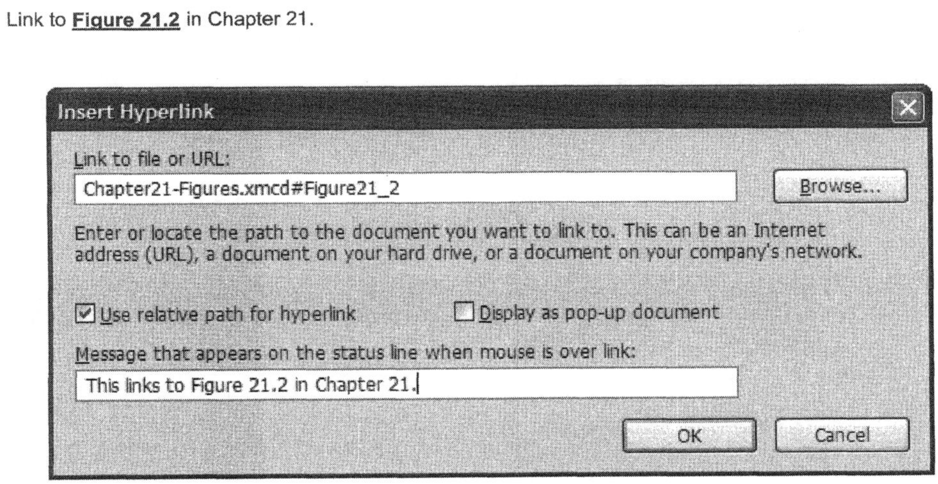

FIGURE 22.3 Hyperlink to another worksheet

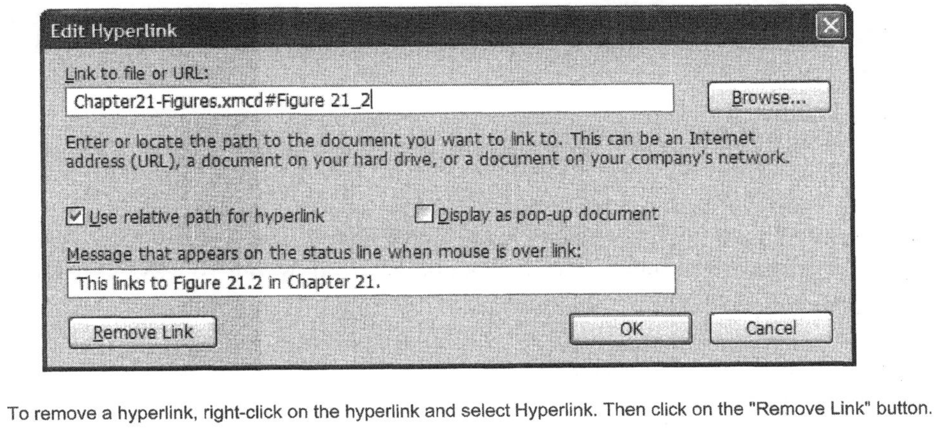

FIGURE 22.4 Removing a hyperlink

Creating a Pop-Up Document

A pop-up document is one that appears in a separate smaller window. A Mathcad worksheet pop-up will display the entire worksheet. You cannot have a pop-up of only one region of the worksheet. To create a pop-up document, click the Display as Pop-Up check box in the Insert Hyperlink dialog box. See Figure 22.5. It is best

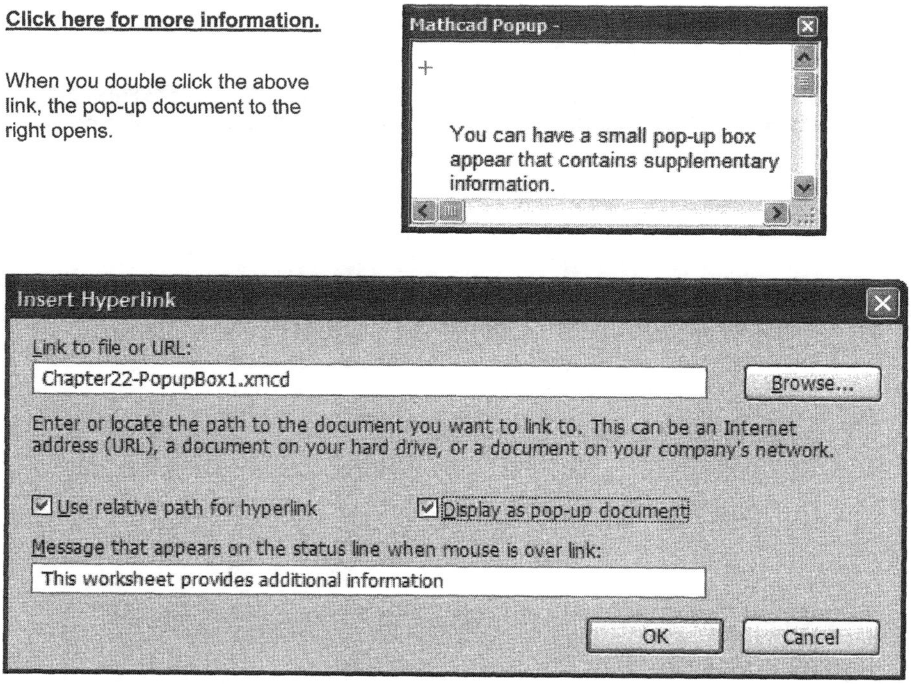

FIGURE 22.5 Pop-up windows

to have the linked pop-up worksheets be small with minimal regions. It is possible to have large worksheets appear in a pop-up window, but you would need to scroll through the window to see all the information.

TABLES OF CONTENTS

Now that you understand how to link to regions in your worksheets, it is possible to create a Table of Contents. We will discuss two types: (1) a Table of Contents for the data within your current worksheet, and (2) a Table of Contents that links to many different files in your project calculations.

To create a Table of Contents in your current worksheet, first examine your worksheet and determine where you would like to have links added. In each of these regions, add a tag to the region. Be sure to write down the tag name for each region.

You can then go to the top of your document and create a text region. The entire Table of Contents can be in a single text region. Type the text of the

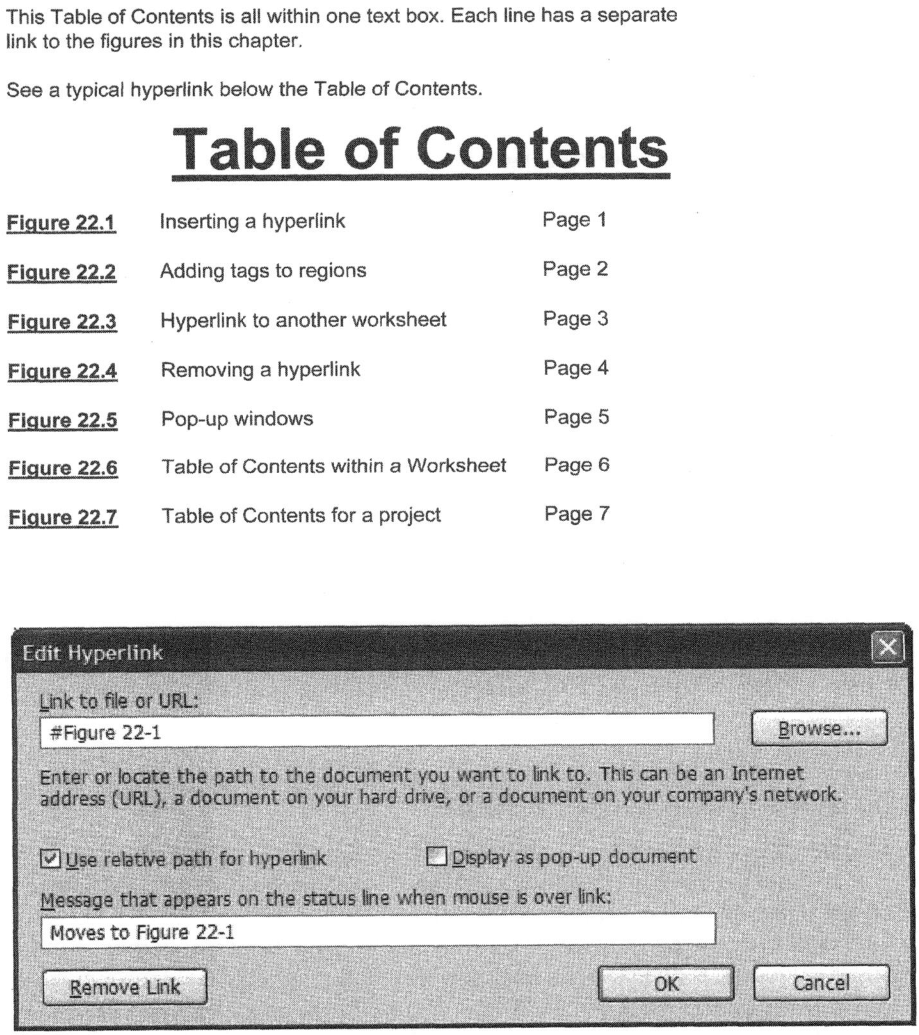

FIGURE 22.6 Table of Contents within a worksheet

hyperlinks, and then go back and add the region hyperlink to each item. See Figure 22.6 for an example.

If you have a project that includes many different worksheet files, it is very useful to have a Table of Contents file that will link to each of the different files. The Table of Contents will be very similar to Figure 22.6. The only difference is that each hyperlink will reference a different file. The project Table of

A project Table of Contents can link all project files together.

Table of Contents

Project Input Variables

Snow Loads

Beams and Joists

Columns

Wind Loads

Seismic Loads

Appendix - Computer Output

FIGURE 22.7 Project Table of Contents

Contents will list each file in the order it should be viewed or printed. It will also provide a hyperlink to each file. See Figure 22.7.

MATHCAD CALCULATION E-BOOK

A Mathcad E-book has all the information of a printed book, plus all the interactive math tools available in Mathcad. These books can be distributed to anyone with the same or higher version of Mathcad.

An E-book is a series of Mathcad files electronically bound together. You can annotate the book, and you can change the math regions to experiment with results. The original copy of the book remains unchanged, but you can save a copy with your annotations and changes. It is also possible to drag and drop regions from the E-book to other Mathcad worksheets.

It takes some effort to create an E-book, but if your calculations are going to be widely distributed, the effort might be worthwhile. The Mathcad Help and Tutorials were created from an electronic E-book.

The details for creating a Mathcad E-book are beyond the scope of this book. The Author's Resources on the **Help** menu have detailed instructions for creating a Mathcad E-book.

SUMMARY

Hyperlinks are a wonderful tool to use in creating project calculations because they allow you to link to other Mathcad calculation files, and to regions in you current worksheet.

In Chapter 22 we:

- Introduced hyperlinks.
- Discussed how to create a hyperlink.
- Showed how to link to files and Web sites.
- Showed how to create a tag in a region by using the Properties dialog box.
- Demonstrated how to link to a region in a current or other worksheet by using the # symbol following the path.
- Told about pop-up windows.
- Learned about creating a Table of Contents for your local worksheet.
- Learned about creating a Table of Contents for a project calculation.
- Briefly introduced the concept of Mathcad E-books.

CHAPTER

23 Conclusion

I hope that you have enjoyed learning the essentials of Mathcad for engineering, science, and math. I hope even more that you have followed along with your own version of Mathcad, and have practiced what you have learned. There is no substitute for hands-on practice and application.

ADVANTAGES OF MATHCAD

You might already have known about the powerful mathematical calculation power of Mathcad. After reading this book, you should now see how useful Mathcad is to do everyday scientific and engineering calculations. You can now use Mathcad for its powerful mathematical capabilities, and for its capability to help you create and organize your calculations. You can now use Mathcad as the primary tool for your project calculations.

Remember some of the key advantages to using Mathcad as the primary tool for your project calculations:

- Units! Mathcad alerts you when you are using inconsistent units. You can easily get results in SI or US units. You can work in both systems, simultaneously displaying the results in both systems.
- Your formulas and equations are visible and can be easily checked. They are not hidden in cells where only the results are visible.
- You can change an input variable, and your results are immediately updated.
- You can create standard calculations worksheets that can be used over and over again.
- Your calculations can be reused in other projects.
- Using Mathcad's metadata you can trace where regions were copied from.
- You can continue using the programs you are comfortable working with, and then store and present the results within Mathcad.
- You can archive the entire project.

CREATING PROJECT CALCULATIONS

Let's review some of the key topics essential to creating project calculations with Mathcad.

- Use the tools in your Mathcad toolbox. Keep them sharp, by reviewing the topics covered in Parts I, II, and III.
- Keep using your existing spreadsheets. Learn how to incorporate them as components in Mathcad so that Mathcad can provide input and receive output from them.
- Use the software programs that make the most sense. Each program has its strengths and weaknesses. Mathcad is not the ideal software program for all applications.
- Keep using your specialty software programs. Link or embed the results in Mathcad. Embed the input and output files in Mathcad.
- Use Mathcad to calculate input and to use the output from your other software programs.
- Use Mathcad to collect and organize the many different software files into one comprehensive project calculation.
- Link all your files together with a Table of Contents file, which contains hyperlinks to each file in the project calculations.
- Use a Table of Contents at the top of each file so that specific areas in the worksheet can easily be found.

ADDITIONAL RESOURCES

We have just scratched the surface. There is so much more that could have been discussed. Our intent was to point out some useful features, and to get you started using Mathcad for all your project calculations.

Here is a list of things you can do to increase your knowledge and Mathcad skills:

- Open the Mathcad Tutorials from the **Help** menu and review each tutorial.
- Open the Mathcad QuickSheets from the **Help** menu and review each sheet.
- Open the Reference Tables from the **Help** menu and explore the wealth of information contained in the tables.
- Review the Mathcad functions from the Insert Function dialog box. Research additional functions that are not discussed in this book.
- Visit the Mathcad User Forums on the Internet. There is a link on the **Help** menu. Explore the many different topics. Ask a question. Look for answers.

- Look for topics at the Mathcad Resource Center located at http://www.ptc.com/appserver/mkt/products/resource/mathcad/index.jsp.
- Review the list of E-books and Mathcad files available in the Mathcad Resource Center.
- Visit the Essential Mathcad website www.elsevierdirect.com/97801237478 and download the Mathcad files used for the figures used in this book.

CONCLUSION

I wish you well, as you begin to use Mathcad in your engineering and technical calculations. I encourage a continued effort to add more tools to your Mathcad toolbox. There are many tools to add that we have not discussed.

It is also important to keep your tools sharp. Tools that are not used very often get rusty. I encourage you to review the chapters in this book and try to incorporate some of the features in your calculations. Review the chapters from time to time and polish your tools so they will be sharp and ready to use when occasion permits.

I wish you the best of success as you enjoy the wonderful world of Mathcad.

Appendix 1: Keyboard Shortcuts

Keystroke Commands

Action	Example	Keystroke
Mathcad Help		[F1]
Context sensitive Help		[Shift] [F1]
Calculate worksheet		[Ctrl] [F9]
Calculate region		[F9]
Redefinition warnings (toggle on and off)		[Ctrl] [Shift] R
Insert text region		[“]
Insert math within text region		[Ctrl] [Shift] A
Addition with line break operator (within a math region)	... + ...	[Ctrl] [Enter]
Character inside brackets as in chemistry notation	[ ]	[Ctrl] [Shift] J
Complex conjugate	$\bar{x}$	["]
Enter special characters into math region		[Ctrl] [Shift] K
Literal subscript	x_1	[.]
Namespace operator		[Ctrl] [Shift] N

Calculator Toolbar

Operator	Example	Keystroke
Absolute value	|x|	[|]
Definition	x: = 5	[:]
Evaluate numerically	x = 5	[=]
Division	x/y	[/]
Multiplication	$3 \cdot 4$	[*]
Inline division	÷	[Ctrl] [/]
Exponentiation	x^2	[^]
Imaginary unit	i	1i
Parentheses	(3 + 4)	[']
Mixed number	$2\frac{1}{2}$	[Ctrl] [Shift] [=]
square root	$\sqrt{x}$	[\]
nth root	$\sqrt[n]{x}$	[Ctrl] [\]
pi	π	[Ctrl] [Shift] P

Boolean Toolbar

Operator	Example	Keystroke
Boolean AND	∧	[Ctrl] [&]
Boolean NOT	¬	[Ctrl] [!]
Boolean OR	∨	[Ctrl] [^]
Boolean XOR	⊕	[Ctrl] [%]
Boolean Equals	=	[Ctrl] [=]
Greater than or equal	≥	[Ctrl] 0
Less than or equal	≤	[Ctrl] 9
Not equal	≠	[Ctrl] 3

Calculus Toolbar

Operator	Example	Keystroke
derivative	$\frac{d}{dt}f(t)$	[?]
nth derivative	$\frac{d^n}{dt^n}f(t)$	[Ctrl] [?]
Indefinite integral	$\int f(x,y)dA$	[Ctrl] I
Definite integral	$\int_a^b f(x)dx$	[&]
Infinity	∞	[Ctrl] [Shift] Z
Iterated product	$\prod_{i=m}^{n} X$	[Ctrl] [#]
Iterated product with range variables	$\prod_{i} X$	[#]
Left-hand limit	$\lim_{x \to 0^-} f(x)$	[Ctrl] [Shift] B
Two-sided limit	$\lim_{x \to 0} \frac{\sin(x)}{x}$	[Ctrl] L
Right-hand limit	$\lim_{x \to 0^+} f(x)$	[Ctrl] [Shift] A
Summation	$\sum_{i=m}^{n} X$	[Ctrl] [$]
Summation with range variables	$\sum_{i} X$	[$]
Gradient	$\nabla_x g(x)$	[Ctrl] [Shift] G

Evaluation Toolbar

Operator	Example	Keystroke
Custom infix operator	x f y	
Custom postfix	x f	[Ctrl] [Shift] X
Custom prefix	f x	
Custom treefix operator	f ⋀ x y	
Definition	$x := 5$	[:]
Evaluation	$x = 5$	[=]
Global definition	$x \equiv 5$	[~]
Evaluate symbolically	$z \to 5$	[Ctrl] [.]

Matrix Toolbar

Operator	Example	Keystroke
Insert matrix	$M := \begin{pmatrix} 1 & 2 \\ 3 & 4 \end{pmatrix}$	[Ctrl] M
Column	$M^{<1>}$	[Ctrl] 6
Determinant	$\lvert x \rvert$	[|]
Vectorize	$\vec{\sin}(M)$	[Ctrl] [-]
Transpose	M^T	[Ctrl] 1
Range variable	1..10	[;]
Cross product	$u \times v$	[Ctrl] 8
Inner (dot) product	$u \cdot v$	[*]
Vector sum	$\sum(123)$	[Ctrl] 4
Insert picture		[Ctrl] T

Programming Toolbar

Operator	Example	Keystroke
Add line	| $z \leftarrow 5$ | ▪	]
break	break	[Ctrl] [{]
continue	continue	[Ctrl] [
for	for i ε 1..10 $V_i \leftarrow i$	[Ctrl] [“]
if	$a \leftarrow 6$ if $z > 3$	[}]
local assignment	$z \leftarrow 5$	[{]
on error	on error	[Ctrl] [’]
otherwise	$z \leftarrow 5$ otherwise	[Ctrl] [}]
return	return	[Ctrl] [|]
while	while	[Ctrl]]

Symbolic Toolbar

Operator	Example	Keystroke
Evaluate symbolically	$z \rightarrow 5$	[Ctrl] [.]
Symbolic evaluation with keywords	$x^2 = 4$ solve, $x \rightarrow$	[Ctrl] [>]

Appendix 2: Keys for Editing and Worksheet Management

Keys for Editing

Enter	Insert blank line. In text, begin a new paragraph.
Delete	Delete blank line. In text or math, remove character to the right of the insertion line.
Shift+Enter	In text, begin a new line within a paragraph.
Ctrl+Enter	Insert a page break. In text, set the width of the text region. In math, insert addition with linebreak operator.
Ctrl+A	In text, select all the text in the text region.
Ctrl+A	In a blank spot, select all regions in the worksheet.
Ctrl+Shift+Enter	In a region, move the cursor out of and below a region.
Ctrl+F	Find a character or string of characters.
Ctrl+H	Replace a character or string of characters.
Ctrl+Z	Undo last edit.
Ctrl+Y	Redo. Reverses the action of Undo.
Ctrl+C	Copy selection to clipboard.
Ctrl+V	Paste clipboard contents into worksheet.
Ctrl+X	Cuts selection to clipboard.
Ctrl+E	Open the Insert Function dialog box.
Ctrl+U	Open the Insert Unit dialog box.
Ctrl+Shift+J	Allows you to type characters inside brackets as for chemistry notation.
Ctrl+Shift+K	Allows you to type characters that usually insert operators.
Ins	Enter insert mode.

Keys for Worksheet Management

The following keys are used for manipulating windows and worksheets as a whole.

Ctrl+F4	Close worksheet
Ctrl+F6	Make next window active
Ctrl+K	Insert or edit hyperlink
Shift+Enter	Exit a text region
Ctrl+N	Create new worksheet
Ctrl+O	Open worksheet
Ctrl+P	Print worksheet
Ctrl+S	Save worksheet
Alt+F4	Quit

Ctrl+R	Redraw screen
F1	Open Help window
F9	Recalculate screen
Shift+F1	Enter or exit context sensitive Help
Esc	Exit context sensitive Help or interrupt a calculation.

Appendix 3: Greek Letters

Greek Toolbar

Enter Roman, then type [Ctrl] G for Greek.

Greek	UC	LC	Roman
Alpha	A	α	A/a
Beta	B	β	B/b
Gamma	Γ	γ	G/g
Delta	Δ	δ	D/d
Epsilon	E	ε	E/e
Zeta	Z	ζ	Z/z
Eta	H	η	H/h
Theta	Θ	θ	Q/q
Theta (alt.)	ϑ		J
Iota	I	ι	I/i
Kappa	K	κ	K/k
Lambda	Λ	λ	L/l
Mu	M	μ	M/m
Nu	N	ν	N/n
Xi	Ξ	ξ	X/x
Omicron	O	ο	O/o
Pi	Π	π	P/p
Rho	P	ρ	R/r
Sigma	Σ	σ	S/s
Tau	T	τ	T/t
Upsilon	Y	u	U/u
Phi	Φ	ϕ	F/j
Phi(alt.)	φ		f
Chi	X	χ	C/c
Psi	Ψ	ψ	Y/y
Omega	Ω	ω	W/w

Appendix 4: Extra Math Symbols

Open QuickSheets by clicking **QuickSheets** from the **Help** menu. Select "Extra Math Symbols" from the list of topics. This opens a list of symbols that can be used for variable names. To use any of these characters, copy and paste them into Mathcad.

Additional characters can be found in the Windows Character Map. From the Start menu select Programs >Accessories>System Tools>Character Map. Choose the Mathcad UniMath font.

°C	°F	‖	/°F	/°C	∑	∏
∀	∃	ℒ	Å	∝	∅	*f*
⊆	⊂	∩	∪	⊃	⊇	∈
∉	×	∝	⊄	≈	÷	±
∟	∢	∠	∡	⊗	⊕	⊙
£	¥	$	€	¢	¤	[illegible]
↔	←	→	↑	↓	⇔	⇐
⇒	⇑	⇓	↲	⇞	↨	⇟
↔	↞	↠	↛	↚	⟻	⟼
È	Æ	℘	ℑ	ℜ	ö	ð
ℵ	Þ	Đ	ϒ	Œ	ℤ	℧
♣	♥	♠	♦	¹	²	³
▬	⟲	↶	∂	•	√	∶
∤	∓	∐	∉	∈	∋	■
⊣	⊤	⊥	⊦	⊨	⊪	⊫
◁	▷	⊶	⊷	∻	⊿	⊾

Appendix 5: Mathcad Predefined Variables

Math Constants

Name	Keystroke	Default Value
∞	**[Ctrl][Shift]z**	10^{307}
e	**e**	Value of e to 17 digits, 2.7182818284590451
π	**[Ctrl][Shift]p** or **p[Ctrl]g**	Value of π to 17 digits, 3.1415926535897931
i or j	**1i** or **1j**	The imaginary unit.
%	%	0.01; multiplying by % gives you the appropriate conversion. You can type [expression]% for inferred multiplication or use it in the unit placeholder.
NaN	NaN	Not a number.

These constants retain their exact values in symbolic calculations. To redefine the value of any of these constants, use the equal sign for definition (:=) as you would to define any variable.

System Constants

Name	Default Value	Use
TOL	.001	Controls iterations on some numerical methods.
CTOL	.001	Controls convergence tolerance in Solve Blocks.
ORIGIN	0	Controls array indexing.
PRNPRECISION	4	Controls data file writing preferences.
PRNCOLWIDTH	8	Controls data file writing preferences.
CWD	Current working directory in the form of a string variable	Can be used as an argument to file handling functions. Also useful to display the current directory path.

(*continued*)

Name	Default Value	Use
FRAME	0	Controls animations.
ERR	NA	Size of the sum of squares error for the approximate solution to a Solve Block.
1L, 1M, 1T, 1Q, 1K, 1C, and 1S	Assigned by the selected unit system	Define custom unit definitions for the base dimensions.

To redefine system constants, use the equal sign for definition (:=) in your worksheet, as you would to define any variable, or choose **Worksheet Options** from the **Tools** menu, and go to the Built-in Variables tab.

Appendix 6: Reference Tables

The Mathcad Reference Tables contain almost 40 tables of scientific and engineering data. You will find information about physics, chemistry, mechanics of materials, mathematics, electronics, and more. You can access these tables from the Resources toolbar.

Mathcad PTC

Reference Tables

Basic Science

Fundamental Constants

Physics - Mechanics

Periodic Table

Calculus

Derivative Formulas

Integral Formulas

Geometry

Areas and Perimeters

Volumes and Surface Areas

Polyhedra

Mechanics

Centroids

Mass Moments of Inertia

Electromagnetics

Capacitance

Oscillators

Properties of Liquids

Density

Viscosity

Specific Gravity

Sound Velocity

Surface Tension

Dielectric Constant

Index of Refraction

Molecular Weight

Properties of Solids

Density

Specific Gravity

Specific Heat

Thermal Conductivity

Dielectric Constant

Properties of Gases

Specific Gravity

Specific Heat

Sound Velocity

Molecular Weight

Properties of Metals

Thermal Conductivity

Specific Gravity

Linear Expansion Coefficient

Electrical Resistivity

Poisson's Ratio

Modulus of Elasticity

Melting Point

Temperature Coefficient of Resistivity

AUTHOR INDEX

A
Adams/BDF method, 294
Adobe Acrobat, software applications, 409
Application programming interface (API), 429
Arrays
  calculation
    addition and subtraction, 114
    division, 117–119
    multiplication, 115–117
  creation of, 19–21
  displaying
    output tables, 111–112
    result format dialog box, 109
    sample output—matrix form, 110
    table display form, 110
  functions
    creation, 120
    extraction, 120–121
    size and lookup, 120
    sorting, 121
  origin, 21
  subscripts, 452–453
  units
    output table, different variable name locations, 112
    single unit system, 113
AutoCAD, software applications, 409
Axes tabs, 172–176

B
Boolean operators
  icons, 205
  logical Mathcad programs, 204
  toolbar, 13, 204–205
Break operator
  and infinite loops, 275
  in Mathcad, 273
Built-in Mathcad functions
  arguments in, 53–55
  categories and names, insert function dialog box, 54
  overwrite, 66
  rules, 53

C
Calculator toolbars
  in-line division, 11
  mixed numbers, 12
Calculus and differential equations
  differentiation
    plot of function, 291
    symbolic and numeric, 290
    symbolic processor and operator in, 289
  functions, 292–293
  integration
    definite and indefinite integrals, 291–292
    live symbolic operator, 292
    ophisticated calculus operations, 291
  shear and moment calcualtions, 298
  slope and deflection calculations, 298–299
Constants
  math style
    for number, 343
    for result, 344
  style change, 349, 350
Constraint tolerance (CTOL) variable, 327
Continue operator, 275, 276
Custom units
  creation, 79
  global definition, 80

D
Data acquisition component (DAC), 428
Data tables, for inputting information, 454–455

E
Engineering equations
  current flow calculation, 256–257
  live symbolic, use of, 247
  lsolve function, 250
  motion, object
  polyroots function, 242
    and polynomial, 246
    usage, 237
  root function, 241–242
  Solve blocks
    functions, use of, 250
    given and find use of, 247–248
    Mathcad forcing, 249
  time solving, 253–256
  TOL, CTOL, and minerr, 252
  units, use of, 253
  water flow, calculation, 257–258

Essential Mathcad metric template, 376
Evaluation toolbar, 13
Explicit Euler procedures
cell 1 lagoon system, 277-280
cell 2 lagoon system, 281-282
well mixed sewage lagoon, 300

F

Factorial operator, 271
Fahrenheit and Celsius temperature functions
adding, 91
in engineering calculations, 91
Feet inch fraction (FIF), 87-89, 92, 93
Find dialog box
Mathcad, 47
text and math regions, 48
For loop functions, 267-271
list and vector, 270
range variables, 267, 269
two local variables, 271

G

Greek toolbar, equal signs, 12-13

H

Highlight region check box, 449-450
Hyperlink
creation from math region or plot region, 467
definition of, 465
inserting, 465-466
linking to regions in worksheet
tag assigning, 466-468
use relative path, 467
removing, 467, 469

I

Input, Mathcad worksheets
Highlight region, 449-450
Show Border box, 450
for standard calculation worksheets, 453, 457
data tables, 454-455
from Mathcad variables, 454
Microsoft Excel, 456
Inserting and deleting lines, 50

L

List box, 461
Literal subscripts, 452
Logic programming
adding lines, 207-211
Boolean
NOT and AND operators, 206
OR and XOR operators, 206
conditional, display statements, 211-212
creation
length checking, 202
operator, benefit of, 201
placeholders, 200
features and programming commands, 199
forms in, 207-211
if statements, 202-204
return operator, 205
toolbar, operators, 199-200

M

Manning
equations, 225
roughness coefficient, 213
wetted perimeter, 214
Mathcad
arrays and subscripts
built-in variables tab, 21
creation of, 18-20
origin, 20
range variables, 22-23
vectors and range variables comparison, 23-24
automatic calculation, 336
basics, 4-5
calculation worksheets, 379
compressed XML document, 326, 392
controls
check box, 460
list box, 461
list/combo box, 462
radio box, 461
scripted language, 457
text box control, 458, 459
web controls, 458
default, 330
editing lines, 6-8
expressions editing
deleting and replacing operators, 9
deleting characters, 8-9
selecting characters, 8
features of, 33
floating point calculations, 228
functions
built-in, 14
user-defined, 14-15
importing files
copying, 406
drawing tools, 408
insert object dialog box, 407
live symbolics, user-defined functions, 224, 225
math expressions, creation, 5-6

my site, 30–31
preferences dialog box, 319
  file locations, 321
  general, 320–321
  HTML options, 322–323
  language, 326
  save, 326–327
  script security, 325
  warnings, 323–325
programming, symbolic calculations, solving, and calculus, 29
redefinition warnings, 394
resources toolbar, 30
result format dialog box
  display options, 333–334
  fraction format, 331
  general format, 331
  individual result formatting, 335–336
  number format, 332–333
  tolerance, 335
  unit display, 334–335
sample database, 416
spell check, 45
subscripts, 21–23
symbolic processor, 67
toolbars
  calculator, 11–12
  greek, 12–13
  standard, formatting, and math, 10
toolbox
  editing, 315
  hand tools, 316
  power tools, 316
  settings, 316
  technical calculations, 317
  templates, customization, 316
  units, 315
  user-defined functions, 315
  variables, 314–315
tutorials, 3–4
units
  evaluating and displaying, 17–18
  to numbers, 16–17
version, 380
worksheet options dialog box
  built-in variables, 327–328
  calculation, 328–330
  compatibility, 331
  display, 330
  unit system and dimensions, 331
worksheets
  highlighting, 450–451
  summarizing, 457
wrapping equations, 9–10
X-Y plots, plotting
  equation quickplot, 26
  functions quickplot, 27
  graphing toolbar, 25
  ranges setting, 28–29
Mathcad, components and communication
  data
    acquisition, 428
    file input, 424–425
    file output, 426–427
    import wizard, 421, 424
    input and output, 428
    read and write functions, 427–428
    table, 425–426
  insertion to application, 429, 430
  mathworks MATLAB®, 418–420
    script writing, 423
    variables, naming, 422
  Microsoft access and open database connectivity
    choose driver, 417
    component insertion, 419
    configuration, 421
    data source, 418
    select table and fields, 419
    windows, configure data source, 416
  Microsoft excel, 414
  scriptable object, 428–429
  wizard dialog box, 425
Mathcad, customization
  additional styles
    math, 343–348
    text, 347
  default styles
    math and text, 342, 344
  headers and footers
    creation, 357
    dialog box, 358
    information, 359
  margins and page setup, 361–362
    dialog box, 361
  math styles, 358
    creation, 351
    default, 345
    new function and variables, 352
    renaming, 351
    variable and constants change, 349
  text styles, 351
    basing, 353
    changing, 352–355

Mathcad, customization (*Continued*)
creation, 355-356
default, 348
define style dialog box, 353
dialog box, 353
paragraph format dialog box, 354
text format dialog box, 354
toolbar, 362
Mathcad functions
angle, 155
built-in functions, 130
complex numbers and polar coordinates, 154
curve fitting, regression, and data analysis, 145-146
error, 151, 153
files, reading and writing, 156
linterp, 146, 148-150
mapping functions, 153
max and min
arrays, 132
complex numbers, 134
with list, 131
string variables, 134
units, 133
variables, 131
mean and median, 132, 135-136
namespace operator, 151, 152
picture and image processing, 153-154
polar notation, 154-155
string, 151
summation operators, 144
range sum, 141, 144-145, 147
vector sum, 140, 141, 143-144
toolbar, 130
truncation and rounding, 137-141
Matrices
creation, 107
element by element multiplication, 125
toolbar, 19
Microsoft excel, 414
arguments, hiding, 436, 440
files, insertion, 446
cell block, creation, 445-446
embedding*vs.* linking, 444
Mathcad data, 445
mechanics, 439, 441-442
print, 444
properties dialog box, 445
for inputting values into Mathcad, 456
and Mathcad, 432
multiple input and output, 436, 439-44
properties, 447
software application, 409
spreadsheet
blank worksheet, insertion, 434-435
cells and, 435
output range, change, 435, 438
usage, 436
units and
continuation, 438, 442
input, 437-438
output, 439
usage, 437, 441, 443
Microsoft word / Corel WordPerfect, software application, 409

N

Normal.xmct template
drawbacks, 377
manual selection, 376

O

Object linking and embedding (OLE), 411, 413
automation, 428
embedded and linked object, 406
software applications
Adobe Acrobat, 409
AutoCAD, 409-410
data files, 410
Microsoft Excel, 409
Microsoft Word/Corel WordPerfect, 409
multimedia, 410
use and limitations of, 408
On error operator, 271
Open database connectivity (ODBC)
components, 417, 421
and Microsoft access, 415-418
Operator notation
infix, 63
postfix, 63
prefix, 63-34
treefix, 64
Operators
break, 273, 257
continue, 273, 276
on error, 277
factorial, 271
return, 276
Ordinary differential equations (ODEs)
half-life, 297
Odesolve function, arguments, 294
salt solution, 295
ORIGIN, built-in variable, 327, 329

Output tables, arrays
font size and properties, 111-112
narrowed columns, 111
variable name location, 112

P

Page setup dialog box, 361
Partial differential equations (PDEs)
arguments, 294
Mathcad quicksheets, 296
Pdesolve function, 294
water flow, 296
Plotting
data points
matrix, 182
numeric display, 180-184
range variables, 179-180
vectors, 180, 181
formatting, 172
axes tab, 172
defaults tab, 176
labels tab, 174-175
number format tab, 174
plot types, 176
traces tab, 173, 175-176
graphing with units, 168-169
log scale, 184
conics, 185
curves family, 185-186, 187
uniform*vs.* variable ranges, 186
variable range, 185
multiple functions graphing, 169-171
with multiple variables, 171
using secondary Y-Axis, 172
parametric, 183-184
polar plot, creation
QuickPlot, functions, 163-164
simple, 162
range setting, 162, 165, 168
range variables, usage, 162-163
plot range setting, 165, 166, 168
solutions finding, usage, 183
three dimensional
after formatting modifications, 191
format dialog box, 190
shear, moment, and deflection, 192
types, 189
X-Y quickplot, creation, 161
zooming, 177-178
Pop-up document, 469-470
Programming, advanced
break and*continue* operators, 273-275
and infinite loops, 275
usage, 276
on error operator, 271
local definition, 265
local variables, 266, 267
looping, 266
for loop, 267-270
while loop, 271-273
return operator, 276
Project calculation input
project variables, 451-452
specific variables, 452

R

Radio button, 461, 463
Range variables
argument for function, 102
data points, plotting, 178-179
increments calculation, 108
matrix creation, 107
plot range setting, 163, 164
Range2Vec function, 122
units and, 107
user-defined functions
multiple range variables, 104
vector, 103
vector creation, 105-106
vs. vectors, 24-25
Regions
aligning and alignment
features, 40
guidelines, 40-41
selecting and moving, 39-40
tabs, 39
worksheet ruler, 39
Replace dialog box, 49
Result highlight text style
characteristics
font, 355
paragraph, 356
style, 356
usage, 357
Return operator, 202-204, 273

S

Show Border box, 450
Software development kit (SDK), 414
Standard calculation worksheets
copying regions
provenance metadata, 384-385
XML and metadata, 381-383
creation, 385-387
file properties dialog box, metadata
annotate selection, 383
annotation views, 383-384

Standard calculation worksheets (*Continued*)
comments, adding, 384-385
custom, 382
properties, 384
summary, 381
find and replace, 403-404
information protection
area boundary, 390
levels of, 391
locking areas, dialog box, 388-390
properties, 389
locking areas*vs.* protecting worksheets, 392
potential problems and solutions, 393
redefined warnings, 394
reference function, 400-403
types of, 387
user-defined functions
inserting multiple times, unique results, 398
potential problem, multiple times, 399
variables and equations, 397
String variables, 37
Structured query language (SQL), 416
Submit box, 462, 464
Subscripts
array, 22
literal, 21
Symbolic calculations
concepts, 237
evaluate, 222
expand, simplify and factor
expanding expressions, 230
floating point calculations, 228-229
numeric evaluations, 228
simplifying expressions and factoring equations, 228
explicit
keyword, 233, 237
variables, 232
float
Mathcad display, 226
symbolic evaluation*vs.* numeric evaluation, 226
numeric evaluation, 234
polynomial expression
solve for numeric results, 210
solve for variable, 220, 221
using units, 236
Symbolic processor
Symbolic processors, 226

T

Table of contents
for data within current worksheet, 470-471
linking files in project calculation, 471-472
Tabs
axes, 172
defaults, 176
labels, 174-179
number format, 174
traces, 173, 175-176
Templates
creation, 367-368
EM metric, 368
headers and footers
page numbers, 373
tabs, 375
margins
notes and nonprinting information, 373
page setup, 372, 374
math styles, 368
page setup settings, 374
saving, 257
text styles
characteristics, 371-372
dialog box, 370
worksheet options
calculation settings, 369
concepts, 368
new dialog box for, 366
opening, 367
Text box control, 458-459
Text regions
click properties, 41
font characteristics, 41
Greek symbols, 42
math regions
graphic image, 47
variable assignment, 46
moving below, 42-43
paragraph properties
hanging indent, 44
word processor, 43
ruler and spell check, 45
Three dimensional (3D) plots
after formatting modifications, 191
format dialog box, 190
shear, moment, and deflection, 192
types, 189
Trace dialog box
plotted points, numeric display, 180
solutions finding, approximate, 183
Truncation and rounding functions
lowercase, 137
units, 139
vectors, 138

uppercase, 140
units, 142
vectors, 141
Two-cell lagoon system, 274

U

Units
assigning, numbers
examples, 17
insert unit dialog box, 16–17
benefits of, 15
custom scaling
degrees minutes seconds (DMS), 89, 92
Fahrenheit and Celsius, 87
feet inch fraction (FIF), 88, 92, 93
hours minutes seconds (hhmmss), 90, 93
postfix notation, 89
temperature change, 88, 90
default unit system, 76–78
display, 76
units formatting result comparison, 77
definitions, 72
derived, 76
dimensionless, 95–96
in empirical formulas, 83–86
in equations
default unit system to US, 82
kinetic energy formula, 80–81
mixed units, 82
redefine built-in units, 81
redefinition warning, 81
evaluating and displaying
results in, 18
solid black box, 17
force and mass, 78–79
limitation of, 96–97
measurement built in, 16
money, 94
custom scaling function, creation, 94
in user-defined functions, 83
using and displaying, 73–75
addition and placeholder, 74
changing and balancing displayed, 75
combining, 75
User-defined functions
arguments, types, 56, 58
examples, 56, 60–61
naming variables and rules, 55
overwrite, 65
samples of, 57
variables in, 59–60
warnings, 65

V

Variables
boolean logic, 33
dialog box, 350
math style
for unit, 343
for variable name, 342
names, style sensitive, 347
naming guidelines, 312–313
descriptive, 312
prime symbol, 314
subscripts, 312–314
underscores, use of, 313
upper and lowercase, combined use, 313
naming rules
case and font, 34
chemistry notation, 36
literal subscripts, 35
special text mode, 35–36
style change, 349
types, 33
uses in calculations, 37–38
Vectors. *See* Mathcad
creation, 105–106
in expressions, 124
user-defined function and, 123
Visual basic programming, 414

W

Web controls, 458
While loop, 271, 273
avoid infinite loops, 275
factorial function, 274
Wrapping equations, 9–10

X

XML document, compressed, 326–327

Appendix 3: Greek Letters

Greek Toolbar

Enter Roman, then type [Ctrl] G for Greek.

Greek	UC	LC	Roman
Alpha	A	α	A/a
Beta	B	β	B/b
Gamma	Γ	γ	G/g
Delta	Δ	δ	D/d
Epsilon	E	ε	E/e
Zeta	Z	ζ	Z/z
Eta	H	η	H/h
Theta	Θ	θ	Q/q
Theta (alt.)	ϑ		J
Iota	I	ι	I/i
Kappa	K	κ	K/k
Lambda	Λ	λ	L/l
Mu	M	μ	M/m
Nu	N	ν	N/n
Xi	Ξ	ξ	X/x
Omicron	O	o	O/o
Pi	Π	π	P/p
Rho	P	ρ	R/r
Sigma	Σ	σ	S/s
Tau	T	τ	T/t
Upsilon	Y	υ	U/u
Phi	Φ	ϕ	F/j
Phi(alt.)	φ		f
Chi	X	χ	C/c
Psi	Ψ	ψ	Y/y
Omega	Ω	ω	W/w

Lightning Source UK Ltd.
Milton Keynes UK
UKHW031005200320
360651UK00021B/1079